Regulation of Functional Foods and Nutraceuticals

A Global Perspective

The *IFT Press* series reflects the mission of the Institute of Food Technologists—advancing the science and technology of food through the exchange of knowledge. Developed in partnership with Blackwell Publishing, *IFT Press* books serve as essential textbooks for academic programs and as leading edge handbooks for industrial application and reference. Crafted through rigorous peer review and meticulous research, *IFT Press* publications represent the latest, most significant resources available to food scientists and related agriculture professionals worldwide.

- *Biofilms in the Food Environment* (Hans P. Blaschek, Hua Wang, and Meredith E. Agle)
- *Food Carbohydrate Chemistry* (Ronald E. Wrolstad)
- *Food Irradiation Research and Technology* (Christopher H. Sommers and Xuetong Fan)
- *Microbiology and Technology of Fermented Foods* (Robert W. Hutkins)
- *Multivariate and Probabilistic Analyses of Sensory Science Problems* (Jean-Francois Meullenet, Hildegarde Heymann, and Rui Xiong)
- *Preharvest and Postharvest Food Safety: Contemporary Issues and Future Directions* (Ross C. Beier, Suresh D. Pillai, and Timothy D. Phillips, Editors; Richard L. Ziprin, Associate Editor)
- *Regulation of Functional Foods and Nutraceuticals: A Global Perspective* (Clare M. Hasler)
- *Sensory and Consumer Research in Food Product Development* (Howard R. Moskowitz, Jacqueline H. Beckley, and Anna V.A. Resurreccion)
- *Water Activity in Foods: Fundamentals and Applications* (Gustavo V. Barbosa-Canovas, Anthony J. Fontana Jr., Shelly J. Schmidt, and Theodore P. Labuza)

Regulation of Functional Foods and Nutraceuticals

A Global Perspective

Edited by Clare M. Hasler

Clare M. Hasler is currently the Founding Executive Director of the Robert Mondavi Institute for Wine and Food Science at the University of California, Davis (www.robertmondaviinstitute.ucdavis.edu) She is one of the world's foremost authorities on functional foods and served as founding Executive Director of the Functional Foods for Health Program at the University of Illinois from 1992-2000. Dr. Hasler received a dual Ph.D. in environmental toxicology and human nutrition from Michigan State University, an M.S. in nutrition science from The Pennsylvania State University and an M.B.A. from the University of Illinois at Urbana-Champaign.

©2005 Blackwell Publishing
All rights reserved

Blackwell Publishing Professional
2121 State Avenue, Ames, Iowa 50014, USA

Orders: 1-800-862-6657
Office: 1-515-292-0140
Fax: 1-515-292-3348
Web site: www.blackwellprofessional.com

Blackwell Publishing Ltd
9600 Garsington Road, Oxford OX4 2DQ, UK
Tel.: +44 (0)1865 776868

Blackwell Publishing Asia
550 Swanston Street, Carlton, Victoria 3053, Australia
Tel.: +61 (0)3 8359 1011

Authorization to photocopy items for internal or personal use, or the internal or personal use of specific clients, is granted by Blackwell Publishing, provided that the base fee of $.10 per copy is paid directly to the Copyright Clearance Center, 222 Rosewood Drive, Danvers, MA 01923. For those organizations that have been granted a photocopy license by CCC, a separate system of payments has been arranged. The fee code for users of the Transactional Reporting Service is 0-8138-1177-5/2005 $.10.

Printed on acid-free paper in the United States of America

First edition, 2005

Library of Congress Cataloging-in-Publication Data

Regulation of functional foods and nutraceuticals : a global perspective / edited by Clare M. Hasler.—1st ed.
    p. cm.
  Includes bibliographical references and index.
  ISBN 0-8138-1177-5 (alk. paper)
  1. Food industry and trade. I. Hasler, Clare M.

TP370.R44 2005
363.19′2—dc22

2004017179

The last digit is the print number: 9 8 7 6 5 4 3 2 1

Table of Contents

Contributors	vii
Preface	xiii
Clare M. Hasler (University of California, Davis)	

1. **The Impact of Regulations on the Business of Nutraceuticals in the United States: Yesterday, Today, and Tomorrow**   3
   Kathie L. Wrick (The Food Group)

2. **The Regulatory Context for the Use of Health Claims and the Marketing of Functional Foods: Global Principles**   37
   Michael Heasman (Food for Good)

3. **Regulation of Quality and Quality Issues Worldwide**   55
   Joy Joseph (Pharmavite Corporation)

4. **Organic Food Regulations: Part Art, Part Science**   69
   Kathleen A. Merrigan (Tufts University)

5. **Health Claims: A U.S. Perspective**   79
   Victor Fulgoni (Nutrition Impact, LLC)

6. **Food and Drug Administration Regulation of Dietary Supplements**   89
   Stephen H. McNamara (Hyman Phelps & McNamara, PC)

7. **Tropicana Pure Premium and the Potassium Health Claim: A Case Study**   101
   Carla McGill (Florida Department of Citrus)

8. **The Importance of the Court Decision in *Pearson v. Shalala* to the Marketing of Conventional Food and Dietary Supplements in the United States**   109
   Elizabeth Martell Walsh, Erika King Lietzan, Peter Barton Hutt (Covington & Burling)

9. **Dietary Supplements and Drug Constituents: The *Pharmanex v. Shalala* Case and Implications for the Pharmaceutical and Dietary Supplement Industries**   137
   Daniel A. Kracov, Paul D. Rubin, Lisa M. Dwyer (Patton Boggs, LLP)

10. **The Role of the Federal Trade Commission in the Marketing of Functional Foods**   149
    Lesley Fair (Federal Trade Commission)

11. **Functional Foods: Regulatory and Marketing Developments in the United States** — 169
    Ilene Ringel Heller (Center for Science in the Public Interest)

12. **The Nutraceutical Health Sector: A Point of View** — 201
    Stephen L. DeFelice (Foundation for Innovation in Medicine)

13. **Regulatory Issues Related to Functional Foods and Natural Health Products in Canada** — 213
    Kelley Fitzpatrick (University of Manitoba, Richardson Centre for Functional Foods and Nutraceuticals)

14. **The Regulation of Functional Foods and Nutraceuticals in the European Union** — 227
    Peter Berry Ottaway (Berry Ottaway & Associates, Ltd.)

15. **Functional Foods in Japan: FOSHU ("Foods for Specified Health Uses") and "Foods with Nutrient Function Claims"** — 247
    Ron Bailey (California Functional Foods)

16. **Chinese Health (Functional) Food Regulations** — 263
    Guangwei Huang and Karen Lapsley (Almond Board of California)

17. **Report of ILSI Southeast Asia Region Coordinated Survey of Functional Foods in Asia** — 293
    E-Siong Tee (International Life Sciences Institute, SE Asia)

18. **Germany and Sweden: Regulation of Functional Foods and Herbal Products** — 303
    Joerg Gruenwald and Birgit Wobst (Phytopharm Research, Analyze & Realize Ag)

19. **Functional Foods: Australia/New Zealand** — 321
    Jane L. Allen, Peter J. Abbott, Sue L. Campion, Janine L. Lewis, Marion J. Healy (Australian/New Zealand Food Authority)

20. **Regulation of Functional Foods in Spain** — 337
    Luis García-Diz and Jose Luis Sierra Cinos (Universidad Complutense de Madrid)

21. **Functional Food Legislation in Brazil** — 367
    Franco M. Lajolo (Universidade de São Paulo)

22. **Codex and Its Competitors: The Future of the Global Regulatory and Trading Regime for Food and Agricultural Products** — 377
    Mark Mansour (Keller and Heckman, LLP)

Index — 389

Contributors

Dr. Peter J. Abbott holds the position of Principal Toxicologist at Food Standards Australia New Zealand. His primary responsibility is the provision of scientific advice in relation to food safety, particularly chemicals in food. Dr. Abbott's academic training and research background is in the area of mechanisms of chemically induced cancer. Since 1996, Dr. Abbott has participated as a technical expert to the World Health Organization on the Joint FAO/WHO Expert Committee on Food Additives.

Jane L. Allen is a senior nutritionist at Food Standards Australia/New Zealand. In this role her particular interests are in functional foods and risk analysis. Ms. Allen's background is in public health nutrition and dietetics, and academic and research activities focusing on food-related consumer behaviors.

Ron Bailey has been an independent consultant for the past 19 years, focusing on the transfer of food technology between the United States and Japan for U.S. and Japanese clients. Mr. Bailey has traveled to Japan more than one hundred times over the past 28 years and has lived in Japan twice. He received his B.S. and M.S. in Chemical Engineering from Iowa State University. Mr. Bailey is President of California Functional Foods, a start-up company conducting research and development on food ingredients with potential health-related benefits.

Sue L. Campion has a Masters degree in Human Nutrition from the University of Otago in New Zealand. Ms. Campion has spent 13 years working for the Australian government focusing on public health aspects of nutrition, including obesity prevention, food and nutrition policy, and food regulatory issues, particularly those relating to food labeling.

Dr. Jose Luis Sierra Cinos is a pharmacist, specialist in nutrition, and former associate professor of Nutrition in the department of Madrid's Complutense University. His expertise is in topics related to foods and nutrition.

Stephen L. DeFelice, M.D., is an internationally recognized authority on clinical research and development of both drugs and natural substances and is also widely recognized as being responsible for coining the term **nutraceuticals**. Dr. DeFelice is a graduate of Jefferson Medical College and served as chief of Clinical Pharmacology, Division of Medicinal Chemistry, at the Walter Reed Army Institute of Research. Dr. DeFelice is also the Founder and Chairman of the Foundation for Innovation in Medicine (FIM) whose goal is to accelerate medical discovery. FIM proposed the Nutraceutical Research and Education Act, which was introduced in Congress by Frank Pallone in 1999. FIM also recently proposed the Doctornaut Act, which would permit physicians to more freely volunteer for clinical studies than nonphysician volunteers. Both acts can be found on the FIM Web site at www.fimdefelice.org.

Lisa M. Dwyer is a sixth-year associate at Patton Boggs LLP, in the Food and Drug Law practice group. She advises clients on issues relating to the development, promotion, and sale of products regulated by the Food and Drug Administration (FDA). Ms. Dwyer handles a variety of regulatory and litigation matters before the FDA, the Federal Trade Commission (FTC), and federal and state courts, and she lobbies Congress and federal agencies in connection with the laws governing FDA-regulated products and other health-related issues.

Lesley Fair is a senior attorney with the Federal Trade Commission's Division of Advertising Practices, where she has represented the Commission in numerous investigations of deceptive advertising both in traditional media and online. She specializes in the marketing of drugs, dietary supplements, foods, and other health-related products. A graduate of the University of Notre Dame and the University of Texas School of Law, Ms. Fair is a member of the adjunct faculty of the Columbus School of Law of the Catholic University of America and holds the title of Distinguished Lecturer.

Kelley Fitzpatrick is the Marketing and Research Development Manager for a new $25 million research and development initiative, the Richardson Centre for Functional Foods and Nutraceuticals, based at the University of Manitoba. Previously, Ms. Fitzpatrick was the President of the Saskatchewan Nutraceutical Network (SNN), an organization that she established in early 1998. Under her direction, the SNN, one of the first networks of its kind in Canada, became recognized nationally and internationally as a superior information resource for the Canadian nutraceutical and functional food sector and a leader in industry representation. Ms. Fitzpatrick holds a Master of Science degree in Nutrition and a Bachelor of Arts degree.

Dr. Victor Fulgoni, III, is currently Senior Vice President of Nutrition Impact, LLC, a consulting firm that helps food companies develop and communicate science-based claims about their products and services. Prior to joining Nutrition Impact, Dr. Fulgoni worked for the Kellogg Company as Vice President of Food and Nutrition Research. At Kellogg he helped develop the company's long-term research program and was intimately involved in the company's research and regulatory efforts to gain health claim approval from the U.S. FDA regarding soluble fiber from psyllium and the risk of heart disease. Dr. Fulgoni completed his Bachelors degree at Rutgers University and his Ph.D. at the University of Tennessee with a major in animal nutrition and a minor in statistics.

Luis García-Diz is Professor of Nutrition at the Complutense University of Madrid, where he is responsible for nutrition and dietetics, nutritional epidemiology, and computer science applied to the sciences of the health. His areas of expertise include nutrition, anthropometry, alimentary behavior, and new technologies related to nutrition. For six years, Dr. García-Diz was assistant manager for Consumer Protection in the Government of Madrid (Spain).

Dr. Joerg Gruenwald is president and founder of Analyze & Realize, Inc., with its subsidiaries Phytopharm Consulting and Phytopharm Research, a specialized consulting company for natural products based in Berlin, Germany, with a strong network in Europe. The activities of the consulting company supports the natural products industry in questions of strategy development, new product development, regulatory affairs, clinical trials, international partner finding and, especially, entering the European markets. Associated research activities have resulted in the book projects *Physicians Desk Reference (PDR) for Herbal Medicines* and *Plant-Based Ingredients for Functional Foods*, as well as the performance of dozens of clinical trials with natural products. Dr. Gruenwald is a leading international expert in the field of botanicals and natural ingredients.

Dr. Marion J. Healy holds the position of Chief Scientist at Food Standards Australia/New Zealand. In this role, she is responsible for the scientific advice that underpins the development of food standards. Dr Healy's research training and experience is in the field of molecular genetics. However, her current interests are focused on food-related health risks.

Dr. Michael Heasman is an internationally recognized writer, researcher, communicator, and opinion-leader in food and health with more than 15 years of experience gained at UK universities. He is a leading commentator on functional foods, serving as co-editor of the international business newsletter New Nutrition Business from 1995–2002, and Director of Studies at the Centre for Food & Health Studies Ltd, a London-based think tank on functional foods and nutraceuticals, from 1999–2002. He is co-author of the first book on business and policy in functional foods called *The Functional Foods Revolution: Healthy People, Healthy Profits?* (London: Earthscan, 2001) as well as *Food Wars: The Global Battle for Minds, Mouths and Markets* (Stylus Pub Llc, 2003), a book on global food and health policy.

Ilene Ringel Heller is a senior staff attorney with the Center for Science in the Public Interest in Washington, D.C., where she has worked extensively on U.S. and international food labeling issues, as well as dietary supplement matters. Mrs. Heller co-authored an extensive report on functional foods entitled "Functional Foods—Public Health Boon or 21st Century Quackery." Prior to joining CSPI, she was an associate at the law firms of Arnold & Porter and Keller and Heckman. While in private practice, Mrs. Heller specialized in counseling clients in the food, drug, and medical device industries on compliance with the Federal Food, Drug, and Cosmetic Act and related laws. Mrs. Heller was awarded a Juris Doctor degree, *cum laude*, from Georgetown University and is a member of the District of Columbia Bar.

Guangwei Huang has been a Technical Manager for the Almond Board of California since September 2001. In that capacity, he provides support to the Food Quality Safety Program, the International Program, and other technical projects. Mr. Huang has eight years of firsthand experience working in the California almond industry, most recently as a Quality Assurance Director for a California company in charge of managing and implementing quality procedures for almond processing. Before immigrating to the United States, Mr. Huang worked for the Chinese Ministry of Health on development of health standards and establishment of testing procedures. Mr. Huang holds a Master of Science in Food Science from the University of California, Davis, and Bachelor of Science in Sanitary Technology from the Wes-China University of Medical Science.

Peter Barton Hutt is a partner in the Washington, D.C., law firm of Covington & Burling, specializing in food and drug law. He graduated from Yale College and Harvard Law

School and obtained a Master of Laws degree in Food and Drug Law from New York University Law School. Mr. Hutt served as Chief Counsel for the Food and Drug Administration during 1971–1975. He is the co-author of the casebook used to teach food and drug law throughout the country. He teaches a full course on this subject during Winter Term at Harvard Law School and has taught the same course during Spring Term at Stanford Law School. Mr. Hutt has been a member of the Institute of Medicine since it was founded in 1971. He was named by *The Washingtonian* magazine as one of Washington's 50 best lawyers (out of more than 40,000) and as one of Washington's 100 most influential people; by the *National Law Journal* as one of the 40 best health care lawyers in the United States; and by *European Counsel* as the best FDA regulatory specialist in Washington, D.C. *Business Week* recently referred to Mr. Hutt as the "unofficial dean of Washington food and drug lawyers."

Dr. Joy Joseph is the Vice President of Quality, Research and Development, and Scientific Affairs for Pharmavite LLC, a leading manufacturer of Dietary Supplements. She has more than 40 years in quality control and GMP training and is a recognized industry expert in "Standard Operating Procedures" that are compliant with pharmaceutical and food GMPs. Dr. Joseph is the chair of the USP Council of Experts on Non-Botanical Dietary Supplements.

Daniel A. Kracov is Deputy Director of Patton Boggs' Public Policy and Regulatory Department and a partner in the firm. He concentrates his practice on matters relating to the Food and Drug Administration's regulation of drugs, biologics, foods, and medical devices and related policy and legislative matters. Mr. Kracov's clients include start-up companies, trade associations, and large manufacturing companies.

Dr. Franco M. Lajolo is a pharmacist and biochemist with a Ph.D. in Food Science from the Massachusetts Institute of Technology, where he was also a post-doctoral fellow. His major areas of investigation have covered food biochemistry and molecular biology and, more recently, functional foods. He has authored more than 200 scientific papers and books and has advised 65 M.S. and Ph.D. students. Dr. Lajolo is a member of the International Academy of Food Science and Technology and of the Functional Foods Committee of the Sanitary Surveillance Agency in Brazil.

Dr. Karen Lapsley is Director of Scientific Affairs for the Almond Board of California where, since August 1999, she has managed contracted-out nutrition and food safety research programs with more than 15 universities and research centers worldwide. As a member of the Health Canada Advisory Panel on Health Claims for Foods (1996–1999), she was considered one of Canada's experts on functional foods. Between 1996 and February 1999, Dr. Lapsley was Vice President for Ceapro, Inc., an Edmonton-based start-up life-sciences company, where she was responsible for the establishment of a Functional Foods Division, which developed oat-based ingredients. From 1983 to 1996, Dr. Lapsley was with Agriculture and Agri-food Canada, most recently in charge of the Food Research Program in Ottawa, which focuses on the safety, quality, and nutritional value of food and feed products. Dr. Lapsley received her Doctorate in Food Science from ETH, Zurich, Switzerland, in 1989.

Janine L. Lewis holds the position of Principal Nutritionist at Food Standards Australia/New Zealand. Her primary responsibility is the provision of scientific advice in relation to nutrition-related food standards. Ms. Lewis's previous role was the development of national food composition data.

Erika King Lietzan is Assistant General Counsel of the Pharmaceutical Research and Manufacturers of America (PhRMA) in Washington, D.C. She works primarily to support its federal legislative team and covers issues such as drug importation, Hatch-Waxman issues, pediatric exclusivity and labeling, product liability, and bioterrorism. Before joining PhRMA in 2002, she was a senior associate in the food and drug practice at Covington & Burling. She advised clients in the food, drug, biologic, medical device, dietary supplement, and cosmetic industries on a variety of issues arising under the Public Health Service Act, the Federal Food, Drug, and Cosmetic Act, the Freedom of Information Act, the Trade Secrets Act, the Federal Advisory Committee Act, and the Administrative Procedure Act.

Mark Mansour is an FDA/Healthcare Regulation Practice partner, resident in the Washington, D.C. office of Morgan Lewis. His practice focuses on national and international food and drug regulation and issues relating to pharmaceutical, agribusiness, biotechnology, and consumer products. He is a graduate of the Georgetown University Law Center and holds a master's degree from Harvard University and a bachelor's degree from Georgetown University. Mr. Mansour currently serves as a member of several biotechnology-related task forces, including the State Department International Economic Policy Task Force on Biotechnology and the Food Industry Codex Coalition. He is admitted to practice in the District of Columbia, Michigan, and the U.S. District Court, Eastern District of Michigan.

Dr. Carla McGill is scientific research director for the Florida Department of Citrus. Previously, she was a Tropicana Fellow, Nutrition Science, for Tropicana Products, Inc. where her responsibilities included overseeing clinical research and nutritional analysis of products, and providing nutrition expertise for product development and marketing. Dr. McGill has experience in both teaching and clinical settings as an educator, researcher, and lecturer. Dr. McGill received her Ph.D. from the Department of Nutrition and Food Science at Auburn University.

Stephen H. McNamara is a member of the law firm of Hyman, Phelps & McNamara, PC, a Washington, D.C.-based law firm that concentrates on Food and Drug Administration (FDA)–related matters. Mr. McNamara formerly served as FDA Associate Chief Counsel for Food, the agency's senior supervisory lawyer for food-related matters. He twice received the FDA's highest award for service, the FDA Award of Merit, and in 2004 he received the Food and Drug Law Institute (FDLI) Distinguished Service Award.

Dr. Kathleen A. Merrigan is Director of the Agriculture, Food and Environment Program at the Gerald J. and Dorothy R. Friedman School of Nutrition Science and Policy at Tufts University. Her teaching and research focus is on sustainable agriculture, interest group politics, and negotiation theory.

Peter Berry Ottaway, CSci, FIFST, is a food scientist and technologist with considerable experience in food law, specializing in the food legislation of the European Union. He is the author, co-author or editor of a number of books on aspects of food science and food law.

Paul D. Rubin is a partner in the Washington, D.C., office of Patton Boggs LLP. He specializes in food and drug law, including the regulation of foods, dietary supplements, and drugs. He is a frequent speaker and author on FDA regulatory issues. Mr. Rubin received his J.D., *cum laude*, from the University of Pennsylvania Law School and his B.S. in Economics, *magna cum laude*, from the Wharton School of the University of Pennsylvania.

Dr. E-Siong Tee is Scientific Director of the International Life Sciences Institute (ILSI), Southeast Asia Region, as well as a nutrition consultant with TES NutriHealth Strategic Consultancy. Dr. Tee is Chairman of the following committees of the Ministry of Health: (a) Nutrition, Health Claims and Advertisement; (b) Classification of Food-Drug Interphase Products; and (c) National Codex Sub-Committee on Nutrition & Foods for Special Dietary Uses. He is also a member of the National Technical Advisory Committee on Food Regulations and the National Coordinating Committee on Food and Nutrition.

Elizabeth Martell Walsh is an associate with the food and drug practice group at Covington & Burling. She primarily provides FDA regulatory advice to the firm's pharmaceutical and medical device clients. Ms. Walsh received her J.D. in 2000 from the University of Virginia, where she was an editor of the *Virginia Law Review* and was elected to the Order of the Coif.

Dr. Birgit Wobst is a biologist who has worked for Phytopharm Research since 1999 as a scientific consultant and medical writer. Her responsibilities are regulatory affairs concerning herbal medicines, dietary supplements, and functional foods. Dr. Wobst writes expert reports, scientific texts, articles, and books about phytopharmaceuticals and other natural products.

Dr. Kathie L. Wrick has more than 26 years of professional experience, more than half of which have been with food companies such as Quaker Oats, H.J. Heinz, Stop and Shop Manufacturing, M&M/Mars, and McNeil Consumer Products. Dr. Wrick also spent 12 years consulting with the Arthur D. Little (now TIAX LLC) Food Industries Practice. For more than a decade she has specialized in market and technology analysis and strategic planning for food ingredient and nutritional product companies, and has maintained a special interest in nutraceutical opportunities. She currently consults with The Food Group, a boutique consulting firm based in Concord, MA, comprised of former senior staff of the Arthur D. Little Food Industries Practice.

Preface

Functional foods and nutraceuticals have been a leading trend in the food industry for well over a decade. Of the principal factors driving the interest in this food category, including aging demographics (instrumental in the establishment of Foods for Specified Health Uses in Japan) and scientific discovery regarding how foods or food components can optimize health, regulatory issues have been the most controversial.

In the United States, the regulation of functional foods and nutraceuticals has been viewed as pivotal to the marketing success of nutritional products for more than 20 years. Indeed, it was in 1984 that Kellogg's, in collaboration with the National Cancer Institute, began marketing All-Bran cereal bearing label statements describing the colon cancer chemoprevention benefits of fiber, thus stimulating the passage of the Nutrition Labeling and Education Act of 1990 (NLEA). The decade between the promulgation of the NLEA regulations (January 1993) and the summer of 2003 witnessed several pivotal changes in food regulations, including the passage of the Dietary Supplement Health and Education Act of 1994 and the FDA Modernization Act of 1997; *Pearson v. Shalala* and issues of free speech protection; and, most recently, qualified health claims. These regulatory developments have provided a rather liberal environment for the marketing of functional foods and nutraceuticals in the United States. As Wrick outlines in the first chapter of this book, "the U.S. regulatory structure . . . has been a powerful force defining the industries that manufacture the products regulated by the FDA" and has "directly or indirectly shaped the structure, key success factors, barriers to entry, rules of competition, and manufacturing operations for conventional foods, various nutritional products, and pharmaceuticals."

Outside the U.S. there has been a similar loosening of restrictions on the use of health claims on foods stimulated by food industry demands to meet marketing goals. However, enormous variation still exists among countries regarding the regulatory tolerance of nutrition and health messages on foods and beverages. Indeed, only Japan has a regulatory structure specifically pertaining to functional foods. In most other countries, the use of disease risk reduction claims is very limited or nonexistent. This is thought to have stifled innovation, competition, and investment in the industry in some cases, and in others has made it almost impossible to directly transfer products from markets such as the United States or Japan.

Of major concern to regulatory authorities in many countries are safety assessment and the need for quality control of products by the industry. Regulatory agencies must ensure that functional foods are regulated in a manner that maximizes health benefits and minimizes health risk for all consumers.

The development and use of health claims is still an evolving area around the world, and the future regulatory framework and climate, especially in the context of the global food marketing and trade, is far from clear. Although numerous books and reports have summarized the latest developments on the science and marketing of functional foods and nutraceuticals, this is the first comprehensive review of how they are regulated from a global perspective. It is my sincere hope that a wide variety of readers from academia, industry, and government agencies around the world will find this text to be a valuable addition to their library on the ever-evolving topic of functional foods and nutraceuticals.

Clare M. Hasler

Regulation of Functional Foods and Nutraceuticals

A Global Perspective

1 The Impact of Regulations on the Business of Nutraceuticals in the United States: Yesterday, Today, and Tomorrow

Kathie L. Wrick

More than a decade has passed since the term **nutraceuticals** was coined by Dr. Stephen DeFelice (1991), founder of the Foundation for Innovation in Medicine. His original definition was very broad:

> A nutraceutical is any substance considered a food or part of a food which provides medical or health benefits including the prevention or treatment of disease and include isolated nutrients, dietary supplements, diets and dietary plans, genetically engineered foods, herbal products and processed foods such as cereals soups and beverages.

These products represented distinctly different types of businesses and markets. For example, isolated nutrients such as vitamins were a distinct segment of the specialty and fine chemicals business with defined end-use markets including animal feeds, human food, cosmetics, and parenteral and enteral solutions, and pharmaceuticals. Diets included product lines such as formulated meal replacements for weight loss or general nutritional supplementation, and dietary plans are a part of the weight loss services business, which guides consumers who join these programs to change their eating habits. Genetically engineered foods were primarily herbicide-resistant soy and other agricultural commodities sold to growers. At the time, herbs were legally considered as something used to season and flavor food; any claims suggesting a use for disease treatment made them an unapproved drug in the United States and many European countries.

Despite disagreement and confusion over the definition, the term **nutraceuticals** captured the spirit of products at the food/drug interface: "Let food be thy medicine." Nutraceuticals could be derived from food and used for medicinal reasons. For example, the beta glucan component of oat bran is recognized by the scientific and U.S. regulatory community as being responsible for lowering elevated blood cholesterol levels (FDA, 2002). Certain dairy peptides, some of which have a mechanism of action similar to ACE inhibiting drugs, have demonstrated a powerful capacity to lower moderately elevated blood pressure at only milligram levels of intake (Seppo et al., 2003; Hata et al., 1996).

A decade beyond its first appearance in the food technology and marketing vocabulary, **nutraceuticals** still has no agreed upon definition or regulatory standing. The emerging world of nutritional products, including nutraceuticals (however defined), took on a life, language, and segmentation of its own, catapulted by a grass-roots consumer curiosity in alternative food and medicines. By 1990, consumers began to experiment in increasing numbers with herbs, vitamins, and other supplements, natural and health foods, and eventually with mainstream food products marketed for unique health benefits. About the time of the passage of the Dietary Supplement Health and Education Act (DSHEA) in 1994, the nutritional products industry was growing by double digits. In only a few years, industry publications, syndicated store-level tracking services, and consumer research companies sprang up to monitor product introductions, company activities, product movement through mainstream and emerging distribution channels, and trends and attitudes of the health-conscious consumers; they also provided thoughtful industry analysis. Today, the *Nutrition Business Journal* (NBJ, New Hope Natural Media, a division of Penton Media, Inc., San Diego) is the industry's lead paid-subscription publication monitoring the financial performance of nutritional products and the activities of their manufacturers. *Nutraceuticals World* and *Nutritional Outlook* are among several trade journals that report on new science, technologies, and market activities. NBJ uses **nutraceuticals** as an umbrella term for anything sold at retail (outside of OTC medications) that is consumed primarily or partially for health reasons. SPINS, a San Francisco–based market research firm, works in conjunction with AC Neilsen to track consumer purchases of natural foods and other health and wellness products through their primary retail channels. These include natural food, health food, and mainstream grocery stores and supermarkets, drug stores, and mass merchandisers. Market-research companies, including HealthFocus International (Atlanta, GA) and The Hartman Group (Bellevue, WA) specialize in the demographics, consumer attitudes, and purchasing patterns of the new consumer segment that spends heavily on food and health-oriented products outside mainstream channels. The Freedonia Group (Cleveland, OH) occasionally reports on the business outlook for nutraceutical chemicals, such as herbal and nonherbal extracts, vitamins and minerals, and healthful food ingredients utilized for their health benefits such as soy, fiber, probiotics, and highly polyunsaturated fatty acids. This latter approach to a market definition is similar to that proposed in the first market-outlook report on the topic (Wrick, 1994) prior to the passage of DSHEA. There, nutraceuticals were defined as the bioactive substance in food agricultural commodities (both plant and animal) that showed demonstrable health benefits beyond satisfying the basic nutritional functions of growth, tissue repair, and energy for daily living. This definition had the advantage of allowing the nutraceutical, or "bioactive," to be discussed as a separate, chemically identifiable substance (or family or combination of substances) that could be isolated or extracted from the tissues in which it naturally occurred or even synthesized. It could then be commercialized through the appropriate regulatory pathway as a component of food, as a dietary supplement, or even as a drug. The definition allowed for relatively easy market analysis, making nutraceuticals the raw material that could be followed down the chain to the end users.

However, the marketplace ignored the conundrum over a definition and cautiously wove a path between communicating the messages implied by the emerging science and the regulatory framework governing product claims for foods and drugs. In the United States and many other countries, the claims made for a product determine its status as a food or a drug. The emerging science of food phytochemicals as a potentially safer means to cure, treat, or prevent disease meant their commercialization in the United States under drug law with its associated high development costs, now estimated at an average of $802 million, up from $318 million in 1980 (Tufts, 2002). This barrier to market entry led to the passage of DSHEA, spurred on by a small but zealous industry group convinced that many of the products it wanted to sell for treating or preventing health problems were safe based on their history of use in the Far East and other countries.

Despite the lack of regulatory standing or a universally agreed upon definition, the term

nutraceuticals is still the subject of market research reports and is occasionally interchanged with some or all of the product segments in what is now known and accepted as the nutrition industry. NBJ, launched in 1995, was the first publication to systematically monitor the activities, trends, and financial performance, mergers, and acquisitions of the industry whose products were purchased for nutrition and health reasons, as distinct from mainstream food products purchased to satisfy hunger, taste, and family tradition. Today, NBJ's definition of the nutrition industry includes dietary supplements (pills, powders, tablets, tonics, extracts, and supplemental beverages), meal replacements, functional foods, natural/organic foods, sports and performance foods and beverages, and, more recently, natural personal care (NBJ, 2002a, b, c; 2003a, b).

In an attempt to bring focus to this update, this chapter reviews the nutritional product segments utilizing nutraceutical actives in the U.S. marketplace today, including dietary supplements and functional foods. These categories are defined as the consumer products that utilize nutrients, specialized food components, or other naturally occurring chemicals (or mixtures of chemicals) as components of pills and tablets and as ingredients in foods intended to provide a health benefit beyond basic nutrition. Dietary supplements and functional foods comprise about 66 percent of the $62.9 billion U.S. nutrition industry and capitalize on the health benefits of added food ingredients, naturally occurring food components, or dietary ingredients they contain (see Table 1.1, Definitions).

Table 1.1. Definitions

> **Dietary ingredient**—A substance used in a dietary supplement, which, by law is defined as a vitamin, a mineral, an herb or other botanical, an amino acid, a dietary substance for use by man to supplement the diet by increasing the total daily intake, or a concentrate, metabolite, a constituent, extract, or combination of these ingredients. (Source: paraphrased from Food and Drug Law Institute Series (1995) Compilation of Food and Drug Laws, Volume 1. Federal Food, Drug, and Cosmetic Act, Chapter 2, Definitions. Section 201[321].
> **Dietary supplement**—A dietary supplement is a product (other than tobacco) that is intended to supplement the diet and that bears or contains one or more specified dietary ingredients (see definition that follows). It is intended for ingestion in tablet, capsule, powder, softgel, gelcap, or liquid form. It may not be represented for use as a conventional food or as the sole item of a meal or diet; and it must be labeled as a "dietary supplement." (Source: paraphrased from Food and Drug Law Institute Series (1995) Compilation of Food and Drug Laws, Volume 1. Federal Food, Drug, and Cosmetic Act, Chapter 2, Definitions. Section 201[321].
> **DSHEA**—The Dietary Supplement Health and Education Act (DSHEA) was passed by Congress in 1994 and amended the Food, Drug, and Cosmetic Act to authorize the creation of a separate regulatory category for dietary supplements distinct from foods and drugs. DSHEA provided broad definition and guidance for what dietary supplements are, how they should and should not be marketed, what they can contain and what form they might take, and how they are to be regulated. (See preceding definition of dietary supplement.)
> **Food ingredient or additive**—All foods in the United States must be made from ingredients that are Generally Recognized As Safe (GRAS) according to FDA regulatory criteria, or which have been approved by the FDA following an agency review of a submitted food additive petition. The petition must present appropriate data demonstrating that the additive is safe for human consumption at the levels at which they are expected to be used in human food.
> **Functional Food**—A food that provides a health or performance benefit beyond providing basic nutritional value. Some examples include products fortified with certain nutrients or phytonutrients, such as fortified cereals, breads, sports drinks, bars, and fortified snack foods, baby foods, and more. **Functional Foods** is primarily a marketing term and there is no regulatory definition for it. These foods are subject to all food regulations regarding their composition and claims. *(continued)*

Table 1.1. Definitions *(Cont.)*

> **GRAS (Generally Recognized As Safe)**—General recognition of safety of food ingredients and additives may be based only on the views of experts qualified to evaluate the safety of substances added directly (or indirectly) to food. Today, FDA has developed specific procedures that industry must follow to establish GRAS status for new ingredients or additives. The process includes a review by a qualified expert panel of all available data on the product's safety for human consumption, and agreement by the panel that the product is safe.
>
> **Health claim**—A claim for a food or supplement product that describes the relationship between a food substance and a disease or health-related condition. To secure a health claim, clinical evidence substantiating the relationship to FDA's satisfaction must be presented to the agency as part of a formal petition process. Most health claims describe a reduction in disease risk with use of the food substance to distinguish the claim from those allowed only for drugs (that is, ". . . cures, treats, mitigates, or prevents a disease.") Currently, health claims are subject to an evidence-based ranking system (A = highest; D = lowest) as a means to provide FDA's assessment of how well the publicly available scientific evidence supports the claim. Unqualified health claims are those with clinical substantiation that meets FDA's significant scientific agreement (SSA) standard, and are awarded the highest ranking of "A." Qualified health claims are those that have less substantiating data than required for SSA, and they are ranked from "B" to "D." A "D" rank will indicate that FDA concludes that there is little scientific evidence to support the claim, and this statement must be made on package labeling if the claim is used.
>
> **Natural foods**—Foods derived from "natural" as opposed to chemically synthesized ingredients. Producers of natural foods strive to use ingredients and raw materials that have not been exposed to pesticides, and avoid preservatives and "synthetic ingredients" whenever possible. There is no legal or commercial standard that defines the term **natural**.
>
> **Nutraceutical**—Consumer products (foods or dietary supplements) that utilize nutrients, specialized food components (phytonutrients), or other naturally occurring chemicals (or mixtures of chemicals) as ingredients intended to provide a health benefit beyond basic nutrition. (Source: NBJ Functional Foods Report, 2002c).
>
> **Organic food**—Foods derived from farms and growers who embrace certain principles of sustainable farm management, pesticide and fertilizer use, and the humane treatment of animals. Federal laws now set standards for growing, certifying, and labeling organic foods.
>
> **Phytonutrients**—Plant food components other than macronutrients, vitamins, or minerals that have some scientific data supporting a health benefit for humans. Examples include antioxidants from grape seed extract, lycopene (the red pigment in tomatoes), or sitostanol ester, the plant sterol derivative from vegetable oils that lower cholesterol at low levels of intake. Phytonutrients can be used as either dietary ingredients in dietary supplements or food ingredients in Functional Foods as long as they meet the appropriate FDA regulatory requirements.
>
> **Structure/Function Claim**—Structure/function claims are used for dietary supplements and occasionally for food products. Most structure/function claims describe the role that a nutrient or dietary ingredient has on the structure or function of the body, for example, "calcium builds strong bones." If a dietary supplement label includes a structure/function claim, it must, by law, have a disclaimer that states: "This statement has not been evaluated by the Food and Drug Administration. This product is not intended to diagnose, treat, cure, or prevent any disease."

Why discuss the marketplace in a volume on food regulations? Quite simply, the U.S. regulatory structure governing foods and drugs has been a powerful force defining the industries that manufacture the products regulated by the FDA. Our regulatory framework has directly or indirectly shaped the structure, key success factors, barriers to entry, rules of competition, and manufacturing operations for conventional foods, various nutritional products, and pharmaceuticals. Figure 1.1 summarizes the statutory or regulatory definitions of the products FDA regulates, and Figure 1.2 shows how these product categories and their attendant regulations have ultimately shaped the boundaries around industries and the scope of their product offerings. Nutraceuticals cut across most of the categories

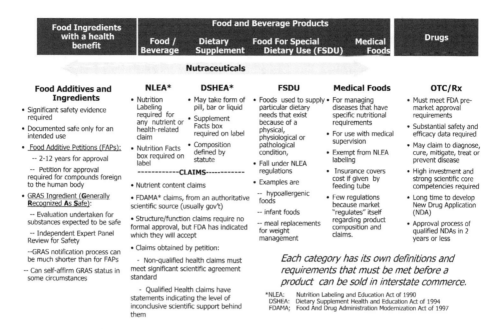

Figure 1.1. U.S. Food, Drug, and Cosmetic Act and its regulations define the product categories for the delivery of health benefits to the American consumer. "Nutraceuticals" cuts across several of these.

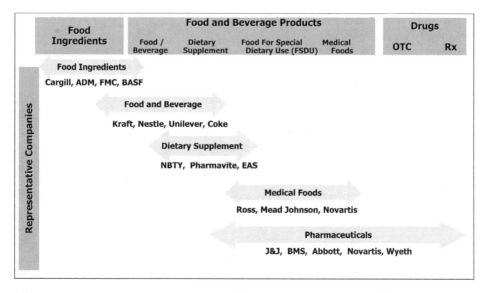

Figure 1.2. Representative global companies in the U.S. nutritional products business. The categories of products created by federal law and regulation for foods and drugs are the single biggest determinant of the industrial and market structures now in place in U.S. nutritional products businesses.

shown. Laws and regulations have been a primary influence on how the companies within each sector have evolved their core competencies, their industry culture, how each markets their products, and the customer segments they serve. Regulations directly shape the cost of entry into a particular market and define the nature of product communications among companies, customers, and end users. The evolution of the U.S. dietary supplement and functional foods markets is interesting in this context.

Dietary Supplements
Background and Market Overview

For decades, dietary supplements were primarily multivitamin/mineral preparations marketed by pharmaceutical manufacturers who built respected brand names that remain today, such as Centrum® (Wyeth), One-A-Day® (Bayer) and Theragran® (Bristol-Meyers Squibb). However, the winds of change began to whisper in the late 1980s as peer-reviewed nutrition research began to show medicinal or disease preventive effects of foods such as berries, garlic, onion and other fruits and vegetables, oats, psyllium, flax and other grains, and certain vitamins or their biological precursors. Around the same time, the National Cancer Institute began research on food phytochemicals and cancer prevention. As China opened its doors to the world following centuries of isolation, Westerners learned of an entirely different approach to health care. The most populous country on earth had not only evolved health care traditions involving acupuncture, herbs, and barefoot doctors but also appeared to be free from many of the chronic diseases that haunted the United States and Europe. Media coverage of these developments combined with growing disenchantment with the escalating costs and perceived insensitivity of health care in the United States and other Western countries made consumer attitudes ripe for alternatives to conventional therapies.

The purveyors of dietary supplements in the growing health-food channel emerged as a segment clearly distinct from the food and pharmaceutical industries in its mission, core competencies, and culture. The segment rapidly took on new, creative products that often did not pass FDA scrutiny when the agency took a close look at the level of substantiation that manufacturers provided for their product claims and, in some instances, data supporting product safety. The fact that the claimed health benefits of many products had been accepted as folklore for centuries in various Asian countries and cultures passed no muster with the agency's regulations, which required safety and efficacy testing by Western scientific methods. Despite periodic market withdrawals of products the FDA judged as unapproved food additives or unapproved drugs, pills and tablets of single-source vitamins, minerals, herbs, and plant extracts had established a significant and very loyal consumer following by the early 1990s. New consumers came to the market because of a growing curiosity about alternative medicines, an increasing suspicion of pharmaceutical products and their side effects, and a growing disenchantment with the U.S. health care system. The latter began as consumers and health care professionals alike began perceiving the industry as an ever more complicated, bureaucratic, indifferent third-party payment system that covered less and less of medical costs. By the mid-1990s as the baby boomers were entering their fifth decade of life, a sizable consumer segment was ripe for a change from what the medical establishment had to offer.

The passage of the DSHEA gave birth to a regulatory environment that provided a legal definition, and therefore credibility, to the fledgling, fragmented, but growing dietary supplement industry, driven by a passionate belief in the health value of its products. The DSHEA was a culmination of efforts by relatively small firms who, unlike larger food and pharmaceutical companies, had nothing to lose and everything to gain by changing the Food, Drug, and Cosmetic Act to their favor. Their goal was to create opportunity for the industry that, until then, could not legally introduce products to market with meaningful claims about their role in health and disease unless the industry invested the time and money into the food additive or drug approval process for claim substantiation. The Act broke down regulatory barriers to market entry by creating a separate class of dietary ingredients distinctly different from food additives and drugs (refer to Table 1.1.).

The new law did not require manufacturers to submit a dossier of preclinical and clinical studies demonstrating product safety or efficacy. Dietary supplements, including vitamins, minerals, herbs, and some specialty products had already grown to almost $5 billion in sales by 1993, primarily through the health and natural foods channels. Natural food stores were growing in number and capturing new consumers interested in health and wellness, organic foods, and beautifully merchandized displays of higher-quality fresh produce, fish, imported cheeses, fancy baked goods, and fresh-prepared foods than could be found in many mainstream supermarkets. Supplement companies seeking to bring alternative therapies to upscale, health-conscious consumers began the joyride of double-digit growth as the industry sought consumer support for the federal legislation during 1993-1994. Congressional passage of DSHEA was virtually assured by the grassroots letter-writing campaign spearheaded at health and natural food stores by the National Nutritional Foods Association, the industry association for health food retailers. At retail outlets nationwide, consumers signed form letters to their representatives in Congress in response to the (unfounded) fear that the FDA was about to cut off consumer access to their favorite products. Congress received more mail and phone calls on this matter than on any of the national issues of the day combined, including NAFTA (the North American Free Trade Agreement), health care reform, and the war in Bosnia. The dietary supplement sales and annual growth rates are shown in Table 1.2. NBJ forecasts a growth rate of about 4 percent for supplements overall from 2004–2008, with vitamins growing at about 3 percent, herbals at 0 percent, minerals at 4 percent, and specialty products in excess of 9 percent during the same period (NBJ, 2004).

Industry Participants

The leading dietary supplement companies are shown in Table 1.3. The companies listed sell pills, tablets, gelcaps, and similar dosing formats. Firms whose main businesses are liquid meal replacements (for use in weight management, diabetes, tube feeding, or general nutritional replenishment) are not included. Table 1.4 shows the highly fragmented nature of U.S. nutritional product companies. There are few leaders and many smaller firms vying for a piece of market share. Some consumer pharmaceutical companies have experimented with herbals and specialty supplements. These companies include Wyeth, with its Centrum® vitamin line, with line extensions with added herbs or phytochemicals, such as added lutein for eye health and soy isoflavones in its product targeted for women. Johnson & Johnson's Personal Products Company has an isoflavone supplement for menopausal women in its Healthy Woman line. McNeil Nutritionals, the newest operating company of J&J and that began in 2001, purchased Viactiv® Calcium Chews from Mead Johnson that

Table 1.2. Dietary supplement sales and growth rates, 1994–2002

Product Category	1994	1995	1996	1997	1998	1999	2000	2001	2002	2003	Forecast 2002–05
Vitamins	3,870	4,250	4,720	5,320	5,620	5,900	5,970	6,025	6,179	6,648	
Growth(%)		9.8	11.1	12.7	5.6	4.9	1.2	0.8	2.6	7.6	2–4
Herbs/Botanicals	2,020	2,470	2,990	3,520	3,960	4,070	4,260	4,397	4,276	4,197	
Growth(%)		22.3	21.1	17.7	12.5	2.8	4.7	3.2	−2.7	−1.8	0–1
Minerals	690	800	890	1,020	1,140	1,290	1,350	1,392	1,527	1,765	
Growth(%)		15.9	11.3	14.6	11.8	13.2	4.6	3.1	9.7	15.6	4–6
Specialty/Other	540	710	860	990	1,210	1,730	2,020	2,230	2,374	2,715	
Growth(%)		31.5	21.1	15.1	22.2	42.9	12.1	10.4	6.5	14.4	5–7
Supplements Total	7,120	8,230	9,460	10,850	11,930	12,990	13,600	14,044	14,356	15,325	
Growth(%)		15.6	14.9	14.7	10.0	8.9	4.7	3.3	2.2	6.7	3–5

Table 1.3. Leading U.S. supplement companies, 2003 (marketers of pills, tablets, and capsules)

Company	$ Million Wholesale Supplement Revenues
NBTY	690
Leiner Health Products*	490
Wyeth	480
Pharmavite*	320
Experimental and Applied Science (EAS)	240
Bayer Corporation	170
Weider Nutrition Group	160
Perrigo*	140
TwinLab Corporation	140
Nutraceutical International	120
Natural Organics (Nature's Plus)	110
GNC Manufacturing	110
Metabolife International	110
Goen Group (TrimSpa)	110
Country Life	100

Source: Compiled from *Nutrition Business Journal,* 2004a
*Companies with a substantial portion of revenues from contract manufacturing, multilevel marketing, or private labeling for retailers

Table 1.4. Universe of U.S. dietary supplement companies in 2003

	Total # of Companies	Revenues ($ Million)	% of Market
Greater than $100 Million	21	4,613	46
$20 Million–$99 Million	72	3,299	33
Less than $20 Million	793	2,028	20
	886	**9,940**	**100**

Source: *Nutrition Business Journal,* 2004a

same year, and a rejuvenated marketing program had a significant impact on sales in 2002 and 2003. Pfizer's Warner Lambert division introduced Quanterra™, an herbal supplement line that was discontinued when sales did not live up to expectations. Bayer's One-A-Day® continues to market line extensions with several herbs and phytochemicals, such as One-A-Day Weightsmart, a multivitamin supplement with added tea catechins from green tea extract, specifically, epigallocatechin gallate (EGCG), which has some data supporting an increase in metabolic rate.

The Products

The products considered here include dietary supplements sold primarily as pills, tablets, capsules, gelcaps, powders, or liquid extracts taken for overall general health and nutritional well-being or for the prevention, delay, or management of chronic disease or its symptoms. These segments include vitamins, herbs/botanicals, minerals, and specialty supplements—the latter being a "catch-all" for products that don't easily fit into the other categories. As can be seen in Table 1.2, the industry's double-digit growth trend declined and started to rebound in 2003. Herbs and botanicals plummeted to single digits and have shown a negative growth rate for the past two years. In general, negative publicity about the effectiveness and labeling accuracy of various herbal products reflected poorly on the entire category. Vitamin usage dropped to pre-DSHEA levels in 2002 but jumped to over 7 percent in 2003. Minerals, supported largely by calcium supplements, have fared considerably better, with 2003 growth in double digits. The specialty category also remains strong, because it is often driven by media coverage of clinical studies reporting product benefits. However, the explosive growth rates of some products (anywhere from 50–200 percent growth over several years) that occurred between 1995–1999 are not likely to be seen again. Examples of specialty supplements, a niche category typically led by blockbuster products that often have wide fluctuations in sales from year to year, include enzymes, hormones, probiotic products, homeopathic remedies, and animal-derived products, to name a few, and often target a specific health condition. Table 1.5 shows individual specialty supplements with the highest sales in 2003. Table 1.6 shows the 2002 rate of growth for condition-specific specialty products in the primary channels in which they are sold. Other channels through which dietary supplements are purchased include multilevel marketers, mail order, and the Internet. Health care practitioners, primarily chiropractors, are an emerging channel through which less than 10 percent of products move. Physicians are generally expected to be slow to enter this channel because the vast majority do not sell products to their patients as part of their practice. Their recommendation level for some dietary supplement products may increase significantly with creative industry outreach programs or compelling efficacy and safety data showing value relative to OTC and prescription medications. Sales growth in the health care channel was strong in 2003, at 8 percent (NBJ, 2004b), but the product volume moving through it is still small.

Table 1.5. Leading products in the $2.39 billion specialty supplement market, 2003

Product	$ Million	% change	% Total Category
Glucosamine/chondroitin	739	9	26.4
Homeopathic remedies	453	13	16.2
CoQ_{10}	258	22	9.2
Plant oils	203	32	7.3
Fish/animal oils	192	43	6.9
Probiotics	177	11	6.3
Digestive enzymes	131	24	4.7
MSM	115	4	4.1
SAMe	92	19	3.3
Bee products	77	13	2.7
Others	355	9	12.7
Total	2,791	15	

Source: *Nutrition Business Journal*, 2004a (adapted)

Table 1.6. Condition-specific dietary supplement sales, 2002(a)

Condition	Overall Growth (%)	Channel Sales	
		Natural Foods (%)	FDM(b) (%)
Arthritis	2.8	9.8	90.2
Eye health	54.6	24.2	75.8
High blood pressure	17.6	11.9	88.1
High cholesterol	17.5	33.3	66.7
Joint health	−1.6	9.1	90.9
Migraine	22.3	38.7	61.3
Osteoporosis	7.4	10.7	89.3
Weight loss	4.7	2.9	97.1
Yeast infections	16.7	53.4	46.6

Source: *Nutrition Business Journal*, 2003c

(a) Data represent growth and channel sales for dietary supplements in pill and tablet form, weight loss beverage products, and natural personal care products
(b) Food, Drug and Mass Merchandisers (excluding Wal-Mart)

Industry Issues and Barriers to Growth

The U.S. dietary supplement industry as defined here now stands at over $15 billion (USD). It faces some serious but potentially resolvable barriers to growth that derive from the current laws and regulations that govern the industry and from noncompliance with some regulations on the part of a small segment of companies.

Like a growing baby, the industry rapidly doubled its size, but is now squarely in adolescence and facing adulthood with all the associated growing pains and tough decisions about the future. Falling growth rates, consolidation, and declining prices signal the maturation of virtually any industry; however, few believed that industry adulthood would arrive so quickly. Some of the reasons for this rapid maturity go beyond the recessionary economic conditions that began in 2000 and are rooted in its own history and culture.

The classic adage "Be careful what you wish for because you just might get it" applies in spades to dietary supplements. The regulatory barriers that prevented growth in the late 1980s and early 1990s gave way with the passage of DSHEA, but not without unanticipated consequences associated with newfound marketplace freedom. When the industry took off, companies found that some of the basics of doing business in consumer products and substantiating claims were still problematic. Some of these problems are discussed next.

Product Claim Regulations Hamper Effective Marketing

A successful marketing program for any product must establish sufficient awareness and interest to initiate consumer trial, and the product must deliver on consumer expectations to generate repeat purchase. Restrictions on product claims under DSHEA effectively undermined the development of successful marketing and promotional programs based on clear communication of product benefits. The ability to communicate scientifically documented product benefits remains a challenge for the industry, which was initially limited to "structure/function claims" by regulation. Structure/function claims allow statements describing how a product affects the structure or function of the (healthy) body; however, the allowed claims are often too broad or too cryptic for meaningful communication to consumers and health professionals. Regardless of the level of recognized scientific support behind a dietary ingredient, claims that promote a product's role in disease management or treatment remain the purview of approved pharmaceuticals. Supplement (and food) companies have the option of filing a health claim petition with FDA's Center for Food Safety and Applied Nutrition (CFSAN) for either an unqualified or qualified health claim. An unqualified health claim would allow a phrase such as "may reduce the risk of (relevant chronic disease)" on package labels; however, this claim must be accompanied by clinical data to satisfy FDA's significant scientific agreement standard. Most supplement firms lack the core competencies and the funds to invest in a comprehensive clinical program to meet this standard. Qualified health claims are a new option for companies to consider. Their purpose is to allow communication of emerging scientific information about food components and disease that has not yet reached the agency's significant scientific agreement standard. As described later in this chapter, these claims are currently of limited marketing value.

In the absence of an approved health claim petition, a product that is effective in lowering serum cholesterol is limited to phrases allowed for in structure/function claims, such as "promotes heart health" or "helps the circulatory system" on package labeling. Similarly, products associated with data demonstrating effectiveness against hypertension are currently limited to phrases such as "reduces blood pressure already within the normal range." Articulating scientifically documented benefits in an accurate message that consumers understand usually establishes the product as an unapproved drug. These communication challenges undermine the industry's ability to effectively brand product lines and develop effective marketing strategies that can be backed with significant advertising and promotion dollars. The exception has been for products with nutrients long accepted by consumers as beneficial, such as calcium supplements for bone health. Nonetheless, some industry participants are willing to risk product claims that do not fall within the letter of the regulations, betting that FDA's severely restrained enforcement resources and focus on food safety and bioterrorism will keep the agency looking the other way.

The challenges of communicating product benefits to consumers through product labeling means that this information has typically been communicated through public relations programs and the media. However, what the media giveth it may also taketh away. The positive media reports of scientifically supported health benefits that prevailed in the early years and helped drive product sales have been overshadowed by recent mainstream news reports on issues of quality, safety, efficacy, and other regulatory infractions by the less responsible industry participants.

Resolution of the marketing challenges will be time consuming, expensive, or both for dietary supplement companies. In theory, one approach is to invest in unqualified health claim development for the supplements that have efficacy data for the well-documented, diet-related chronic diseases, such as heart disease, hypertension, and diabetes. This approach requires that a dietary ingredient first qualifies for as Generally Recognized As Safe (GRAS) according to FDA procedures and then substantiates its health benefit claims through adequately controlled clinical trials. However, no dietary supplement company has ever chosen this route as part of its development strategy, and few are likely to do so because of the cost and nonproprietary nature of the claim. Another approach is to continue to hammer away at FDA by taking it to court when an appropriate case can be made so that claim restrictions may be loosened. This pathway can have unpredictable results, however. The outcome of the *Pearson v. Shalala* case was in the industry's favor in that CFSAN was charged by the court to establish a means to communicate that there is **not** significant scientific agreement behind a diet/disease relationship. Unfortunately, the qualified claims the agency proposed as an outcome of the *Pearson* case (Table 1.7) are very unwieldy and of questionable value from a marketing standpoint. The FDA rationale behind qualified health claims is described later in this chapter.

Table 1.7. Qualified health claims permitted for U.S. foods and dietary supplements

1. Selenium and Cancer Risk (eligible for dietary supplements containing selenium)
    Selenium may reduce the risk of certain cancers. Some scientific evidence suggests that consumption of selenium may reduce the risk of certain forms of cancer. However, FDA has determined that this evidence is limited and not conclusive.
    or,
    Selenium may produce anticarcinogenic effects in the body. Some scientific evidence suggests that consumption of selenium may produce anticarcinogenic effects in the body. However, FDA has determined that this evidence is limited and not conclusive.
2. Antioxidant Vitamins and Cancer (eligible for dietary supplements containing Vitamin E and/or Vitamin C)
    Some scientific evidence suggests that consumption of antioxidant vitamins may reduce the risk of certain forms of cancer. However, FDA has determined that this evidence is limited and not conclusive.
    or,
    Some scientific evidence suggests that consumption of antioxidant vitamins may reduce the risk of certain forms of cancer. However FDA does not endorse this claim because this evidence is limited and not conclusive.
    or,
    FDA has determined that although some scientific evidence suggests that consumption of antioxidant vitamins may reduce the risk of certain forms of cancer, this evidence is limited and not conclusive.
3. Nuts and Heart Disease (eligible for the following whole or chopped nuts that are raw, blanched, roasted, salted and/or lightly coated and/or flavored, for which any fat or carbohydrate added must meet the definition of an insignificant amount [§101.9 (f) (1)]: almonds, hazelnuts, peanuts, pecans, some pine nuts, pistachio nuts and walnuts, or other nuts listed in the petition as long as they do not exceed 4 grams saturated fat per 50 grams
    Walnuts:
        Supportive but not conclusive research shows that eating 1.5 ounces per day of walnuts as part of a diet low in saturated fat and cholesterol may reduce the risk of heart disease. See nutrition information for fat content.
    or,
        Scientific evidence suggests but does not prove that eating 1.5 ounces per day of most nuts, such as walnuts, as part of a diet low in saturated fat and cholesterol may reduce the risk of heart disease. See nutrition information for fat content.
4. Omega-3 Fatty acids and Coronary Heart Disease (eligible for dietary supplements containing omega-3 long chain polyunsaturated fatty acids eicosapentaenoic acid (EPA) and/or docosahexaenoic acid (DHA).
    Consumption of omega-3 fatty acids may reduce the risk of coronary heart disease. FDA evaluated the data and determined that although there is scientific evidence supporting the claim, the evidence is not conclusive.   (*continued*)

Table 1.7. Qualified health claims permitted for U.S. foods and dietary supplements (*Cont.*)

5. B Vitamins and Vascular Disease (eligible for dietary supplements containing vitamins B6, B12, and/or folic acid)
   As a part of a well-balanced diet that is low in saturated fat and cholesterol, Folic Acid, Vitamin B6 and Vitamin B12 may reduce the risk of vascular disease. FDA evaluated the above claim and found that, although it is known that diets low in saturated fat and cholesterol reduce the risk of heart disease and other vascular diseases, the evidence in support of the above claim is inconclusive.
7. Phosphatidylserine and Cognitive Dysfunction and Dementia (eligible for dietary supplements for soy-derived phosphatidylserine)
   Consumption of phosphatidylserine may reduce the risk of dementia in the elderly. Very limited and preliminary scientific research suggests that phosphatidylserine may reduce the risk of dementia in the elderly. FDA concludes that there is little scientific evidence supporting this claim.
   or,
   Consumption of phosphatidylserine may reduce the risk of cognitive dysfunction in the elderly. Very limited and preliminary scientific research suggests that phosphatidylserine may reduce the risk of cognitive dysfunction in the elderly. FDA concludes that there is little scientific evidence supporting this claim.
8. 0.8 mg Folic Acid and Neural Tube Birth Defects (dietary supplements containing folic acid)
   0.8 mg folic acid in a dietary supplement is more effective in reducing the risk of neural tube defects than a lower amount in foods in common form. FDA does not endorse this claim. Public health authorities recommend that women consume 0.4 mg. folic acid daily from fortified foods or dietary supplements or both to reduce the risk of neural tube defects.

Source: FDA (2002b), Summary of Qualified Health Claims Permitted, http://www.cfsan.fda.gov/~dms/qhc-sum.html

Product Quality Problems in the Industry Persist

The early double-digit growth of the supplement industry put pressure on consistent sourcing of raw materials, especially those in herbal products and specialty supplements, for which reliable sources of supply were still being established. The high growth rates enticed foreign suppliers who were not familiar with U.S. standards for product certification or regulatory requirements for labeling. Negative media coverage increased as independent laboratories began testing products to see whether they delivered what package labels promised. One California firm made headlines because some of the raw materials it obtained from China were found to contain undeclared prescription drug ingredients. Though some of these product shortfalls could arise from irresponsible companies trying to cut costs, a very real problem is the absence of validated analytical methods for use in manufacturing controls and finished product testing to assure that label claims, and therefore consumer expectations, are met. In contrast to pharmaceuticals, for which active compound(s) are identified and methods of analysis for finished formulated products are established before a product is approved and marketed, dietary supplements have no similar regulatory requirement. Natural products are some of the most complex matrices found in the world of analytical chemistry. Sometimes they contain thousands of phytochemicals in one plant, which may vary in composition and quantity depending on season and soil. Analytical methods developed to confirm the identity of incoming raw materials often cannot be applied to finished products to confirm dosages (or serving sizes, in the regulatory language of dietary supplements) against label claims. This is especially true of combination products for which other botanicals or excipients and binders are used as part of formulations that may interfere with marker compound analysis. Consumer confidence will return when negative media reports cease and become a distant memory so that they no longer influence purchase decisions.

FDA's strained resources have prohibited the level of DSHEA enforcement that responsible companies would like to see. Companies will have to address their own quality issues internally, and the industry must agree on how to police those members whose misdeeds tarnish its image.

More Rigorous Clinical Substantiation to Support Product Claims for Many Dietary Ingredients Is Needed

Different segments of the supplement industry face different challenges in clinically substantiating product benefits. The biological behavior and nutritional value of vitamins and minerals have been studied for decades, ever since analytical methods were established to synthesize or process pure compounds of a consistent molecular identity for further study in animals and humans. The National Academy of Sciences' Institute of Medicine continues its 60-year program of periodic comprehensive and systematic scientific reviews of the known science behind recommended nutrients. This organization establishes recommended amounts as well as upper limits of intake that should not be exceeded for the healthy, noninstitutionalized members of the U.S. population as a whole (Food and Nutrition Board, 1998–2000). For more than a decade, a growing focus in the nutrition science community has been the epidemiology of nutrient (or phytonutrient) intake and its relationship to chronic disease. Frequent studies have related nutrient intake at multiples of the levels satisfying established nutritional standards to disease prevention or reduction in disease risk. Though inadequate for drawing firm conclusions about metabolic responses by individuals, these studies are often compelling to consumers who follow the media reports about their results.

Herbal, botanical, and some specialty supplement products face different hurdles for clinical substantiation of benefits. The analytical challenges discussed previously that create difficulties in confirming finished product composition also create problems in clinical study design. The absence of a standard chemical fingerprint for a particular active or group of actives means that different investigators may administer products that have the same name and taxonomy but different compositions. Lack of standard preparations for clinical testing is one of the major factors contributing to conflicting outcomes of human studies investigating herbal products and ingredients used in specialty supplements. It is an area of investigation in some of the National Institutes of Health (NIH)–funded Botanical Centers (Table 1.8). The pharmaceutical industry sought to overcome these same challenges decades ago in the search for a plant's active constituent that had a recognized chemical identity and fingerprint. The active could be synthesized in pure form to be free of the inert or other undesirable components that co-existed in the plant tissue. The ability to measure physiologic potency in terms of a pure chemical entity, whose composition and manufacture could be controlled, was one of the key reasons that administering a purified active was advantageous in the first place (NBJ, 2002a; Foster and Tyler, 2000). The enthusiastic and passionate purveyors of herbal and botanical supplements unknowingly inherited these scientific challenges with their newfound marketplace freedom under DSHEA. The responsible companies are trying to resolve these problems; however, well-designed clinical trials for many herbs and botanicals await standardized methods to vali-

Table 1.8. NIH botanical research centers

Center and Principal Investigator	Funding	Research Focus/Project Description
UCLA Center for Dietary Supplements Research: Botanicals. David Heber, MD, PhD.	$1.5 million/yr for 5 years from ODS and NCCAM	Chinese **red yeast rice** and heart disease, **green tea extract** and **soy** for effect on tumor inhibition, and **St. John's Wort** and depression.
University of Illinois at Chicago: NIH Center for Dietary Supplements Research in Women's Health. Norman Farnsworth, PhD.	$1.5 million/yr for 5 yrs from ODS, NCCAM, Office of Women's Health, and NIGMS (General Medical Sciences)	Botanicals (approximately 10) and Women's Health, especially menopause. Projects include: Standardization of botanical preparations, bioassay development for estrogenic agents, *in vitro* and *in vivo* studies of absorption, metabolism and toxicity, and clinical evaluation.
The Arizona Center for Phytomedicine Research at Tuscon. Barbara Timmerman, PhD.	$1.5 million/yr for 5 yrs from ODS, NCCAM	For the botanicals **ginger**, **tumeric** and **boswellia**, considered to have antiinflammatory (including antiarthritic) properties: (1) chemistry, MOA, (2) bioavailability, (3) pharmacokinetics, pharmacodynamics in humans.
Purdue University and the University of Alabama at Birmingham. Botanicals Research Center for Age Related Diseases. Connie M. Weaver, PhD.	$1.5 million/yr for 5 yrs from ODS, NCCAM	Age-related diseases and disorders (including osteoporosis, cancer, cardiovascular disease and loss of cognitive function). (1) **grape polyphenols** and neuroprotection; inflammation, (2) **tea catechins** and cancer, (3) **soy isoflavones** and bone resorption in post-menopausal women.
University of Missouri Center for Phytonutrient & Phytochemical Studies. Dennis Lubahn, PhD.	$6 million over 5 yrs from NIEHS	Molecular mechanisms for several phytonutrients. (1) phytonutrients and prostate tumor progression, (2) **phytoestrogens** and immunity in animals, (3) **plant polyphenols** and neuroprotection and, (4) botanical identification and characterization.
University of Iowa and Iowa State University. Center for Dietary Supplement Research. Diane F. Birt, PhD.	$6 million over 5 yrs from ODS, NCCAM, NIEHS, and other NIH Institutes	Health effects of **echinacea** and **St. John's Wort**.

Source: compiled from www.nih.gov

date plant species and analytical methods to establish a quantitative chemical fingerprint for the preparations under study. This standardization has been achieved or is in progress for some botanicals, animal tissues, and other specialty dietary ingredients such as St. John's Wort, shark cartilage, gingko biloba, glucosamine and chondroitin sulfate, tea extracts, ginger, tumeric, boswellia, grape polyphenols, soy isoflavones, and echinacea. Some of these are under investigation at the six botanical research centers awarded to several universities through competitive grants funded by NIH's National Center for Complementary and Alternative Medicine (NCCAM) and the Office of Dietary Supplements (ODS), as shown in Table 1.8.

The major U.S. pharmaceutical companies that are participants in the dietary supplement industry avoided many of these problems by marketing carefully selected actives with well-understood chemistry (or microbial species identification in the case of probiotic products) and some clinical data supporting their safety and efficacy. Firms such as Wyeth and Bayer with established vitamin supplement brands created line extensions with added specialty ingredients that these companies reviewed for safety and efficacy. Their strong medical science core competency and established marketing expertise will allow these firms to pick and choose selected dietary ingredients that resonate with consumers while minimizing any risk of product liability because of safety issues.

A creative outcome of DSHEA was a mandate for the creation of an Office of Dietary Supplements (ODS) within the NIH to study dietary ingredients. DSHEA framers recognized that scientific stature would be needed for their products and that without patent protection, their margins would not be sufficient over the long term to support expensive scientific investigations. This means that U.S. tax dollars support a sizable portion of the research on dietary ingredients used in dietary supplements. Only a few firms invest in clinical trials, and their objective is usually to substantiate marketing and advertising claims rather than to understand the fundamental aspects of biological behavior.

The ODS resides in the program offices of the NIH Director, under the Office of Disease Prevention. As are its sister program offices, the ODS is responsible for developing research initiatives and for supporting, funding, and coordinating research activities through the 27 institutes and seven centers within NIH. ODS was established in 1995 with the specific mission to support research, evaluate scientific information, and publicize research results on dietary supplements. It also supports conferences and workshops on scientific topics relevant to dietary ingredients and their usage. ODS does not have the authority to directly fund investigator–initiated research grant applications; it can, however, co-fund research either through contracts or by funding grants or awards to investigators in conjunction with the other NIH institutes and centers.

In 1998, Congress authorized the formation of the National Center of Complementary and Alternative Medicine (NCCAM) within NIH to support research, train researchers, and disseminate information to health professionals and the public about complementary and alternative medicine.

The establishment of ODS and NCCAM, combined with the aggressive growth of supplement use, served as a catalyst for scientific investigation of numerous vitamins, minerals, herbs and botanicals, and other dietary ingredients in relation to various disorders. Though investigations are funded and conducted through NIH, a large, prestigious, and highly credible venue, investigators will design their research to expand the knowledge base about these substances, and findings may not necessarily support a marketing claim that the industry wishes to make for its products. Industry will have little to no influence on how results are communicated. Science has shown mild support for the efficacy of some supplements, but results for other studies under way may test the passionately held health-related beliefs of some industry segments and their loyal consumer following. Recent examples of products that did not prove efficacious in clinical trials were St. John's Wort for major depression (Hypericum Depression Trial Study Group, 2002; Shelton et al., 2001) and ginkgo biloba for memory in adults without cognitive impairment (Solomon et al., 2002). During the first quarter of 2003, NCCAM authorized a $4 million, four-year, 300-participant study to determine St John's Wort's effectiveness in mild depression. The multicenter trial will be conducted at Massachusetts General Hospital in Boston, Cedars-Sinai Medical Center in Los Angeles, and the University of Pittsburgh Medical School.

Marketers in the glucosamine/chondroitin sulfate category, a $739 million business for dietary supplement and cosmetic applications in 2003 (NBJ, 2004b), are likely waiting for the results of the study under way by the National Institute of Arthritis, Musculoskeletal and Skin Diseases (NIAMS). This trial is investigating the role of these compounds in the prevention and treatment of osteoarthritis (OA) of the knee. The $14 million Glucosamine/Chondroitin Arthritis Intervention Trial (GAIT) study of more than 1,500 adults with OA is being coordinated by the University of Utah School of Medicine. Results are expected in 2005.

Overall, NIH financial support for research on dietary supplements is substantial. Programs encompass a wide variety of dietary ingredients, so few are under investigation in great depth. Approximately $98 million was awarded in 1999 and $117 million in 2000 according to the ODS. Independent estimates prepared from government and state databases that track funding from ODS and NCAAM suggest that these funding levels were surpassed in subsequent years.

Business and Economic Issues Increased as Industry Growth Declined

DSHEA eliminated the major technical and regulatory barriers to market entry for companies wishing to market dietary supplement products. The post DSHEA era saw a rush of companies to the market because in contrast to food additives and drugs, dietary supplements had no FDA pre-market review mandated for safety and structure/function claim substantiation. In only 10 years, an industry of almost 900 supplement companies is doing business in the United States. Almost 800 of them command less than $20 million in annual revenues (Table 1.4). With growth having reached a plateau in 2001-2002, industry fragmentation led to consolidation followed by an impressive 6 percent overall industry growth in 2003.

Price wars led to price declines, lower profit margins, and the search for alternative markets for dietary ingredients. Though many dietary ingredient suppliers now look to the food industry as an outlet for their products, many firms seek outside scientific and regulatory counsel for establishing GRAS status for their products. Larger firms in the food ingredient and pharmaceutical business that are well recognized for skills in scientific and regulatory affairs have succeeded in establishing GRAS status for some dietary ingredients such as plant sterols and stanols or their esters (McNeil Consumer Healthcare, Cargill, ADM, Novartis Consumer Health), milk-derived lactoferrin (DMV International) lutein and lutein esters (Roche/Kemin Foods, and Cognis Corporation, respectively) and lycopene (Roche, BASF).

When many companies market similar products, as occurs in the dietary supplement industry, the result can be downward pressure on prices. Average gross margins for dietary supplements are now about 40–45 percent, not unlike those of the food industry. In addition, some products that have short life cycles curtail industry growth as manufacturers respond to the latest trend, which often lasts only a few years. The business dynamic and associated economics is not conducive to investment in clinical trials, analytical method development programs, or thoughtfully designed, long-term marketing programs backed by significant advertising and promotional support.

As more dietary ingredients qualify for use in food products, the growth of functional foods should slowly increase. However, establishing GRAS status, successfully formulating ingredients into food products that meet consumer expectations for taste and texture, and securing broad retail distribution will take far more time than some dietary ingredient suppliers are used to. In addition, claim substantiation remains problematic for the benefits these ingredients provide because preparing a successful health claim petition, especially for an unqualified health claim, requires a significant investment of money and time.

Not Enough Consumers Embrace the Regular Use of Supplements

Industry estimates of heavy supplement users generally fall between 35–40 percent, and the occasional and rare users number around a third of the U.S. population. National surveys of supplement use typically take a snapshot-in-time approach, concluding that anywhere from 55 to 75 percent of the U.S. population takes a vitamin or other dietary supplement. Most consumers, however, do not do this regularly, and regular use of multiple products is what sustains the supplement industry. For many, supplement use is a discretionary decision rather than one considered essential for health. The recent economic downturn has shown that the business is vulnerable to these consumer attitudes. According to SPINS (2002), supplements in the food/drug/mass market channel averaged about $12–14 for a 30-day supply in 2001-2002, though some cost as much as $60. Despite the dramatic increase in supplement distribution over the past five years, many consumers doubt that supplements are either necessary or effective. A number of factors may further discourage the large segment of occasional users who could drive the business forward. These factors include the lack of new and intriguing products, inconsistent quality, reports of serious safety problems (including deaths in some ephedra and androstenedione users), a lack of medical endorsement for herbal and specialty products, and press coverage contradicting product claims. There are some bright notes for the industry, however. The American Medical Association now recommends the use of a multivitamin and mineral supplement for the American population at large because our lifestyle changes in recent years have compromised our nutrient intake. On the business front, industry consolidation creates larger companies with more sizable marketing budgets, making airwave and print advertising communication options to consumers. Federal Trade Commission (FTC) rules on claim substantiation differ somewhat from those of FDA, which potentially creates new opportunities for companies to bring new users into the category once a base of clinical data to support claims is established.

The Outlook for Dietary Supplements

As the industry adjusts to slower growth rates, its future depends on successful policing of the companies whose activities generate negative publicity that reflects on all members. In

2004, some members of Congress and key regulators have brought DSHEA under attack following the ban on ephedra alkyloid use in dietary supplements. S-722 is a bill imposing a premarket approval process on some dietary supplements introduced by Senator Richard Durban (D-IL) in 2003. A related piece of legislation was introduced in the House (HR-3377) and is co-sponsored by Susan Davis (D-CA), Henry Waxman (D-CA), and John Dingell (D-MI), who have been critics of the industry in the past. More bills at the national level are expected that may even go so far as to call for DSHEA's repeal or at least some strong modification. The industry is seeking support among its membership to preserve DSHEA as it is through outreach, education, and lobbying of Congress to pressure FDA for more rigorous enforcement of companies who are noncompliant with existing laws and regulations. Many responsible companies, especially the pharmaceutical companies and their suppliers with established, branded supplement products, are usually above the fray. But for those seeking bigger growth and marketplace stature in mainstream channels, the internal issues described will have to be addressed in their business plans and rectified over the near term.

Functional Foods in the United States

Background and Market Overview

As is nutraceuticals, **functional foods** is a marketing term whose definitions have ebbed and flowed over the years. The U.S. Institute of Medicine of the National Academy of Sciences has put forward a definition for functional foods that appears to satisfy marketers as well as nutritionists and food scientists. Simply put, Functional Foods are those in which the concentrations of one or more ingredients have been manipulated or modified to enhance their contribution to a healthful diet (American Dietetic Association, 1999). In practice, a small but growing segment of the retail U.S. food industry is comprised of foods for health or improved physical performance. Sometimes these products are made with added nutrients or phytochemicals. Alternatively, some are promoted for a new health benefit attributable to a food component that has always been present. Perhaps the most salient example is cranberry juice, in which the proanthocyanidins naturally present in the cranberry fruit have been shown to prevent urinary tract infections by inhibiting bacterial adhesion to the urinary tract wall and therefore the cellular reproduction needed for infection to occur. Clearly the new science has supported the old conventional wisdom about this beverage.

These products are now generally accepted as functional foods and are distinct from those that have been enriched with added nutrients for the public health purpose of preventing nutrient deficiencies in the population. These latter examples include iodine in salt, B vitamins and iron in certain cereals, pasta, and flours, or vitamin D in milk. Few of these products are available today without enrichment to meet government standards.

The large, mature food industry serves as a backdrop to the comparatively spritely little segment that comprises functional foods. This segment grew slightly in excess of 7 percent in 2003 to $21.9 billion compared to the $518 billion mainstream food industry's 3 percent growth. The retail value of organic foods was about $10.4 billion in 2003, an increase of over 20 percent from 2002 (NBJ, 2004b). The organic and natural food segments have been the source of many functional food acquisitions by mainstream food companies. Despite the slow growth rates, the food industry is highly competitive, with thin-to-modest margins along the value chain from agricultural commodities to retail food products. Compared to the natural/organic and dietary supplement industries, the mainstream food industry culture can be characterized as risk averse and seeking a harmonious rather than confrontational relationship with its regulators. Against this cultural landscape, larger companies have positioned their products for health benefits much more conservatively, generally with stronger scientific and regulatory support than those in the dietary supplement industry. In the early 1990s the food industry did not take seriously the small companies who took risks with product claims lacking scientific rigor. As a result, mainstream food companies and their industry associations were caught by surprise by DSHEA's passage. However, the growth of nutritional products in general since the 1994 legislation became law has caught the food industry's attention in a big way.

Traditional food industry messages on health and nutrition have centered on the fact that a balanced diet includes a wide variety of foods, including sweets and snacks, all consumed in moderation. For the most part, this nutrition education philosophy has historically served industry participants well, assuring consumers and the manufacturers of all foods, including salty snacks, carbonated beverages, dessert products, and confectionery that these products had a place in the American diet. When consumer interest in low-fat foods materialized during the early 1990s, many companies offered low- or reduced-fat alternatives. Intensive industry efforts reduced the fat content of processed foods wherever it was possible through ingredient substitution and new ingredient technology development, though there has been general agreement that taste qualities were often compromised. The meat and meat-processing industry provided substantially leaner products at retail through traditional breeding techniques, modified animal diets, and increased trimming of visible fat. These efforts not only addressed consumer demand but also helped to exceed the specific objectives to the industry made in 1990 by the U.S. Department of Health and Human Services' "Healthy People 2010" initiative. The assigned food industry target to increase the availability of reduced-fat processed foods was achieved in substantially fewer than five years (National Center for Health Statistics, 2001).

The functional food concept that began to catch on in the early 1990s ran counter to the trend of reducing the less desirable food components such as sugar and fat. The original concept for functional foods allowed a food product to become a delivery vehicle for a particular health benefit that was distinct from traditional mainstream food commodities and brands (Wrick, 1994) though that definition expanded over the 1990s (Wrick and Shaffer, 2002; NBJ, 2002b). In order to capitalize on rapid growth of new products while also protecting existing brand franchises, major food firms such as Kraft and Nestle acquired functional food brands in the energy bar market (Balance Bar and PowerBar®, respectively).

Despite the interest in the functional food concept, most companies entered the market through acquisition because line extensions for many prepared foods positioned for better health would imply that the flagship brand was nutritionally inferior. Only a few food and beverage companies, who had products already associated with good nutrition, adopted a strategy that included a health positioning for line extensions to current brand franchises. Tropicana Orange Juice (PepsiCo) is one of the most notable examples, with two FDA approved health claims to their credit. Their calcium-fortified citrus juices contain Fruitcal®, the brand name for calcium citrate malate, the acid-soluble form of calcium for which they have an exclusive license from the ingredient's inventor, Procter & Gamble. These products provide similar amounts of calcium to a glass of milk without the chalky mouth feel or precipitation of calcium salts typically encountered in acidic products. As a result, they qualify for the FDA unqualified health claim that relates calcium intake to the prevention of osteoporosis, though the company has not consistently used the claim as a marketing tool. Tropicana later filed for the first health claim under FDAMA (the Food and Drug Administration Modernization Act). This statute allowed Tropicana to claim that consumption of its citrus juice products could help reduce the risk of hypertension and stroke by virtue of the authoritative statements made by the National Academy of Sciences National Research Council publication, *Diet and Health* (1988). This publication reviewed the well-accepted relationship between increased potassium and low sodium intake and the reduced incidence of high blood pressure and stroke.

Commodity businesses such as poultry and fish began promoting the health benefits of naturally occurring components that scientific reports indicated were important, such as omega−3 fatty acids in fish, or in eggs from hens on special diets. Cancer-fighting phytochemicals in cruciferous vegetables (that is, glucosinolate compounds found naturally in broccoli, broccoli sprouts, Brussels sprouts, and other vegetables) were occasionally touted in promotional materials, as was the FDA health claim that the increased consumption of fruits and vegetables could reduce the risk of certain forms of cancer. Though scientifically accurate, these health messages did not resonate well with many consumers, and few companies used these messages on package labels. More recently, the National Cancer Institute "5-A-Day" program has been used to try to encourage more produce consumption for its vitamin and phytochemical content.

These conservative approaches to communicating health messages relative to the dietary supplement and natural/organic segments, and the industry's history of conflict-avoidance with the FDA, have resulted in a slower, more gradual start of the functional foods category than was achieved with dietary supplements following the passage of DSHEA. But the cautious cultural environment may be just what is needed to sustain a more gradual and prolonged growth curve than that initially achieved for the dietary supplement segment. Clearly, the 7–9 percent growth rates for functional foods during 2001–2003 (NBJ 2002b, 2003b, 2004) began to turn the heads of industry executives toward a serious evaluation of their companies' opportunities in this market. Historical and projected sales for functional foods are shown in Table 1.9.

Table 1.9. Historical and projected sales of functional foods in the U.S.

	1997	1998	1999	2000	2001	2002	2003	2008e
Sales $B	$13.7	14.8	16.1	17.4	18.9	20.5	21.9	29.7
Growth (%)		8.0	8.8	8.0	8.6	9.0	7.7	Est. 6.8%/yr
% of total food industry sales	3.0	3.2	3.4	3.5	3.8	4.0	4.3	Est. 5–6%

Source: *Nutrition Business Journal*, 2002c, 2003a, 2004a estimates

Nutrition Business Journal defines functional foods as follows: Foods fortified with added (nutrients) or concentrated ingredients to a functional level that improves health and/or performance. They also include products marketed for their "inherent" functional qualities. They include some enriched cereals, breads, sports drinks, bars, fortified snack foods, baby foods, prepared meals, and more.

Participants and Products

The first mainstream U.S. food product that could be called a functional food by today's definition was Kellogg's All Bran®, a wheat bran cereal high in fiber and well known for its laxative properties. In 1984, Kellogg pushed the envelope on FDA labeling regulations and, in conjunction with the National Cancer Institute, described the relationship of a high-fiber diet to cancer prevention, based on the epidemiological evidence of the day, on the back of the All Bran package. Kellogg's strategy with All Bran led to the establishment of the health claim regulations we have in place today. Later, the Quaker Oats Company successfully filed a health claim petition for oats and the reduced risk of heart disease, based on accumulated evidence that oat bran (specifically the beta glucan–rich soluble fiber fraction) helped lower serum cholesterol.

Kellogg began a functional foods division in 1996 based on its patented technologies for soluble fiber derived from psyllium seed husk, which was awarded a health claim for reduced heart disease risk. However, this path was later abandoned because product acceptability of the Ensemble product line with psyllium fiber did not pass the critical taste and texture acceptability test with consumers, despite its health benefits of lowered cholesterol. Today, with its acquisition of Loma Linda, Worthington Foods, and Morningstar Farms brands, Kellogg is well positioned to take advantage of the FDA-approved health claim for soy and its role in reducing the risk of heart disease.

Companies participating in the functional food market today vary considerably and include large, mainstream multinationals such as PepsiCo, General Mills, Kellogg, Coca-Cola, and Kraft, who collectively had about a 34 percent share of the $21.9 billion market in 2003. Another 25 branded manufacturers are responsible for more than $100 million each in U.S. functional food sales (NBJ, 2003a). The remaining companies focus primarily on the natural/organic channel, which has served as a testing ground for many nutritional products that have grown over time to a size attractive enough to be acquired by large food companies. An example of one of the relative newcomers to the United States is Red Bull GmbH, which markets, to young adults, beverages with caffeine, taurine, and gluconolactones for energy and alertness. Red Bull® has been marketed for more than 14 years in Europe and is now found in more than 60 countries, from Angola to Yemen. Since its founding in 1992, Ferrulito, Vultaggio & Sons, owners of AriZona brand beverages, have become leaders in the New Age beverage category, which includes flavored teas and fruit-based beverages with added herbs and botanicals; whether the levels provided of these added ingredients deliver any demonstrable health benefit is not clear, however. No health-related claims are made for the products. Clif Bar® was introduced to the United States in 1992 and grew to a $130 million business in 10 years. Clif Bar, Inc. has also launched line extensions, including one with a salty taste rather than the sweet flavors offered by other energy bar products. All products provide high-carbohydrate, low-fat, portable nutrition initially targeted to the physically active who want to be able to eat immediately before, during, or after their sport. These products have evolved to meet the mainstream need for a quick yet satisfying snack or meal replacement when time is not available for breakfast or lunch.

PepsiCo, with its acquisition of Quaker Oats and its Gatorade® brand in 2001 is now the largest functional food company, though its Pepsi beverage franchise and Frito Lay snacks still make up the lion's share of the business. Pepsi's brands associated with health and nutrition or performance include Quaker cereals, Tropicana® and Dole® Juices,

Lipton® teas (which Pepsi distributes for Unilever), Aquafina® bottled water, SoBe® non-carbonated soft drinks, and Gatorade. In addition to the success of the Quaker and Tropicana health-claim positioning, teas that contribute polyphenols and antioxidants to the diet are gently promoted through print media rather than product labeling and advertising. SoBe beverages are either tea- or fruit juice–based and have added caffeine for "mind-blowing energy," or herbal extracts. Gatorade, a 30-year-old functional beverage, replenishes some of the salt lost through perspiration and is well accepted for its hydrating and rapid fluid replenishment during and after exercise. Today, in addition to the athletes to whom the product was originally targeted many years ago, a broad consumer segment has emerged who appreciates the sweet-salty taste, and the brand has a global presence.

General Mills has been very proactive in repositioning long-standing brands such as Cheerios®, which is now promoted for heart health based on oat content. General Mills was responsible for the FDAMA health claim petition that FDA accepted regarding the role of whole-grain foods in reducing the risk of heart disease and certain cancers. Recently, a joint venture between General Mills and Dupont's Protein Technologies International (Dupont/PTI's Solae) launched a new brand, 8th Continent™, a significant entry into the soymilk market. The 8th Continent™ product is formulated with soy protein isolate, which has successfully overcome many (but not all) of the flavor problems that can occur when the crushed whole bean is used in soymilk manufacture. Most soymilks in the marketplace utilize the heart health claim for soy because it appeals to their health-conscious target market. As a result, these soymilks contain at least 6.25 grams of soy protein per serving, which is the qualifying amount required by regulation for use of the claim.

Though soymilks have had the strongest consumer acceptance as reflected in soyfood sales, other soy products are now considered mainstream following acquisitions of their brands by large food companies. Consumers have been exposed to soy as an ingredient in many processed foods for decades. In the late 1960s, soy protein was sold as a hamburger extender when beef prices temporarily skyrocketed. Long before the Solae joint venture between DuPont and PTI was even conceived, PTI built an extensive scientific database of peer-reviewed studies on the health benefits of soy and soy protein, from feeding milk-intolerant or allergic infants to cholesterol reduction in adults. Work in the latter area gradually built the significant scientific agreement in the nutrition and medical science communities required to support its health claim for reduced risk of heart disease. The market for soymilk is forecast to reach $1 billion by 2005. About six firms command more than 70 percent of the market and include White Wave, Hain, Eden Foods, Imagine Foods, Vitasoy USA, and Pacific Foods. General Mills' 8th Continent™ is expected to expand the category and take shares from others. Soy-based meat alternatives are also a significant category. Today's products are meat-free patties that look like hamburgers but have a grilled, meat-like taste. Large multinationals such as Con Agra (Lightlife), Kraft (Boca Burger) and Kellogg (Worthington, Loma Linda) have acquired soy businesses and are now major players.

Entry of soy foods into the mainstream grocery channel occurred around the time that FDA published its final rule allowing for the soy/heart disease health claim for products containing 6.25 grams of soy protein per serving. Curiosity and interest in soy had already started in consumers, and the claim became a true driver of soy food sales. Increases in purchases of soy-based foods of more than 20 percent were reported within the first year the claim was used. However, not all products can be reformulated to contain enough soy to qualify for the health claim because of the large amount of soy protein required per serv-

Table 1.10. FDA criteria for the health claim relating soy protein intake to the reduction in risk for heart disease

---

Soy protein—Must have at least 6.25 grams of soy protein per RACC (Recommended Amount Commonly Consumed).

Low fat—Each serving or RACC must contain less than 3 grams of fat. (Requirements as stated in 21CFR §101.62.) If the product consists of or is derived from whole soybeans and contains no fat in addition to the fat inherently present in the whole soybeans, the 3-gram limit will not apply.

Low saturated fat—A serving or RACC must contain less than 1 gram of saturated fat. The total amount of saturated fat per serving must be no more than 15% of total calories. (Requirements as stated in 21CFR §101.62.)

Low cholesterol—A serving or RACC of food must also contain no more than 20 milligrams (or less) of cholesterol. (Requirements as stated in 21CFR §101.62.)

Note that the eligibility of the food to bear the soy health claim is based on the soy protein per RACC per eating occasion so that 4 eating occasions would provide a total daily intake of 25 grams soy protein. For many foods, but not all, a RACC equates to one serving. After the food qualifies for the health claim, the label statement for the amount of soy protein per serving is based on the actual serving size of the food. Foods currently on the market that are eligible for the health claim include soy-based beverages, tofu, tempeh, and soy-based meat alternatives.

Source: Food and Drug Administration, 1999, Food Labeling: Health Claims; Soy protein and Coronary Heart Disease: Final Rule. *Federal Register* 64(206): 57732–57733

---

ing as well as other food composition criteria. When the criteria listed in Table 1.10 are met, a product can claim in labeling that "diets low in saturated fat and cholesterol and high in soy protein may reduce the risk of heart disease." The FDA has indicated that 25 grams of soy protein a day (or 4 servings) are needed to positively impact the risk of cardiovascular disease. Soymilk, meat alternatives, and some nutritional bar products can meet the 6.25 grams of protein per serving while meeting the requirements for fat, cholesterol, and sodium content. Other product categories will find formulating to these standards without compromising consumer taste and texture expectations more challenging.

Barriers and Challenges to the Growth of Functional Foods

Today, the functional food industry is poised for growth but faces technical, marketing, intellectual property, and regulatory challenges. The same drivers that helped shape the business environment for nutraceutical products are still in place today; however, incorporating dietary ingredient technologies into food products creates a unique set of challenges.

The Technical Challenges

Not all products can successfully incorporate added nutrients or phytochemicals. Commodity products such as meat, poultry, fish, and eggs can have their composition altered to some extent by the animal's diet, but delivery of truly unique health benefits will probably require genetic engineering, which is not yet accepted by consumers. Should consumer skepticism soften, the future might bring genetic alterations to beef and dairy cow metabolism to produce increased levels of naturally occurring conjugated linoleic acid (CLA) in meat and milk products. Strong evidence supports the claim that increased levels of dietary CLA in animals help to convert fat tissue to muscle; however, this finding awaits substantiation in humans regarding a role in weight management. Should these benefits be confirmed, CLA will likely be sought after for addition to other food products, if it qualifies as GRAS at the levels needed to deliver health benefits.

Some foods, such as milk, chocolate, and mayonnaise, have government regulations or standards of identity governing product composition that make adding novel ingredients difficult. Unless these standards are revised, adding novel ingredients may require renaming these products to something that implies "imitation" to consumers. The compositional standards for foods such as chocolate have strong economic and labeling implications in world trade, so little incentive exists to change them to incorporate ingredients that appeal to a relatively small consumer segment.

Other indulgent foods, such as ice cream and sweet baked goods, have historically not been candidates for this kind of fortification because consumers want to retain their tradition as a dessert or a special treat, though this may change as consumer attitudes and ingredient technologies evolve. The nutrition bars in the U.S. market have been the creative exception, using confectionery or bakery technology but being positioned as healthful, not too sweet, low in fat, satisfying, and delivering energy and a few added nutrients in a portable, convenient format.

Functional food ingredients may offer formulation challenges as well. Unlike most dietary supplements, which are in the form of pills or tablets, functional foods have the challenge of meeting the taste preferences of target consumer groups. Many functional food ingredients, such as isolated nutrients, phytochemicals, and plant extracts, have inherently unpleasant flavors. When added to foods to meet nutritional requirements, levels of vitamins and some minerals are usually low enough that many undesirable flavor notes or chalky textures are not perceived on the palate, or they can be masked with other ingredients. Higher levels can cause flavor problems over the shelf life of some products. Even if qualifying as GRAS, many herbs, plant extracts, or other phytochemicals that have been found to deliver health benefits have strong, unpleasant flavors or textures. At worst, their flavors cannot be masked or hidden, and at best, their use can cause delays in the food formulation process or increased ingredient costs in order to achieve sensory targets. Modifying ingredients that are complex molecular blends to address sensory problems without compromising health benefits can be a major research undertaking without a guaranteed positive outcome. Resolution of these issues can add significantly to product development time.

The Marketing Challenges

Functional foods have different marketing challenges than those of dietary supplements. Some successful products in the functional food category have capitalized on long-known diet-health relationships for their products or their ingredients, such as calcium and strong bones, carbohydrates and physical energy, and caffeine and alertness. Very little "low-hanging fruit" is left for marketers. Ingredients providing health benefits that may have a complicated educational message require much more careful positioning. The experience with plant sterol- or plant stanol-enriched margarine-like spreads has reinforced the fact that not all consumers with high cholesterol are interested in managing it through diet. And of those who are, many don't want to use a higher-priced, though good-tasting, spread multiple times a day even if it makes their cholesterol go down. These products developed a small but very loyal consumer following, but mainstream U.S. consumers who might be at risk for heart disease were not necessarily responsive to a product positioning that reminds them of it. Chronic diseases such as hypertension, diabetes, and cardiovascular disease have no demographic boundaries. Identifying the target consumers within a group of people who suffer from a particular disorder usually requires psychographic (for example, attitudinal) segmentation and knowledge of consumer purchasing patterns in order to estimate potential market size. Market research tools exist to do this, though this research is expensive.

To date, marketing a functional food to address a specific medical condition has not proved successful in the eyes of the mainstream food industry. "Food as medicine" appeals to a relatively small consumer segment and will not generate the revenue needed to justify the necessary resources for development or to satisfy shareholder expectations. Large food companies are generally not interested in products with a market potential of less than several hundred million dollars. Industry is now betting that consumers seek positive nutrition in general for overall health and wellness, outside the context of preventing specific diseases. The health benefits of a high protein intake for weight loss as part of a low carbohydrate diet, calcium for strong bones, and dietary fiber and probiotic bacteria for gastrointestinal health are now promoted in a wide variety of functional food categories that are growing at attractive rates. Collectively, these health and wellness products became about as popular as vitamin-fortified products, with a 49 percent and 46 percent share of 2002 functional food sales, respectively (NBJ, 2003b).

The concept of functional foods with key ingredients responsible for a particular health benefit may shift as marketers change their focus from foods for disease management and prevention to the broader message about health and wellness. Ingredients associated with good health but without a link to disease could be used in products marketed to a larger audience. The soy market gives a clue to this trend. That soy is a healthful alternative to meat and dairy protein may be the only message needed to drive consumer sales and for marketers to avoid regulations that can be cumbersome.

The Regulatory Landscape for Nutraceuticals

The regulatory landscape has always had a strong influence on the evolving market for nutraceutical ingredients whether used in foods or supplements, and it will continue to do so for years to come. In the years after DSHEA, when dietary supplement growth was at double-digit rates, very few ingredients were available that had passed regulatory safety requirements for use in food, and insufficient scientific data was available on most to clear the regulatory hurdle of "significant scientific agreement" for health claims. The high costs of building new brands and the need to protect existing ones created downward pressure on functional food development for many processed-food companies. In addition, formulation with some phytochemical actives in foods and beverages presented technical challenges in meeting consumer preferences for taste and texture when the amounts needed to deliver the promised health benefit were used.

A major regulatory distinction exists between the ingredients used for claimed health benefits for foods and for dietary supplements. Any substance added to food, for whatever reason, must meet FDA standards for GRAS status or be an approved food additive, having received agency approval via a formal petition to the FDA documenting safety under conditions of intended use. The burden of establishing safety clearly rests with the companies that introduce these ingredients into the food supply. Alternatively, dietary supplement companies are free by federal statute to utilize ingredients that have a different (and controversial) safety standard that does not routinely require a formal FDA premarket scientific review or approval. Dietary ingredients must demonstrate a history of use or other evidence of safety that establishes that they can reasonably be expected to be safe when used as specified by the product label. The burden of proving a dietary ingredient unsafe remains the responsibility of the resource-starved FDA, though this element of the DSHEA statute may be poised for serious reconsideration.

Following DSHEA's passage, it was easy to speculate that the time and expense in establishing GRAS status and meeting the significant scientific agreement standard for a potential health claim could be avoided by putting a dietary supplement product in food form. This strategy was tested in 1998 with McNeil Consumer Healthcare's Benecol® spread and salad dressing lines, which contained the patented and clinically proven cholesterol-lowering ingredient plant stanol esters (sitostanol ester). These products were labeled and positioned as dietary supplements in promotional campaigns to health professionals prior to product launch. Some well-respected attorneys in food and drug law insisted that case law would support this strategy should an FDA court challenge ensue. FDA did intervene with the Benecol launch plan and required that plant stanol esters qualify for GRAS status before the product could enter interstate commerce in the United States. McNeil chose to comply with the FDA's request rather than go to court. Fortunately for McNeil, sufficient scientific backup was already on file so that the notification document supporting GRAS status could be prepared in a matter of weeks. Unilever's Take Control® spread, which also lowered cholesterol using a slightly different patented plant sterol–based ingredient with GRAS status, was introduced to the retail market within weeks of Benecol. Both companies submitted health claim petitions, and FDA granted permission to use a claim for reduced risk of heart disease for products utilizing either plant sterol or plant stanol esters. The strength of the scientific data supporting the health benefit of these ingredients allowed a petition review period of only a few months, one of the fastest on record. Numerous plant sterol–derived ingredients are now available for use; however, products utilizing them represent only about 1 percent of functional food sales (NBJ, 2003a).

Intellectual Property and Regulatory Barriers to Growth

Several barriers to growth for the nutraceutical ingredients used in functional foods and dietary supplements exist. They have centered on intellectual property, claim exclusivity, and federal regulations that may evolve more slowly than the technical and scientific developments in the business environment of the products that are regulated.

There is a dearth of proprietary ingredients or manufacturing technologies that create a clear marketplace advantage for most bioactives derived from plant or animal tissues that offer well documented health benefits and are suitable for use in foods or supplement products. Many bioactives from herbs and plants have been used in folklore and do not qualify for patent protection. Vitamins and mineral preparations have been manufactured for decades and are now priced low enough to be considered commodity ingredients. Any industry know-how, trade secrets, or critical patents with remaining life are held in the hands of companies that have been in the business of nutrient manufacture for a long time.

Though providing a strong technology base, proprietary positioning by means of patented ingredients or manufacturing methods generally have not offered strong marketplace advantages for most dietary or food ingredients used in supplements or foods as they have for pharmaceutical companies. Where patents and licensing agreements exist in the supplement industry, infringements also exist and are costly for smaller companies to fight in court. New chemical forms of traditional nutrients, such as vitamin or mineral chelates, continue to be developed and sometimes patented. However, the majority of supplement companies are too small to defray the cost and often do not have the internal technical expertise to design, manage, and oversee improvements over existing products. Though larger firms could make these investments, anticipated benefits often are not sufficiently attractive to consumers to bring the needed sales or justify the price premium needed to recover clinical costs as well as the strong marketing and communications program needed. An additional disincentive to supplement companies to conducting clinical work is that the trials take a long time to design, conduct, analyze data and draw conclusions, followed by many months of peer review before publication in a respectable, peer-reviewed journal. The timeline from study design to publication is often longer than the product life cycle for some dietary supplement products, though not necessarily for functional foods.

The Raisio Group in Finland had a patented technology in the production and use of stanol esters (sitostanol ester) for lowering cholesterol and found tremendous success in its home country when it launched Benecol margarine. In an unfortunate but classic case of underestimating the competitive landscape for the technology in foreign markets, Raisio and its U.S. marketing partner, McNeil Consumer Products (now McNeil Nutritionals), executed a business plan that did not consider the competitive ingredient technologies already patented or in the development pipeline. Raisio's patent protection offered little marketplace advantage because other food phytochemicals also lower serum cholesterol. McNeil's entry into the retail margarine case with national ambitions prompted Unilever to dust off its plant sterols technology, manufacture a comparable product at lower cost and greater profit, and equally share the market space with McNeil. Even if plant stanol esters could be determined to be more effective at lower levels of intake than any other cholesterol-reducing food, ingredient, or drug, FDA regulations would preclude comparative claims. The approved health claim is for a reduction in the risk of heart disease, not for lowering cholesterol by a certain amount.

Current federal statute requires that products making claims for disease treatment or prevention are drugs and can be sold in interstate commerce only when they meet strict premarket approval requirements for safety and efficacy. This remains the case even when abundant and compelling peer-reviewed science supports a role for amelioration or treatment of disease or its symptoms that should support a "truthful and non-misleading" standard in many cases. In addition, there has been no clear route for companies to communicate this science to consumers in an understandable manner in order to drive sales. Further, razor-thin margins in the food and supplement industry discourage investment in preclinical or clinical work in the face of these restraints. Approved health claims for foods and supplements are generic, not exclusive to the petitioner, and may not always reflect what a product actually does, should studies indicate that it is effective in treating or managing disease. Technology development for new ingredients that are sufficiently safe for use in our food supply and could contribute public health benefits against chronic disease will remain curtailed unless some established regulatory barriers are eased.

One possible approach to this dilemma would have the FDA set a regulatory framework to allow a period of claim exclusivity to the company that invests in the ingredient development, safety testing, and claim substantiation, a model similar to that of the U.S. orphan drug regulations. A fledgling industry group called the Research-based Dietary Ingredient Association (RDIA) was beginning to evaluate workable routes to ultimately propose to FDA. However, this organization never quite achieved the critical mass of industry support and funding required to create a force for change. In addition, the 9/11 terrorist attacks in New York initiated a refocus of FDA's mission and strategy to work with industry to ensure the safety of the U.S. food supply from deliberate attempts to undermine its safety through terrorism.

In light of the new national priorities, legislative or regulatory change designed to foster food and dietary ingredient innovation is getting little if any attention. This absence will likely mean only sporadic development for new nutraceutical ingredients for the foreseeable future. New ingredients that are developed will generally find their way into dietary supplements before functional foods unless the business opportunity is clearly greater for a food application. DHA is one of the n-3 fatty acids with a legacy of clinical research supporting its role in infant retinal and cognitive development. Martek Biosciences developed the methods to produce purified DHA from algal fermentation, made the business decision to seek the support of the infant formula industry worldwide, and proactively sought approval from governments for inclusion in infant formula products. The company's less attractive alternative was to enter the dietary supplement market with its less stable market dynamics. The technical hurdles with the inherent instability of n-3 fatty acids are slowly being overcome, and DHA is now used in many infant formulas internationally.

Recent Developments at the U.S. FDA Liberalize Health Claim Regulations

A major revision to the process for making product claims about diet and health for conventional foods and dietary supplements was initiated in 2002. The revision was based on accumulated experience with product claims in labeling and advertising, and especially the requirements of the 1999 Court of Appeals decision regarding the *Pearson v. Shalala* case. Briefly, the outcome of this ruling mandated that FDA allow qualified health claims on dietary supplements that did not meet the rigorous "significant scientific agreement" standard in place since health claims for foods were first authorized. In practice, this means that a claim can be made for a diet-disease relationship that had not reached significant scientific agreement as long as the consumer is apprised that the data supporting the claim are not yet conclusive.

An FDA guidance document described a process for systematically evaluating and ranking the scientific evidence for a qualified health claim. The ranking system uses an A, B, C, or D grading system. A Grade A claim would meet significant scientific agreement standards for a traditional (now called "unqualified") health claim, and a B Grade would be assigned to those petitions for which good scientific evidence exists supporting the claim but for which the evidence is not entirely conclusive. A C Grade would apply to claims for which the evidence is limited and inconclusive; a D grade would be given to claims with little scientific support.

A petition must be submitted for unqualified and qualified health claims so that FDA can review each potentially qualified claim on a case-by-case basis. Petitions will be accepted for both conventional foods and dietary supplements. Qualified claims will be reviewed according to a new agency policy that will utilize a "weight of the scientific evidence" standard, as has long been used by the (FTC) for claims made in print and airwave media advertising. Petitioners must specifically request a review for a qualified health claim and provide a body of data that is sufficiently scientifically persuasive to demonstrate to expert reviewers that the weight of the evidence supports the proposed claim. The data they provide need not rise to the level of significant scientific agreement.

In evaluating whether claims used in package labeling are truthful and not misleading, FDA will use a "reasonable consumer" standard (as opposed to the "ignorant, unthinking, or credulous consumer"). The reasonable consumer standard is consistent with FTC deception analysis, which means that its use by FDA will contribute to the rationalization of the legal and regulatory environment for food and supplement promotion. This standard also reflects the agency's belief that consumers are active partners in their own health care and behave in health-promoting ways when given health information (FDA, 2002b).

The qualified claims allowed as of September 2003 are listed in Table 1.7, which appears earlier. Each claim listed has some conditions for use that are specific to each. All must meet the health claim requirements of 21 CFR 101.14 except for the significant scientific agreement standard. Other conditions for use include claim placement on the label, package recommendations conforming to the National Academy of Sciences/Institute of Medicine's (NAS/IOM) Tolerable Upper Intake Levels that should not be exceeded, and/or particular product nutrient content specifications. As can be seen in Table 1.7, the allowed claims are not ones that most marketers will immediately embrace. However, as has happened over time with structure/function claims, the specific wording on package labels might be negotiated with the agency on a case-by-case basis. This would inch marketers closer to label-claim strategies to which consumers are more receptive despite the built-in disclaimer regarding inconclusive or lack of scientific support. It will be interesting to see whether the new FDA policy toward claim liberalization will encourage additional clinical research to reach a significant scientific agreement standard for some products, especially foods.

Long-Lasting Drivers That Favor Growth of Functional Foods and Nutraceuticals

There are some industry drivers favoring gradual, long-term growth of functional foods; these factors should provide some patience to consumer product companies that find the longer product development process that some functional foods can take to be frustrating.

For one, consumers in the "baby boom" category now comprise about 27 percent of the U.S. population, and this group, which in general is interested in good health, will be around for a long time. For this group, the food-health connection is already made, though some subsegments may need more convincing of the link between food and disease prevention.

Second, Americans of any age will always eat food for energy and basic nutritional needs. For many, dietary supplements beyond the multivitamin taken for "nutritional insurance" are viewed as expendable. This means that there will likely be a place for food products to be the source of new health benefits. The time pressures of our twenty-first century American lifestyles have led us to become a nation of "grazers" who grab the opportunity to eat when they can, whether while walking to work or riding in a car. Nonetheless, while the supplement industry addresses the marketing challenge of winning over new consumers, most Americans don't yet take supplements regularly, but they all eat food every day.

Third, though some dietary supplements may claim to suppress appetite, supplements in general will never satisfy hunger when someone has gone for extensive periods without eating. Only food can do that.

Functional foods have the potential to be a major growth driver for the food industry; however, the right mix of market research, marketing, science and technology, branding, distribution, pricing, product taste and convenience, and adequate development time before launch has been elusive for the pioneers in the business. Kellogg (Ensemble), Novartis (Aviva line), and McNeil (Benecol) all had mainstream ambitions for a business that appears for now to be destined to start out as a niche. Nonetheless, functional foods might yet become premium brands, even within product segments at or approaching commodity status. Tropicana's brands are a preeminent example. NBJ (2002b) reports that Tropicana's venture into functional foods has brought its citrus beverages to far greater than a $1 billion brand. If Soyatech Inc.'s forecasts are correct, soy food sales will exceed that figure by 2005. A "health and wellness" marketing trend for mainstream foods appears to be under way now that will position products with health benefits as "good-for-you" or "better-for-you" rather than be designed to prevent specific diseases. If successful, this trend will continue to help functional foods become one of the larger-growing segments in the mainstream food industry.

The Future Outlook

Consumer interest in finding real or perceived solutions to health problems outside the traditional Rx and OTC routes will continue to grow as consumers pick up an ever-increasing share of their health care costs and as subsequent activism about quality of health care increases. Continued pressure on physician productivity within managed-care institutions will further compromise the quality of doctor-patient communication. Any further publicity about research studies questioning the claimed benefits of blockbuster pharmaceutical products, as has occurred with hormone replacement therapy (HRT), will continue to shake the confidence of a significant consumer segment that will look for alternatives. A positive outcome of the NIH glucosamine trial, anticipated by about 2005, will help strengthen the credibility of specialty supplements. The dietary supplement industry has some challenging times ahead. It must plan to adjust its businesses to new regulations governing current good manufacturing practices, for which FDA published a proposed set of standards in March of 2003. In addition, the industry must prepare to deal with potential legislative initiatives in Congress, which has the potential to eliminate some of the industry's freedom in selecting products for market in the interest of product safety.

The near-term outlook for functional foods should be brighter as more dietary ingredients begin to qualify for GRAS status and can therefore be used in foods. Time will be required for food or dietary ingredient firms and consumer product companies to establish safety and then substantiate health claims, and the new avenue created for qualified health claims created by FDA in early 2003 may mean increased opportunities for responsible product labeling claims made under standards that are harmonized with those of the FTC.

One might envision a significant, perhaps double-digit growth for functional foods, which now comprise less than 5 percent of retail foods sold in the U.S., if incentives for ingredient technology development, such as claim exclusivity, eventually become part of the regulatory framework. Such initiatives would lead to a greater search, exploration, and clinical study of potential ingredients with health benefits safe enough for use in foods. One possible scenario could entail the development of food or supplement products that are readily accepted by consumers, show comparable efficacy to some medications in well-designed clinical trials, and are also low enough in cost relative to medications to make managed-care companies begin to take notice. Currently, however neither the FDA nor the pharmaceutical industry is interested in further blurring the lines between foods and drugs, and the managed-care companies are focused on far more near-term industry structure, economic issues, and congressional attitudes toward health care regulation at a national level. But thoughtful and creative pharmaceutical companies could begin the strategy now to identify a bioactive that is safe enough for GRAS status and provides a similar benefit as an established class of pharmaceuticals. These companies have access to and credibility with the medical and scientific community, and they are capable of rapid clinical trial design and execution for building the scientific data and support product claims. A partnership with the right food company would link them to food marketing expertise and manufacturing, and capabilities for nationwide access. The time may be right again for the pharmaceutical companies to revisit functional foods as a strategic option and capitalize on different times and lessons learned.

References

American Dietetic Association. 1999. Functional Foods. A Position Paper of the American Dietetic Association. *J. Am. Diet. Assoc.* 99:1278–1285.

DeFelice, Stephen. 1991. The Nutraceutical Initiative: A Proposal for Economic and Regulatory Reform. The Foundation for Innovation in Medicine. Cranford, New Jersey, 07016.

Food and Drug Administration. 2002. http://www.cfsan.fda.gov/~dms/guidance.html; or, http://www.fda.gov/ohrms/dockets/default.htm.

Food and Drug Administration. 2002b. Guidance for Industry: Qualified Health Claims in the Labeling of Conventional Foods and Dietary Supplements. http://www.cfsan.fda.gov/~dms/hclmgui2.html.

Food and Drug Administration. 1999. Food Labeling: Health Claims; Soy Protein and Coronary Heart Disease: Final Rule. *Federal Register* 64(206):57700–57733

Food and Drug Administration. 2002. Food Labeling: health claims: soluble dietary fiber from certain foods and coronary heart disease. Interim Final Rule. *Federal Register* 2002, 67(191):61773–61783.

Food and Nutrition Board, Institute of Medicine. 2004. Dietary Reference Intakes for Water, Potassium, Sodium, Chloride and Sulfate. National Academy Press. Washington, D.C.

Food and Nutrition Board, Institute of Medicine. 2002. Dietary Reference Intakes for Energy, Carbohydrate, Fiber, Fat, Fatty Acids, Cholesterol, Protein and Amino Acids. Parts I and II. National Academy Press. Washington, D.C.

Food and Nutrition Board, Institute of Medicine. 2001. Dietary Reference Intakes for Vitamin A, Vitamin K, Arsenic, Boron, Chromium, Copper, Iodine, Iron, Manganese, Molybdenum, Nickle Silicon, Vanadium and Zinc. National Academy Press. Washington, D.C.

Food and Nutrition Board, Institute of Medicine. 2000. Dietary Reference Intakes for Thiamine, Riboflavin, Niacin, Vitamin B_6, Folate, Vitamin B_{12}, Pantothenic Acid, Biotin and Choline. National Academy Press. Washington, D.C.

Food and Nutrition Board, Institute of Medicine. 2000. Dietary Reference Intakes for Vitamin C, Vitamin E, Selenium and Carotenoids. National Academy Press. Washington, D.C.

Food and Nutrition Board, Institute of Medicine. 1997. Dietary Reference Intakes for Calcium, Phosphorus, Magnesium, Vitamin D and Fluoride. National Academy Press. Washington, D.C.

Foster, S. and V.E. Tyler. 2000. *Tyler's Honest Herbal. A Sensible Guide to the Use of Herbs and Related Remedies*, Haworth Press, New York, p. 1–19.

Hata Y., M. Yamamoto, M. Ohni, K. Nakajima, and Y. Nakamura. 1996. A placebo-controlled study of the effect of sour milk on blood pressure in hypertensive subjects. *Am. J. Clin Nutr.* 64:767–771.

Hypericum Depression Trial Study Group. 2002. Effect of *hypericum perforatum* (St. John's Wort) in major depressive disorders. A randomized controlled trial. *J.A.M.A.* 287:1807–1814.
NAS/NRC. 1988. Diet and Health. National Academy Press: Washington, D.C.
National Center for Health Statistics. 2001. *Healthy People 2000*, Final Review. Hyattsville, MD, Public Health Service. Library of Congress catalog card number 76-641-496, Table 2b.
National Research Council (U.S.), committee on Diet and Health. 1989. *Diet and Health: Implications for Reducing Chronic Disease Risk*. Food and Nutrition Board, Commission on Life Sciences, National Research Council. National Academy Press: Washington, D.C.
Nutrition Business Journal. 2002a. *Dietary Supplement Business Report*. 2001. Penton Media: San Diego, CA.
Nutrition Business Journal. 2002b. *Annual Industry Overview* VII, 7(5/6):1–30.
Nutrition Business Journal. 2002c. *Functional Foods Report 2002*. Penton Media: San Diego, CA.
Nutrition Business Journal. 2003a. *Functional Foods VI*, 8(2/3):1–31.
Nutrition Business Journal. 2003b. *NBJ's Annual Industry Overview VIII*, 8(5/6):1–47.
Nutrition Business Journal. 2003c. *Niche Markets VI: Specialty Supplements,* 8(7):1–31.
Nutrition Business Journal 2004a. *NBJ's Annual Industry Overview IX,* 9(5/6):3.
Nutrition Business Journal 2004b. *NBJ Newport Summit 2004 Market Overview Snapshot.* Coronado Island Marriott Resort. July 21–23, 2004. info@nutritionbusiness.com
Seppo L., T. Jauhiainen, T. Poussa, and R. Korpela. 2003. A fermented milk high in bioactive peptides has a blood pressure–lowering effect in hypertensive subjects. *Am. J. Clin. Nutr.* 77(2):326–330.
Shekelle P., S. Morton, M. Maglione, et al. 2003. Ephedra and Ephedrine for Weight Loss and Athletic Performance Enhancement: Clinical Efficacy and Side Effects. Evidence Report/Technology Assessment No. 76 (Prepared by Southern California Evidence-based Practice Center, RAND, under Contract No 290-97-0001, Task Order No. 9). AHRQ Publication No. 03-E022. Rockville, MD: Agency for Healthcare Research and Quality. February 2003.
Shelton, R.C., M.B. Keller, A. Gelenberg, D.L. Dunner, et al. 2001. Effectiveness of St. John's Wort in major depression. *J.A.M.A.* 285:1978–1986.
Solomon, P.R., F. Adas, A. Silver, J. Zinna, et al. 2002. Gingko for memory enhancement. A randomized controlled trial. *J.A.M.A.* 288:835–840.
SPINS, Inc. 2002. San Francisco, personal communication.
SPINS/Soyatech, in association with Arthur D. Little. 2002. Soy foods: The U.S. Market 2002. Soyatech, Inc. Bar Harbor Maine.
Tufts Center for the Study of Drug Development. 2002. Outlook 2002. Boston, MA, www.tufts.edu/med/csdd.
Wrick, K.L.. 1994. The Business Prospects for Nutraceuticals in the United States. Decision Resources, Inc. Waltham, MA. Nov., 165. 02453
Wrick, K.L. and J. Shaffer. 2002. A Nutraceuticals Update: Business Gains and Growing Pains of the U.S. Dietary Supplements Sector. Spectrum/Life Sciences, Pharmaceutical Industry Dynamics, Decision Resources, Inc. Waltham, MA 02453.

2 The Regulatory Context for the Use of Health Claims and the Marketing of Functional Foods: Global Principles

Michael Heasman

This chapter considers the distinct and highly contentious area of global functional food and nutraceutical regulatory activity that is the statutory and voluntary use of health claims on foodstuffs. Health claims and their regulation and use are some of the more complex and controversial issues facing policy makers, public health specialists, and the food and drink industry nationally and internationally in the area of functional foods (Lawrence and Raynor, 1998). The development and use of health claims is still an evolving area, and the future regulatory framework and climate, especially in the context of the global marketing and trade in foodstuffs, is far from clear. But the value and use of health claims on foodstuffs and in particular their use in food marketing are seen by many in the food industry as central to the future marketing success of nutrition-based products.

The aim here is to examine, in terms of general principles, the evolution of the regulation and use of health claims, using examples from around the world. The marketing waters are muddied further by the fact that enormous variation exists from country to country in the regulation (and regulatory tolerance) of nutrition and health messages on foods and drinks (Childs, 1998). So, to provide a generalized framework in which to consider regulation and food marketing activity with respect to health claims, which by necessity entails simplifying many complex issues, this chapter concludes with a series of checklists detailing general observations about the state of health claims use internationally.

The core health claims regulatory principle can, in general terms, be characterized by the need to reconcile the historic regulatory prohibition on foods that claim to prevent, treat, or cure disease with how such prohibition is restricting the modern communication of the role of a healthful diet in maintaining good health and reducing the risk of disease. Such a regulatory challenge has arisen from growing scientific understanding about the role of diet and various nutrients and bioactive substances in foods and the development of the concept of optimum nutrition, that is, foods that benefit an individual, including a role in the prevention of disease, beyond the normal provision of energy and nutrients (Heasman and Mellentin, 2001).

Added to this regulatory conundrum has been the increasing pressure by the international food industry to shift the policy and regulatory agenda to a loosening of restrictions on the use of health claims on foodstuffs driven to meet marketing and competitive goals, against a public health perspective which looks to meet the health needs of populations and "at risk" groups in society (Truswell, 1998). Such groups include those parts of the population facing diet-related ill-health who are socially and economically disadvantaged and so far largely excluded from the functional food and nutraceutical marketing practices of food and beverage producers.

The importance of the ability to develop and use health claims on products gained growing momentum during the 1990s on public health grounds, from an industry perspective, and from an international scientific standpoint, but, crucially, the regulatory environment to a large extent remains incomplete, inconclusive, and confusing. This is despite the fact that by the early 2000s most developed countries had proposed or introduced some sort of regulatory proposal or framework for the use of health claims. Apart from Japan, no country has developed regulations that specifically defines functional foods or nutraceuticals and, in the case of Japan, the term **functional food** was dropped (to avoid confusion with pharmaceutical regulations) and the term **Foods for Specified Health Uses**—the FOSHU system—was adopted. Another key issue and principle not resolved in the regulatory arena is the protection of intellectual property with respect to products and ingredients with health-promoting properties, including issues of exclusivity and confidentiality to encourage product innovation in addition to investments in research and development.

General Principles in the Regulation of Health Claims

In the United Kingdom, a voluntary Code of Practice for the approval of health claims on foods has been developed by the Joint Health Claims Initiative (2001) set up in June 1997 among consumer organizations, enforcement authorities, and industry bodies. In the preamble to the Code, the JHCI neatly summarizes key general principles for the use of health claims that could also apply in the regulatory context. For the development of the Code, the following objectives were taken into account:

- To protect and promote public health
- To provide accurate and responsible information relating to food to enable consumers to make informed choices
- To promote fair trade and innovation in the food industry
- To promote consistency in the use of health claims internationally

But what is a health claim? The numerous formal and legal definitions vary by country. The international food standards authority, Codex Alimentarius, has defined a health claim as follows: ". . . any representation that states, suggests or implies that a relationship exists between a food or a nutrient or other substances contained in a food and a disease or health-related condition."

A health claim is, therefore, fundamentally different in principal from a "nutrition claim," a "nutrient function claim," or a "nutrient content claim" (for example, that a food is low in fat, reduced in cholesterol, or high in fiber) and in many countries falls into that gray regulatory area between food and medicinal claims (that is, that a food is capable of curing, treating, or preventing a human disease or any reference to such properties) (see Clydesdale, 1997).

Further, the term "health claim" can be subdivided:

- Claims that refer to possible disease risk factors (for example, "can help lower blood cholesterol")
- Claims that refer to nutrient function (for example, "calcium is needed to build strong bones and teeth")
- Claims that refer to recommended dietary practice (for example, "eat more oily fish for a healthy lifestyle")

Positive messages that make links between foods and health may help consumers to maintain or move toward a healthy diet. Many health claims in use on food products today are in effect "nutrient function" claims—very much like the regulated "structure/function" claims in the United States. Examples include:

- "Calcium aids in the development of strong bones and teeth"
- "Protein helps build and repair body tissues"
- "Iron is a factor in red blood cell formation"
- "Folic acid contributes to the normal growth of the foetus"

Health claim regulation is further complicated by the need for two types of health claim: "generic" health claims, that is, claims that can be applied across a range of foodstuffs and open for anyone to use (see Tables 2.1 and 2.2 for U.S. examples of generic health claims); and "product specific" health claims, which apply to a single product or product range. Further distinctions can be made between a "direct" health claim, that is, for the food itself rather than its ingredients or category (for example, food X has been shown to maintain healthy skin) and an "indirect" claim, that is, applying to the food ingredients rather than the food itself (for example, X is good for the heart; this food is high in X) (adapted from JHCI, 2001). Tables 2.1 and 2.2 illustrate such distinctions by detailing examples of the nutrition and dietary scope of regulated health claims approved by the Food and Drug Administration (FDA) (Table 2.1) and examples of "model" health claims statements suggested by the FDA (Table 2.2).

Scientific Substantiation of Health Claims: Developments in the United States and European Union

A further key general principle for the international community in developing regulation for health claims is the need to clarify and strengthen the requirements for evidence to substantiate health claims. But the level of scientific substantiation and its relationship to health claim regulation is probably the most contentious issue in the regulation of functional foods and nutraceuticals.

Table 2.1. FDA-approved generic health claims (as of December 2002)

1. Potassium and the risk of high blood pressure and stroke
2. Whole grain foods and risk of heart disease and certain cancers
3. Plant sterol/stanol esters and risk of coronary heart disease
4. Soy protein and risk of coronary heart disease
5. Soluble fiber from certain foods and risk of coronary heart disease
6. Dietary sugar alcohol and dental caries
7. Folate and neural tube defects
8. Fruits and vegetables and cancer
9. Fruits, vegetables and grain products that contain fiber, particularly soluble fiber, and risk of coronary heart disease
10. Fiber-containing grain products, fruits, and vegetables and cancer
11. Dietary saturated fat and cholesterol and risk of coronary heart disease
12. Dietary fat and cancer
13. Sodium and hypertension
14. Calcium and osteoporosis

Table 2.2. Examples of FDA model health claims and statements

---
Calcium and osteoporosis: "Regular exercise and a healthy diet with enough calcium help teens and young adult white and Asian women maintain good bone health and may reduce their high risk of osteoporosis later in life."
Saturated fat: "While many factors affect heart disease, diets low in saturated fat and cholesterol may reduce the risk of this disease."
Folate and neural tube defects: "Healthful diets with adequate folate may reduce a woman's risk of having a child with a brain or spinal cord defect."
Plant sterol-based ingredients: "Foods containing at least 0.65 gram per serving of vegetable oil sterol esters, eaten twice a day with meals for a daily total intake of at least 1.3 grams, as part of a diet low in saturated fat and cholesterol, may reduce the risk of heart disease. A serving of [*name of food*] supplies __ grams of (vegetable oil sterol esters)/(plant stanol esters)."
Soy protein: "25 grams of soy protein a day, as part of a diet low in saturated fat and cholesterol, may reduce the risk of heart disease. A serving of [*name of food*] supplies __ grams of soy protein."
FDAMA approved statements—Whole grains
  Required wording of the claim: "Diets rich in whole grain foods and other plant foods and low in total fat, saturated fat, and cholesterol may reduce the risk of heart disease and some cancers."
FDAMA approved statements—Potassium
  Required wording for the claim: "Diets containing foods that are a good source of potassium and that are low in sodium may reduce the risk of high blood pressure and stroke."

Source: U.S. Food and Drug Administration, 2002

For example, in the United States, the scientific substantiation of health claims on foods and beverages has turned into a free speech (First Amendment) regulatory issue. Following a series of court rulings, the practical outcome of this principle was the FDA's issuing, on December 18, 2002, of new guidance on the scientific substantiation of health claims, which many commentators believed would open the market for their use on foods. In announcing its new initiative, the FCA changed its previous position by saying that it would allow future health claims to be supported by "qualified" instead of "significant" scientific data, as required previously.

The FDA says that its Consumer Health Information for Better Nutrition initiative is designed to foster two complementary goals concerning the labeling of food and dietary supplements: (1) to encourage makers of conventional foods and dietary supplements to make accurate, science-based claims about the health benefits of their products; and (2) to help eliminate bogus labeling claims by taking on those dietary supplement marketers who make false or misleading claims.

The FDA said that by putting credible, science-based information in the hands of consumers, it hopes to foster competition based on the real nutritional value of foods rather than on portion size or spurious and unreliable claims, and to help empower consumers to make smart, healthful choices about the foods they buy and consume.

To meet the criteria for making a new, qualified claim on a conventional food, the manufacturer must provide a credible body of scientific data supporting the claim. The company must demonstrate, based on a fair review by scientific experts of the totality of information available, that the "weight of scientific evidence" supports the proposed claim. All qualified health claims will require review by FDA before they may be used on the food label.

At the same time, the FDA announced the establishment of a Task Force on Consumer Health Information for Better Nutrition. The task force, which came into being in January

2003, will develop a framework to help consumers obtain accurate, up-to-date, and science-based information about conventional food and dietary supplements. This includes the development of additional scientific guidance on how the "weight of the evidence" standard will be applied, as well as the development of regulations that will give these principles the force and the effect of law.

The European Union Struggles to Find a Health Claim Regulatory Framework

In the European Union, the approach to the scientific substantiation of health claims and the path to a regulatory framework has been less bloody in public arenas than in the United States, but still fiercely contested. Sweden was the first European country to develop a program for the use of generic health claims in the labeling and marketing of food products. The program first came into effect in August 1990 but was revised in August 1996. The code allows eight connections between diet related diseases, or risk factors, namely for obesity, cholesterol level in the blood, blood pressure, atherosclerosis, constipation, osteoporosis, dental caries, and iron deficiency (Swedish Nutrition Foundation, 1996).

In the UK, the JHCI, the self-regulatory body referred to earlier, has "approved" six food-related health claims, the first on October 12, 2001. However, of these six, the JHCI warns that three should not be used on foods or food advertisements because the JHCI regards these claims as illegal under current UK food labeling regulation. The three claims approved (all on February 15, 2002) but considered illegal by the JHCI are the following:

- "Eating more vegetables as part of a healthy lifestyle may help reduce the risk of bowel cancer."
- "Eating more fruit may help reduce the risk of lung cancer. This does not overcome the adverse effects of smoking on lung cancer."
- "Eating more fruit and vegetables may help reduce the risk of stomach cancer."

The "approved" claims that the JHCI considers may be used in food marketing are the following:

- "People with a healthy heart tend to eat more wholegrain foods as part of a healthy lifestyle" (approved February 4, 2002).
- "Decreasing dietary saturates (saturated fat) can help lower blood cholesterol" (approved November 12, 2001).
- "The inclusion of at least 25g of soya protein per day, as part of a diet low in saturated fat, can help reduce blood cholesterol levels" (approved July 30, 2002).

During the late 1990s, Europe-wide moves to develop the scientific basis for health claims regulation were led by the International Life Sciences Institute (ILSI) Europe. But much of the scientific push toward allowing health claims and the communication of health benefits of foodstuffs has been developed with little empirical evidence or consideration of what is known about the communication of health messages. For example, ILSI Europe, although producing outstanding reviews of the science and technologies behind functional foods (Bellisle et al., 1998; Knorr, 1998) in its 1999 "Functional Foods in Europe Consensus Document," mapped out a policy and communications strategy for the public for health claims and functional foods without referring to a single reference or other evidence on the labeling and communication of health messages on products (Diplock et al., 1999).

More recently, ILSI Europe has initiated an EU Concerted Action called PASSCLAIM, which commenced in April 2001 and is due to conclude with a consensus document in 2005. The EU-funded PASSCLAIM initiative is collating current and potential claims and describing scientific principles needed to support these claims, developing criteria to substantiate claims and basic scientific support, and assessing the utility and validity of biomarkers to support claims. Areas of health being investigated in this respect are as follows: diet-related cardiovascular disease; bone health and osteoporosis; physical performance and fitness; insulin sensitivity and diabetes risk; diet-related cancer; mental state and performance; and gut health and immunity. Results of the first stages of this work were published in a special supplement to the *European Journal of Nutrition* in March 2003 (PASSCLAIM, 2003). General issues raised as important during the early stages of this work (up to the end of 2002) include:

- The demarcation between functional foods and pharmaceuticals
- The real distinction between what the project calls claims for proposed enhanced function and those for the reduction of disease risk (in particular, identifying the distinction between the two)
- The need to have an evidence-based approach to the substantiation process
- The standard of evidence required for generic as opposed to product-specific claims.

To what extent the work of the PASSCLAIM project and other country specific initiatives such as the JHCI in the UK and others within European countries will have on the final European regulatory framework is a moot point. Unlike other parts of the world, the EU has lagged behind in the regulatory field on health claims, with the Commission of the European Communities publishing only a draft proposal for regulation of nutritional, functional, and health claims in July 2002 (European Commission, 2002). Views on the proposal were gathered from industry, consumer groups, regulatory authorities in European member states, and other interested parties, and a second draft of the proposed regulation was in the public arena in March 2003. At the time the draft proposals were published there was still room for modification, which was likely because in some areas the food industry considered the regulatory proposals particularly restrictive, such as the Commission's proposal to prohibit certain claims (set out in Article 6 of the draft). The proposals should be finalized in 2003.

As part of the general goal to keep European consumers properly informed and in an effort to fight food-related diseases such as obesity, the Commission has also requested a detailed study, due out in 2003, to look at how consumers react to information and claims provided on product labels and packaging.

The Growing Importance of Health Claims in Food Marketing

The 1990s saw an almost remarkable convergence and rethinking about regulatory initiatives in the arena of health claims. For example, the United States and Japan have both put into place formal regulations that allow, if a company so chooses, a process for the approval of a generic and product-specific health claim on foods and drinks. In the United

States, there are now 14 Food and Drug Administration–approved health claims, and in Japan, by the end of 2002, more than 300 products had achieved FOSHU status that allows a product-specific health claim. In Europe, regulatory proposals are in place that will allow health claims of some sort to be made sooner rather than later.

Therefore, a significant shift in regulatory principle has occurred, something Geiger (1998) captures by saying that during the twentieth century, food labeling laws and regulations evolved from protecting consumers from economic harm to protecting consumers from health risk. The more recent momentum in this direction reflects advances in food science and technology, research in nutrition and health, and consumer demand for "health benefit" information. There has also been concentrated action by the industry and its representatives to influence the policy environment for health claims.

But global regulatory developments in the United States, Japan, and Europe have to be set in the context of a food economy that has seen the marketing of hundreds of new products or repositioning of established products all sending out health messages and nutrition-based claims to consumers. Many of these unregulated health messages are still supported by good, even groundbreaking science. At the same time, unfortunately, many products are of dubious merit, including those making illegal or misleading health claims.

The Marketing Benefits of Regulated and Voluntary Health Claims

On the plus side for health claims use and their appropriate regulation, some empirical evidence shows that the marketing of foodstuffs using explicit health claims actually sells more product, thereby imparting a health benefit directly to consumers. The classic early study in this respect was by Ippolito and Mathios (1990), who examined the effect on fiber consumption after the Kellogg Company used advertisements and explicit package labeling claims on its All-Bran® breakfast cereal in the United States in 1984 to inform consumers that a high-fiber/low-fat diet could reduce the risk of developing certain forms of cancer.

Adding to the credibility of the Kellogg promotion was a highly respected third-party reference—the National Cancer Institute (NCI)—which reviewed Kellogg's fiber-cancer claims prior to the launch of its All-Bran cereal promotion. The NCI was used in promotional materials to endorse the Kellogg campaign. At the time, the Kellogg promotion was also illegal, being in direct contradiction to FDA policy that had banned the use of health claims to promote food products (Hutt, 1986). The Kellogg campaign, in terms of public policy, was the stimulus for a full review of the use of health claims in food promotion (Hutt, 1986; Ippolito and Mathios, 1990) and resulted in the ban on health claims in the United States being suspended while this took place. But just as important, the Kellogg campaign was shown to have a significant nutritional as well as policy impact. After 1984, per capita consumption of fiber in cereals increased to double that of their previous levels, mainly because manufacturers started to increase the fiber content of their cereal products. As Ippolito and Mathios (1990) say:

> Cereals introduced between 1985 and 1987 were significantly higher in fiber than the average cereal on the market in 1984 . . . . Thus, the evidence on new product developments in the cereal market is consistent with the hypothesis that the ability to use health claims to advertise new products is a significant factor in stimulating the development and introduction of more nutritious products in the market. (p.424)

They estimate, using a simplified model of the cereal market, that the revenues for high-fiber cereals increased $280 million **over** market projections in 1987, which implied an increase of approximately two million households eating high-fiber cereals due to the advertising. The authors point out, however, that it was not calculated how important the role of the NCI was as an "authoritative source" in Kellogg's success. Second, they highlight how it was not known how market structure, with few firms accounting for the majority of the U.S. cereal market, produced the movement toward more healthful products. In other words, the market structure and key company players may be just as important as the health message itself.

More recently, since the late 1990s, major food companies in the United States have been showing the rest of the world how to effectively use regulated health claims to promote nutrition messages to consumers. Key players have been the Quaker Oats Company and General Mills in exploiting FDA approved health claims for oats and whole grains, respectively, in relation to heart health. Another successful company developing a strategy using regulated health claims has been PepsiCo-owned Tropicana Products Inc., which clearly demonstrates how "good health sells." The Tropicana nutrition marketing success stories are for its orange juice products: Pure Premium Calcium and Pure Premium Double Vitamin C and 100% E, but it is how the company developed the potassium health claim for orange juice that serves as a model for the principle of using regulated health claims in marketing. The process and result are briefly described here.

Tropicana submitted notification for a claim in relation to potassium and a reduced risk of high blood pressure and stroke to the FDA on July 3, 2000, under the provisions of the FDA Modernization Act of 1997 (FDAMA). The company's notification was based on an authoritative statement published by the U.S. National Academy of Sciences in a publication called *Diet and Health: Implications for Reducing Chronic Disease*. An important point to note is that this was published in 1989—so Tropicana did not need up-to-the-minute science or have to undertake human clinical trials to submit a dossier for the regulation of a health claim.

The rationale behind the potassium claim was that 50 million Americans have high blood pressure, stroke is the third leading cause of death, and orange juice is a good source of potassium, having 450 mg in 8 ounces, or 13 percent of the U.S. daily value (DV). Tropicana also had data showing that 82 percent of Americans do not consume the recommended amount of potassium and that awareness about the relationship between potassium and blood pressure was low.

The health claim was allowed after October 31, 2000, with the claim stating: "Diets containing foods that are good sources of potassium and low in sodium may reduce the risk of high blood pressure and stroke."

Even though the health claim is a generic one (that is, open to anyone to use), Tropicana calculated it could "own the claim because we were first." To ensure its ownership, the company had in place a multichannel $10 million support plan to put into action immediately after the claim became public. The Tropicana support plan included a new health logo, packaging graphics, point-of-sale material, advertising, PR, employee communications, a launch event, customer communications, health professional communication, and Web site development (www.tropicanahealthnews.com). Tropicana had the new graphics on the shelf within three weeks; the company distributed through retail outlets more than two million point-of-sale materials in the first few months and had the company's heart logo integrated into retailers' holiday ads. The company claims that the potassium health claim support plan accelerated Pure Premium growth by 16 percent in the first few months from the FDA approval (McGill, 2002).

But, as a caveat, other companies have been equally successful at this nutrition marketing game without using FDA-approved health claims—notably, Heinz with its lycopene campaign (lycopene being a bioactive substance found in tomatoes and associated with the reduced risk of certain cancers) and Ocean Spray with its cranberry juice. Ocean Spray built a highly successful PR campaign surrounding the use of cranberry juice for urinary tract infection, starting with just one peer-reviewed study published in 1994 (Heasman and Mellentin, 2001).

In Europe, the probiotic "little bottle" market for fermented milk products has grown from nothing in 1995 to several hundred million dollars and has involved communicating a complicated health message to consumers (gut health and the role of "good" or "friendly" bacteria) in an environment without health-claims regulation. Cholesterol-lowering spreads in the UK (created by McNeil—a division of Johnson & Johnson—and Unilever) have created a $100+ million market without health-claims regulation. The energy bar market in the United States has grown from a humble beginning in the early 1990s into a $1.1 billion giant without relying on health-claims regulation (Health Strategy Consulting, 2003). Energy drinks have become a $1+ billion global market, again without using regulated health claims. More generally, functional soft drink consumption across the United States, Japan, and 16 West European countries passed the 12 billion litre mark in 2002, representing 6 percent of soft drinks in these markets—compared to just 4 percent in 1998 (Zenith International, 2003), all without a consistent health claims regulatory structure in place.

Public Health and Consumer Advocacy Concerns about Marketing Using Health Claims

As the momentum to establish, regulate, and implement health claims grows and the marketing of nutrition and products with health benefits becomes widespread, there has, at the same time, been mounting public health concern about the health effects of current dietary practices, prompted in particular by the alarming growth in the number of overweight and obese individuals. A revised assessment of the prevalence of obesity worldwide published in 2003 estimates the numbers of overweight and obese at 1.7 billion people, 50 percent higher than earlier estimates (International Obesity Task Force, 2003). Of particular concern is the rise in obesity and being overweight among children. The world of public health has become increasingly vocal in criticizing the food industry. For example, on the launch in March 2003 of a major report it commissioned (WHO, 2003) and produced by a panel of experts from around the world, the World Health Organization (WHO) criticized the food industry for the "heavy marketing practices of energy-dense micronutrient-poor foods." The WHO called on the food industry to reduce amounts of certain types of fats as well as salt and sugar in snacks and processed foods. The report particularly singled out the prevention of obesity in children as a priority, including recommending restricting the consumption of packaged snacks and sugar-sweetened soft drinks (Fleck, 2003). The report will form the basis of a major new WHO global strategy on diet and physical activity to be unveiled in 2004 (WHO, 2003).

In relation to health-claim regulatory activity, it is not clear what long-term public health impact that health claim policy for foodstuffs, if turned into regulation, will have on the health and well being of the populations experiencing diet-related ill-health. Health claim labeling, even if regulated, may confuse some consumers about food and health by increasing the health message load in the market place. A general principle largely absent from regulatory activity around health claims is long-term evaluation of health claims and the marketing of nutrition-related claims on changes to eating behavior leading to favorable health outcomes. In addition, relatively few new resources have been put into enforcement efforts for the prevention and control of misleading or fraudulent health claims in relation to foods and ingredients.

But even if the regulation of health claims is an appropriate marketing route, the evidence for their effectiveness is often contradictory. For example, although surveys clearly indicate that a substantial majority of consumers wish to be informed about the health benefits of foodstuffs or believe that food and nutrition play a "great role" in maintaining or improving overall health (Schmidt et al., 1997), other research suggests that the proliferation of health messages on products, both approved and unofficial, can leave consumers cynical about health messages and confused, rather than enlightened, about nutrition (McNutt, 1997; Patterson et al., 2001). Moreover, it has been argued that the purpose of the food label is to sell the product and, by itself, is not an adequate means of nutrition education (Kessler, 1999).

The public health arguments for health claims are that they enable consumers to develop and maintain healthful dietary practices and enable consumers to be informed promptly and effectively of important new knowledge regarding the nutritional and health benefits of food. But for many public health professionals, a basic problem exists with the way the health claims debate has been framed; they argue that a fundamental public health nutrition principle is that the *total diet*, not individual food products, determines health. Some health professionals see singling out foodstuffs by health claims as undermining that principle. A further major concern is achieving the right balance between consumer choice and consumer protection. Much of the marketing of food and drinks with health messages is seen as misleading, even for food manufacturers. For example as Richardson (1998) writes:

> . . . what could be extremely frustrating to reputable food manufacturers is that functional foods backed by considerable research efforts and investment could be undermined by, and appear to be the same as, those that are crude, carelessly made, lacking substantiating evidence, and, at worse, fraudulent. If the consumer did not believe or trust the ability of a product to provide the stated benefits, the long-term credibility of the industry would soon be damaged (p. 205).

As well as public health concerns about the health status of the food supply, functional foods and the claims being made for them have been directly criticized by consumer advocacy groups. The headline in the September 25, 1989, edition of *Business Week* sums up the problem:

Snap, Crackle, Stop—States crack down on misleading food claims

The article continued: "Stroll down a supermarket aisle these days. You could swear you're in a drug store" (*Business Week*, 1989).

The article was prompted by the fact that at the end of the 1980s, food marketing was pursuing with a vengeance the promotion of food and drink products to help prevent disease to such an extent it was causing concern to consumer advocacy groups and regulators. Around 40 percent of food products introduced in the first half of 1989 bore health messages. In the late 1980s, Kellogg, Quaker Oats, General Mills, and the Campbell Soup Company were all under legal attack about the health claims they were making for their products (*Business Week*, 1989). The years 1984–1990 have been described as a health claims free-for-all in the United States—or a golden age, depending on your point of view (Silverglade, 1995).

During the 1990s, consumer advocacy groups saw developments in functional foods and nutraceuticals as leading to a new concern about the health messages appearing on foods and beverages. For example, in a 1999 report by The International Association of Consumer Food Organizations, they argued that functional foods might amount to little more than "21st century quackery" (International Association of Consumer Food Organizations, 1999). Across the globe, a general principle of the consumer advocacy movement has been to call for strict national and international regulation of health claims.

And the campaign against misleading labeling, especially in relation to health messages on foods, has continued. In February 2003, for example, the UK's National Consumer Council (NCC) published research showing that logos on food labels are more likely to confuse and mislead consumers than inform them (National Consumer Council, 2003).

The report highlights how the sheer number of labeling schemes has caused confusion among consumers who do not know what the labels mean. The problems in food labeling, the report says, can be illustrated by the term "healthful eating," which would appear, at face value, to be clear and well defined. But in fact, the study found that it is open to interpretation. There are huge inconsistencies between the different healthful eating schemes, and manufacturers do not necessarily publish their criteria for entry into the scheme, nor do they explain clearly to consumers what their logos actually mean. The research highlights how a fragmented approach to food labeling has led to more confusion for consumers and that a proliferation of labels and logos has caused information overload among consumers.

Also in the UK, the Consumers' Association (CA) has called for better rules to guide manufacturers and to protect and inform consumers specifically regarding functional foods. In a June 2000 report, the CA said there needed to be tighter controls on ingredients, nutrition labeling, and health claims for foods with added nutrients. They called for all health claims on functional foods to be proven before the products are marketed (Consumers' Association, 2000).

The Effectiveness of Health Claims in Australia, Japan, and the United States

The limitations of health claims per se in food marketing and consumer impact can also be demonstrated by the results of a health-claim pilot study in Australia for intakes of folic acid and the reduction in the risk of neural tube defects in newborn infants. This was the first government evaluation and assessment of a health claim and was an initiative by the Australian and New Zealand Food Authority (ANZFA) in March 1998.

The folate health claim was permitted on eligible foods from November 1998 to February 2000. Companies participating in the pilot, such as Kellogg and Sanitarium, were able to carry a direct health claim for folate-fortified foods stating that if women consume at least 400 micrograms of folate a day, the risk of neural tube defects in pregnancy, which can lead to babies being born with spina bifida, is greatly reduced.

ANZFA was keen to stress that eating a combination of folate-rich foods is vital, rather than taking a supplement or eating fortified cereals or breads. More than 100 foods were approved by ANZFA to carry the health claim, including eggs, legumes, various fruits and vegetables, together with a wide range of breads, cereals and juices—a total of 28 primary foods and 72 processed foods. Processed foods making the claim also had to meet criteria for fat, salt, and sugar. Part of the trial evaluation was to assess whether health claims make any difference to sales of products using the claim, in addition to the main public health goal of increasing folate intake.

However, the evaluation of the folate health claim pilot published in September 2000 concluded that it is hard to attribute any changes in knowledge, attitudes, and product sales solely to the health claim. The process of evaluation for the pilot health claim included the experiences and perceptions of key stakeholders from government, industry, consumer, community, and professional organizations collected during two rounds of consultation. Around 100 consultations took place (Australia New Zealand Food Authority, September 2000).

The Japanese Experience with Health Claims

Even in Japan, with its unique FOSHU system that has been in force since 1991 and which allows product specific health claims, it has been reported that the food industry has mixed opinions on the marketing effectiveness of health claims as approved under the FOSHU system. And for consumers, the FOSHU system has had a slow start. According to surveys conducted by the Japan Health Food and Nutrition Food Association (JHNFA), in 1996, only 2 percent of consumers were aware of FOSHU status, but by 1999 awareness had increased to 20 percent, and in the new century 25 percent of Japanese consumers could recognize the FOSHU approval logo on packaging.

One of the few market analyses available in English on the effectiveness of FOSHU health claims is by Yamaguchi (2002). In this report, Japanese companies gave a mixed review of the success of using FOSHU health claims. Examples of comments noted in the report include:

> I don't think our products attract consumers because of FOSHU [being] approved or even the printed symbol. Our products have been in the market for three years before they had approval for FOSHU but, since then, the sales have grown 5 percent compared to 4 percent before they became FOSHU (representative of Calpis).

> We emphasize Econa Cooking Oil is "FOSHU approved" on our advertisements, but we are unable to measure how much it [FOSHU] contributed to sales.

> After we got FOSHU approval some stores reported the sales increased 200 percent (Ajinomoto AGF, on its product Bitahot).

Other key factors identified in Japan for market success for products with a health claim were media coverage, well-coordinated publicity, and good distribution networks. Successful products were also seen by consumers as innovative and good tasting. Some Japanese food companies in this study complained that the requirements to obtain FOSHU health claims are expensive because of the volume of research necessary to establish the scientific validity of claims. They also pointed out that allowable statements of health claims are vague and ineffective in attracting consumers. But many also say that FOSHU status helped attract consumers.

Use of Health Claims in U.S. Food Advertising

The mixed use and impact of nutrition messages and health claims on foods used in food advertising in the United States has been documented in a unique study undertaken by the Federal Trade Commission's (FTC) Bureau of Economics (Ippolito and Pappalardo, 2002). The report examines a wealth of data on the content of 11,647 food print advertisements during the years 1977 to 1997 taken from eight leading magazines. The data make it clear that nutrition-related claims have become a major feature of food advertising and an important focus of competition. The evidence also makes it clear that regulatory rules and enforcement policy matter: The study found that the content of food advertising shifts markedly as the regulatory policies toward nutrition and health claims vary over these years. Interestingly, the study found that the evidence shows no increased advertising on "good foods" and in fact advertising for fruits and vegetables had fallen significantly since the Nutrition and Labeling Education Act (NLEA, 1990 [implemented 1994]).

From a public health perspective, an important finding from the FTC study is the way it shows how nutrition-based claims in U.S. magazine food advertising rise and fall over very short periods of time. Thus, nutrition advertising claims would not be consistent with long-term health education about diet and food. But the authors of this study do argue that the rise and fall of nutrition and health claims has been prompted by changes in U.S. regulatory rules; for example, the passage of the NLEA. But again, the impact of regulatory change on the use of nutrition messages in ads appears to be short lived.

However, the study shows that the number and range of nutrition claims on advertising are so widespread and general as to offer a mass of information to the consumer. For example, in 1977, 50 percent of all ads had a general nutrition claim. Their use rose to nearly 70 percent of ads by 1983 and held steady through the 1990s before falling back to 56 percent of ads in 1997. The same fluctuations in use of more health claims in food advertising were seen in the survey; for example, heart and serum cholesterol claims were used very little in the late 1970s and then began again in 1983, rising substantially to a peak of 8.2 percent of all ads in 1989 before falling dramatically in the early 1990s. In 1997, 3.4 percent of ads included a heart or serum cholesterol claim, 41 percent of the peak use. The authors explain the rise and fall of certain nutrition claims as follows:

> This evidence is consistent with the view that the FDA label rules have an important influence on producers' willingness to make advertising claims (2002).

The FTC report raises another important issue over the effectiveness of health claims or other nutrition information: Such claims must be seen in the context of competing messages on foodstuffs. The FTC survey found that most food ads make multiple informative-type claims; more than 50 percent of ads surveyed include information about different varieties of the product, such as flavor. Around one-third make an explicit claim about the product's convenience; 20 percent claim that the product is new or has been improved. Approximately 80 percent of the ads make claims about use, texture, or aroma of food. The study authors point out that, assuming that the nutrition label is credible to consumers, most of these claims involve what they call "search or experience" characteristics, that is, consumers need to bring some prior knowledge or interpretive skills to make full use of the labeling information (Ippolito and Pappalardo, 2002).

Lessons for Health Claim Use and Food and Drink Marketing from Regulatory and Market Activity

What general principles in relation to the regulation of health claims and the marketing of functional foods/nutraceuticals can be learned from the literature, regulatory developments to date, and marketplace activity? These principles are summarized here in a series of checklists and a consideration of the advantages and disadvantages of using health claims. These lists draw on both regulatory developments and voluntary use of health claims.

Checklist 1: Market Activity—Some General Observations

- Larger companies now dominate regulated health claim market activity and have been the drivers for securing specific health claim regulation.
- Companies start to build the consumer market for health benefits through advertisements and other publicity during the approval process and use a regulated health claim to leverage public awareness.
- Big companies look to lock-in generic claims for themselves; that is, they claim ownership of a particular health claim regulated area, for example, whole grains.
- Big ingredient companies are starting to move toward branded ingredient strategies to try to capture advantages from health claim regulation, for example, over soy ingredients.
- Health claims are used to help to build an overall perception of healthfulness for a company's products and to establish consumer trust.
- Health claims, regulated or otherwise, do not guarantee success. Companies build in regulated health claims to develop and implement an integrated communications and marketing strategy, often requiring substantial marketing expenditure beyond the reach of most smaller- and medium-sized companies.
- "Authoritative sources" (such as health organizations) might be as important in marketing as health claims, and companies increasingly look to link with third-party endorsers.

Checklist 2: Targeting the Right Audience

Health claims are used to address two distinct audiences:

- Disease conditions (for example, high cholesterol; people who have experienced an illness)
- Optimal health (helping healthy people stay healthy; prevention rather than treatment)
- A key principle overlooked in much functional food marketing and directly relevant to regulators is what the health claim target audience will find most appealing and moti-

vating from a desired behavior change. Regulated health claims must be persuasive to the target, even though health messages have clear-cut scientifically based benefits; for example, "reduce your risk of cancer." These public health benefits may not be persuasive to the target audience.

Checklist 3: The Context in Which Health Claims and Regulation Are Used
- Clinical approach (one-to-one, personalized)
- Public health approach (getting populations to change; setting national food and nutrition policy; promoting health and disease prevention)
- Commercial approach (used purely as a marketing tool to create new markets; keep up with competitors)

Checklist 4: Advantages of Regulated Health Claims
- Consumers want to learn more about the health benefits of foods and drinks. Companies can become an authoritative source through developing products and marketing around approved health claims.
- Regulated health claims enable producers to communicate sound science.
- Consumers will switch brands or start buying a food product because of something they read on the label.
- Providing compelling information can change consumer behavior.
- Consumers who already had more knowledge about the health benefits of an ingredient had more positive attitudes toward claims related to diet and disease on food labels (that is, prior beliefs about product healthfulness also appear to override claim information and add something of value in the minds of the consumer [Geiger, 1998]); health claims can reinforce this perception.
- Regulated health claims offer a range of health message labeling options. For example, for soy protein in the United States, it is possible to use the FDA-approved health claim in relation to heart health, a nutrient-content claim, a statement of quantity, or a structure/function claim.
- Claims establish a product's health-related credibility and reputation.
- Claims help to establish long-term support for a brand or company.
- Claims can be used to distinguish and differentiate companies using approved health claims from those using misleading nutrition-related messages.

Checklist 5: Disadvantages of Regulated Health Claims
- Most claims will be generic rather than product specific (that is, open for anyone to use so that others can benefit from another company's nutrition marketing without making the same commitment to resources and marketing).
- Approved health claims are generally written in consumer "unfriendly" language that is hard for consumers to understand.
- Potential escalation in the use of regulated (and unregulated) health claims may lead to an increase in nutritional uncertainties (consumer confusion, nutritional message overload).
- Consumer advocacy groups call for stricter regulation of functional foods (arguing that claims are misleading); we are unlikely to see a wide range of health claim approval.

- It is often difficult for consumers to distinguish between approved and unapproved product claims.

    Marketing claims sometimes involve selective use of science, and preliminary studies might be refuted by new studies; companies might therefore be seen as "over-interpreting" nutrition research in their marketing to consumers in the sense that the science is far from complete and consumers can also raise concerns about there being too much "hype" in the use of health messages on products.
- Consumer backlash can occur against diet and health messages.
- As health messages become increasingly complex and more numerous in the market place, it becomes more challenging for consumers to keep up-to-date with new information and more challenging for companies to find innovative and compelling ways to differentiate their "health claims" from competing health messages.
- Many sources of health messages exist, often with markedly different agendas (such as popular media, consumer advocacy groups, health professionals, food industry marketers, and regulators themselves).
- Cost and methodological challenges of providing scientific evidence.

    Claims are nonproprietary.

    Claims compete among many in the market (less, reduced, no, low, healthy, natural, organic, and so on).

    The growth of "self-regulation" health claim labeling programs in the United States that compete with one another (for example, those run by organizations such as the American Heart Association or educational campaigns developed by industry, such as the soy industry).
- Effects of health claims may not last (increasing sales can be difficult to sustain, as with the Kellogg All-Bran campaign in 1980s).
- Claims create a bandwagon approach. After some companies start to show success with claim, many others look to join in—thus there is the business risk that the product innovators or ingredient pioneers quickly find their marketing advantage eroded by "me, too" products and the activity of competitors.

Checklist 6: Issues for Policy and Regulatory Processes

- Is it the ingredient or the food as consumed that needs regulating?
- Should policy be directed toward disease/illness conditions or to promoting optimal health/health prevention of populations (issues include assessing the relationship between targeting "high risk" individuals *versus* population risk reduction, calculating the costs of benefits over potential risks, the safety of products and ingredients, long-term health goals in relation to short-term objectives, and so on)?
- Post-market surveillance (monitoring/evaluation; public health benefit).
- How can health claim regulation be more flexible in wording, placement, and length of claims?
- What is the required level of scientific substantiation for health claims; should this be different for generic health claims versus product specific health claims?
- How can intellectual property and commercial confidentiality be protected?
- What are the food safety issues; can too much of a "healthful" ingredient have safety issues?

What Does the Future Hold?

For the future, the regulation and use of health claims will become more complex. It will require marketing and communications sophistication and consistency to succeed in marketing food and drinks with health messages. To succeed with health claims, the industry needs to develop more inclusive, transparent, and workable partnerships among health organizations, consumer advocacy groups, government agencies, food industry, and media to advance nutrition education and consistent messages. For the food and drink industry, the only general principle that truly matters is understanding and implementing genuine innovation in the area of health in relation to food and beverages, within or outside a health claim regulatory framework.

References

Australia New Zealand Food Authority (ANZFA). (September 2000) *Evaluating the Folate-Neural Tube Defect Claim Pilot. Outcome Evaluation*, Canberra: ANZFA.

Bellisle et al. (1998) Functional Food Science in Europe. *British Journal of Nutrition*, vol. 80 (Suppl. 1), p. S1–S193.

Business Week. (1989) September 25, 1989, p. 42–43.

Childs, N. (1998) Public policy approaches to establishing health claims for food labels: an international comparison. *British Food Journal*, 100/4 p. 191–200.

Clydesdale, F. (1997) A proposal for the establishment of scientific criteria for health claims for functional foods. *Nutrition Reviews*, vol. 55, No. 12, p. 413–422.

Consumers' Association. (2000) *Functional Foods—Health or Hype?* London: Consumers' Association.

Diplock et al. (1999) Scientific concepts of functional foods in Europe: Consensus document. *British Journal of Nutrition*, vol. 81, (Suppl. 1) p. S1–S27.

European Commission. (2001) Discussion paper on nutrition claims and functional claims. SANCO/1341/2001. Prepared by Directorate General Health and Consumer Protection (SANCO D4).

Fleck, F. (2003) WHO challenges food industry on report on diet and health. *B.M.J.*, vol. 326, March 8, p. 515.

Geiger, C. (1998) Health Claims: history, current regulatory status, and consumer research. *J. of the American Dietetic Association*, vol. 98, no. 11, p. 1312–1322.

Health Strategy Consulting, personal communication, www.health-strategy.com.

Heasman, M. and J. Mellentin. (2001) *The Functional Foods Revolution: Healthy People, Healthy Profits?* London: Earthscan.

Hutt, P.B. (1986) Government Regulation of Health Claims in Food Advertising and Labeling. *Food, Drug Cosmetic Law Journal*, vol. 41, no. 1, p. 3–73.

Ippolito, P. and A. Mathios. (1990) Information, advertising and health choices: a study of the cereal market. *RAND Journal of Economics*, vol. 21, no. 3, p 459–480.

Ippolito, P. and J.K. Pappalardo. (2002) *Advertising Nutrition & Health: Evidence from Food Advertising 1977–1997*. Washington, D.C.: Federal Trade Commission.

International Association of Consumer Food Organizations. (1999) *Functional Foods—Public Health Boon or 21st Century Quackery?* Washington, D.C.: IACFO.

International Obesity Task Force. (2003) *Global Burden of Disease*. Geneva: WHO/Harvard.

JHCI. (2001) *Joint Health Claims Initiative Code of Practice and Guidelines for the Substantiation of Health Claims*. www.jhci.org.uk.

Kessler, D. (1989) The Federal regulation of food labeling. *The New England Journal of Medicine*, vol. 321, no. 11, p. 717–725.

Knorr, D. (1998) Special Issue: Functional Food Science in Europe. *Trends in Food Science and Technology*. vol. 9, p. 293–344.

Lawrence, M. and M. Raynor. (1998) Functional Foods and Health Claims: A Public Health Policy Perspective. *Public Health Nutrition*, vol. 1, no. 2, p. 75–82.

McGill, C. (2002) Good Health Sells: Tropicana Success Stories. Presentation at Wellness, Energy & Functional Beverages Conference, ACI, Orlando, Florida, January 28–29.

McNutt, K. (1997) Why some consumers don't believe some nutrition claims. *Nutrition Today*, vol. 32, no. 6, p. 252–256.

National Consumer Council (NCC). (2003) *Bamboozled, Baffled and Bombarded: Consumers' Views on Voluntary Food Labeling*. London: NCC.

PASSCLAIM. (2003) *European Journal of Nutrition*, vol. 42 (Suppl. 1), p 96–111.

Patterson et al. (2001) Is there a consumer backlash against the diet and health message? *J. of the American Dietetic Association*, vol. 10, no.1, p. 37–41.

Richardson, D. (1998) Scientific and Regulatory Issues About Foods Which Claim to Have a Positive Effect on Health in Functional Foods: *The Consumer, the Products, the Evidence* edited by M. Sadler London: Royal Society of Chemistry.

Truswell, S. (1998) Practical and realistic approaches to healthier diet modifications. *Am. J. Clin. Nutr.* vol. 67 (suppl):583S–590S.

Schmidt et al. (1997) Communicating the benefits of functional foods. *CHEMTECH*. December, p 40–44.

Silverglade, B. (1995) *Facilitating Nutritional Research—Is Liberalizing Health Claims Regulation the Answer?* The Foundation For Innovative Medicine, Fifth Nutraceutical Conference, Washington, D.C., November 15.

Swedish Nutrition Foundation. (1996) *Health Claims in the Labelling and Marketing of Food Products, The Food Industry's Rules (Self-Regulating Programme)*. Revised programme, August 28, 1996, Applicable from January 1, 1997.

World Health Organization (WHO). (2003) *Diet, Nutrition and the Prevention of Chronic Diseases*. WHO Technical Report Series 916. Geneva: WHO.

Yamaguchi & Associates, Inc. (2002) *FOSHU 2002: Japan's Functional Foods with Health Claims*, Yamaguchi & Associates. Tarrytown, NY, USA.

Zenith International. (2003) *Functional Soft Drinks 2003*. Bath: Zenith.

3 Regulation of Quality and Quality Issues Worldwide

Joy Joseph

For the purposes of this chapter, nutraceuticals are defined as food products that contain ingredients that have been recognized in the Dietary Supplement Health and Education Act of 1994 (DSHEA) as dietary ingredients. These products can be in tablet, capsule, liquid, or conventional food forms such as bars, drinks, soft chews, or hard candy types of products. These ingredients are generally recognized as safe or must have been exempted due to use in the food chain as a dietary ingredient prior to 1994. These ingredients also have a proven science-based nutritional benefit. The terms **nutraceutical** and **dietary supplement** are used interchangeably throughout this chapter.

In most countries other than the United States, nutraceutical products in pharmaceutical dosage forms such as compressed tablets, two-piece gelatin capsules, soft elastic gelatin capsules, and liquids are regulated as drugs. Some even require preapproval prior to marketing and have a quasi-drug status. In most of those countries, only those vitamins or nutritional products that contain doses of nutrients much lower than the U.S. Recommended Daily Allowances can be marketed as foods. In the United States, nutraceutical products or dietary supplements, as this category has been officially named by Congress (DSHEA), have always been regulated as foods.

Food quality for most of the world is regulated according to the principles of sanitation laws. Some countries have stricter requirements for testing, verification, and record keeping of sanitation practices. Hygienic practices for employees play a major role in providing safe food for human consumption, and those procedures are clearly defined for all countries.

In the United States, food sanitation regulations include some specific requirements for certain foods, such as processing and filling of bottled water; Hazard Analysis and Critical Control Points (HACCP). Procedures for the safe and sanitary processing and importing of juice, fish, and fishery products; the Infant Formula Quality Control Procedures; and the processes for thermally processed low-acid canned foods packaged in hermetically sealed containers. All other foods fall under 21CFR, Part 110, CGMPs (Current Good Manufacturing Practices) for manufacturing, packing, and holding human food, for which extensive record keeping is not a requirement.

Nutraceutical/Dietary Supplement Quality Issues

Most consumers believe that fortified foods and nutraceuticals are necessary to ensure health-sustaining nutritional intake. In the United States alone, 100 million people take dietary supplements because they believe nutrient benefits are lost through food processing. Consumers have concerns about the erosion of the food chain due to the use of pesticides and preservatives. There are concerns around the nutritional values of genetically engineered foods. Some cooking processes also destroy valuable nutrients. Others believe that their busy lifestyles prevent them from getting proper nutrition. Those consumers who do endeavor to eat a balanced diet also support the use of nutraceuticals to provide that extra benefit to assure good health and prevent nutrient deficiency.

Unfortunately, the dietary supplement/nutraceutical industry has not enjoyed a reputation of trust among consumers or regulatory agencies. More often than not, news of the health benefits of supplements has been overshadowed by a continuous flow of negative publicity because of a lack of appropriate regulations and knowledge among industry members regarding suitable steps to take to ensure that products deliver what is stated in the labeling. In addition, the industry has been plagued with a few bad actors that have over-promised benefits and results. Some of them have labeled or advertised products to have drug-like benefits, and others have knowingly taken advantage of the public by providing products that have delivered no value at all or have resulted in injury or other unsafe consequences when used.

There have been cases of products contaminated with drug products that render them unsafe for food use. Reports have also arisen of unsafe side effects from products that have been improperly tested or not tested at all and sold as dietary supplements. Products have undergone post-market surveillance by third-party laboratories and have been found to be void of claimed nutrients. Cases of both subpotent and superpotent products have occurred.

Other industries have felt threatened by some of the more beneficial products and have attacked these products because they tend to infringe upon some of the tested and proven claims made by their products. Only recently have a small number of nutraceuticals actually undergone experimental studies or clinical trials to prove efficacy.

Because nutraceuticals and dietary supplements cannot, for the most part, enjoy the benefits of patents and trademarks (and, therefore, exclusivity), the industry has lacked the enthusiasm to support testing and clinical studies. (A good example is the finding that folic acid prevents neural tube defects. Any and all dietary supplement manufacturers can and do market this product without having participated in the research that identified this benefit).

Several issues made it clear to the industry and the regulatory agencies that the Good Manufacturing Practices (GMPs) promulgated for foods were inadequate to assure the quality of nutraceutical or dietary supplement products. Those issues are the following:

1. Products are in the market that do not deliver what they purport to provide.
2. Labeled ingredients have been nonexistent when tested.
3. Product purity has been compromised with adulterated ingredients.
4. Product potencies have not met label claims.
5. Products are on the market that contain ingredients that are unsafe for the products' intended use.

Missing from food GMPs, but appropriate for dietary supplements, were such requirements as:

- Raw material qualifications and testing
- In-process product testing
- Finished product testing
- Cleaning and sanitizing
- Record keeping

- Standard operating procedures
- Stability studies

The 1994 DSHEA called for a GMP requirement based upon food regulations, thus giving FDA the authority to develop separate regulations for dietary supplements.

In 1995, the industry members submitted to FDA an outline for GMPs based on practices from some of the more responsible ones and their trade associations.

In 1997, the FDA issued an Advance Notice of Proposed Rule Making based upon the industry submission.

In 1999, the FDA conducted a survey that found strong public support for increased regulations to ensure that these products are safe and do what they claim to do.

A different set of GMPs for dietary supplements was **found to be** necessary to assure that consumers get products with the strength, quality, and purity they expect.

These GMPs should:

- Ensure that every dietary supplement has the safety, identity, strength, and quality its label purports
- Include ingredient identity and purity testing
- Ensure that dietary supplement products are produced using standard procedures under sanitary conditions
- Require the manufacturers to document the manufacturing process and ensure its consistency from lot to lot
- Require testing of dietary ingredients for harmful contaminants
- Require expiration dating for products based upon testing, particularly those subject to deterioration
- Require manufacturers to keep records of complaints and to report those events that are deemed serious or life threatening

In March 2003 the FDA published the proposed "Current Good Manufacturing Practice" regulations for dietary ingredients and dietary supplement products. These rules provide the minimum standards necessary to ensure that these products would not be adulterated or misbranded.

How Implementation of Good Manufacturing Practices Will Help to Ensure Good-Quality Products

Organization

To ensure the quality of the products manufactured, the first requirement is a commitment to an established quality control program. This first requirement must have full support and a commitment to quality from the entire organization, beginning with and endorsed by executive management.

The second requirement is to have appropriately trained employees who understand their processes and are well trained and experienced with manufacturing practices that accomplish the intended outcomes of safety, quality, potency, and purity. Adequately trained employees are the cornerstones for building quality into the product.

The quality control unit must be separate from the manufacturing unit, preferably reporting directly to executive management. All functional groups must understand in enough detail the criteria and line of authority that contribute to acceptance or rejection decisions for the products, ingredients, and processes. Functional organizational structure with delineation of authority must be understood by all employees. Personnel from both functions should participate in establishing procedures, training employees, and conducting product failure investigations, but the quality control unit should have the ultimate responsibility for release and reject decisions.

The 2003 proposed CGMPs includes provisions for the following:

- Personnel
- Facilities
- Equipment and utensils
- Production and process controls
- Holding and distribution
- Consumer complaints
- Records and record keeping

Although these CGMPs as of this writing are in "proposed status," this chapter focuses on those practices that will ensure quality products even if the proposal undergoes change by the date of final publication.

Personnel

Supervisors and employees should be qualified to take measures to exclude from operations any persons who might present a risk because of microbial contamination. They also must be trained to use hygienic practices to prevent product contamination.

In addition to the hygienic practices, personnel must be trained in the performance of the tasks to which they are assigned. This training must be in detailed procedures and quality requirements of the actual job performed in order to ensure worker familiarity at the conclusion of the training program. Records of training must be maintained and kept current. Training must be on a continuous basis with retraining schedules and reviews. Standard operating procedures and GMPs must be reviewed and updated as procedures and processes change.

Employee training is best accomplished when it is the subject of standard operating procedures with schedules, rotations, and a dedicated staff. Training is one of those areas that have not been consistent or routinely practiced in this industry.

Scientific personnel should be qualified through appropriate academic studies for their specific function. Laboratory and testing professionals need education and training in analytical chemistry or microbial analyses to perform their assigned tasks. They need to be able to interpret test results and take responsibility for their deliverables.

Personnel engaged in the manufacture, processing, packing, or holding of a product will wear clean protective clothing to prevent contamination of dietary supplement products.

Companies must provide a sufficient number of uniforms to personnel and specify in writing maximum intervals between changes for each function. Employees should understand that uniforms are provided to protect the products from contamination.

Personnel must be instructed to report any health conditions that may adversely affect the dietary supplement products.

Management must be sensitive to employee needs and assign them to other areas where product is not exposed wherever possible, in order to promote honesty and trust between management and line workers. When employees are often sent home and salary reduced or eliminated, employees may not be open and honest about health conditions.

Physical Plant

Physical plants must be of suitable size and dimensions to allow for orderly placement of equipment, and allow for the separation of components, dietary ingredients, finished products, packing materials, and labels that will be used during manufacture. The plant must be suitable in design to facilitate maintenance, cleaning, and sanitizing procedures and have adequate space for laboratory analyses and the storage of laboratory supplies.

Floors, walls, and ceilings must be smooth and of a hard surface to facilitate cleaning and repair.

The facility must provide for adequate ventilation or environmental controls and equipment that control temperature and humidity. All such equipment must be installed and operated in a manner to prevent contamination of products.

Plants should be designed and constructed in a manner to protect products from becoming adulterated during manufacturing, packing, and holding. Plants should be constructed to facilitate a clean and sanitary environment. Plants should be kept in sufficient repair to prevent contamination of products and components.

The proposed CGMPs are extremely comprehensive in regard to physical facility requirements, including but not limited to the use of light bulbs, protective coverings, the size of aisles, working spaces, and much more. Because this chapter only *touches upon the minimum requirements, it is suggested that the reader do a follow-up after the document becomes final to assure full compliance.*

Cleaning compounds and sanitizing agents must be free from microbes and safe for their intended use.

Toxic materials must not be held or used unless they are necessary to maintain sanitary conditions, or for use in the testing laboratory or in the operation of the plant or equipment.

Animals and pests must not be allowed in any area of the physical plant. Measures must be taken to exclude pest and prevent contamination of products. Pesticides, rodenticides, fumigants, and fungicides can be used only if procedures for use prevent product contamination.

It is recommended that plant sanitation schedules and procedures for handling cleaning compounds, pesticides, and rodenticides be written and documented when used or performed in order to facilitate investigation of use if some unforeseen event takes place regarding proper use of plant sanitation products.

Water supply must be safe and sanitary for use in manufacturing and for use in cleaning contact surfaces. Water that comes in contact with ingredients or products must be in compliance with drinking water regulations. Compliance to these standards must be documented.

Plumbing must not present a source of product contamination and must not allow for back flow or cross connection between discharge systems and manufacturing water use systems.

It is recommended that water used in manufacturing be tested for chemical and microbial purity and records be kept and maintained.

Bathroom and hand-washing facilities must be adequate for the number of employees and must be kept clean and sanitary. Employees must be required to wash and sanitize hands before handling products and components.

Trash bins and trash disposal must be handled and managed in a sanitary manner and must not be a source of product contamination. Hazardous waste must be controlled to prevent product contamination.

Sanitation procedures must be assigned to and managed by supervisors qualified in training and experience.

Equipment and Utensils

Equipment and utensils should be made of nontoxic material and have an appropriate design, construction, and workmanship for their intended use. They will be appropriately cleaned and maintained. Instruments must be calibrated for accuracy and precision to ensure that automatic, mechanical, and electronic equipment works as intended. Equipment and utensils must be designed so that their use does not result in contamination with manufacturing adjuncts, such as lubricants, fuel, or particulate matter.

Freezers and cold storage compartments must be fitted with thermometers and temperature measuring and recording devices that show temperature accurately.

Instruments or controls used in the manufacturing, packing, or storage of dietary supplements, including instruments used to measure, regulate, or record temperature, hydrogen-ion concentration, water activity, or other conditions that control or prevent the growth of microorganisms or other contamination, must be calibrated before use for accuracy and precision. Calibration results must be documented with all pertinent details including measures and standards. Documentation must be signed or initialed by the person performing the calibration.

Calibration procedures must be written with appropriate accuracy and precision limits and approved by the quality control unit.

Instruments out of specification must be repaired and replaced.

In the proposed rule, calibration of utensils and equipment was very specific and prescriptive in nature. It was one of a few sections (other than in the quality control documentation) in which signatures were spelled out as a requirement. Serious attention to detailed procedures is recommended for this section.

Automatic, Mechanical, and Electronic Equipment

Automatic, mechanical, and electronic equipment must be designed to ensure that specifications are consistently met. Suitability checks must be performed to determine whether automatic equipment is capable of operating satisfactorily within the limits of the process. Specifications and verification of capability must be approved by the quality control unit.

Schedules must be established for the routine checks, inspections, and calibration of automatic equipment. These schedules must be documented and followed. Records must be written and maintained.

The quality control unit must approve all changes related to equipment use in all functions.

Back-up files of software programs and data entered into computer systems must be documented and maintained. These files must be exact and complete records of the data entered and must be protected from alterations, erasures, or loss.

It is recommended that machinery product profiles be utilized that have been established through a design of experiments used to validate that the process and equipment are within the range of variability expected for the processing times and conditions for the specific products produced. This can be accomplished by running a sufficient number of production size lots and collecting numerous samples for testing for relevant chemical, microbial, and physical parameters. Records should be maintained and serve to reduce calibration schedules and testing frequency for like products.

Production and Process Controls

Use of a quality control unit is required in manufacturing packaging and labeling operations. The production and in-process control systems must be reviewed and approved by the quality control unit.

Specifications must be established for any step in the manufacturing process for which control is necessary to prevent adulteration.

Specifications must be established for the identity, strength, quality, and composition of components, dietary ingredients, and dietary supplement products. These specifications must be tested to ensure that they are met. Test methods must be scientifically validated methods.

Specifications must be established for in-process controls in the manufacturing process, and these control points must be monitored to ensure that specifications are met.

Specifications must be established for labels and packaging components and must be examined or monitored for compliance to specifications.

Corrective action plans must be established for use when an established specification is not met. A documented material review must be conducted to determine the cause of the out-of-specification occurrence, and the quality control unit must conduct a separate review and approve any reprocess or disposition decisions.

Rejected materials must not be reprocessed unless approved by the quality control unit.

Dietary supplements may not be reprocessed if they are rejected because of contamination with microorganisms or other contaminants.

Material review data, test results, and disposition decisions must be recorded and maintained as a part of the batch record.

Master manufacturing records and batch control records must be established and used to assure batch-to-batch consistency.

The proposed rule goes into great detail on production and process controls, with particular emphasis placed upon product failures and quality control unit responsibilities. The rule is also specific to rejection of adulterated products without the possibility of reprocessing.

Quality Control Unit

The quality control unit must approve or reject all processes, specifications, controls, tests, and examinations, and deviations from or modifications to them, that may affect the identity, purity, quality, strength, and composition of a dietary supplement.

The quality control unit must:

- Determine specifications for components, dietary ingredients and products, and packaging and labels.

- Approve or reject components, dietary ingredients and products, and packaging and labels.
- Review and approve all master manufacturing, batch production and production-related records, and any and all modifications to each.
- Review and approve all processes for calibrating instruments or controls and all records for calibration of instruments and gauges.
- Review and approve all laboratory control processes and testing results.
- Review and approve all packaging and label records, approval for repackaging and re-labeling, and approval for release for distribution.

The quality control unit is responsible for the sampling and testing of each lot of components, dietary ingredients, dietary supplements, and packaging and labels received to determine compliance to specifications.

The quality control unit is responsible for ensuring that all in-process control steps are followed to ensure product quality.

The quality control unit is responsible for the assurance that the correct packaging components and labeling has been applied to each batch of finished dietary supplement product.

The quality control unit is responsible for review and evaluation of all processing and testing for each lot of product prior to release for distribution.

The quality control unit is responsible for keeping reserve samples of each batch distributed. The sample must be an amount equal to two times the amount required for full testing. It must be maintained for three years from the date of manufacture for use in appropriate investigations associated with consumer complaints or product failures.

The quality control unit must review and approve the reprocessing or distribution of any returned dietary supplements.

The requirements for the Quality Control Unit for the proposed Dietary Supplement regulations have been expanded to include specificity not found in the CGMPs for finished pharmaceuticals. It is believed that this expansion has been promulgated because no requirement exists for process validation. Industry has proposed a more process-controlled approach to FDA in order to reduce the burden of quality control oversight as proposed. Some dietary supplement manufacturers already utilize a process control format similar to that used in the pharmaceutical industry as standard practice and would like to be allowed to perform some process control verification in lieu of extensive testing.

Components, Dietary Ingredients, Packaging, and Labels

Raw Materials

Containers of raw materials and components must be examined upon receipt for the appropriate label, container damage, broken seals, and foreign substances on the surface that might indicate tampering or possible contamination.

Examination must also include supplier's invoice, guarantees, or certificates that accompany shipments to assure that the correct product and information have been received.

Components and raw materials must be quarantined from use, and sampled and tested by the quality control unit to determine whether predetermined specifications have been met.

Components and raw materials must be assigned a unique identifier that allows traceability to the supplier, the date received, and the name and status of the component throughout the manufacturing cycle and distribution.

Components and raw materials must be held under conditions that will protect them against contamination and deterioration and will avoid mix-ups.

Raw materials and dietary ingredients have been a major source of contamination and adulteration in the dietary supplement industry. Ingredients of particular concern are herbal ingredients and other natural ingredients for which validated test methods are nonexistent or in developmental stages. Sophisticated testing equipment and methodologies are becoming more prevalent, resulting in efficient and accurate methods for isolation of synthetic chemicals that are being added to herbals to meet superficially imposed specifications. Also, the USP, AOAC, and some industry trade associations have active methods development projects under way to improve the speed at which validated methods can become available to the industry. In lieu of compendia validated test methods, the existence of in-house testing laboratories with competent scientist and chemist is a must-have to provide a high level of confidence that products meet purity and potency standards.

Packaging and Labeling

Packaging and labels must be visually examined for appropriate content label, damaged containers, and broken seals to determine whether the container condition has resulted in contamination or deterioration of the packaging or labels.

Packaging and labels must be held in quarantine until examined by the quality control unit and released for use.

Packaging and labels must be assigned a unique identifier that allows for traceability back to the manufacturer and must be used whenever a disposition is recorded for each shipment received.

Packaging and labels must be held under conditions that will prevent mix-ups, contamination, and deterioration.

Records must be created at the time of performance and maintained for all receipts, sampling, testing/investigations, and dispositions of packaging and labels. These records should include dates and the signature of the person performing the requirements.

New to the dietary supplement industry is the requirement to assign a unique identifier to each lot of packaging components and labels and to keep records of issuance and use in order to trace these components and labels back to the product lots where used, and back to the manufacturer.

Master Manufacturing Records

Master manufacturing records must be prepared for each type of dietary supplement manufactured and for each batch size to ensure batch-to-batch uniformity.

Master manufacturing records must include specifications for the points, steps, and stages in the manufacturing process where control is necessary to prevent adulteration.

Master manufacturing records must establish controls and procedures to ensure that each batch of dietary supplement manufactured meets those specifications.

Master manufacturing records must contain the name of the supplement to be manufactured and the strength, concentration, weight, or measure of each ingredient for each batch size.

Master manufacturing records must contain:

- A complete list of components to be used
- An accurate statement of the weight of each ingredient declared in the label
- A statement of any intentional excess amount of a dietary ingredient
- A statement of theoretical yield expected at each stage of processing, including maximum and minimum limits of variability beyond which a deviation investigation must be performed
- A description on packaging components and a copy of the label or labeling
- Written instructions for each processing step
- Sampling and test procedures
- Special actions required to verify steps necessary to prevent adulteration, including the number of steps, checks, special notations, and precautions to be followed
- Corrective action plans to be taken when specifications are not met
- Review and approval of master manufacturing records by the quality control unit

Under food GMPs there is no requirement for master manufacturing records. However, master formulas specifying ingredients and amounts to be weighed or measured are commonly used in the dietary supplement industry. The proposed rule requirements for corrective action plans and explanations for the intentional excess or overages and the identity of control points to prevent contamination have not been common practice. Even those dietary supplement manufacturers who already employ the use of master manufacturing records will be required to make changes and upgrade requirements to comply with the proposed rule.

Batch Production Records

Batch production records for each lot to be manufactured must be prepared, must be an accurate reproduction of the master manufacturing record, and each processing step included therein must be performed as directed.

Batch production records must include:

- Batch or lot control number
- Documentation at the time of performance that each step was performed, showing dates and initials of persons performing each step
- The identity of equipment and processing lines
- The date and time on the maintenance, cleaning, and sanitizing of equipment and processing lines used in processing
- The unique lot identifier and weight or measure of each ingredient or component used
- The initials at the time of performance or at the completion of the batch of the person responsible for verifying the addition of components to the batch
- A statement of the actual yield and a statement of the percentage of theoretical yield at the appropriate stages of processing
- The actual test results for any test performed
- Documentation that the dietary supplement meets specifications

- Copies of all container labels used and the results of examinations conducted during the label operation
- Material review documentation with disposition decisions, including the signature of the quality control unit

Batch record documentation requirements in the proposed rule mimic exactly what is required for finished pharmaceuticals. For many dietary supplement companies, some of this data is nonexistent, particularly with reference to material review and depth of requirements for quality control investigation. Some of the required steps are performed as a matter of good business practice, such as maintaining copies of labeling, statements of actual yields, and identity, weights, or measures of ingredients, to name a few. The challenge will be to increase the level of detail and to compile this data as part of the batch record for each lot manufactured.

It is recommended that qualified and experienced quality control professionals be assigned to the retraining of line employees to assist in developing standard operating procedures that address the procedures for incorporating all data into the batch record as tasks are performed. Collecting data after the fact is not a recommended practice when the requirements are as comprehensive as in the proposed rule.

Laboratory Operations

Adequate laboratory facilities must be used to perform testing and examinations to ensure that dietary supplements meet specifications established in the master manufacturing records.

Laboratory control processes must be approved by the quality control unit and must include:

- Establishing and using criteria for selecting and developing appropriate test methods and for selecting reference standards
- Establishing and using criteria for appropriate specifications
- Establishing and using sampling plans for obtaining representative samples for testing of raw ingredients, components and dietary supplement products, and packaging and labels
- Establishing and using criteria for appropriate test method validations
- Documentation of tests and examinations at the time of performance by the person performing the tests

Laboratory records must be signed, dated, and maintained for a period specified in the Standard Operating Procedures in accordance with the final rule or CGMPs.

Even though the proposed rule calls for the use of laboratory facilities, it is recommended in this chapter that manufacturers equip facilities with in-house laboratories or be in control of outside testing laboratories to ensure that test methods used are appropriate and valid for their products. Outside testing laboratories have standard test methods that may not be applicable to every formula variation for a given nutrient. Although some label claims may be the same based on regulations or science-based recommendations, the excipient base (other components in the product) may be different and pose assay problems that affect accuracy of test results. If outside services must be used, it is beneficial to establish a relationship of trust with the testing labs that will allow manufacturers to share proprietary information about the formulation and enable the testing lab to design a valid test method. Industry-recognized official compendia test methods such as USP and AOAC are usually the best general methods to begin a product evaluation because they are based upon a large sampling of products. They, too, however, are not always valid for a specific product that may be unique in the excipient base.

For plant-based and animal-derived materials, this chapter recommends plant audits, either by the manufacturer or a third-party auditor who is trained and experienced in the discipline of auditing. It is important to inspect for those specifications that cannot be tested, such as GMPs, cleaning and sanitation practices, use of pesticides, and identity and origin of starting materials in order to assure that the deliverable is what was requested and is not adulterated, contaminated, or deteriorated.

Manufacturing and Packaging Operations

Manufacturing and packaging processes must be designed to ensure that dietary supplement specifications are consistently met and are conducted with adequate sanitation principles.

Necessary precautions must be put in place to prevent contamination. These precautions must include:

- Performing operations in an environment that protects against the growth of microorganisms or contamination
- Washing and cleaning components that contain soil or other contaminants, and the cleaning and sanitizing of all filling and packaging equipment, utensils, and dietary supplement containers
- Sterilizing, pasteurizing, freezing, refrigerating, controlling hydrogen ion concentration, controlling humidity, controlling water activity, and employing any other means of control against contamination or decomposition
- Holding ingredients that support growth of microorganisms of public health significance in a manner to prevent adulteration
- Adequate storage procedures for withholding and preventing mix-ups of materials under review until a quality control decision has been made
- Effective measures that protect against inclusion of metal or other foreign materials, including the use of traps, filters, strainers, magnets, or electronic devices
- Segregation of batches according to phase of manufacturing, along with identification of processing lines and major equipment
- Physical and spatial separation of packaging and label operations to prevent mix-ups
- Examination of representative samples of each batch of dietary supplements to ensure that each product meets specification and that the label specified in the master record has been applied
- Use of packaging components and labels that are issued and reconciled according to written procedures and documented in the batch production record
- Maintenance of packaging and labeling records in accord with the requirements of the final rule

Many of the requirements in this section are overlapping with other sections and have already been commented upon. The requirement for reconciliation of packaging compo-

nents is new to the industry and will present the most challenge for manufacturers. Bottle and cap suppliers need to be made aware of this requirement and can be helpful in establishing handling procedures. They already have to accommodate these requirements for the pharmaceutical industry, which already performs lot control and reconciles packaging components. Most of the container closure systems used are the same as those for the pharmaceutical industry.

The proposed rule also addresses:

- Holding and distribution
  – Records and recordkeeping
  – Rejected materials
- Returned goods
- Consumer complaints

These topics are not addressed in this chapter.

Expiration Dating

Other topics that are extremely important to nutraceuticals or dietary supplements are finished product stability studies and expiration dating.

The proposed rule does not require shelf life or expiration dating for dietary supplements. Recommended in this chapter is the inclusion of expiration dating supported by studies or scientific back-up data to support dating. Nutrition labeling regulations require that products provide at least 100 percent of labeled ingredients throughout the shelf life of the product. In addition, DSHEA declares dietary supplement products to be misbranded if they fail to have the identity and strength that they are represented to have. These two regulations alone support a requirement for expiration dating and, ultimately, stability studies. Most of the industry already assigns expiration dates to products because of customer demand. However, the methods of assignment are many and variable. It is recommended that the industry propose a standard method that will be acceptable to the scientific community for determining dating order to preclude a prescriptive method from the regulatory agencies resulting from a product failure.

Summary

Even though GMPs specify extensive testing of ingredients and finished products, it is imperative that quality be built into the product. Testing alone will not guarantee that products will indeed meet all specifications for quality, potency, and purity. A controlled and documented manufacturing process goes a long way in ensuring product quality. Testing should be performed to verify that the process is in control and that the end product from the controlled process can be guaranteed to meet the preset specifications.

With the passage of the Dietary Supplement CGMPs, the manufacturers of dietary supplements or nutraceuticals will be bound by law to take the responsibility for control.

The proposed rule is based upon present knowledge and can be expected to change as emerging problems are resolved and superior methods of testing and controls are devised.

Although the current proposal is not perfect and pleasing to the entire industry, it is in fact the first step in the right direction. It is in the direction that responsible manufacturers want to take in cleaning up the industry and getting products that support healthful living to the consumers without the uncertainty of the quality, potency, purity and safety to which we have been subjected. It is believed that global acceptance of these new beginnings will help to improve product quality worldwide.

4 Organic Food Regulations: Part Art, Part Science

Kathleen A. Merrigan

Food products labeled organic are grown in 100 countries, with countless more countries offering organic items in the marketplace.[1] In less than two decades, the organic sector grew from an amalgamation of tiny, localized markets denounced by critics as "hippy dippy agriculture" to a booming industry with major multinational corporations launching organic divisions and worldwide sales estimated at $26 billion in 2003.[2] What brought about this phenomenal growth? Did government regulations help or impede this economic sector? This chapter reviews the development of organic regulations in the United States and compares the regulations with organic standards in other parts of the world.

Defining Organic Food

When asked for a quick definition of **organic**, most people respond by listing the prohibitions that accompany an organic claim—that is, organic farmers do not use: pesticides, fertilizers, GMOs (genetically modified organisms), sewage sludge (a.k.a. biosolids), irradiation, antibiotics, or hormones. While this list is basically true, organic agriculture also incorporates an equally long list of positive requirements, such as crop rotation, humane animal treatment, water conservation, and wildlife habitat preservation. When a simple phrase is demanded, organic agriculture can be described as being environmentally sound. But that characterization does not begin to address the complexities in organic production, or clearly differentiate organic food from that produced via other sustainable agriculture practices, such as integrated pest management.

The United States took its first step in formalizing a federal definition of organic food when Congress passed the Organic Foods Production Act of 1990.[3] The law defines the term **organically produced** in a circular fashion; anything that meets the terms of the 1990 law and subsequent regulations can be labeled **organic**. The bill drafters adopted this somewhat unsatisfying definition because they realized that no succinct definition existed and because the debate was politically volatile. Organic opponents argued that an organic label would confuse consumers because its meaning would depart from common usage of **organic** in chemistry as well as perpetuate consumers' mistaken impression that organic food was safer than nonorganic food to consume. In the end, the legislators sidestepped controversy by delegating the task of forming a detailed organic definition to the administrative agency responsible for implementing the law.

Yet a fundamental decision made by Congress with regard to the meaning of organic has been pivotal in the formation of the National Organic Program (NOP).[4] Congress determined that the organic label would represent a production claim rather than a product claim. For example, an organic label on an apple provides a guarantee about how that apple was produced on the farm, transported, stored, and, in the case of apple products such as pie, processed. The organic label does not, however, indicate that the organic apple differs in quality or safety from the nonorganic apple. The legislators noted that because it was difficult, and often impossible, to distinguish organic from nonorganic products when inspecting the product itself, the focus of the NOP would necessarily be that of establishing a federal system of inspection, recordkeeping, and audits to verify organic production claims.

After passage of the law, the organic definition debate shifted to the Agricultural Marketing Service (AMS), an agency of the U.S. Department of Agriculture (USDA)[5] given the task of developing the detailed rules necessary to implement the NOP. The rulemaking process was complicated by scarce public consensus over the details that would formulate a NOP, and AMS continued to receive pressure from organic opponents who wanted to derail implementation of the law. A proposed rule was finally published in 1997 but was exceedingly unpopular, generating a record-breaking 275,603 letters and postcards from the public to AMS denouncing one or more aspects of the proposal.[6] Among the controversies was the proposed definition of **organic**.

The proposed rule included a circular definition of organic not unlike that in the 1990 law, but the public protested. Consumers need a real definition, the public argued, and it gave specific suggestions. In particular, the public wanted a definition that referenced a positive relationship between organic agriculture and biodiversity. In the time between passage of the U.S. law and the proposed U.S. rule, private and international organizations had moved forward in specifying organic production standards, and several popular definitions were available as models. The International Federation of Organic Agriculture Movements (IFOAM),[7] an international nongovernmental organization (NGO) and the international CODEX Alimentarius[8] guideline established by the Food and Agriculture Organization (FAO) of the United Nations and the World Health Organization[9] (WHO), had organic definitions that referenced organic agriculture's contribution to biodiversity. AMS regulators, particularly those with significant expertise in rule writing, were hesitant to mirror a definition with reference to biodiversity in the U.S. regulation because they worried about the responsibilities such a definition would place on AMS and organic certification agents to measure and ensure compliance with the law. How would conservation of biodiversity be determined?

Organic advocates had operated for years with private and state standards phrased as principles of organic agriculture. Organic advocates argued that regulators needed to fuse art with science in drafting the NOP and create standards stated as principles rather than prescriptive measures that could get "stuck in time" and hamper innovation. For example, a typical so-called standard at the time was "organic farmers should respect animal welfare." A U.S. rule written this way, regulators argued, would provide inspectors too much discretion in interpreting the meaning of animal welfare. Regulators explained that an animal welfare standard needed to be detailed; otherwise there would be no way to consistently enforce the law. In this case, the final NOP standard published in 2000 details requirements for animal medications and outdoor access as well as specifying when, and if, common livestock practices such as tail docking can be performed.[10,11]

Today, the NOP is a mishmash of principles and measurable standards. The strong demand for an organic definition referencing biodiversity and other hard-to-measure aspects of organic production trumped regulator concerns. The NOP defines organic as "site-specific conditions that integrate cultural, biological, and mechanical practices that foster cycling of resources, promote ecological balance, and conserve biodiversity."[12] Yet there are many more areas of the NOP in which the United States avoided the principles and vague terminology found in other countries' regulations. The European Union (EU) prohibits "factory farms" and manure from factory farms in organic production. Although the public made a similar request in the United States, regulators argued that there is no working definition of factory farms or ways to readily determine sources of manure. This tension between the U.S. style of regulations, which require specificity and err on the side of science, and the EU/IFOAM style of standards, which are composed of looser principles that err on the side of art, explains many current disputes over what constitutes organic food.

Overview of the U.S. Regulatory Structure

The challenge of developing national standards for organic production was great. When Congress debated the organic law in 1990, 49 state and private organic standards were already in place in various regions of the United States.[13] Indeed, the existence of these myriad standards was among the justifications for federal action, because conflicts between standard-setting organizations had created interstate commerce problems and confused consumers. A federal rule was clearly needed, but there was reluctance among organic advocates and policymakers to disband existing organic organizations because they were the source of organic production expertise and consumer confidence in the organic label. As a result, Congress structured the NOP as a public-private partnership. The responsibility of the federal government is to establish organic standards and accredit private and state certification organizations to ensure their competency as private inspectors working on behalf of the USDA. The responsibility of state and private organizations is to certify that farming and handling operations meet NOP standards.

The goal then and now is to have one national standard. Private organizations are prohibited from enacting standards that differ from or are in addition to the federal standards. States are allowed to enact additional organic standards provided that they are: (1) deemed environmentally essential due to particular ecological attributes of that state; (2) not used to imply superiority over products from other states; (3) not imposed on imported products; (4) enforced by the state; and (5) approved by the USDA. So far, no state standards exist.

Is having one national standard possible? The European Union has one regulation for member states that must be adhered to by producers and handlers who label their product as organic. But the EU regulation is considered the minimum that must be done to qualify for the organic label.[14] Member states and private certification organizations working within the EU may and do enact additional requirements. The United States recognized the impossibility of enacting detailed national standards that could account for the extreme ecological and cropping variability in the country but rather than allow for additional "standards," Congress required that all organic farmers and handlers develop, implement, and update an annual organic production or handling system plan and submit it to a certifying agent for approval. The plan must describe practices to be performed and maintained, substances used as production or handling inputs, monitoring practices, the recordkeeping system, all separation measures, and any other measures necessary to ensure the organic integrity of the product. Thus, the mechanism of the organic plan allows producers and certification agents to tailor site-specific requirements but still comply fully within the national standards.

Congress was also motivated to enact national organic standards because of concern over fraudulent organic claims. In 1989, the television show *60 Minutes* focused on Alar, a pesticide used in apple production. Actress Meryl Streep, speaking on behalf of a national environmental organization, claimed that children were being exposed to dangerous pesticide residues through their consumption of apple juice. A "panic for organic" ensued, and people who had never before purchased organic food rushed to their supermarkets to stockpile organic apple juice and other organic products. Suddenly the profits in the organic marketplace soared, attracting many new entrepreneurs who wanted to catch the trend. Indeed, until the NOP was implemented, there was no requirement that any particular standard be followed, and many organic farmers and handlers placed the organic label on their products with no oversight. Today, the NOP requires all farmers and handlers of organic food to adhere to NOP standards, and all but the smallest operators (those with organic sales of less than $5,000) must be certified.

Beyond setting the standards, the central role of the USDA is to inspect the inspectors—in other words, the federal government runs an accreditation program for certifying agents. AMS accredits both domestic and foreign certifying agents in the areas of crops, livestock, wild crops, and handling. Those who seek accreditation must apply to the NOP and demonstrate their expertise in organic production techniques, provide certification information to interested parties, and maintain records and confidentiality of clients. Accreditation is valid for a period of five years. Accredited agents may produce and use a seal or logo to denote certification (see Figure 4.1). AMS conducts site evaluations to determine compliance of accredited certifying agents and requires such agents to submit annual fees and reports to AMS.

The accredited certifying agents in turn inspect the farmers and handlers. At a minimum, this includes annual on-site inspection of all production facilities and fields and review of records and organic plans to ensure compliance with the NOP. Certifying agents may also perform residue testing of organic products as an additional way to ascertain compliance with the regulation. The costs of certification are paid directly by farmers and handlers to the certifying agent.

U.S. Organic Standards

The standards for organic production are proposed by the NOP and, after public comment and debate, are considered final and posted at www.ams.usda.gov/nop. Proposals to amend the standards are accepted anytime and, given widespread support, such amendments can be adopted. This chapter lays out the broad stipulations of the standards that, in their entirety, comprise hundreds of pages.

To pass inspection, farmers must demonstrate that their land has not had prohibited substances (for example, synthetic pesticides and fertilizers) applied to it for three years be-

Figure 4.1. California Certified Organic Farmers (CCOF) certification seal (http://www.ccof.org/).

fore planting their first organic crop. All organic fields need to have distinct boundaries and buffer zones to any conventional fields. Cultivation practices must improve soil and minimize erosion. Manure must be applied such that it does not contribute to contamination of crops, soil, or water. Raw manure must be composted unless it is used on crops not for human consumption (such as animal feed) or if it is incorporated into the soil 120 days before harvest if the edible plant part is in contact with soil (such as lettuce) or 90 days before harvest if the edible plant part does not touch the soil (such as sweet corn). Composted plant and animal materials must be produced according to specified time, temperature, and turning requirements. Weeding by means of heat and flame is allowed. Seeds treated with pesticides or antibiotics may not be used except if required by federal or state phytosanitary regulations. The use of inputs used as disinfectants and sanitizers in the production system (such as in irrigation) is restricted.

All animals for meat production, with the exception of chickens, must be managed organically from the last third of gestation or hatching. Milking cows require at least one year under organic management before milk can be sold as organic. Poultry must be under organic management from the second day of life. Organic production requires that animals be allowed exercise, freedom of movement, and reduction of stress. Most of the year, this means that animals must be given access to the outdoors. Feed must be 100 percent organically produced. Livestock bedding needs to be organic if typically consumed by the animals. No antibiotics or hormones are allowed at any time. If animals get sick and need treatment, they must be culled from the organic herd and treated with antibiotics for ethical reasons, but the resulting animal product cannot be sold as organic.

Organic has traditionally meant that synthetic materials are prohibited and, conversely, that all natural materials are allowed. United States organic standards, similarly to IFOAM guidelines and standards in other countries, include a list of exceptions to this synthetic/natural rule. Called "The National List," these allowed synthetics are specified (for example, soap) as well as prohibited naturals (for example, arsenic). The National List is developed by the National Organic Standards Board (NOSB), a panel of citizens including environmental and consumer group representatives, that evaluates proposed materials for organic production based on criteria established by law. No synthetic input can be used in organic production without this board's recommendation, no matter how benign.

At no time after the food product leaves the farm can it be exposed to prohibited substances, such as chemical fumigants. Audit trails must be maintained that follow the organic food product from the farm, through various handlers (such as processors and food re-packagers), and to the retail shelf to ensure that organic integrity is maintained. All packaging material must be clean and pose no risk of contamination. If residue testing detects prohibited pesticides at levels that are greater than 5 percent of the Environmental Protection Agency's tolerance level for that pesticide residue, the agricultural product must not be sold, labeled, or represented as organically produced even if the farmer or handler can prove that all NOP standards have been followed. This is a concession to consumer demand that organic food be essentially pesticide free.

The "Big Three"

In 1997, the USDA proposed that organic foods be produced using genetic engineering, sewage sludge, and irradiation. This was contrary to existing U.S. state and private organic standards, IFOAM guidelines, and organic regulations in most European nations. The public backlash was swift and overwhelming: The vast majority of public comments to the USDA focused exclusively on the need to prohibit what became known as the "big three." The Secretary of Agriculture promised the public that the USDA would write an entirely new proposal that would ban genetic engineering, sewage sludge, and irradiation consistent with consumer expectations.

Banning irradiation and sewage sludge was easy enough. But policymakers feared that a prohibition on genetic engineering would signal U.S. concerns over the technology just as the United States was in a heated trade dispute with the European Union over its refusal to accept genetically engineered American corn and soybeans. Policymakers also struggled over whether to establish a threshold for some trace amount of genetically engineered material that would be considered unavoidable contamination, and not disqualify a product for a "GMO free" designation. Indeed, the European Union had enacted a 1 percent so-called "adventitious presence" threshold and Japan a 5 percent level.[15,16] But the United States was not ready to resolve the GMO threshold issue even though it was important in finalizing national organic standards. Instead, an effort to isolate the organic standard from the larger political debate over genetic engineering took place. The NOP redefined genetic engineering as "excluded methods" and stated that the mere presence of GMO, at any level, does not disqualify a product from being labeled organic, unless the farmer or handler is found to have violated the standards. In this way, unintentional genetic drift or other means of contamination does not eliminate organic products from entering the market.

This political resolution means that current NOP standards do not perfectly match consumer expectations that organic food be GMO-free. But it is a practical solution. A 2002 survey of American organic farmers found that nationally, 30 percent of organic farmers characterized their risk as high to very high for contamination by GMO crops, and 48 percent had carried out some measure to reduce GMO impacts by increasing buffer zones, discontinuing certain inputs, adjusting planting times, and other measures. Nationally, 27 percent had a test for GMO presence demanded by a certifier or buyer, with 11 percent of these tests finding evidence of GMO presence.[17]

Reading the Organic Label

The NOP established four organic label categories. The four label designations provide consumers with important information about the overall content of the organic product and create an incentive for food processors to maximize their organic ingredients. The highest level is "100 percent organic," which means that the entire product, including any additives, must be certified organic substances. The second highest organic content level is "organic," which designates products consisting of at least 95 percent organically produced ingredients. Many products fall into this category because essential inputs that are nonorganic, such as baking soda, must be used. Products meeting the requirements for "100 percent organic" and "organic" may display the USDA Organic seal (see Figure 4.2). Products labeled "made with organic" must contain at least 70 percent organic ingredients and list up to three of the organic ingredients or food groups on the principal display panel. Products that contain less than 70 percent organic ingredients cannot use the term "organic" other than to identify the specific ingredients that are organically produced in the ingredients statement.

Figure 4.2. U.S.D.A. Organic certification seal (Agricultural Marketing Service, National Organic Program http://www.ams.usda.gov/nop/indexIE.htm).

Health and Safety

Consumers assume that organic food is safer and more nutritious. According to numerous surveys, health remains the number one reason people buy organic food.[18,19,20] Yet there is little scientific evidence to support this conclusion. The link between health and organic production is a nascent area of scientific inquiry, primarily because governments have not significantly invested in any kind of research related to organic agriculture. The Organic Farming Research Foundation searched the database of USDA-sponsored research for money spent on inquiries "directly pertinent" to organic farmers. The conclusion of the 1997 study: less than one-tenth of 1 percent of the USDA's total research investment had been so dedicated.[21]

Upon enactment of the organic law and regulation, policymakers were careful to stress in every press release and public meeting that all food is safe and wholesome if it meets basic federal laws and regulations of the Food and Drug Administration, EPA, and USDA. Organic does not imply that other food is inferior, policymakers said; rather, organic is simply a matter of marketing and personal preference. Yet, NOP standards do restrict the use of toxic materials and antibiotics and hormones, all of which raise health concerns. Organic food is, at least, different from conventionally produced food.

Because of historic concerns with the use of manure in agriculture, Congress stipulated that manure must be used safely. Yet the federal government, despite concerns about *E-coli* 0157 and other food pathogens, has not developed a manure standard for any aspect of agriculture. After much public debate over what constituted safe use, the NOP imposed a rule, as described earlier, that requires 60- and 120-day waiting periods from application to harvest. In some northern climates, this rule essentially eliminates the opportunity to use raw manure because the growing season can be less than 120 days. Composted manure must have an initial C:N ratio between 25:1 and 40:1; be maintained at 55–77°C for three days (in-vessel/aerated system) or for 15 days (windrow composting system) and the compost must be turned at least five times. Does all this provide a guarantee of safety? At this time, no scientific consensus exists on safe use of manure, but organic standards do require safeguards that are absent in most production systems.

As more research is conducted on organic food, making health claims in the future may be possible. New and tantalizing research suggests that in comparisons of organic and conventionally grown food, certain organic food products may be preferable due to lesser pesticide residues, no antibiotic use, increases of some secondary plant metabolites that may play a cancer-fighting role, and increased omega-3 fatty acids.[22,23]

How U.S. Standards Compare Worldwide

No two countries have identical standards. IFOAM standards and Codex guidelines are adhered to for the most part by all traders. The major rift between organic programs concerns the aforementioned articulation of goals as either principles or measurable outcomes. That aside, crop and processing standards are largely consistent. All organic standard programs recognize human health as a concern. All programs have a material list very much like the National List in both its construction and contents. Yet even the smallest differences in standards can create trade friction. The European Union, for example, does not allow the use of sodium (Chilean) nitrate in organic production, whereas the United States allows this natural material to be used under certain conditions. This difference has generated additional talks between U.S. and EU policymakers as well as complicated contractual arrangements by private companies trading in both markets.

The greatest standards discrepancies are in the area of animal production. Codex, the EU, and IFOAM standards allow use of most animal drugs, including antibiotics, whereas the United States prohibits almost all animal drugs except specified vaccines. Animal welfare rules differ across organic programs and are constantly evolving. As mentioned earlier, manure rules differ, with Codex and the European Union banning manure from factory farms and the United States establishing highly prescriptive standards for raw manure and compost use. Japan has not yet adopted any animal standards.

At the time the U.S. Congress acted in 1990, organic farming was a very small enterprise, described by participants purposely as a community and not an industry. Organic farming was almost synonymous with small-scale family farming. Farm labor was not a particularly big issue because most organic farms were owner operated and there were few organic processed food products. Times have changed and social issues have become contentious within U.S. organic constituencies. The debate over organic standards at the turn of the century is whether to include social justice criteria among the standards. IFOAM has extended its standards to include not only animal welfare but also producer welfare and the social justice implications of all aspects of production.[24] Should organic foods come from overseas, big corporations, and farms or should organic foods be linked with local foods and community food security? Should such foods be produced in accordance with fair trade principles that address wages and worker conditions? United States advocates for social justice standards are urging consumers to buy food labeled fair trade as well as organic while they seek to make NOP standards consistent with IFOAM.[25]

Did Government Regulations Help or Hurt?

I used to begin my speeches building on the old punch line that ended many jokes: "I'm from the government and I'm here to help." Critics of the U.S. NOP predict that the intervention of the federal government into what had essentially been a nonprofit, farmer-driven community enterprise will be the downfall of organic agriculture. Perhaps 15 long years of struggle over formation of the NOP that began with early bill drafting in 1988 and concluded with implementation in 2003 has generated cynicism. But the lesson I draw from the experience is that an activated public can determine the content of organic standards. The U.S. organic regulatory system is structured to heavily weigh citizen input and, to the extent that the public continues to exercise its voice, organic standards should adhere to consumer expectations and needs.

Federal government recognition of the organic label has provided the legitimacy, consistency, and predictability necessary to attract more than the small band of dedicated, ideologically driven farmers to the organic market. This, in turn, has meant more land in ecologically responsible management. It has also meant a larger and more powerful clientele for organic services, and the U.S. government is responding by extending traditional government programs, such as crop insurance and market reporting, to assist this burgeoning sector.

Notes

1. Willer, H. and M. Yussefi (Eds.) 2004. *The World of Organic Agriculture 2004: Statistics and Emerging Trends*, 6th ed. Koenigstein, Germany: Verlagsservice Wilfried Niederland.

2. Jones, D. 2003. Organic Agriculture, Sustainability, and Policy. In *Organic Agriculture Sustainability, Markets, and Policies, OECD Organization for Economic Development*, p. 9. Wallingford, UK: CABI Publishing.

3. 7 U.S.C. Chap. 94, Sec. 6501.

4. 7 C.F.R. Part 205.

5. *U.S. Department of Agriculture, Agriculture Marketing Service* Home Page. U.S. Dept. of Agriculture. http://www.ams.usda.gov/.

6. 65 Fed. Reg. 13,511-13,560 March 13, 2000 (proposed rule).

7. *International Organization for Organic Agriculture Movements* Home Page. IFOAM. http://www.ifoam.org/.

8. *CODEX Alimentarius* Home Page. Food and Agriculture Organization of the United Nations and the World Health Organization. http://www.codexalimentarius.net/.

9. *CODEX Alimentarius* Home Page. Food and Agriculture Organization of the United Nations and the World Health Organization. http://www.codexalimentarius.net/.

10. 65 Fed. Reg. 80,547-80,596 December 21, 2000 (final rule).

11. 7 C.F.R. Part 205. Sec. 236–239.

12. 7 U.S.C. Chap. 94, Sec. 6501.

13. U.S. Congress. Senate. Committee on Agriculture, Nutrition and Forestry. 1990. *Food, Agriculture and Trade Act of 1990*. 101st Cong., 2d sess., S. Rept. 101–357.

14. Countries are allowed some discretion in applying EU Organic Standards upon consideration of their local conditions. Specific areas where discretion is allowed on the national level are defined within the standards.

15. Japanese Agricultural Standards for Organic Agricultural Products. Notification No. 59 of the Ministry of Agriculture, Forestry, and Fisheries, January 20, 2000.

16. Council Regulation (EC) No 2092/91 of 24 June 1991: Organic production of agricultural products and indications referring thereto on agricultural products and foodstuffs. Official Journal L 198, 22/07/1991, p. 1.

17. Walz, E. 4th National Organic Farmers' Survey: Sustaining Organic Farms. Organic Farming Research Foundation. May 16, 2003. http://www.ofrf.org/publications/survey/GMO.SurveyResults.PDF.

18. Batt, P.J. and M. Giblett. 1999. A pilot study of consumer attitudes to organic fresh fruit and vegetables in Western Australia. *Food Australia* 51(11):549–550.

19. Makatouni, A. 1999. The consumer message: What motivates parents to buy organic food in the UK? Results of a Qualitative Study. In *Quality & communication for the organic market*. W. Lockeretz and B. Geier (Eds.). 6th IFOAM Organic Trade Conference, October 20th–23rd, 1999, Florence/Italy.

20. Williams, P.R.D. and J.K. Hammitt. 2000. A comparison of organic and conventional fresh produce buyers in the Boston area. *Risk Analysis* 20(5):735–746.

21. Walz, E. Final Results of the Third Biennial National Organic Farmers' Survey. Organic Farming Research Foundation, 1999 http://www.ofrf.org/publications/survey/Final.Results.Third.NOF.Survey.pdf.

22. Benbrook, C. M. Minimizing Pesticide Dietary Exposure Through the Consumption of Organic Food. May 2004. The Organic Center. http://www.organic-center.org/.

23. Is Organic Better? PowerPoint presentation from the Organic Center. 20 May 2004 <http://www.organic-center.org/about4.htm>.

24. Section B, chapter 8 of IFOAM Basic Standards for Organic Production and Processing, IFOAM General Assembly, Victoria, CA, 2002.

25. Henderson, E., R. Mandelbaum, O. Mendieta, and M. Sligh. *Towards Social Justice and Economic Equity in the Food System: A Call for Social Stewardship Standards in Sustainable and Organic Agriculture.* The Rural Advancement Foundation International, USA, final draft, October 2003. http://www.rafiusa.org/pubs/SocialJustice_final.pdf.

5 Health Claims: A U.S. Perspective

Victor Fulgoni

Health claims can be defined as a statement regarding a food or food ingredient and its ability to reduce the risk of a disease or health-related condition. Structure function claims are very different from health claims. The former refer to the effects of a food or food ingredient (typically a nutrient) on the structure or function of the human body. Structure function claims must not explicitly mention or imply that the food or food ingredient will diagnose, cure, mitigate, treat, or prevent disease.

Interestingly, although structure function claims such as "calcium builds strong bones" have been used for decades, health claims are a relatively recent phenomenon. This chapter provides a short history of health claims and then discusses some of the scientific requirements necessary to persuade the government that a health claim should be authorized.

Brief History of Health Claims

In 1984, health claims took on a new perspective. Prior to that time, any mention that a food or food ingredient could have a role in disease prevention resulted in the Food and Drug Administration's (FDA) deeming the product an unapproved drug/misbranded food. Serious regulatory action, including product seizure and possibly even criminal charges, were possible.

In 1984, the Kellogg Company introduced its groundbreaking collaboration with the National Cancer Institute (NCI) to communicate on their All-Bran® cereal packages the message that a "low fat, high-fiber diet may reduce the risk of certain cancers" (Fig. 5.1). This campaign, which included television advertising and public relations efforts, was a "lightning rod" for health claims. This was the first time any major food manufacturer had ever even used the word **cancer** on a food package. Kellogg's strategy was brilliant and the food and dietary supplement industry owe those Kellogg executives much gratitude for taking the regulatory risk to initiate what can be called the "health claims process." Although the FDA seriously considered regulatory action against Kellogg, having NCI support for the health claim eventually caused the FDA to reconsider taking regulatory action on Kellogg products. Kellogg eventually expanded this campaign to other cereal brands, such as Bran Flakes and Raisin Bran. The company worked with NCI to develop other communication materials. By almost all accounts, this campaign was very successful. Tens of thousands of consumers contacted NCI for more information about cancer prevention, and Kellogg sales increased significantly. A report by former Federal Trade Commission employees found that the campaign was very successful in reaching population groups that usually do not see health and nutrition information[1]. This campaign became the "gold standard" of health claims.

In the late 1980s, other companies tried to replicate the Kellogg approach. Probably the most successful endeavor was an effort to communicate the health benefits of consuming oat bran to reduce cholesterol levels and thus the risk of heart disease. At that time, a significant body of literature existed indicating that oats/oat bran could reduce blood

Figure 5.1. All Bran®—health claims leader

cholesterol levels as part of a diet low in saturated fat and cholesterol[2-5]. These claims proliferated, and many food manufacturers made significant revenues by making and promoting oats and oat bran products. Then the "oat bran boom" burst with the release of a study that indicated that, in nurses with normal cholesterol, oat bran was not effective[6]. Simultaneously, food companies, trade associations, and consumer groups were actively petitioning the FDA to formally recognize health claims or to ban health claims. All the debate ended when the Nutrition Labeling and Education Act was passed in 1990. This act, among other actions, mandated that the FDA create regulations authorizing health claims and even specifically required FDA to review the relationship between several food ingredients and certain diseases.

The *Federal Register* containing the final rules resulting from the act was published in January 1993. Thus, nine years after the Kellogg/NCI campaign, the FDA formally authorized health claims. As part of the final rules, the FDA authorized nine (two claims, folic acid's relation to neural tube defects and sugar alcohols' relation to dental caries, were authorized a short time after the initial NLEA final rules) health claims and declined to authorize five health claims[7-16]. Additionally. The FDA developed a regulation on a petition process for new health claims[17].

The NLEA-Authorized Health Claims

The authorized claims with sample language resulting from NLEA are presented in Figure 5.2. Five health claims were specifically not authorized as part of NLEA[7,8,10,12,14]:

1. Antioxidant vitamins and cancer
2. Dietary fiber and cancer
3. Dietary fiber and cardiovascular disease
4. Omega-3 fatty acids and coronary heart disease
5. Zinc and immune function in the elderly

- **Calcium and Osteoporosis - 21 CFR 101.72**
Regular exercise and a healthy diet with enough calcium helps teen and young adult white and Asian women maintain good bone health and may reduce their high risk of osteoporosis later in life.
Regular exercise and a healthy diet with enough calcium helps teen and young adult white and Asian women maintain good bone health and may reduce their high risk of osteoporosis later in life. Adequate calcium intake is important, but daily intakes above about 2,000 mg are not likely to provide any additional benefit.

- **Dietary Fat and Cancer – 21 CFR 101.73**
Development of cancer depends on many factors. A diet low in total fat may reduce the risk of some cancers.
Eating a healthful diet low in fat may help reduce the risk of some types of cancers. Development of cancer is associated with many factors, including a family history of the disease, cigarette smoking, and what you eat.

- **Fiber-Containing Grain Products, Fruits, and Vegetables and Cancer – 21 CFR 101.76**
Low fat diets rich in fiber-containing grain products, fruits, and vegetables may reduce the risk of some types of cancer, a disease associated with many factors.
(2) Development of cancer depends on many factors. Eating a diet low in fat and high in grain products, fruits, and vegetables that contain dietary fiber may reduce your risk of some cancers.

- **Fruits and Vegetables and Cancer - 21 CFR 101.78**
Low fat diets rich in fruits and vegetables (foods that are low in fat and may contain dietary fiber, vitamin A, and vitamin C) may reduce the risk of some types of cancer, a disease associated with many factors. Broccoli is high in vitamins A and C, and it is a good source of dietary fiber.
(2) Development of cancer depends on many factors. Eating a diet low in fat and high in fruits and vegetables, foods that are low in fat and may contain vitamin A, vitamin C, and dietary fiber, may reduce your risk of some cancers. Oranges, a food low in fat, are a good source of fiber and vitamin C.

- **Sodium and Hypertension -21 CFR 101.74**
Diets low in sodium may reduce the risk of high blood pressure, a disease associated with many factors.
(2) Development of hypertension or high blood pressure depends on many factors. [This product] can be part of a low sodium, low salt diet that might reduce the risk of hypertension or high blood pressure.

- **Dietary Saturated Fat and Cholesterol and Risk of Coronary Heart Disease - 21 CFR 101.75**
While many factors affect heart disease, diets low in saturated fat and cholesterol may reduce the risk of this disease.
Diets low in saturated fat, cholesterol, and total fat may reduce the risk of heart disease. Heart disease is dependent upon many factors, including diet, a family history of the disease, elevated blood LDL-cholesterol levels, and physical inactivity.

- **Fruits, Vegetables and Grain Products that contain Fiber, particularly Soluble Fiber, and Risk of Coronary Heart Disease - 21 CFR 101.77**
Diets low in saturated fat and cholesterol and rich in fruits, vegetables, and grain products that contain some types of dietary fiber, particularly soluble fiber, may reduce the risk of heart disease, a disease associated with many factors.
(2) Development of heart disease depends on many factors. Eating a diet low in saturated fat and cholesterol and high in fruits, vegetables, and grain products that contain fiber may lower blood cholesterol levels and reduce your risk of heart disease.

- **Folate and Neural Tube Defects - 21 CFR 101.79**[1]
Healthful diets with adequate folate may reduce a woman's risk of having a child with a brain or spinal cord birth defect.
Women who consume healthful diets with adequate folate throughout their childbearing years may reduce their risk of having a child with a birth defect of the brain or spinal cord. Sources of folate include fruits, vegetables, whole grain products, fortified cereals, and dietary supplements.

- **Dietary Sugar Alcohol and Dental Caries - 21 CFR 101.80**[1]
Frequent between-meal consumption of foods high in sugars and starches promotes tooth decay. The sugar alcohols in [name of food] do not promote tooth decay.
Example of the shortened claim for small packages:
(i) Does not promote tooth decay.

[1]These claims were finalized shortly after NLEA regulations

Figure 5.2. Original health claims authorized

In 1997, the FDA Modernization Act (FDAMA) was passed and authorized FDA to allow certain health claims without having to go through the health claim petition process. Such a health claim must be the subject of an "authoritative statement" issued by a "scientific body" of the U.S. government with official responsibility for public health protection or research directly relating to human nutrition. The FDA identified the National Institutes of Health, the Centers for Disease Control and Prevention, the National Academy of Sciences or any of its subdivisions, and the USDA as a qualified "scientific body." Health claims under the FDAMA process must provide notification to the FDA of the desire to make a health claim based on an "authoritative statement" and provide all relevant information to the agency (for example, the claim, the "authoritative statement, scientific summary, and so on). FDA must respond to the notification in 120 days; otherwise, the claim will be automatically allowed[18]. The three health claims allowed via the FDAMA process are as follows:

1. "Diets rich in whole grain foods and other plant foods and low in total fat, saturated fat, and cholesterol, may help reduce the risk of heart disease and certain cancers." This claim was based on a notification from General Mills, Inc., which used the following "authoritative statement:" "Diets high in plant foods—i.e. fruits, vegetables, legumes, and whole-grain cereals—are associated with a lower occurrence of coronary heart disease and cancers of the lung, colon, esophagus, and stomach," from the Executive Summary of the 1989 National Academy of Sciences (NAS) report, *Diet and Health: Implications for Reducing Chronic Disease Risk*[19].
2. "Diets containing foods that are good sources of potassium and low in sodium may reduce the risk of high blood pressure and stroke." This claim was based on a notification from Tropicana Products, Inc., which used the following "authoritative statements:" "Epidemiological and animal studies indicate that the risk of stroke-related deaths is inversely related to potassium intake over the entire range of blood pressures, and the relationship appears to be dose dependent. The combination of a low-sodium, high potassium intake is associated with the lowest blood pressure levels and the lowest frequency of stroke in individuals and populations. Although the effects of reducing sodium intake and increasing potassium intake would vary and may be small in some individuals, the estimated reduction in stroke-related mortality for the population is large." The second statement was: "Vegetables and fruits are also good sources of potassium. A diet containing approximately 75 mEq (i.e., approximately 3.5g of elemental potassium) daily may contribute to reduced risk of stroke, which is especially common among blacks and older people of all races. Potassium supplements are neither necessary nor recommended for the general population." Both of these statements came from the 1989 NAS report *Diet and Health: Implications for Reducing Chronic Disease Risk*[20].
3. "Diets rich in whole grain foods and other plant foods, and low in saturated fat and cholesterol, may help reduce the risk of heart disease." This claim, although the same as the first claim described previously, was the result of a notification by Kraft Foods and allowed claims for whole grain foods with less than 6.5 grams total fat, less than 1.0 g saturated fat, and 0.5 gram or less *trans* fat per reference amount customarily consumed. Again, the authoritative statements came from the NAS report. Kraft Foods supplied numerous statements from the NAS report that focused fat recommendations on saturated fat but not total fat[21].

Since the adoption of the NLEA health claim petition process and the FDAMA notification process, FDA has authorized nine additional health claims for foods[19–27]. As can be seen in Figure 5.3, FDA has authorized a new claim just about every year since the original NLEA claims. It takes considerable effort and resources to complete the review of a new health claim petition. Submissions must evaluate the safety of the food or ingredient and must provide a balanced summary of the scientific evidence regarding the food or ingredient and the disease or health-related condition. For most new health claims a considerable body of scientific evidence has accompanied the petition. Figure 5.4 shows the number of studies submitted, the number of studies reviewed (some studies did not meet minimum eligibility standards), and the number of studies given special weight for several of the new health claims (for example, especially strongly designed studies, typically randomized control trials, conducted in relevant population groups using realistic levels of the food or ingredient). As few as 15 studies and as many as 43 studies have been submitted for the health claims listed in Figure 5.4. Interestingly, as few as eight studies received "special weight" to support authorization of a new health claim[25]. More recently, even fewer studies were used to authorize a health claim for oatrim (purified source of beta-glucan from oats), though a considerable body of relevant studies for whole oats and oat bran were previously reviewed[26].

"Qualified Health Claims"

In the late 1990s, the FDA lost a series of court battles that greatly transfigured the "health claim process." The major court case, *Pearson v. Shalala,* forced the FDA to allow health claims for certain ingredient/disease relationships even if there was not significant scientific agreement about the relationship of the food or ingredient and the disease/health-related condition. The court ordered the FDA to consider using qualifying language to modify the health claim rather than deny the health claim. Thus, "qualified health claims" were created. Initially, the FDA tried to interpret the court's decisions very narrowly and limit qualified health claims to dietary supplements only. Eventually, in December 2002, as part of the report of the results of an internal FDA task force (Consumer Health Infor-

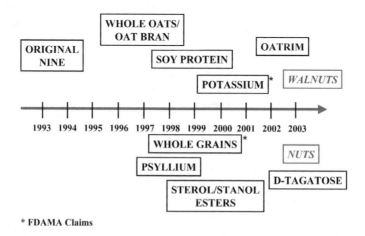

* FDAMA Claims

Figure 5.3. Health claims for foods post NLEA

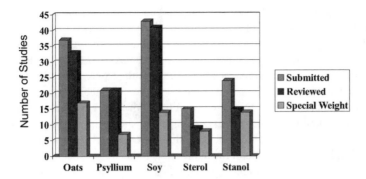

Figure 5.4. Science submitted in FDA health claim petitions

mation for Better Nutrition), FDA agreed to allow "qualified health claims for food." A very interesting aspect of these "qualified health claims" is that instead of a regulation authorizing the health claim, FDA has decided to use its "enforcement discretion" and has agreed not to take any action regarding the use of a properly qualified claim if the claim is consistent with the items outlined in the letter in response to the health claim petition. To date, including dietary supplements, FDA has issued eight "qualified health claims"[28–35] (Figure 5.5). Only two of these claims were for foods—walnuts and other tree nuts[30, 31]. However, in the future more "qualified health claims" for foods can be expected.

Summary of the Current Health Claim Process

Much has happened in the almost 20 years of health claims. This area is likely to continue to evolve as we learn more about the impact of food or food ingredients on health. Figure 5.6 attempts to describe the existing health claim process in a concise manner.

Basically, the starting point (from left to right in Figure 5.6) is a *public health issue* (for example, coronary heart disease, hypertension). Following that must be a body of **science**, which deals with both the **safety** and **effectiveness** of the food or ingredient with respect to the disease or health-related condition. More likely than not, the food or ingredient will affect a biomarker, rather than the disease itself, of the disease or health-related condition. One reason most of the new health claims authorized by the FDA are for effects on coronary heart disease is the fact that an accepted biomarker of this disease exists, namely, blood low-density lipoprotein cholesterol levels. Although a variety of study types can be used to support a new health claim, the "gold standard" remains the randomized clinical trial. Although the FDA has authorized health claims with only epidemiological data, claims using only these types of studies will be difficult to reach the "significant scientific agreement" standard and, consequently, may be eligible only for a "qualified health claim." As indicated in Figure 5.6, with an adequate body of science, a "qualified health claim" might be allowed. However, FDA has indicated its desire to ensure that "qualified health claims" would eventually achieve the "significant scientific agreement" standard, that is, eventually more science would allow for "significant scientific agreement."

It is important to ensure that relevant health professionals know about the science regarding a food or ingredient. On one hand, publication of the science in peer-reviewed journals will help to disseminate the information regarding the effects of a food or ingre-

Claims for Dietary Supplements

Selenium & Cancer

(1) Selenium may reduce the risk of certain cancers. Some scientific evidence suggests that consumption of selenium may reduce the risk of certain forms of cancer. However, FDA has determined that this evidence is limited and not conclusive.

(2) Selenium may produce anticarcinogenic effects in the body. Some scientific evidence suggests that consumption of selenium may produce anticarcinogenic effects in the body. However, FDA has determined that this evidence is limited and not conclusive.

Antioxidants and Cancer

(1) Some scientific evidence suggests that consumption of antioxidant vitamins may reduce the risk of certain forms of cancer. However, FDA has determined that this evidence is limited and not conclusive.

(2) FDA has determined that although some scientific evidence suggests that consumption of antioxidant vitamins may reduce the risk of certain forms of cancer, this evidence is limited and not conclusive.

Omega-3 Fatty Acids & Coronary Heart Disease

Consumption of omega-3 fatty acids may reduce the risk of coronary heart disease. FDA evaluated the data and determined that, although there is scientific evidence supporting the claim, the evidence is not conclusive.

B Vitamins & Vascular Disease

As part of a well-balanced diet that is low in saturated fat and cholesterol, Folic Acid, Vitamin B6 and Vitamin B12 may reduce the risk of vascular disease. FDA evaluated the above claim and found that, while it is known that diets low in saturated fat and cholesterol reduce the risk of heart disease and other vascular diseases, the evidence in support of the above claim is inconclusive.

Phosphatidylserine & Cognitive Dysfunction and Dementia

1) Consumption of phosphatidylserine may reduce the risk of dementia in the elderly. Very limited and preliminary scientific research suggests that phosphatidylserine may reduce the risk of dementia in the elderly. FDA concludes that there is little scientific evidence supporting this claim.

(2) Consumption of phosphatidylserine may reduce the risk of cognitive dysfunction in the elderly. Very limited and preliminary scientific research suggests that phosphatidylserine may reduce the risk of cognitive dysfunction in the elderly. FDA concludes that there is little scientific evidence supporting this claim.

Folic Acid (0.8 mg) & Neural Tube Birth Defects

0.8 mg folic acid in a dietary supplement is more effective in reducing the risk of neural tube defects than a lower amount in foods in common form. FDA does not endorse this claim. Public health authorities recommend that women consume 0.4 mg folic acid daily from fortified foods or dietary supplements or both to reduce the risk of neural tube defects.

Claims for Foods

Nuts & Heart Disease

Scientific evidence suggests but does not prove that eating 1.5 ounces per day of most nuts [such as *name of specific nut*] as part of a diet low in saturated fat and cholesterol may reduce the risk of heart disease. [See nutrition information for fat content.]

Walnuts & Heart Disease

(1) Supportive but not conclusive research shows that eating 1.5 ounces per day of walnuts as part of a diet low in saturated fat and cholesterol may reduce the risk of heart disease. See nutrition information for fat content.

(2) Scientific evidence suggests but does not prove that eating 1.5 ounces per day of most nuts, such as walnuts, as part of a diet low in saturated fat and cholesterol may reduce the risk of heart disease. See nutrition information for fat content.

Figure 5.5. Original health claims for dietary supplements.

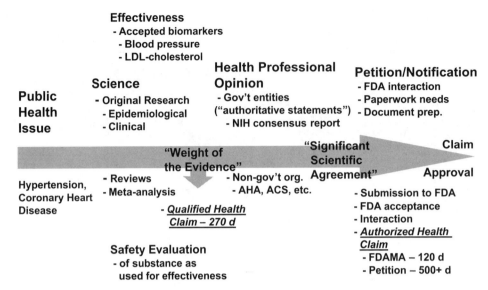

Figure 5.6. Schematic of health claim process

dient on a biomarker of disease. On the other hand, more proactive approaches may be necessary to ensure that health professionals are aware of the science. These proactive activities may include organizing symposia at relevant meetings, preparing review articles for specific audiences (for example, focusing on dietitians, primary care physicians, and others) or meeting one-on-one with key individuals and groups. Without these efforts, "significant scientific agreement" will take longer to occur.

Finally, after considerable effort on the aspect of science and health professional communication, a formal notification or petition can be prepared requesting FDA to authorize a new health claim. The petitioner or notifier can expect the FDA staff to take a constructive attitude to responding to the petition or notification, asking relevant scientific questions. These interactions, although at times challenging, have always led to a better or faster outcome.

Conclusions

Health claims are one way to formally recognize the potential that proper nutrition, especially that which exceeds meeting basic nutrient needs, can have on public health via reduction of risk for several chronic disease states. Although they have existed for only approximately 20 years, health claims have certainly provided numerous opportunities for food and dietary supplement manufacturers to participate in the prevention of certain chronic diseases while simultaneously making money for their shareholders and stakeholders.

References

1. Ippolito, P, Mathios, A. 1989. *Health Claims in Advertising and Labeling: A Study of the Cereal Market.* Bureau of Economics, Federal Trade Commission, Washington, DC.

2. Kirby RW, Anderson JW, Sieling B, Rees ED, Chen WJ, Miller RE, Kay RM. 1981. Oat-bran intake selectively lowers serum low-density lipoprotein cholesterol concentrations of hypercholesterolemic men. *Am J Clin Nutr* 34:5 824–829.

3. Anderson JW, Story L, Sieling B, Chen WJ, Petro MS, Story J. 1984. Hypocholesterolemic effects of oat-bran or bean intake for hypercholesterolemic men. *Am J Clin Nutr* 40:6 1146–1155.

4. Van Horn LV, Liu K, Parker D, Emidy L, Liao YL, Pan WH, Giumetti D, Hewitt J, Stamler J. 1986. Serum lipid response to oat product intake with a fat-modified diet. *J Am Diet Assoc* 86:6 759–764.

5. Gold KV, Davidson DM. 1988. Oat bran as a cholesterol-reducing dietary adjunct in a young, healthy population. *West J Med* 148:3 299–302.

6. Swain JF, Rouse IL, Curley CB, Sacks FM. 1990. Comparison of the effects of oat bran and low-fiber wheat on serum lipoprotein levels and blood pressure. *NEJM* 322:3 147–52.

7. Food and Drug Administration. Food labeling: Health claims and label statements: Antioxidant vitamins and cancer. Final rule. Federal Register 58:2622–2660.

8. Food and Drug Administration. Food labeling: Health claims; Zinc and immune function in the elderly. Final rule. Federal Register 58:2661–2664.

9. Food and Drug Administration. Food labeling: Health claims; Calcium and Osteoporosis. Final rule. Federal Register 58:2665–2681.

10. Food and Drug Administration. Food labeling: Health claims and label statements: Omega-3 fatty acids and coronary heart disease. Final rule. Federal Register 58:2682–2738.

11. Food and Drug Administration. Food labeling: Health claims and label statements: Dietary saturated fat and cholesterol and coronary heart disease. Final rule. Federal Register 58:2739–2786.

12. Food and Drug Administration. Food labeling: Health claims and label statements: Dietary fat and cancer. Final rule. Federal Register 58:2787–2819.

13. Food and Drug Administration. Food labeling: Health claims and label statements: Sodium and hypertension. Final rule. Federal Register 58:2820–2849.

14. Food and Drug Administration. Food labeling: Health claims and label statements: Dietary fiber and cancer. Final rule. Federal Register 58:2537–2551.

15. Food and Drug Administration. Food labeling: Health claims and label statements: Dietary fiber and cardiovascular disease. Final rule. Federal Register 58:2552–2605.

16. Food and Drug Administration. Food labeling: Health claims and label statements: Folic acid and neural tube defects. Final rule. Federal Register 58:2606–2620.

17. Food and Drug Administration. Food labeling: General requirement for health claims for food. Final rule. Federal Register 58:2478–2536.

18. Food and Drug Administration. Guidance for Industry: Notification of a Health Claim or Nutrient Content Claim Based on an Authoritative Statement of a Scientific Body. http://www.cfsan.fda.gov/~dms/hclmguid.html

19. Food and Drug Administration. FDAMA Health claim notification for whole grains. http://www.cfsan.fda.gov/~dms/flgrains.html.

20. Food and Drug Administration. FDAMA Health claim notification for potassium containing foods. http://www.cfsan.fda.gov/~dms/hclm-k.html.

21. Food and Drug Administration. FDAMA Health claim notification for whole grains with moderate fat content. http://www.cfsan.fda.gov/~dms/flgrain2.html.

22. Food and Drug Administration. Food Labeling: Health claims; Oats and coronary heart disease. Final rule. Federal register 62:3584–3601.

23. Food and Drug Administration. Food Labeling: Health claims; Soluble fiber from certain foods and coronary heart disease. Final rule. Federal register 63:8103–8121.

24. Food and Drug Administration. Food Labeling: Health claims; Soy protein and coronary heart disease. Final rule. Federal register 64:57700-57733.

25. Food and Drug Administration. Food Labeling: Health claims; Plant sterol/stanol esters coronary heart disease. Interim final rule. Federal register 65:54686–54739.

26. Food and Drug Administration. Food Labeling: Health claims; Soluble fiber (beta glucan) from certain foods and coronary heart disease. Interim final rule. Federal register 67:61733–61783.

27. Food and Drug Administration. Food Labeling: Health claims; D-tagatose and dental caries. Interim final rule. Federal register 67:71461–71470.

28. Food and Drug Administration. Qualified health claim: Selenium and cancer. http://www.cfsan.fda.gov/~dms/ds-ltr32.html and http://www.cfsan.fda.gov/~dms/ds-ltr35.html.

29. Food and Drug Administration. Qualified health claim: Antioxidants and cancer. http://www.cfsan.fda.gov/~dms/ds-ltr34.html.

30. Food and Drug Administration. Qualified health claim: Nuts and heart disease. http://www.cfsan.fda.gov/~dms/qhcnuts2.html.

31. Food and Drug Administration. Qualified health claim: Walnuts and heart disease. http://www.cfsan.fda.gov/~dms/qhcnuts.html.

32. Food and Drug Administration. Qualified health claim: Omega-3 fatty acids and heart disease. http://www.cfsan.fda.gov/~dms/ds-ltr11.html and http://www.cfsan.fda.gov/~dms/ds-ltr28.html.

33. Food and Drug Administration. Qualified health claim: B vitamins and heart disease. http://www.cfsan.fda.gov/~dms/ds-ltr12.html and http://www.cfsan.fda.gov/~dms/ds-hclbv.html.

34. Food and Drug Administration. Qualified health claim: Phosphatidylserine and cognitive dysfunction and dementia. http://www.cfsan.fda.gov/~dms/ds-ltr33.html and http://www.cfsan.fda.gov/~dms/ds-ltr36.html.

35. Food and Drug Administration. Qualified health claim: Folic acid (0.8 mg) and neural tube defects. http://www.cfsan.fda.gov/~dms/ds-ltr7.html and http://www.cfsan.fda.gov/~dms/ds-ltr22.html.

6 Food and Drug Administration Regulation of Dietary Supplements

Stephen H. McNamara

Introduction

This chapter reviews the key aspects of the law governing the composition and labeling of "dietary supplement" products.

During the early 1990s, much pressure was on Congress to amend the Federal Food, Drug, and Cosmetic Act (FDC Act) to reduce the regulatory burdens on dietary supplements. Many members of the House of Representatives and Senate stated that they were receiving more mail, more phone calls, and generally more constituent pressure on this subject than on anything else—including health care reform, abortion, or the deficit.

Not surprisingly, given all this pressure, Congress eventually passed the Dietary Supplement Health and Education Act of 1994 (DSHEA).[1] The House of Representatives approved the measure by unanimous consent at approximately 3:00 a.m. on October 7, 1994; and the Senate approved it, also by unanimous consent, shortly after midnight on the morning of October 8, 1994. With such broad, bipartisan support for the legislation, President Clinton signed it on October 25, 1994. (Senator Orrin Hatch, Republican of Utah, was the leading proponent of the legislation in the Congress.)

The new law, as with virtually all legislation, was a compromise. It does not include all the restraints on U.S. Food and Drug Administration (FDA) regulation of dietary supplements that the sponsors had originally wanted. Furthermore, it imposes some significant new requirements for such products. Nevertheless, viewed as a whole, there can be little doubt that the legislation has been a very favorable development for those who want to sell dietary supplements of vitamins, minerals, herbs, other botanicals, amino acids, and other similar substances.

A review of key provisions of DSHEA is set forth in the following sections.

Why Was DSHEA Enacted?

At the outset, one should consider why Congress approved DSHEA by unanimous consent.[2] DSHEA was enacted because FDA was viewed as distorting the law that existed before DSHEA to try improperly to deprive the public of safe and popular dietary supplement products.[3] FDA's authority needed to be better defined and controlled. In its official report about the need for DSHEA to curtail excessive regulation of dietary supplements by FDA, the Senate Committee on Labor and Human Resources stated explicitly, "in fact, FDA has been distorting the law in its actions to try to prevent the marketing of safe dietary supplement substances."[4] The Senate Committee also concluded, "FDA has attempted to twist the statute [that is, the provisions of the FDC Act as it then existed] in what the Committee sees as a result-oriented effort to impede the manufacture and sale of dietary supplements."[5]

Among the examples of FDA excesses described by the Senate Committee was FDA's campaign to prevent the marketing of safe dietary supplements of black currant oil (from the same fruit used to make jam).[6] FDA argued that the addition of black currant oil to a gelatin capsule caused the black currant oil to become a "food additive" within the meaning of the FDC Act, and that as a "food additive," the substance could not be marketed as a dietary supplement without first obtaining FDA issuance of an approving "food additive" regulation.[7] In this regard, the Senate Committee noted the prohibitive costs and delays for dietary supplement manufacturers that were inherent in FDA's attempt to require such "food additive" approval: "The cost to a manufacturer to prepare a food additive petition can run to $2 million. FDA approval of a food additive petition typically takes from 2 to 6 years."[8]

As the Senate Committee also noted, the federal courts had rejected repeatedly FDA's allegations of "unapproved food additive" status and consequent illegality for dietary supplements of black currant oil, although FDA was still persisting to attack the product. For example, the U.S. Court of Appeals for the Seventh Circuit, in a unanimous three-judge opinion, stated as follows:

> The only justification for this Alice-in-Wonderland approach [that is, FDA's "food additive" allegation] is to allow the FDA to make an end-run around the statutory scheme. . . . We hold that [black currant oil] encapsulated with glycerin and gelatin is not a food additive. . . . FDA has not shown that [black currant oil] is adulterated or unsafe in any way. . . .[9]

The U.S. Court of Appeals for the First Circuit, also in a unanimous three-judge opinion, ruled similarly to the Seventh Circuit:

> FDA's reading of the [FDC] Act is nonsensical. . . . The proposition that placing a single-ingredient food product into an inert capsule as a convenient method of ingestion converts that food into a food additive perverts the statutory text, undermines legislative intent, and defenestrates common sense. We cannot accept such anfractuous reasoning.[10]

Another example noted by Congress of FDA excess was provided by the agency's similar campaign to try to eradicate dietary supplements of evening primrose oil. Although evening primrose oil was (and remains today) widely available around the world as a safe and popular dietary supplement, FDA asserted that this substance, too, was an "unapproved food additive."[11] As an illustration of the extent of the agency's massive campaign, evidence was presented before the House of Representatives and the Senate in 1993 that the FDA Commissioner had awarded the Commissioner's Special Citation to 61 FDA personnel, who at that time comprised the "Evening Primrose Oil Litigation Team."[12]

Nevertheless, although two U.S. district courts and two three-judge U.S. courts of appeals had all unanimously rejected FDA's regulatory program, Congress concluded that unless it stepped in and passed DSHEA, the facts showed that FDA would continue to try to prohibit marketing of safe and proper dietary supplements by using its own interpretation of the then-existing "food additive" law:

> Although a fair reading of the current statute [i.e., the "food additive" provisions of the FDC Act], as most recently interpreted by two United States courts of appeal, should make . . . amendment [of the FDC Act by DSHEA] unnecessary, the committee has heard testimony that the FDA has rejected these [judicial] holdings. The committee is therefore concerned that the FDA will persist in such litigation, and thereby continue to subject small manufacturers to the choice of abandoning production and sale of lawful products, or accepting the significant financial burden of defending themselves against baseless lawsuits [brought by the FDA].[13]

So, when FDA last believed that it had comprehensive authority to preclear the use of dietary ingredients in dietary supplement products (under its "food additive" theory), both the federal courts and Congress concluded that the agency repeatedly had abused its authority to try to prevent the marketing of safe and proper dietary supplement products.[14]

It is important to remember this history, both because it helps to explain why, under current law, FDA is not entrusted with blanket preapproval authority over the marketing of all dietary supplement products, and because it illustrates the possibility of excessively restrictive regulation that might be presented if FDA were to be given such comprehensive preclearance authority in the future.

Broad, Expanded Definition of "Dietary Supplement"

For many years, there has been a substantial business in the selling of "dietary supplements" in the United States.[15] These products, usually consisting of tablets or capsules, have provided not only vitamins or minerals viewed as "essential" by the mainstream community of nutritionists, such as vitamin A or iron, but also other substances that FDA personnel often have regarded as being of dubious usefulness—substances ranging from rutin and other bioflavonoids to herbs to shark cartilage.

Until DSHEA was passed, agency personnel often maintained that it was "misleading" and improper to distribute as a "dietary supplement" a substance that the agency regarded as lacking in nutritional usefulness. Agency personnel sometimes also asserted that a substance that did not provide "taste, aroma, or nutritional value" in its dietary supplement (pill-type) form could not properly be sold as a food product.[16]

DSHEA deals with this fundamental definitional matter by unequivocally providing a broad, **expansive** definition of "dietary supplement" that includes a

> Product . . . intended to supplement the diet that bears or contains . . . a vitamin; . . . a mineral; . . . an herb or other botanical; . . . an amino acid; . . . a dietary substance for use by man to supplement the diet by increasing the total dietary intake; or . . . a concentrate, metabolite, constituent, extract, or combination [of any ingredient described above].[17]

The practical effect of this definition is to be clear that the protections provided by DSHEA for dietary supplements apply broadly to a wide class of products, including even products that FDA nutritionists might regard as having no nutritional value.

Exemption of Dietary Ingredients in Dietary Supplements from "Food Additive" Status

One of the most important provisions of DSHEA from the perspective of manufacturers and consumers of dietary supplement products is the explicit amendment of the FDC Act to be clear that the term **food additive** does **not** apply to a dietary ingredient in, or intended for use in, a dietary supplement.[18]

Previously, FDA had argued that substances added to dietary supplement products were much like substances added to any other food product, that is, that if such a substance was not "generally recognized as safe" (not GRAS), by "experts," based on published scientific literature, it would be subject to regulation as a **food additive**.[19] Under the FDC Act, if a substance is a **food additive**, it may not be added to food products unless a **food additive** regulation, issued by FDA, explicitly permits such addition.[20] Typically, preparation of a food additive petition, including the conduct of needed research (often including extensive animal feeding studies) and participation in the ensuing administrative proceedings, can cost a petitioner $1 million or more, and it sometimes takes FDA five years or more after receiving even a well-founded food additive petition before the agency issues an approving food additive regulation.[21] The bottom line of all of this for dietary supplements had been that alleged food additive status and the absence of a food additive regulation approving use as a dietary supplement had meant alleged **illegality**, and the end or curtailment of marketing, for many products. Based on allegations of **unapproved food additive** status before DSHEA was enacted, FDA had pursued regulatory actions against numerous once-popular dietary supplement ingredients, including calcium acetate, orotate compounds such as magnesium orotate, evening primrose oil, black currant oil, borage seed oil, linseed/flaxseed oil, chlorella, lobelia, St. John's wort, and coenzyme Q10.

DSHEA therefore frees dietary supplement ingredients from the previously existing continuing risk that, at any time, FDA might assert that its scientists did not believe that a particular dietary ingredient was GRAS for use in dietary supplements and, therefore, that the ingredient was an **unapproved food additive** and illegal.

New Safety Standards

As a quid pro quo, the new law replaces the voided food additive provisions with some new safety standards.

In general, DSHEA provides that a dietary supplement will be deemed to be adulterated if it "presents a **significant or unreasonable risk of illness or injury** under . . . conditions of use recommended or suggested in labeling, or . . . if no conditions of use are suggested or recommended in the labeling, under ordinary conditions of use (emphasis added).[22] The new law is also clear, however, that FDA shall bear the burden of proof in court if it asserts that a dietary supplement is adulterated under this standard.[23]

There are additional requirements with respect to a **new** dietary ingredient, that is, an ingredient that "was not marketed in the United States before October 15, 1994."[24] A dietary supplement that contains a **new** dietary ingredient is deemed to be adulterated unless, **either** (1) the supplement "contains only dietary ingredients which have been present in the food supply as an article used for food in a form in which the food has not been chemically altered," **or** (2) there is a "history of use or other evidence of safety establishing that the dietary ingredient when used under the conditions recommended or suggested in the labeling . . . will reasonably be expected to be safe."[25] In the latter case, also, "at least 75 days before being introduced or delivered for introduction into interstate commerce," the manufacturer or distributor of the dietary ingredient or dietary supplement must provide FDA "with information, including any citation to published articles, which is the basis on which the manufacturer or distributor has concluded that a dietary supplement containing such dietary ingredient will reasonably be expected to be safe."[26]

In addition, the new legislation provides that the Secretary of Health and Human Services may declare a dietary supplement "to pose an imminent hazard to public health or safety," in which case it immediately becomes illegal to market the product, although the Secretary must thereafter promptly hold a formal hearing to assemble data "to affirm or withdraw the declaration."[27] (The new law provides that the authority to declare a dietary supplement to be an "imminent hazard" must be exercised by the Secretary him- or herself and **may not be delegated to the FDA**.)

In sum, under the new law, with respect to safety, the FDA-asserted **food additive** requirement for agency **preclearance** of dietary ingredients not believed by FDA to be GRAS has been deleted, but FDA and the Secretary have been granted substantial new policing authority to stop the distribution of a dietary supplement if government personnel believe they can show that the product is not safe.

Rights for Sellers to Convey Information about Usefulness of Dietary Supplements

Some of the most intriguing, and potentially important, changes brought about by DSHEA relate to new freedoms for sellers of dietary supplements to use published literature to convey information to potential customers about the usefulness of dietary ingredients.

Prior to enactment of DSHEA, FDA routinely would assert that if a book, article, or other publication was used in connection with the sale of a dietary supplement to customers, such a publication could be regulated as "labeling" for the product, and that if the publication included information to the effect that an ingredient present in the product might be useful in the cure, mitigation, treatment, or prevention of any disease, the product itself could be subject to regulation as a **drug**. For example, FDA would assert that a seller of a dietary supplement product could not promote the product to customers by showing the customers, in connection with the sale of the product, copies of books or articles that asserted disease prevention benefits for nutrients provided by the supplement.[28]

DSHEA greatly restricts the FDA's ability to object to the use of books and other publications in connection with the sale of dietary supplement products. In a significant new provision, the legislation provides that a "publication," including "an article, a chapter in a book, or an official abstract of a peer-reviewed scientific publication that appears in an article and was prepared by the author or the editors of the publication," "shall not be defined as labeling" and may be "used in connection with the sale of a dietary supplement to consumers" **if** the publication is "reprinted in its entirety" **and** meets certain specific criteria. Among the criteria that must be met are (1) the publication must not be "false or misleading," (2) it must not "promote a particular manufacturer or brand of a dietary supplement," (3) it must be "displayed or presented . . . so as to present a balanced view of the available scientific information," (4) **if** "displayed in an establishment," it must be "physically separate from the dietary supplements," and (5) it must **not** "have appended to it any information by sticker or any other method."[29]

As long as these criteria are met, DSHEA would appear to enable a salesperson to call to the attention of a potential customer published nutritional or other scientific literature that describes the health benefits of a dietary supplement's ingredients—**without** enabling FDA to regulate the literature as labeling.

Furthermore, there is also an additional, explicit provision in the new legislation that it shall **not** "restrict a retailer or wholesaler of dietary supplements in any way whatsoever in the sale of books or other publications as a part of the business of such retailer or wholesaler."[30] This is the first affirmative provision in the FDC Act that protects the right of sellers of dietary supplements also to sell "books or other publications"—publications that would, of course, be expected to describe the health-related benefits of nutrients and to help support sales of dietary supplements.

Statements of Nutritional Support

FDA regulations published pursuant to the Nutrition Labeling and Education Act (NLEA) provide that, generally, **no** "health claim" may appear on the label or in other labeling (including brochures) for food products, **including dietary supplements**, unless FDA has first approved use of the claim in a final regulation.[31]

As a limited exception to the general requirement for FDA approval of "health claims" in food labeling, under DSHEA, dietary supplements are **allowed** to make four types of "statements of nutritional support" on labels or in other labeling **without** obtaining FDA approval. The four types of statements are as follows:

1. A "statement [that] claims a benefit related to a classical nutrient deficiency disease and discloses the prevalence of such disease in the United States."
2. A statement that "describes the role of a nutrient or dietary ingredient intended to affect the structure or function in humans."
3. A statement that "characterizes the documented mechanism by which a nutrient or dietary ingredient acts to maintain such structure or function."
4. A statement that "describes general well-being from consumption of a nutrient or dietary ingredient."[32]

The legislation provides that such a statement may be made in labeling for a dietary supplement **if** all the following are met:

1. The manufacturer "has substantiation that such statement is truthful and not misleading."
2. The labeling contains, prominently displayed, the following additional text: "This statement has not been evaluated by the Food and Drug Administration. This product is not intended to diagnose, treat, cure, or prevent any disease."
3. The manufacturer notifies FDA "no later than 30 days after the first marketing of the dietary supplement with such statement that such a statement is being made."

Summary of Requirements for Dietary Supplements under DSHEA

The following sections present a summary of the requirements for dietary supplements established under DSHEA.

Requirements for Dietary Supplements

Prohibition of Dietary Supplements That Present "Significant or Unreasonable Risk"

DSHEA provides that a dietary supplement shall be deemed to be "adulterated" (and therefore illegal) if it presents, or if it contains a dietary ingredient that presents, "a significant or unreasonable risk of illness or injury under . . . conditions of use recommended or suggested in labeling, or . . . if no conditions of use are suggested or recommended in the labeling, under ordinary conditions of use."[33] This provision does not require FDA to prove that a product will harm anyone; instead, a dietary supplement is deemed to be "adulterated" (illegal) if it simply presents a "significant or unreasonable risk" of illness or injury.[34]

Prohibition of Any "Poisonous or Deleterious Substances" in Dietary Supplements

DSHEA retains and incorporates the section of the FDC Act that provides that a food (this includes a dietary supplement"[35]) shall be deemed to be "adulterated" (and accordingly illegal) if it "bears or contains any poisonous or deleterious substance which may render it injurious to health . . . under the conditions of use recommended or suggested in the labeling of such dietary supplement."[36] This provision does not require FDA to prove that a substance will be injurious, only that it may render a product injurious.

Prohibition of Dietary Supplements That Are "Unfit for Food"

Another section of the FDC Act, which was not modified by DSHEA and remains in full effect, provides that a food (again, this includes a dietary supplement[37]) shall be deemed to be "adulterated" (and accordingly illegal) if it is "unfit for food."[38] This extremely expansive provision gives FDA plenary authority to take action to stop the marketing of any dietary supplement that the agency believes is not a fit item for human consumption.

Prohibition of "Drug" Claims for Dietary Supplements

DSHEA does not allow a dietary supplement to bear labeling claims that represent that the product is intended for use in the diagnosis, cure, mitigation, treatment, or prevention of disease; claims that suggest such an intended use subject a dietary supplement to regulation as a drug.[39] Such a product becomes illegal if it fails to comply with all drug requirements, including the requirement for FDA approval of a "new drug" prior to marketing.

Requirement for Truthful and Informative Labeling

The FDC Act, as amended by DSHEA and as implemented by FDA's regulations, requires extensive informative labeling for dietary supplements, including an informative name for each product, the net quantity of contents, detailed information about the nutrients provided by the product, including the name and quantity of the dietary ingredients presented in a standard format (the "Supplement Facts" labeling), information about any other ingredients, and the name and place of business of the responsible company.[40] Failure to comply with any of these requirements causes a dietary supplement to be deemed "misbranded" (and accordingly illegal).

The FDC Act also provides that a dietary supplement is "misbranded" (and accordingly illegal) if any of its labeling is "false or misleading in any particular."[41]

Sanctions for Violations

There are substantial and severe sanctions for violations of the aforementioned requirements. A dietary supplement that is "adulterated," "misbranded," or in violation of applicable drug requirements is subject to seizure, condemnation, and destruction by the U.S. district courts.[42] A person who is responsible for the shipment of such a product in interstate commerce is subject both to an injunction action[43] and to criminal prosecution.[44] Moreover, such a violative product is subject to FDA's regulations concerning requests for recall,[45] and the importation from foreign countries into the United States of any product that simply "appears" to be violative in any manner described previously is also prohibited.[46]

New Dietary Ingredients

DSHEA also includes additional safety-related requirements with respect to the introduction into commerce in the United States of *new* dietary ingredients. A **new dietary ingredient** is defined as "a dietary ingredient that was not marketed in the United States before October 15, 1994.[47] DSHEA provides that, in addition to all the other safety-related requirements described previously in this article, a dietary supplement that contains a new dietary ingredient shall be deemed to be "adulterated" (and subject to all the enforcement sanctions described previously) unless, **either:** (1) the supplement contains "only dietary ingredients which have been present in the food supply as an article used for food in a form in which the food has not been chemically altered,"[48] **or**, (2) there is a

> history of use or other evidence of safety establishing that the dietary ingredient when used under the conditions recommended or suggested in the labeling of the dietary supplement will reasonably be expected to be safe and, at least 75 days before being introduced or delivered for introduction into interstate commerce, the manufacturer or distributor of the dietary ingredient or dietary supplement provides the [FDA] with information, including any citation to published articles, which is the basis on which the manufacturer or distributor has concluded that a dietary supplement containing such dietary ingredient will reasonably be expected to be safe.[49]

In essence, if a dietary ingredient that is to be used in a dietary supplement was not marketed in the United States before October 15, 1994, and if the ingredient has not otherwise been present in the food supply as an article used for food in a form that has not been chemically altered, the dietary ingredient may not be marketed in a dietary supplement product unless one first submits evidence of safety to FDA and waits at least 75 days for a response from the agency about whether the ingredient may be marketed.

FDA has published regulations to govern the submission to the agency of "new dietary ingredients notifications,"[50] and the agency has processed numerous notifications of this type and has notified a number of companies not to market particular new dietary ingredients in the United States.[51]

It also should be noted that this requirement for FDA notification before marketing a new dietary ingredient for use in dietary supplements is more stringent than the requirements that apply to the marketing of new ingredients in other food products (that is, for conventional foods, unlike dietary supplements, as long as a manufacturer determines for itself that a new food ingredient is "generally recognized as safe" (GRAS), the manufacturer is free to market that ingredient for food use without any notice to, or preclearance by, FDA).[52]

Authority Immediately to Stop Marketing of Any Dietary Supplement That Presents an "Imminent Hazard"

In addition to all the other provisions described previously, DSHEA authorizes the Secretary of the Department of Health and Human Services immediately to stop shipment of any dietary supplement product by declaring it "to pose an imminent hazard to public health or safety."[53] If the Secretary declares a dietary supplement or dietary ingredient "to pose an imminent hazard," the government must promptly thereafter conduct an administrative proceeding to review the merits of the "imminent hazard" conclusion. During the proceeding, however, the product cannot be sold to the public.

This "imminent hazard" authority over dietary supplements is a power that the government does not have over other food products. The FDC Act does not provide any similar imminent hazard authority to the Secretary with respect to conventional foods or ingredients in conventional foods (that is, under DSHEA, the government has more power to immediately stop the marketing of a dietary supplement than it has for any conventional food).

Good Manufacturing Practices

DSHEA also authorizes FDA to issue regulations to "prescribe good manufacturing practices for dietary supplements," and the law provides that a dietary supplement shall be deemed to be "adulterated" (and subject to all the sanctions described above) if "it has been prepared, packed, or held under conditions that do not meet current good manufacturing practice regulations . . . issued by [FDA]."[54]

More than two years after the passage of DSHEA, FDA published an "advance notice of proposed rulemaking" concerning good manufacturing practice (GMP) regulations for the dietary supplement industry."[55] FDA published proposed dietary supplement GMP regulations in March 2003.[56] FDA expects to publish final dietary supplement GMP regulations by the end of 2004.[57] The final regulations will have the status of law and will be subject to enforcement in the courts.[58]

Conclusion

In summary, FDA has substantial policing powers with respect to the regulation of dietary supplement products and their ingredients. Although the agency is not given preclearance authority over all dietary ingredients in dietary supplements, "new" ingredients that were not marketed prior to October 15, 1994, must be the subject of a notification to the agency, documenting safety, to be submitted at least 75 days before marketing begins.

Although there has been some controversy as to whether FDA should be given greater authority over dietary supplement ingredients, including the authority to preclear all such ingredients, the agency's history of suppressing marketing even of safe dietary supplement substances before DSHEA was enacted probably explains why the agency has not been given comprehensive preclearance authority (see "Why Was DSHEA Enacted?"), and the authority that was granted to the Secretary by DSHEA to immediately stop the marketing of any ingredient deemed to present an imminent hazard is a power that the agency does not possess with respect to other food products and ingredients.

Although controversy will surely continue over how pervasively and how aggressively dietary supplements should be regulated by the FDA, it is clear that the agency does possess substantial regulatory authority at present.

Notes

1. Public Law 103-417; 108 Stat. 4325-4335; 103d Congress, 2d Sess. (October 25, 1994).
2. 140 Cong. Rec. Sl1, 173-79 (daily ed. Oct. 6, 1994); 140 Cong. Rec. S14, 798-800 (daily ed. Oct. 7, 1994).
3. *See infra* notes 6-15 and accompanying text.
4. S. Rep. No. 103-410, at 16 (Oct. 8, 1994).
5. *Id.* at 22.
6. *United States v. Two Plastic Drums, More or Less of an Article of Food, Labeled in Part, Viponte Ltd. Black Currant Oil Batch No. BOOSF 039*, 984 F.2d 814 (7th Cir. 1993).
7. *Id.*
8. S. Rep. No. 103-410, *supra* note 6, at 21.
9. *Black Currant Oil*, 984 F.2d at 819, 820.
10. *United States v. Oakmont Investment Co.*, 987 F.2d 33, 37, 39 (1st Cir. 1993).
11. *United States v. 45/194kg. Drums of Pure Vegetable Oil*, 961 F.2d 808 (9th Cir. 1992).
12. *Dietary Supplements, Before the House Committee on Appropriations, Subcommittee on Agriculture, Rural Development, Food and Drug Administration, and Related Agencies*, 103d Cong., 1st Sess. 208 (Oct. 18, 1993). *See also* S. Rep. No. 103-19, at 128 (Oct. 21, 1993).
13. S. Rep. No. 103-410, *supra* note 6, at 21.
14. Accordingly, one of the provisions of DSHEA amended the definition of "food additive" in the FDC Act to provide explicitly that the term does not include dietary ingredients in dietary supplements. 21 U.S.C. § 32l(s)(6); FDC Act § 201(s)(6).
15. FDA has long regarded dietary supplements as a type of **food** intended for "special dietary uses." Section 403(j) of the FDC Act, which has been part of the Act since its original passage in 1938, provides that a food shall be deemed to be "misbranded" (and therefore illegal) "[i]f it purports to be or is represented for special dietary uses, unless its label bears such information concerning its vitamin, mineral, and other dietary properties as the Secretary determines to be, and by regulations prescribes as, necessary in order fully to inform purchasers as to its value for such uses." 21 U.S.C. § 343(j). This section of the FDC Act recognizes that a product that is intended "for special dietary uses" because of "its vitamin, mineral, and other dietary properties" is a type of **food**. Note that section 403(j) authorizes but does not require the issuance of regulations to prescribe mandatory label information. The authority of the Secretary of Health and Human Services to issue regulations under the FDC Act has been delegated to the Commissioner of Food and Drugs, who directs the FDA. 21 C.F.R. § 5.10(a)(1). FDA regulations issued pursuant to section 403(j) of the Act state that the term "special dietary uses" includes "[u]ses for **supplementing** or fortifying the ordinary or usual **diet** with any vitamin, mineral, or other dietary property." 21 C.F.R. § 105.3(a)(1)(iii) (hence "dietary supplement"). (Emphasis added.)
16. Judicial rulings that FDA personnel cited as supportive of this proposition include *Nutrilab, Inc. v. Schweiker*, 713 F.2d 335 (7th Cir. 1983), and *American Health Products Co. v. Hayes*, 574 F. Supp. 1498 (S.D.N.Y. 1983), aff'd, 744 F.2d 912 (2d Cir. 1984). However, perhaps inconsistently, FDA also has long accepted that "Ingredients or products such as rutin, other bioflavonoids, para-amino-benzoic acid, inositol, and similar substances which have in the past been represented as having nutritional properties but which have not been shown to be essential in human nutrition . . . may be marketed as individual products or mixtures thereof: Provided, That the possibility of nutritional, dietary, or therapeutic value is not stated or implied . . . ." 21 C.F.R. § 101.9(i)(5) (April 1, 1993 ed.). Tablets or capsules of such products might have no taste, aroma, or nutritional

value, but nevertheless the agency acknowledged that they could properly be sold as foods (provided that no misleading nutritional or therapeutic claims were made). (This particular statement no longer appears in FDA's revised regulations on nutrition labeling. 21 C.F.R. § 101.9 [April 1, 2004 ed.]. However, when the FDA revised its nutrition labeling regulations to delete this statement, it explicitly said that it was not intending to change its policy that permitted the sale, as food, of a substance that offered no nutritional value [in FDA's judgment]. For example, *see generally* 58 Fed. Reg. 2166 [January 6, 1993].)

17. Section 201(ff)(1) of the FDC Act, 21 U.S.C. § 321(ff)(1). There are additional criteria. The product must either be intended for ingestion in tablet, capsule, powder, softgel, gelcap, or liquid droplet form, or, if not intended for ingestion in such a form, not be "represented for use as a conventional food or as a sole item of a meal or the diet." Section 201(ff)(2) of the FDC Act, 21 U.S.C. § 321(ff)(2). In addition, it must be labeled as a "dietary supplement." Section 201(ff)(2)(C) of the FDC Act, 21 U.S.C. § 321(ff)(2)(C); section 403(s)(2)(B) of the FDC Act, 21 U.S.C. § 343(s)(2)(B).

The new definition includes some highly technical provisions about the situation in which an ingredient in a supplement is also approved by FDA for use as a drug. In general, if "an article" has been "marketed as a dietary supplement or as a food" **before** it is approved by FDA as a new drug, certified by FDA as an antibiotic, or licensed by FDA as a biologic, it may continue to be marketed as a dietary supplement unless FDA publishes a prohibitory regulation (which would be subject to judicial review). Section 201(ff)(3)(A) of the FDC Act, 21 U.S.C. § 321(ff)(3)(A).

On the other hand, in general, if, **before** "an article" is "marketed as a dietary supplement or as a food," **either** (1) FDA has approved the article as a new drug, certified the article as an antibiotic, or licensed the article as a biologic, **or** (2) FDA has authorized the article for investigation as a new drug, antibiotic, or biologic, **and** "substantial clinical investigations have been instituted," **and** "the existence of such investigations has been made public," under these circumstances the article may **not** be marketed as a dietary supplement unless FDA first issues an approving regulation. Section 201(ff)(3)(B) of the FDC Act, 21 U.S.C. § 321(ff)(3)(B).

18. Section 201(s)(6) of the FDC Act, 21 U.S.C. § 321(s)(6).

19. Section 201(s) of the FDC Act, 21 U.S.C. § 321(s). As an example of FDA's use of the "food additive" allegation for dietary ingredients in dietary supplement products prior to enactment of DSHEA, **see** FDA Compliance Policy Guide No. 7117.04, "Botanical Products for Use as Food" (as issued on July 1, 1986). Even before enactment of DSHEA, however, some courts had expressed the view that FDA was overreaching in its attempts to regulate dietary ingredients in dietary supplement products as "food additives." *See* the discussion in Section II of this chapter.

20. 21 U.S.C. §§ 342(a)(2)(C), 348(a).

21. Kutak, Rock & Campbell, "FDA Safeguards Against Improper Disclosure of Financially Sensitive Information: The Product Approval Centers," Final Report (November 14, 1991) at 162; *33 Food Chem. News* 67 (November 4, 1991).

22. Section 402(f)(1)(A) of the FDC Act, 21 U.S.C. § 342(f)(1)(A).

23. Section 402(f)(1) of the FDC Act, 21 U.S.C. § 342(f)(1).

24. Section 413(c) of the FDC Act, 21 U.S.C. § 350b(c).

25. Section 413(a) of the FDC Act, 21 U.S.C. § 350b(a).

26. *Id.*

27. Section 402(f)(1)(C) of the FDC Act, 21 U.S.C. § 342(f)(1)(C).

28. For a judicial ruling upholding FDA assertions of this type, *see United States v. Articles of Drug . . . Honey*, 344 F.2d 288 (6th Cir. 1965) (ruling that a jar of honey became subject to regulation as a drug when a book that made drug claims for honey was used in "immediate connection . . . with the sale of the product" by the retailer). Cf. *United States v. "Sterling Vinegar and Honey" . . . Balanced Foods*, 338 F.2d 157 (2d Cir. 1964) (ruling that a book that made drug-type claims for a vinegar-honey combination did **not** create drug status for a vinegar-honey product, although the book was available for sale in the same store as the product, when there was "no evidence of any joint promotion" of the book and the product).

29. Section 403B(a) of the FDC Act, 21 U.S.C. § 343-2(a).

30. Section 403B(b) of the FDC Act, 21 U.S.C. § 343-2(b).

31. 21 C.F.R. § 101.14(e)(1).

32. Section 403(r)(6) of the FDC Act, 21 U.S.C. § 343(r)(6).

33. 21 U.S.C. § 342(f)(1)(A); FDC Act § 402(f)(1)(A).

34. On February 11, 2004, FDA issued a final rule declaring all dietary supplements that contain ephedrine alkaloids under the "significant or unreasonable risk" standard. 69 Fed. Reg. 6788. FDA announced in this rule that the government's burden of proof of adulteration under 21 U.S.C. § 342(f)(1)(A) is met when a product's risks

outweigh its benefits. This is a new "risk/benefit" standard for dietary supplements that may become important in future FDA assessments of adulteration. Both the new standard and the scientific basis for FDA's final rule have been challenged in two lawsuits. See *Nutraceutical Corp. v. Crawford*, No. 2:04-cv-00409 (D. Utah filed May 3, 2004); *NVE, Inc. v. Dep't Health & Human Servs.*, No. 2:04-cv-00999 (D.N.J. Aug. 4, 2004).

35. 21 U.S.C. § 321(ff); (FDC Act § 201(ff).
36. 21 U.S.C. § 342(a)(1), (f)(l)(D); FDC Act § 402(a)(1), (f)(1)(D).
37. 21 U.S.C. § 321(ff); FDC Act § 201(ff).
38. 21 U.S.C. § 342(a)(3); FDC Act § 402(a)(3).
39. 21 U.S.C. §§ 321(g)(1)(B), 343(r)(6)(C); FDC Act §§ 201(g)(l)(B), 403(r)(6)(C); 21 C.F.R. § 101.93(f) and (g).
40. 21 C.F.R. §§ 101.3, 101.4, 101.5, 101.36, 101.105.
41. 21 U.S.C. § 343(a)(1); FDC Act § 403(a)(1).
42. 21 U.S.C. § 334; FDC Act § 304.
43. In *United States v. Lane Labs-USA, Inc.*, 324 F. Supp. 2d 547 (D.N.J. 2004), *appeal docketed*, No. 04-3592 (3d Cir. Sept. 10, 2004), the court found that "monetary equitable remedies beyond injunctive relief are available pursuant to the" FDC Act. *Id.* At 578. The court found that the defendants were making drug claims for their dietary supplements products and that there was a reasonable likelihood that this behavior would continue. *Id.* At 574. The court ordered that in addition to being permanently prohibited from selling and marketing their products, the defendants had to pay restitution to consumers who had bought the product. *Id.* At 581-82. The court's ruling is inconsistent with opinions publicly expressed by food and drug lawyers. *Id.* At 576 n.16.
44. 21 U.S.C. §§ 331, 332, 333; FDC Act §§ 301, 302, 303.
45. Recalls, 21 C.F.R. pt. 7, subpt. C.
46. 21 U.S.C. § 381(a); FDC Act § 801.
47. 21 U.S.C. § 350b(c); FDC Act § 413(c).
48. 21 U.S.C. § 350b(a); FDC Act § 413(a).
49. *Id.*
50. 21 C.F.R. § 190.6.
51. FDA actions are on public display in FDA Docket No. 95S-0316, FDA Dockets Management Branch, 5630 Fishers Lane, Rm. 1061, Rockville, MD.
52. 62 Fed. Reg. 18,938, 18,941 (Apr. 17, 1997) ("a manufacturer may market a substance that the manufacturer determines is [generally recognized as safe] GRAS without informing the agency [FDA]").
53. 21 U.S.C. § 342(f)(l)(C); FDC Act § 402(f)(l)(c).
54. 21 U.S.C. § 342(g) (FDC Act § 402(g).
55. 62 Fed. Reg. 5700-09 (Feb. 6, 1997).
56. 68 Fed. Reg. 12,158 (Mar. 13, 2003).
57. 69 Fed. Reg. 37,452 (June 28, 2004).
58. 21 U.S.C. § 342(g)(1); FDC Act § 402(g)(1).

7 Tropicana Pure Premium and the Potassium Health Claim: A Case Study

Carla McGill

Introduction

On November 21, 1997, President Clinton signed the FDA Modernization Act of 1997 (FDAMA)[1]. Prior to enactment of FDAMA, manufacturers could not use a health claim in food labeling unless the Food and Drug Administration (FDA) published a regulation authorizing such a claim. Provisions of FDAMA permit food manufacturers to use claims if such claims are based on current, published, authoritative statements from certain federal scientific bodies such as the National Institutes of Health, The Centers for Disease Control and Prevention, and the National Academy of Sciences.

Tropicana Products, Inc., utilized the FDAMA process in 2000 to author a health claim linking foods that are good sources of potassium and low in sodium to reduced risk of high blood pressure and stroke. The potassium health claim was allowed on food labels after October 31, 2000 (Fig. 7.1). The integrated communication plan to drive awareness of the new health claim strengthened Tropicana Pure Premium's position as the category leader and as a healthful brand with intrinsic benefits.

Health Claims

Health claims are among the various types of claims allowed on food labels. They describe the relationship between a nutrient or other food substances and a disease or health-related condition. Health claims are intended to inform consumers about a product's potential to promote health by stating that certain foods or food substances may reduce the risk of certain diseases. Other types of claims allowed on food labels include nutrient content claims, such as "low fat," and structure/function claims. Structure/function claims make statements about a food substance's effect on the structure or function of the body but do not state disease risk reduction.

Attention to health claims on food labels increased in the 1980s when marketing strategies began to reflect increased recognition of the role of nutrition in promoting health and reducing the risk of disease. In 1990 the Nutrition Labeling and Education Act (NLEA) was passed[2]. Part of the intent of NLEA was to prevent the use of misleading claims and to require that claims be supported by sufficient scientific evidence. FDA initially authorized seven health claims in 1993 under the provisions of NLEA. Under NLEA, companies petition FDA to consider new health claims through rule making. This process can require more than a year to complete due to the necessary scientific review and the need to allow for comment to a proposed rule. In an effort to decrease the time required to make health claim information available to consumers, FDAMA includes a provision intended to expedite the process.

The intended purpose of health claims is to provide information to consumers on healthful eating patterns that may help reduce the risk of heart disease, cancer, osteoporo-

Tropicana's health claim was allowed after October 31, 2000

"Diets containing foods that are good sources of potassium and low in sodium may reduce the risk of high blood pressure and stroke."

Qualifying foods:

- Good source of potassium
- Low sodium
- Low in fat, saturated fat and cholesterol

Figure 7.1. Health claim language, qualifying foods and example print ad.

sis, hypertension, or certain birth defects. Health claims are intended to be interpreted in the context of the total diet so that no one food or nutrient is perceived to be a "magic bullet." Health claims also include language that references **disease risk reduction**, not **disease prevention**.

The FDA Modernization Act (FDAMA)

The FDAMA became law on November 21, 1997[1]. Section 303 of the FDAMA amends section 403 of the Food, Drug, and Cosmetic Act to add new subparagraphs authorizing food labels to include health claims without approval by a FDA regulation. The health claim must be the subject of a published authoritative statement, which is currently in effect, issued by:

> A scientific body of the United States Government with official responsibility for public health protection or research directly relating to human nutrition (such as the National Institutes of Health or the Centers for Disease Control and Prevention) or the National Academy of Sciences or any of its subdivisions . . . .

Published authoritative statements that qualify for these types of health claims must be official statements in a report that has completed the institutional report review process.

The health claim must be stated "in a manner so that the claim is an accurate representation of the authoritative statement" and "so that the claim enables the public to comprehend the information provided in the claim and to understand the relative significance of such information in the context of a total daily diet." The claim must not be "false or misleading" and the food bearing the claim cannot exceed limits for fat, saturated fat, cholesterol, or sodium per reference amount customarily consumed.

The petition submitted to FDA must contain the specific authoritative statement that is the basis for the health claim. In addition, a review of the scientific literature and information on nutrient intakes should be included. If FDA does not act to prohibit or modify the claim, it may be used 120 days after receipt of the notification.

Tropicana Strategic Analysis

Orange juice has been ranked as the most nutritious juice by the Center for Science in the Public Interest[3]. It provides more nutrients per serving than do other juices. According to the 1999 HealthFocus Trend Report, 45 percent of consumers ranked orange juice as the most nutritious juice and it surpassed broccoli as the food consumers seek most often to reduce risk of disease[4]. In the same survey, consumers rated Tropicana as the second most healthful brand. In 2000 consumers perceived Tropicana Pure Premium as the most nutritious orange juice, according to Millward Brown Tracking [5].

Tropicana Pure Premium orange juice, a premium not from concentrate juice, has long enjoyed the perception as a healthful brand. Orange juice is a good source of potassium. An 8-ounce serving contains as much potassium as a medium banana—450 mg. An internal consumer survey conducted by Tropicana in 2000 found that consumer awareness about potassium was low. Ninety-four percent of those surveyed did not know that orange juice is a good source of potassium, and 75 percent did not link potassium intake to reduced risk of hypertension or stroke [6].

Cardiovascular disease is the leading cause of death in the United States. Statistics from the American Heart Association document that one in four American adults has high blood pressure[7]. More than four million Americans suffer from stroke annually, and it is the leading cause of permanent disability [8]. Consumers listed cardiovascular disease as the number one health concern in the 1999 HealthFocus Trend Report[4].

Action Plan: The decision was made to reinforce Pure Premium as the category leader in 100 percent orange juice products providing the perfect combination of taste and nutrition. Key decision makers in Tropicana Research and Development along with marketing approved the plan to submit a FDAMA petition to FDA linking foods containing potassium to reduced risk of hypertension and stroke. The goal was to communicate the news about potassium and cardiovascular disease risk reduction to consumers and to link the health claim to the Pure Premium brand.

The Tropicana Process

On July 3, 2000, Tropicana Products, Inc. submitted a notification to FDA containing a proposed health claim about the relationship of potassium-containing foods to reduced risk of high blood pressure and stroke[9]. The notification cited statements from the National Academy of Sciences report, *Diet and Health: Implications for Reducing Chronic Disease Risk* as authoritative statements for the claim[10]. The two statements are from the Executive

Summary of the *Diet and Health* report, which integrates all the evidence reviewed in the total report, and are based on consensus of the Committee of Diet and Health. The two statements are from pages 11 and 15 of the report and state:

> Epidemiological and animal studies indicate that the risk of stroke-related deaths is inversely related to potassium intake over the entire range of blood pressures, and the relationship appears to be dose dependent. The combination of a low-sodium, high potassium intake is associated with the lowest blood pressure levels and the lowest frequency of stroke in individuals and populations. Although the effects of reducing sodium intake and increasing potassium intake would vary and may be small in some individuals, the estimated reduction in stroke-related mortality for the population is large.

> Vegetables and fruits are good sources of potassium. A diet containing approximately 75 mEq (i.e., approximately 3.5g of elemental potassium) daily may contribute to reduced risk of stroke, which is especially common among blacks and older people of all races. Potassium supplements are neither necessary nor recommended for the general population.

The submitted petition contained a 31-page body that included an Executive Summary, Introduction, Background, Foods Eligible to Make the Claim, Scientific Literature Review, and Conclusions. Attached appendices contained copies of pages from the *Diet and Health* report, data on potassium intakes in the United States, potassium content of foods based on USDA Standard Reference 13, menu plans, both historical and current literature reviews of clinical studies involving potassium and its role in blood pressure and stroke, a bibliography, and letters of support.

Of the 36 studies presented in the appendix of the petition, 28 suggest that increasing potassium intake is associated with a decrease in blood pressure and reduced risk of stroke. Five studies suggest that fruits and vegetables or their components are an important determinant of hypertension, risk of stroke, or both. Many of the authors suggest that an increase in consumption of foods rich in potassium would constitute a relatively low-cost, low-risk public health measure to decrease risk of high blood pressure and stroke.

The Dietary Approaches to Stop Hypertension (DASH) trial deserves special mention because it directly examined the impact of increasing either fruits and vegetables or fruits, vegetables, and low-fat dairy foods on blood pressure[11]. The DASH study included 459 people, randomized into three diet groups: (1) a control diet (typical American diet, 37 percent of calories from fat); (2) the control diet with the addition of 8–10 servings of fruits and vegetables; and (3) the DASH diet (8–10 servings of fruits and vegetables and two servings of low-fat dairy products). The diets were maintained for eight weeks with all foods prepared by the study centers. After two weeks, the fruits and vegetables and DASH diets resulted in significant reductions in blood pressure that were maintained for the remaining six weeks of the study. Adding fruits and vegetables to the typical American diet resulted in a 2.8 mm Hg drop in systolic blood pressure and a 1.1 mm Hg drop in diastolic blood pressure compared to controls. The DASH diet resulted in a 5.5 mm Hg drop in systolic blood pressure and a 3.0 mm Hg drop in diastolic blood pressure compared to controls. Potassium intakes were significantly greater in the fruits and vegetables and DASH diets compared to the control diet[11].

As part of the petition submitted to FDA, Tropicana analyzed potassium intakes in the United States based on the USDA Continuing Survey of Food Intake by Individuals. Mean intake of potassium for various age-sex-race groups was below the Daily Value (DV) of 3,500 mg/d and the level of potassium NAS reports to decrease risk of stroke (75 mEq or about 3,500 mg/d). Only about 18 percent of the American population consumes the recommended level of potassium. Foods containing potassium (certain fruits and fruit juices, some vegetables, certain grain products and low-fat dairy products) are recommended as components of a healthful diet. Very few foods provide potassium content of foods on the Nutrition Facts panel because it is not required as part of mandatory nutrition labeling.

Tropicana technical personnel conducted several phone conversations with FDA personnel in the Office of Nutrition Products, Labeling and Dietary Supplements between July and October, 2000. Tropicana submitted several amendments to the original petition in writing. FDA reviewed the authoritative statements from *Diet and Health* in their context and in light of existing authorized health claims for hypertension and cardiovascular disease. FDA did not act to prohibit or modify the claim as amended. After October 31, 2000, manufacturers could use the claim on the label and in labeling of any food product that meets the eligibility criteria described in the notification. Under the provisions of FDAMA, the claim must appear exactly as stated in the notification. The allowed claim states:

> Diets containing foods that are good sources of potassium and low in sodium may reduce the risk of high blood pressure and stroke.[9]

Communicating the Message

During the early phases of the project until shortly after submission of the petition to the FDA, knowledge of the project within Tropicana was kept to the minimum core team to ensure confidentiality and competitive advantage. As plans for the announcement began, the team was expanded to obtain organizational alignment. Key decision makers from Research and Development, Marketing, Communications, and Business Development were enlisted to maximize the impact of the announcement of a new health claim and to link the Pure Premium brand to the claim. An integrated campaign to drive awareness and consumption was developed (Fig. 7.2). On October 31, 2000, Tropicana launched a $10 million promotional campaign that included public relations, advertising, and marketing components.

Consumer communication included new package graphics for Pure Premium orange juice 16-, 32-, and 64- ounce containers that included a new heart-health version of the trademark straw-in-orange icon along with the claim language (Fig. 7.3). New graphics were on the shelf within three weeks of the announcement. Advertising included full-page print ads that ran in national daily publications in November, 2000, with a sustained print campaign that included holiday-themed ads. One holiday spot featured Santa watching his blood pressure. A television commercial featured an elderly softball team and the song "You Gotta Have Heart." Trade and field support included floor graphics, point-of-sale materials, ad slicks, and trade advertising. The health messages with the new heart-health icon were incorporated into retailers' holiday ads.

An aggressive public relations campaign was launched to generate awareness. Public relations efforts focused on the business and financial community at the time of the announcement. Tropicana's president and CEO conducted interviews in New York with the business press. Along with sampling at the New York Stock Exchange, the core team rang

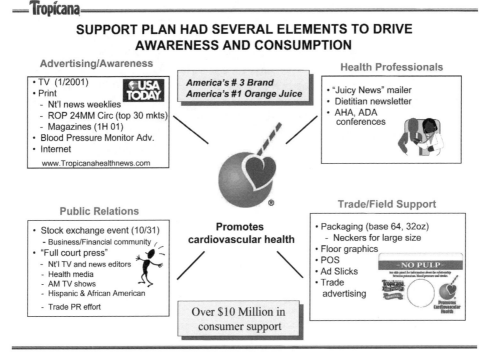

Figure 7.2. Support plan had several elements to drive awareness and consumption.

the bell at the closing of the Exchange on October 31, 2000. The business story drew coverage from CNN on "Market Sweep." Marketing representatives conducted interviews with the trade media. Continuing public relations efforts included an audio news release and targeted communications to African-American and Hispanic consumers.

Consumer communication also included a medical spokesperson on the "Good Morning America Show" on the day of the announcement. In-depth information about the new health claim and the benefits of potassium were available on a new Web site, www.tropicanahealthnews.com, with separate sections for health care professionals and consumers. Health professional outreach also included a "Juicy News" mailer to registered dietitians that included materials for use in advising patients. A print ad also ran in the *Journal of the American Dietetic Association.*

Results

The integrated communication plan coupled with the announcement of the potassium health claim strengthened Tropicana Pure Premium's stature as the leader in the orange juice category and maximized opportunities to "own" the claim. The successful strategy involved layering intrinsic benefits of a product in a category in which Pure Premium already had established brand equity and significant market share. In 2000 Tropicana boosted its share of the U.S. orange juice market to 35 percent. Total sales of Pure Premium reached nearly $2.4 billion. Prior to the health claim announcement, Pure Premium volume was growing at a rate of 11 percent. From the announcement on October 31, 2000, to

Figure 7.3. Potassium health claim packaging.

February 1, 2001, Pure Premium volume grew at 16 percent and gained 2.5 share points (Fig. 7.4). Sales for the last four weeks of December 2000 increased by 54.4 percent over the same period in 1999. The communication campaign generated 895 million media impressions reaching target consumers and health professionals. A content analysis of available coverage established that Tropicana's name was mentioned in every story that covered the claim and in 62 percent of all headlines.

Several components of Tropicana's health claim communication strategy have won peer recognition awards. The health professional outreach to registered dietitians that included a print ad won an annual advertising award from the American Dietetic Association. The public relations campaign that included the New York Stock Exchange event won the prestigious Silver Anvil Award for Integrated Communications from the Public Relations Society of America. According to one industry analyst, Tropicana's successful health claim approval for potassium "caught everyone flat footed" and he called the claim a "brilliant case study"[12].

Tropicana Products, Inc., a division of PepsiCo, Inc., is the world's leading producer

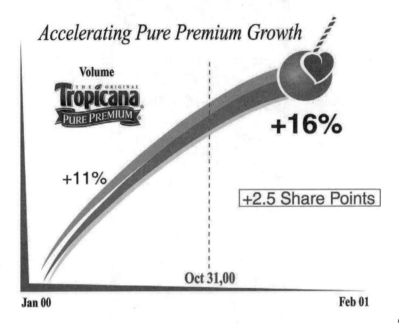

Figure 7.4. Share volume increase following health claim announcement.

and marketer of branded fruit juices. Tropicana markets its products in the United States under a number of brand names, including Tropicana Pure Premium orange and grapefruit juices and Pure Premium calcium-fortified juices. The Tropicana Nutrition Center is committed to the progress and dissemination of credible science in the areas of health and nutrition and sponsors independent research into the health benefits of citrus.

References

1. Food and Drug Administration Modernization Act of 1997. PL 105-115. 111 Stat 2296.
2. Nutrition Labeling and Education Act of 1990. PL 101-535 104 Stat 2353.
3. Center for Science in the Public Interest. *Nutrition Action Health Letter.* Vol. 26 No. 5; June 1999.
4. The 1999 HealthFocus Trend Report. HealthFocus, Inc., Des Moines, IA; 1999.
5. Millward Brown IMS Tracking Report. Millward Brown IMS, Inc. Dublin, Ireland; 2000.
6. Data on file, Tropicana Products, Inc., Bradenton, FL.
7. American Heart Association Heart Disease and Stroke Statistics. American Heart Association, Inc., Dallas, TX; 1999.
8. American Stroke Association Statistics. American Stroke Association Inc., Dallas, TX: 1999.
9. Food and Drug Administration Dockets Management 00Q-1582. Health Claim Notification for Potassium Containing Foods. October 31, 2000.
10. National Research Council. *Diet and Health: Implications for Reducing Chronic Disease Risk.* National Academy Press, Washington, DC; 1989.
11. Harsha, D., Lin, P.H., Obarzanek, E., Karanja, N.M., Moore, T.J. and Caballero, B. (1999). Dietary approaches to stop hypertension: a summary of study results. *JADA*; 99(suppl):S35–S39.
12. *Nutrition Business Journal*, Vol. VI, No. 10. *Functional Foods V. San Diego, CA*. Penton Media Inc. 2001.

8 The Importance of the Court Decision in *Pearson v. Shalala* to the Marketing of Conventional Food and Dietary Supplements in the United States

Elizabeth Martell Walsh, Erika King Lietzan, and Peter Barton Hutt

Introduction

In December 2002, the Food and Drug Administration (FDA) established two major policies affecting the marketing of food and dietary supplements in the United States.[1] First, FDA determined that, in accordance with the free speech clause of the First Amendment to the United States Constitution, disease claims may be made for conventional food and dietary supplements if the "weight of the scientific evidence" supports the claim and the claim is appropriately qualified so that consumers are not misled. This policy reversed an earlier FDA decision that refused to apply the First Amendment to disease claims for conventional food. Second, FDA stated that it would use a "reasonable consumer" standard in determining whether a claim is misleading. The reasonable consumer standard replaced the standard of "the ignorant, the unthinking, and the credulous" consumer, used by courts at the request of FDA in the past, which FDA concluded was not consistent with the First Amendment. Both of these agency policies were the direct result of the landmark decision by the United States Court of Appeals for the District of Columbia Circuit three years earlier in the case of *Pearson v. Shalala*.[2]

In *Pearson v. Shalala* the D.C. Circuit (and in subsequent proceedings, the D.C. District Court as well[3]) held that FDA is required to apply to dietary supplement labeling the same principles of free speech under the First Amendment as the United States Supreme Court has applied to other commercial speech. This chapter explores the importance of this decision to the marketing of conventional food (including functional food) as well as dietary supplements.

Section I explains the governing law and the FDA decisions to permit qualified disease claims[4] on dietary supplement labeling under the Federal Food, Drug, and Cosmetic Act (FD&C Act) of 1938[5] pursuant to an "enforcement discretion" policy announced on October 6, 2000[6] as a result of the *Pearson* decision. Section II explains why FDA was required to permit these qualified disease claims on conventional food labeling, pursuant to the *Pearson* decision. Failure to approve the qualified disease claims for conventional food labeling would have contravened the decision in *Pearson* and would have violated the First Amendment to the United States Constitution. Section III explains that FDA has the obligation to interpret and apply the statute in a constitutional manner, that the exercise of enforcement discretion in this instance lies within the Agency's discretion, and that it has precedent in the Agency's past practices. Section IV briefly explores the impact of the two policies established by FDA in December 2002.

Background

The Nutrition Labeling and Education Act of 1990 (NLEA)[7] amended the FD&C Act to permit the use of disease claims in food labeling. Congress defined a disease claim as a claim in the label or labeling of a food that "expressly or by implication . . . characterizes the relationship of any nutrient . . . to a disease . . . ."[8]

The NLEA "Significant Scientific Agreement" Standard

Under the NLEA, FDA must approve a disease claim for conventional food if it finds that "based on the totality of the publicly available scientific evidence . . . there is significant scientific agreement, among experts, . . . that the claim is supported by such evidence."[9] A conventional food manufacturer may not use an NLEA disease claim in its labeling unless and until FDA promulgates a regulation authorizing that claim.[10] (Use of a disease claim in the absence of an authorizing regulation subjects the product to the drug provisions of the Act.[11]) The NLEA did not define "significant scientific agreement."

Although the NLEA prescribed a standard for FDA's review of disease claims for conventional foods, it did not prescribe a standard for disease claims in dietary supplement labeling. Instead, Congress provided that such a claim would be "subject to a procedure and standard" established by FDA in a regulation.[12]

FDA Rulemakings on the "Significant Scientific Agreement" Standard

FDA split the rulemaking on conventional food disease claims from the rulemaking on dietary supplement disease claims as a result of the enactment of the Dietary Supplement Act of 1992.[13]

Conventional Food

In January 1993, FDA adopted final regulations implementing the NLEA with respect to disease claims for conventional food.[14] In these regulations, FDA explained briefly what was meant by the "significant scientific agreement" standard and how it would assess conformity to that standard.[15] In particular, FDA stated that it would authorize a disease claim only if it determined:

> based on the totality of publicly available scientific evidence (including evidence from well-designed studies conducted in a manner which is consistent with generally recognized scientific procedures and principles) that there is significant scientific agreement, among experts qualified by scientific training and experience to evaluate such claims, that the claim is supported by such evidence.[16]

Rather than flesh out the evidentiary requirement, FDA announced that it would "make case-by-case determinations."[17] FDA stated that it would not permit disease claims "based only on preliminary data," even if those claims accurately disclose the preliminary nature of the data.[18] FDA claimed that it lacked the authority to permit preliminary disease claims "that are qualified by an explanation that a difference of scientific opinion exists."[19]

Dietary Supplements

In the rulemaking addressing the general requirements for disease claims in dietary supplement labeling, FDA decided to use the same "significant scientific agreement" standard—and associated procedures—as applied by statute to conventional food.[20] FDA has consistently characterized that decision as a decision to adopt "the same standard" for both types of food.[21] Indeed, to support its decision to use the same standard, FDA cited both the need to eliminate consumer confusion[22] and the need for "fairness" as between dietary supplement and conventional food manufacturers.[23]

The Food and Drug Administration Modernization Act "Authoritative Statement" Standard

In the Food and Drug Administration Modernization Act of 1997 (FDAMA),[24] Congress created an alternative to the NLEA process for approval of disease claims in conventional food labeling. The new disease claims provision permits conventional food manufacturers to make disease claims based on "authoritative statements" of qualified federal scientific bodies. So-called "authoritative statement disease claims" may be made after premarket notification to FDA, rather than approval by FDA. FDA is not required to prescribe the language of the permitted claim, nor is it required to promulgate a regulation authorizing the claim.[25]

In June 1998, FDA by regulation "overruled" the Congressional mandate of FDAMA by declaring that it would not permit disease claims on the basis of an "authoritative statement" alone.[26] Instead, it wrote, it would incorporate the "significant scientific agreement" standard into the "authoritative statement" premarket notification process. Specifically, FDA stated that it intended "to determine whether the standard of significant scientific agreement is met by a health claim based on an authoritative statement."[27] This standard, FDA contended, would not allow for a claim based on "findings characterized as preliminary results, statements that indicate research is inconclusive, or statements intended to guide further research."[28]

Although the FDAMA "authoritative statement" standard applies only to conventional food, FDA has proposed extending it to dietary supplements.[29]

The "Significant Scientific Agreement" Standard after *Pearson*

The *Pearson* case established unequivocally that FDA regulation of food labeling is subject to the First Amendment commercial free speech doctrine. A disease claim that does not satisfy the "significant scientific agreement" standard is not inherently false and misleading. The First Amendment does not permit FDA to ban such claims categorically.

The *Pearson* case arose from FDA's decision not to approve four disease claims for dietary supplements. (The claims were among the ten as to which Congress had mandated a decision in the NLEA.[30])

- 0.8 mg of folic acid in a dietary supplement is more effective in reducing the risk of neural tube defects than is a lower amount in foods in common form.
- Consumption of omega-3 fatty acids may reduce the risk of coronary heart disease.
- Consumption of antioxidant vitamins may reduce the risk of certain kinds of cancers.
- Consumption of fiber may reduce the risk of colorectal cancer.

In January 1993, FDA rejected all four claims for conventional food labeling, based on the lack of significant scientific agreement.[31] In October 1993, FDA proposed not to authorize three of the four claims for the labeling of dietary supplements.[32] It proposed to authorize a claim relating folic acid to a reduced risk of neural tube defects, for dietary supplements and for food, although not the comparative claim requested.[33] On December 31, 1993, both proposals became final.[34] The folic acid regulation, applicable both to food and to dietary supplements, was modified in 1996.[35] The final regulation provides that:

> The claim shall not state that a specified amount of folate per serving from one source is more effective in reducing the risk of neural tube defects than a lower amount per serving from another source.

Following the January 1994 denial of the four original claims, the *Pearson* plaintiffs brought suit in the District Court for the District of Columbia. They alleged that FDA's final regulations (denying all four claims) were unconstitutional prior restraints in violation of the First Amendment, that they violated the First Amendment commercial free speech doctrine, and that they were overbroad in violation of the First Amendment. The plaintiffs also argued that the final regulations were void for vagueness under the Fifth Amendment. Finally, the *Pearson* plaintiffs argued that FDA had violated the Administrative Procedure Act[36] by failing to adopt a defined standard for "significant scientific agreement" and by arbitrarily and capriciously denying all four claims.

The District Court granted FDA's Motion to Dismiss and denied the plaintiffs' Motion for Summary Judgment.[37] In a strongly worded opinion, the U.S. Court of Appeals for the District of Columbia Circuit reversed the District Court and confirmed that FDA's labeling and advertising regulations are subject to the First Amendment commercial free speech doctrine.[38] It is "undisputed," the court wrote, "that FDA's restrictions on appellants' health claims are evaluated under the commercial speech doctrine."[39] FDA conceded as much, but argued in the alternative that (1) disease claims lacking "significant scientific agreement" are inherently misleading and thus entirely outside the protection of the First Amendment or (2) even if such claims are only potentially misleading, under the three-part test set forth by the United States Supreme Court in the case of *Central Hudson Gas & Electric Corp. v. Public Service Commission of New York*[40], the government is not obligated to consider requiring disclaimers in lieu of an outright ban.

The D.C. Circuit dismissed the first argument as "almost frivolous."[41] "We reject it," the court wrote.[42] As to the second, the court stated, protection of public health and prevention of consumer fraud—the cited bases for the ban—are admittedly "substantial" government interests.[43] Nevertheless, "the government's regulatory approach" fails the final two parts of the *Central Hudson* test.[44]

Although suppression of disease claims might protect consumers from fraud, it does not directly advance the government's interest in protecting the public health, because FDA does not claim the products themselves are harmful.[45] And "the difficulty with the government's consumer fraud justification," the court wrote, "comes at the final *Central Hudson* factor."[46] There is not a reasonable fit between the government's stated goal (prevention of fraud) and the means chosen to advance it (outright suppression of any disease claims). FDA argued that the commercial speech doctrine does not embody a preference for disclosure over outright suppression. "Our understanding of the doctrine," the court wrote, "is otherwise."[47] Under *Central Hudson*, FDA must consider a disclaimer in lieu of an outright ban.[48] The court therefore invalidated the regulations governing all four of the disease claims at issue.[49]

The court also held that the Administrative Procedure Act requires FDA to "give some definitional content" to the phrase "significant scientific agreement."[50] The court stated that "FDA must explain what it means by significant scientific agreement, or, at minimum, what it does not mean."[51]

The Court of Appeals denied rehearing,[52] and FDA did not seek review in the Supreme Court.

FDA's Slow Implementation of the Pearson Ruling

FDA did not quickly implement the *Pearson* decision and initially refused to apply the First Amendment aspects of the ruling to conventional food.

Announcement of Strategy

On December 1, 1999, more than eleven months after the Court of Appeals decision, FDA announced a "strategy" to implement *Pearson*.[53] First, FDA would update the scientific evidence on the four claims at issue in *Pearson*. Second, FDA would issue a guidance document clarifying the "significant scientific agreement" standard. Third, FDA would hold a public meeting to solicit input on changes to FDA's general disease claim regulation for dietary supplements[54] that might be warranted in light of *Pearson*. Fourth, FDA would initiate a rulemaking to reconsider the general disease claim regulation for dietary supplements that might be warranted in light of *Pearson*. Fifth, FDA would initiate a rulemaking on each of the four *Pearson* claims. FDA also stated that it would deny all other pending disease claims without prejudice if they failed to meet the "significant scientific agreement" standard, until the disease claim regulation was revised.[55] FDA made no mention of conventional food.

Significant Scientific Agreement Guidance

Later in December 1999, FDA published a guidance document addressing the meaning of "significant scientific agreement."[56] This document defined the phrase as it applies to disease claims for both dietary supplements and conventional food.[57] Key to this guidance is FDA's insistence that there be significant scientific agreement about the substance/disease relationship rather than significant scientific agreement about the actual claim being made. Section 403(r)(3)(B)(i) of the FD&C Act, however, requires only that a disease claim be supported by significant scientific agreement. It does not require that the relationship between the food substance and the disease condition be established by significant scientific agreement (except to the extent that the claim characterizes the relationship). FDA applied this guidance to both dietary supplements and conventional foods.

Letter to Congress

In the spring of 2000, FDA took the position that it would not apply the *Pearson* ruling to conventional food absent a direct court order. In a letter to Representative David M. McIntosh dated May 16, 2000, FDA wrote:

The claims that were the subject of *Pearson* were for dietary supplements. The court's mandate did not direct FDA to reconsider any health claims for conventional foods. There is a statutory requirement that FDA authorize health claims for conventional foods only when there is significant scientific agreement that the nutrient-disease relationship is valid. Therefore, absent a court ruling finding the statute unconstitutional, FDA does not have authority to authorize health claims for conventional foods when such a claim would require a disclaimer to render it truthful and nonmisleading. For these reasons, the *Pearson* implementation strategy announced in the December 1, 1999, *Federal Register* did not address health claims for conventional foods.[58]

Until December 2002, FDA adhered to this position.

Public Meeting

In April 2000, FDA held a public meeting to solicit comments on two topics pertaining to disease claims in food and dietary supplement labeling. The first issue was whether a disease claim relating to an existing disease (not simply a claim of risk reduction) could properly be authorized under the NLEA disease claim process. As to this issue, FDA wrote, its decision would apply to dietary supplements and to conventional food.[59] The second issue was how to implement the aspect of *Pearson* requiring FDA to consider the use of qualified disease claims. As to this issue, FDA wrote, its decision would apply only to dietary supplements:

> Unlike the statutory provision for the use of health claims on dietary supplements, Section 403(r)(3)(B)(i) of the act provides that FDA may authorize health claims on conventional foods only when there is significant scientific agreement among qualified experts that the totality of publicly available scientific evidence supports the claim. As a result of this statutory requirement for conventional foods and because the *Pearson* case involved only dietary supplements, this portion of the public meeting will be restricted to health claims on dietary supplements.[60]

Attempts to broaden the public meeting to include consideration of qualified disease claims for conventional food were unsuccessful.

New Interim Strategy

On October 3, 2000, FDA revoked its regulations codifying its decision not to authorize the four *Pearson* claims.[61] On October 6, 2000, FDA announced a new strategy for disposition of pending dietary supplement disease claims.[62] FDA announced that it would use its "enforcement discretion" to decline to take action against a dietary supplement disease claim provided that the following conditions are met: (1) the disease claim petition meets FDA requirements for such petitions; (2) the scientific evidence supporting the claim outweighs the scientific evidence against the claim; (3) consumer health and safety are not threatened; and (4) the claim meets the general requirements for a disease claim (that is, except for the "significant scientific agreement" standard and the requirement that the claim be made in accordance with an authorizing regulation). If these criteria are satisfied, FDA will send a letter to the petitioner outlining the rationale for its determination that the evidence does not meet the "significant scientific agreement" standard and stating the conditions under which the Agency will ordinarily expect to exercise enforcement discretion regarding the claim.[63] FDA stated that this implementation of the *Pearson* mandate would apply only to disease claims for dietary supplements.

Application of Interim Standard

Since October 2000, FDA has applied this "interim standard" three times—in each case, in response to a petition or lawsuit from a dietary supplement manufacturer.

i. Fiber. On October 10, 2000, it denied the fiber claim.[64] This decision has not been challenged.

ii. Folic Acid. On October 10, 2000, FDA concluded that the folic acid claim was "inherently misleading" and it declined to authorize the claim even with clarifying disclaimers. Instead, it stated that it would exercise "enforcement discretion" as to the following four alternative claims, each of which recommends that women capable of becoming pregnant consume 0.4 mg (400 mcg) folate daily to reduce the risk of neural tube defects.[65]

> Example 1: Healthful diets with adequate folate may reduce a woman's risk of having a child with a brain or spinal cord birth defect. The Institute of Medicine of the National Academy of Sciences recommends that women capable of becoming pregnant consume 400 mcg folate daily from supplements, fortified foods, or both, in addition to consuming food folate from a varied diet.

> Example 2: Healthful diets with adequate folate may reduce a woman's risk of having a child with a brain or spinal cord birth defect. The scientific evidence that 400 mcg folic acid daily reduces the risk of such defects is stronger than the evidence for the effectiveness of lower amounts. This is because most such tests have not looked at amounts less than 400 mcg folic acid daily.

> Example 3: Healthful diets with adequate folate may reduce a woman's risk of having a child with a brain or spinal cord birth defect. Women capable of becoming pregnant should take 400 mcg folate/day from fortified foods and/or a supplement, in addition to food folate from a varied diet. It is not known whether the same level of protection can be achieved by using only food that is naturally rich in folate. Neither is it known whether lower intakes would be protective or whether there is a threshold below which no protection occurs.

> Example 4: Healthful diets with adequate folate may reduce a woman's risk of having a child with a brain or spinal cord birth defect. Women capable of becoming pregnant should take 400 mcg of folate per day from a supplement or fortified foods and consume food folate from a varied diet. It is not known whether the same level of protection can be achieved by using lower amounts.

The petitioners challenged FDA's decision in court. On February 2, 2001, the United States District Court for the District of Columbia concluded in a sharply worded opinion that FDA's denial of the folic acid claim violated the First Amendment.[66] The court observed that "FDA simply failed to comply with the constitutional guidelines in *Pearson*. Indeed, the Agency seems to have at best, misunderstood, and at worst, deliberately ignored, highly relevant portions of the Court of Appeals opinion."[67] The court ordered FDA to draft "one or more short, succinct, and accurate alternative disclaimers" to accompany the folic acid claim.[68]

When FDA moved for reconsideration of this decision, the court denied the motion[69] and reiterated its conclusion:

> Defendants again seem to ignore the thrust of *Pearson I*. While that decision might leave certain specific issues to be fleshed out in the course of future litigation, the philosophy underlying *Pearson I* is perfectly clear: That . . . First Amendment analysis . . . applies in this case, and that if a health claim is not inherently misleading, the balance tilts in favor of disclaimers rather than suppression.[70]

The court stated that the two earlier decisions in this litigation "established a very heavy burden which Defendants must satisfy if they wish to totally suppress a particular health claim" and quoted from the Court of Appeals that FDA must "demonstrate with empirical evidence that disclaimers . . . would bewilder consumers and fail to correct for deceptiveness."[71]

In a letter dated April 3, 2001, FDA reversed itself and stated it would allow the following claim and disclaimer.

> 0.8 mg folic acid in a dietary supplement is more effective in reducing the risk of neural tube defects than a lower amount in foods in common form. FDA does not endorse this claim. Public health authorities recommend that women consume 0.4 mg folic acid daily from fortified foods or dietary supplements or both to reduce the risk of neural tube defects.[72]

On this basis, the case was dismissed.

iii. Omega-3 Fatty Acids. On October 31, 2000, FDA determined that there was no significant scientific agreement as to the relationship between omega-3 fatty acids in dietary supplements and lowered risk of coronary heart disease, and announced that it would exercise "enforcement discretion" as to certain qualified claims describing that relationship.[73] It offered the following as a sample qualified claim.

> The scientific evidence about whether omega-3 fatty acids may reduce the risk of coronary heart disease (CHD) is suggestive, but not conclusive. Studies in the general population have looked at diets containing fish and it is not known whether diet or omega-3 fatty acids in fish may have a possible effect on a reduced risk of CHD. It is not known what effect omega-3 fatty acids may or may not have on risk of CHD in the general population.

The petitioners asked the Agency to revisit its October 31 decision in lieu of further litigation. In a response dated February 8, 2002, FDA reversed itself and agreed to the following modified language:

Consumption of omega-3 fatty acids may reduce the risk of coronary heart disease. FDA evaluated the data and determined that, although there is scientific evidence supporting the claim, the evidence is not conclusive.[74]

iv. Antioxidants. On May 4, 2001, FDA denied the fourth *Pearson* claim, relating to antioxidant vitamins and cancer, and stated that it would not permit qualified claims under *Pearson*.[75]

Once again, the District Court overruled FDA and ordered the agency to prepare one or more short, succinct, and accurate alternative disclaimers for use by manufacturers.[76] After reviewing the Court of Appeals decision and the two District Court decisions in the *Pearson* litigation, the District Court found that the proposed antioxidant vitamin claim is not inherently misleading and that FDA therefore erred in not considering disclaimers to accompany the claim. The court determined that FDA failed to carry its burden of showing that suppression of the antioxidant claim is the least restrictive means of protecting consumers against the potential of being misled by the claim. It relied heavily on the statement by the Supreme Court in the *Western States* decision that "if the Government could achieve its interest in a manner that does not restrict speech, or that restricts less speech, the Government must do so."[77]

The court stated that the *Pearson* opinion identified only two situations in which a complete ban would be reasonable: where no evidence supports a disease claim or where the evidence in support of the claim is qualitatively weaker than evidence against the claim, for example, where the claim rests on only one or two old studies.[78] Even in those two situations, a complete ban would be appropriate only when "the government could demonstrate with empirical evidence that disclaimers . . . would bewilder consumers and fail to correct for deceptiveness."[79] Accordingly, the district court concluded that under the *Pearson* decision:

> . . . any complete ban of a claim would be approved only under narrow circumstances, i.e., when there was almost no qualitative evidence in support of the claim and where the government provided empirical evidence proving that the public would still be deceived even if the claim was qualified by a disclaimer.[80]

Of particular importance, the court explicitly rejected the FDA "weight-of-the-evidence" standard—which FDA had relied upon strongly in its guidance announced only a few days before the court decision.[81] First, the court concluded that the FDA standard did not comply with the *Pearson* decision:

> The Court of Appeals established clear guidelines for the FDA in determining whether a particular health claim may be deemed "inherently misleading" and thus subject to total suppression. The court implied, though it did not declare explicitly, that when "credible evidence" supports a claim, that claim may not be absolutely prohibited."[82]

* * *

> The FDA has banned the Plaintiffs' claim by concluding that the evidence in support of it was weaker than evidence against it, but it is clear that *more than*

60 recent studies reviewed by the FDA supported the claim. This hardly constitutes the "one or two old studies" that the Court of Appeals contemplated might support a total ban.[83]

Second, even if the FDA standard were accepted, the agency failed to present empirical evidence that a disclaimer or other explanatory information would be inadequate to make a qualified disease claim truthful and nonmisleading:

> . . . even if the FDA's decision to ban the Claim could be justified by finding that the evidence in support of it was clearly qualitatively weaker than the evidence against it, the FDA has failed to provide empirical evidence that an appropriate disclaimer would confuse consumers and fail to correct for deceptiveness.[84]

As a result of this decision, FDA has prepared three acceptable qualified claims:

> 1. Some scientific evidence suggests that consumption of antioxidant vitamins may reduce the risk of certain forms of cancer. However, FDA has determined that this evidence is limited and not conclusive.

> 2. Some scientific evidence suggests that consumption of antioxidant vitamins may reduce the risk of certain forms of cancer. However, FDA does not endorse this claim because this evidence is limited and not conclusive.

> 3. FDA has determined that although some scientific evidence suggests that consumption of antioxidant vitamins may reduce the risk of certain forms of cancer, this evidence is limited and not conclusive.[85]

Whitaker accepted these qualified health claims and thus this litigation was concluded.

v. Folic Acid/B Vitamins. On November 28, 2000, FDA determined that there was no significant scientific agreement about a disease claim submitted after the four *Pearson* claims were submitted—concerning the relationship between folic acid, vitamin B_6, and vitamin B_{12}, in dietary supplements, and the risk of heart disease and other vascular disease. FDA announced that it would exercise "enforcement discretion" as to qualified claims describing that relationship.[86] In announcing its approach, FDA offered the following as a sample qualified claim.

> It is known that diets low in saturated fat and cholesterol may reduce the risk of heart disease. The scientific evidence about whether folic acid [folate], vitamin B_6, and vitamin B_{12} may also reduce the risk of heart disease and other vascular diseases is suggestive, but not conclusive. Studies in the general population have generally found that these vitamins lower homocysteine, an amino acid found in the blood. It is not known whether elevated levels of homocysteine may cause vascular disease or whether high homocysteine levels are caused by other factors. Studies that will directly evaluate whether reducing homocysteine may also reduce the risk of vascular disease are not yet complete.

After litigation, FDA and the plaintiffs in a companion case to *Pearson* filed a joint notice of dismissal in which FDA once again reversed itself and agreed to permit the following claim.

> As part of a well-balanced diet that is low in saturated fat and cholesterol, Folic Acid, Vitamin B6 and Vitamin B-12 may reduce the risk of vascular disease. FDA evaluated the above claim and found that, while it is known that diets low in saturated fat and cholesterol reduce the risk of heart disease and other vascular diseases, the evidence in support of the above claim is inconclusive.

vi. Saw Palmetto. In May 1999, the same individuals who had pursued the other disease claims for dietary supplements submitted a petition for a claim that saw palmetto extract may reduce the symptoms of mild benign prostatic hyperplasia (BPH). Because this was the first petition to request a claim relating to an effect on an existing disease, rather than prevention of disease, FDA decided to hold a public hearing on the matter[87] and then determined that the disease claim provisions in the FD&C Act authorize only disease prevention and not disease treatment claims.[88]

On appeal, the same District Court that has handled the other dietary supplement claims ruled in favor of the FDA interpretation of the FD&C Act and held that it does not violate the First Amendment.[89] The court concluded that the statute is ambiguous with respect to this issue and, because the FDA interpretation is not unreasonable, it must be upheld. This decision was upheld on appeal in the D.C. Circuit.[90]

FDA's Initial Refusal to Apply the Pearson Ruling to Conventional Food

The GMA Citizen Petition

In April 2000, the Grocery Manufacturers of America (GMA) submitted a citizen petition arguing that FDA must apply the *Pearson* ruling to all food, not just to dietary supplements. The *Pearson* decision, GMA pointed out, arose under the same standard for approval of disease claims as applies to all food under the NLEA. GMA maintained that FDA's implementation strategy inhibits GMA members from disseminating truthful and nonmisleading nutrition and health information to consumers. GMA argued that FDA must conform its regulation of food labeling to *Pearson*'s First Amendment standards by taking the following six actions.

1. FDA must immediately withdraw and revise its proposed strategy to implement the *Pearson* decision.
2. FDA must apply *Pearson* to all food, including but not limited to dietary supplements, because the *Pearson* case interpreted the NLEA standard for approval of disease claims for food (which FDA extended without change to dietary supplements).
3. FDA must withdraw the significant scientific agreement guidance because it does not permit FDA to authorize all truthful, nonmisleading claims (including claims for which the level of scientific support can be set forth meaningfully in disclaimers or other explanatory information).

4. FDA must withdraw the authoritative statement guidance because it indicates that FDA will use its unconstitutional interpretation of "significant scientific agreement" to determine whether a statement is "authoritative."
5. FDA must amend all existing disease claim regulations (both procedural and substantive) in 21 C.F.R. § 101.14 and 21 C.F.R. Part 101, Subpart E, to comply with *Pearson*.
6. FDA must immediately suspend all enforcement action against claims that are truthful, accurate, and not misleading.

FDA did not respond to this petition. Because the FDA policies announced in December 2002 effectively grant the action requested, GMA has now withdrawn this petition.

FDA Refused GMA's Disease Claim Petition

Because FDA had not responded to the first GMA petition, on March 14, 2001, GMA submitted a disease claim petition pursuant to Section 403(r)(4) of the FD&C Act and 21 C.F.R. 101.70 seeking approval for conventional food labeling of the specific qualified claims permitted by FDA in dietary supplement labeling pursuant to *Pearson*. GMA incorporated by reference the entire docket for each original disease claim at issue and conceded that each claim lacked significant scientific agreement, as FDA defined the standard. GMA explained that both *Pearson* and the First Amendment require FDA to treat all food similarly—permitting the same qualified claims for conventional food as for dietary supplements.

The claims at issue were the following:

- "Healthful diets with adequate folate may reduce a woman's risk of having a child with a brain or spinal cord defect. The Institute of Medicine of the National Academy of Sciences recommends that women capable of becoming pregnant consume 400 mg of folate daily from supplements, fortified foods, or both, in addition to consuming food folate from a varied diet."[91]
- "0.8 mg folic acid in a dietary supplement is more effective in reducing the risk of neural tube defects than a lower amount in foods in common form. FDA does not endorse this claim. Public health authorities recommend that women consume 0.4 mg folic acid daily from fortified foods or dietary supplements or both to reduce the risk of neural tube defects."[93]
- "The scientific evidence about whether omega-3 fatty acids may reduce the risk of coronary heart disease (CHD) is suggestive, but not conclusive. Studies in the general population have looked at diets containing fish and it is not known whether diets or omega-3 fatty acids in fish may have a possible effect on a reduced risk of CHD. It is not known what effect omega-3 fatty acids may or may not have on risk of CHD in the general population."[94]
- "Consumption of omega-3 fatty acids may reduce the risk of coronary heart disease. FDA evaluated the data and determined that, although there is scientific evidence supporting the claim, the evidence is not conclusive."[94]
- "It is known that diets low in saturated fat and cholesterol may reduce the risk of heart disease. The scientific evidence about whether folic acid [folate], vitamin B_6, and vitamin B_{12} may also reduce the risk of heart disease and other vascular diseases is sugges-

tive, but not conclusive. Studies in the general population have generally found that these vitamins lower homocysteine, an amino acid found in the blood. It is not known whether elevated levels of homocysteine may cause vascular disease or whether high homocysteine levels are caused by other factors. Studies that will directly evaluate whether reducing homocysteine may also reduce the risk of vascular disease are not yet complete."[95]

- "As part of a well-balanced diet that is low in saturated fat and cholesterol, Folic Acid, Vitamin B6 and Vitamin B12 may reduce the risk of vascular disease. FDA evaluated the above claim and found that, while it is known that diets low in saturated fat and cholesterol reduce the risk of heart disease and other vascular diseases, the evidence in support of the above claim is inconclusive."[96]

FDA responded to the petition on June 22, 2001, raising what were essentially technical objections and refusing to address the petition on the merits.[97] FDA first asserted that GMA's incorporation by reference of materials in the dietary supplement disease claim dockets was inadequate insofar as GMA did not make "specific reference" to the precise "location" of required information. FDA also suggested that the scientific considerations for dietary supplements and conventional foods are not identical.

GMA's First Amendment Comments

In May 2002, the Supreme Court handed down a decision involving FDA in a First Amendment case, *Thompson v. Western States Medical Center*.[98] Shortly after, FDA requested public comment to ensure that its policies comply with the governing First Amendment case law.[99] GMA submitted extensive comments to FDA arguing that the First Amendment requires both that the *Pearson* decision be applied to conventional food in the same way as it is applied to dietary supplements and that the standard for determining whether a claim is misleading should be the reasonable person rather than the ignorant, unthinking, and credulous person. Because the FDA policies established in December 2002 grant these two requests, GMA has now also withdrawn its second petition.

The Pearson Ruling, the First Amendment, and Conventional Food Labeling

Both the *Pearson* decision and the First Amendment require FDA to permit conventional food manufacturers to make qualified disease claims in their labeling just as the Agency permits dietary supplement manufacturers to make those claims in their labeling. Neither *Pearson* nor the First Amendment permits FDA to treat the speech of dietary supplement manufacturers differently from the speech of conventional food manufacturers.

The Ruling in Pearson Required FDA to Permit the Proposed Claims in Conventional Food Labeling

The Court of Appeals in *Pearson* applied the First Amendment commercial speech doctrine to FDA regulation of product labeling. It is "undisputed," the court wrote, "that FDA's restrictions on appellants' health claims are evaluated under the commercial speech doctrine."[100] The claims involved are commercial speech, and FDA was therefore obliged under *Pearson* to conform its regulation of these claims to the *Central Hudson* doctrine.

The Court of Appeals in *Pearson* unambiguously held that FDA's application of the significant scientific agreement standard to bar disease claims was unconstitutional. Indeed, FDA's argument that claims lacking significant scientific agreement are inherently misleading was deemed to be "almost frivolous." The significant scientific agreement standard applies to all foods, whether in conventional form or in dietary supplements.

The regulatory schemes for conventional food and dietary supplements are identical. Dietary supplements are food under the FD&C Act. Rules that apply to dietary supplement disease claims also must apply to disease claims for food. The FDA disease claims regulation makes this clear when it states:

> The requirements of this section apply to foods intended for human consumption that are offered for sale, regardless of whether the foods are in conventional food form or dietary supplement form.[101]

The standard that FDA applies is the same. FDA recognized that the same standard applies to disease claims for dietary supplements and conventional food when it issued the significant scientific agreement guidance following the *Pearson* decision and when it issued the guidance on authoritative body claims under the Food and Drug Administration Modernization Act of 1997.

The D.C. Circuit's holding is not limited to dietary supplements. The court expressly invalidated the disease claim regulations that apply to both dietary supplements and conventional foods. Nor is the court's reasoning limited to statutory standards rather than regulatory standards. Under the *Pearson* court's reasoning, if it is frivolous for FDA to argue that dietary supplement disease claims lacking in significant scientific agreement are inherently misleading, it is equally frivolous for FDA to argue that conventional food disease claims lacking in significant scientific agreement are inherently misleading. Similarly, if suppression of disease claims on dietary supplements would not directly advance the government's interest in protecting the public health, suppression of disease claims on conventional food would not directly advance the government's interest in protecting the public health. Finally, if there is no reasonable fit between the prevention of fraud and the outright suppression of disease claims on dietary supplements, there is no reasonable fit between the prevention of fraud and the outright suppression of disease claims on conventional food.

The commercial speech doctrine embodies a preference for disclosure over outright suppression. This is no less true as to conventional food labeling than it was as to dietary supplement labeling. Nothing in the *Pearson* ruling limits its holding to dietary supplements. Accordingly, under *Pearson*, FDA had no alternative but to consider other methods of assuring that disease claims in conventional food labeling are truthful and nonmisleading—such as disclaimers, explanatory statements, and the like.

The First Amendment Required FDA to Permit the Proposed Claims in Conventional Food Labeling

The D.C. Circuit held in *Pearson* that FDA's regulations were an unconstitutional restriction on commercial speech. Regardless of the D.C. Circuit holding, however, the First Amendment principles firmly established by the Supreme Court independently required FDA to permit truthful qualified disease claims in conventional food labeling.

The Supreme Court's First Amendment Cases Protect Commercial Speech

The First Amendment protects "commercial speech," including food and dietary supplement labeling. Disease claims in food labeling also impart vital noncommercial information to consumers, such as the health risks and benefits of consuming a particular product. Food labels and labeling bearing a hybrid of commercial and noncommercial speech are entitled to a heightened form of intermediate scrutiny (that is, an even more rigorous application of *Central Hudson*).[102] As explained below, however, even conventional commercial speech doctrine requires that FDA approve the proposed disease claims.

Under conventional commercial speech doctrine, the government may not prohibit or restrict commercial speech unless it satisfies the criteria in *Central Hudson Gas & Electric Corp. v. Public Service Commission*.[103] Under this decision, the government may prohibit commercial speech only if the speech is inherently false or misleading or proposes an unlawful transaction. Otherwise, it may regulate commercial speech only if it has a significant interest in doing so, the regulation in question directly furthers that interest, and there is no less restrictive means of furthering that interest.

The *Central Hudson* test can be distilled into two principles. First, "only false, deceptive or misleading commercial speech may be banned."[104] Second, commercial speech that is not false, deceptive, or misleading may be restricted, but only if the government shows that there is a "reasonable fit" between its objectives and the degree of restriction that it uses to achieve its objectives.[105]

As to the first principle, FDA has the burden to establish that a disease claim is false or misleading before it may ban that claim.[106] As to the second principle, FDA has the burden "of identifying a substantial interest and justifying the challenged restriction."[107] FDA may not satisfy its burden with speculation. It must present proof that its feared harm is real and that the intended statement will indeed harm the public.[108]

The Supreme Court has repeatedly and emphatically rejected what it calls the "paternalistic" suppression of commercial speech. As the Court has explained:

> The First Amendment directs us to be especially skeptical of regulations that seek to keep people in the dark for what the government perceives to be their own good. That teaching applies equally to state attempts to deprive consumers of accurate information about their chosen products.[109]

To the contrary, the Supreme Court clearly directs the government to give consumers information on which they can base their own decisions:

> Information is not in itself harmful . . . people will perceive their own best interest if only they are well enough informed . . . the best means to that end is to open the channels of communication rather than to close them.[110]

The Court made the same point in *Central Hudson*:

> "Even when advertising communicates only an incomplete version of the relevant facts, the First Amendment presumes that some accurate information is better than no information at all.[111]

Finally, the restriction must be "narrowly tailored."[112] The "cost" of the restriction—that is, the burden it imposes on the speech—must be "carefully calculated."[113] That cost/benefit assessment in turn requires that "the regulation not 'burden substantially more speech than is necessary to further the government's legitimate interests.'"[114]

The Western States Decision

In April 2002, the Supreme Court had occasion to apply the *Central Hudson* principles in a case involving advertising of FDA-regulated products. In *Thompson v. Western States Medical Center*,[115] a group of pharmacies engaged in the practice of compounding prescription drugs challenged a provision of FDAMA that allowed compounding only in response to an "unsolicited" prescription and prohibited a pharmacy, pharmacist, or physician from advertising that it could compound any particular drug or category of drugs.[116] The pharmacies argued that the FDAMA provision violated their First Amendment right to advertise their services in a truthful and nonmisleading manner. FDA responded that advertising was "a fair proxy for actual or intended large-scale manufacturing,"[117] an activity viewed by Congress as violating FDA's new drug approval process.

The Supreme Court agreed with the pharmacies and held that the provision unconstitutionally limited legitimate commercial speech. FDA conceded—and all nine justices agreed—that the First Amendment applies to FDA. Thus, in the wake of *Western States*, it is clear that any speech restriction imposed by FDA on commercial speech must be assessed within the *Central Hudson* framework.

The Court flatly rejected FDA's contention that it had satisfied the final part of the *Central Hudson* test. The Court's precedent established that "if the Government could achieve its interests in a manner that does not restrict speech, or that restricts less speech, the Government must do so."[118] Here, the Court found that the advertising restriction was not narrowly tailored to advance the claimed interest. As the Court stated,

> If the First Amendment means anything, it means that regulating speech must be a last—not first—resort. Yet here it seems to have been the first strategy the Government thought to try.[119]

The Court characterized the dissent's arguments as "a fear that people would make bad decisions if given truthful information about compounded drugs,"[120] a rationale for speech restrictions the Court had rejected in prior cases:

> We have previously rejected the notion that the Government has an interest in preventing the dissemination of truthful commercial information in order to prevent members of the public from making bad decisions with the information.[121]

The Court noted that, even if FDA had a legitimate fear that advertising would create patient confusion about the risk of compounded drugs as the dissent implied, "this interest could be satisfied by the far less restrictive alternative of requiring each compounded drug to be labeled with a warning that the drug had not undergone FDA testing and that its risks were unknown."[122]

Application of These Principles to the Use of Qualified Disease Claims in Conventional Food Labeling

The qualified disease claims that GMA has sought to use were truthful and nonmisleading. FDA conceded this by permitting them in dietary supplement labeling.[123] Under *Central Hudson* and *Western States*, FDA could not categorically ban such claims for conventional food. Rather, it would be required to satisfy a heavy burden of justifying any restriction on the claims, and it could not rely on paternalistic assumptions about the ability of consumers to interpret qualified claims. Nor could it arbitrarily argue that consumers may understand qualified claims on dietary supplements but not the same claims on conventional food.

A public health justification would not support suppression of the qualified disease claims. The conventional foods at issue are concededly safe. Nor would a consumer fraud justification support suppression of the qualified disease claims. The claims, as qualified, are accurate and nonmisleading, and the First Amendment does not permit FDA to assume that consumers are incapable of understanding qualifications and caveats. The Supreme Court's commercial speech cases supported the conclusion reached by the *Pearson* court and required that FDA consider qualified disease claims in conventional food labeling.

FDA Can and Should Apply the FD&C Act in a Way that Protects the First Amendment Rights of Conventional Food Manufacturers

FDA claimed in its May 2000 letter to Representative McIntosh that it could not apply the *Pearson* ruling to conventional food absent a court order. The agency reasoned that it is required to apply the significant scientific agreement standard to conventional food because the food standard is embodied in a statute, whereas the dietary supplement standard is merely embodied in a regulation. However, a federal statute is subject to the same constitutional standard as an agency regulation. If FDA may not by regulation categorically ban from food labeling disease claims lacking significant scientific agreement, neither may Congress do so by statute. Moreover, it is incumbent on FDA to interpret Section 403(r)(4) of the FD&C Act in a way that comports with the Constitution. As an instrument of the federal government, whose officers are sworn to uphold the Constitution, FDA may not knowingly adopt a statutory interpretation that violates the Constitution.

Clarification of the "Significant Scientific Agreement" Standard

FDA is not bound to any particular interpretation of the significant scientific agreement standard in the FD&C Act. The point at which scientific agreement becomes significant is inherently ambiguous and an insufficient guideline for judicial review under the Administrative Procedure Act (APA). It is the policies expressed in FDA regulations or guidance documents that clarify the meaning of the term, and FDA may amend those policies.[124]

The statute instructs FDA to issue a regulation permitting a manufacturer to make a disease claim only when the claim meets the statutory requirements, as articulated by FDA. The APA requires that agencies give content to their enforcement policies so as to prevent arbitrary and capricious enforcement decisions.[125] The language in Section 403(r)(3)(B)(i) of the FD&C Act is clear in one regard. FDA may not promulgate a new regulation for a claim that lacks significant scientific agreement. As the D.C. Circuit has suggested, however, the statutory language alone may not create a sufficiently clear standard to guide a court's review of FDA's exercise of enforcement discretion.[126]

The D.C. Circuit suggested in *Pearson v. Shalala*[127] that the operative statutory language on significant scientific agreement, standing alone, may not pass muster under the APA. FDA argued in *Pearson* that its regulation requiring significant scientific agreement for dietary supplements was justified merely because Congress chose the same term in the statute. The D.C. Circuit squarely rejected that claim and suggested a broader implication:

> We are quite unimpressed with the government's argument that the Agency is justified in using this standard without definition because Congress used the same standard in [the statute]. Presumably—we do not decide—the FDA in applying that statutory standard would similarly be obliged under the APA to give it content.[128]

Federal statutes are "to be construed so as to avoid serious doubts as to their constitutionality."[129] FDA's initial interpretation and enforcement of the "significant scientific agreement" standard infringed food manufacturers' First Amendment right to make nondeceptive claims about their products. FDA could remedy this—and has now done so—by adopting a constitutionally valid interpretation of the "significant scientific agreement" standard. For example, scientific agreement may be "significant" even though the scientific community continues to research and debate various details of a claim. Such a standard would ensure the dissemination of accurate consumer information while not infringing on manufacturers' legitimate speech concerns.

In short, on its own and without further explication from FDA, the statutory provision does not contain sufficient "law to apply" to guide FDA's enforcement actions.[130] Under Supreme Court commercial speech precedent, the Agency's initial interpretation of the statutory standard represented an unconstitutional infringement on food manufacturers' commercial speech rights. For APA purposes, an FDA regulation or guidance on "significant scientific agreement" was needed to flesh out the meaning of the term and give content to the statutory prohibition before FDA could enforce it fairly and in a manner that does not unconstitutionally restrict food manufacturers' speech rights.

FDA Has an Obligation to Interpret and Apply the FD&C Act Constitutionally

Federal agencies have an independent obligation to uphold the Constitution, which is the supreme law of the land.[131] Cases dating from the earliest years of the republic establish that a congressional enactment that conflicts with the Constitution is not a "law." As such, an executive branch agency is not required to enforce it.

In *Marbury v. Madison*, Justice Marshall explained the proposition that courts have an obligation to overturn statutes and other official acts that conflict with the Constitution. The theory of a constitutional government must be that "an act of the legislature repugnant to the constitution is void."[132] *Marbury* addressed the power of the judiciary to invalidate a congressional enactment on the basis of a conflict with the Constitution. The principle underlying *Marbury v. Madison* also leads to the logical conclusion that the executive branch has an identical obligation to uphold the superior source of law in the United States. In short, an executive branch agency must uphold the Constitution even when it conflicts with a statutory directive.[133]

This point was made during the debates that led to the adoption of the Constitution. At the Philadelphia Convention in 1787, James Wilson argued that the Constitution imposed significant restraints on the power of the legislature.[134] In his view, the power of the Constitution is paramount to the power of the legislature. Just as a judge may consider constitutional principles in assessing the legitimacy of a legislative enactment, in "the same manner, the President of the United States could shield himself, and refuse to carry into effect an act that violates the Constitution."[135]

In the present context, of course, FDA is acting on behalf of the President.[136] Upon taking the oath of office, the President vows to "preserve, protect and defend the Constitution of the United States."[137] As agents of the President, the FDA Commissioner and staff likewise have an obligation to uphold the tenets set forth in the Constitution, including the First Amendment.

FDA Can Decline Enforcement of the Misbranding Prohibition on First Amendment Grounds

FDA's Enforcement Authority Is Discretionary

FDA has the discretion to decline to bring enforcement action against a nonmisleading disease claim that lacks an authorizing regulation. The FD&C Act states that the Secretary "shall" promulgate regulations authorizing disease claims,[138] and further states that disease claims "may only be made" if, among other things, they meet the requirements of those regulations.[139] Nowhere does the Act state that FDA "shall" enforce violations of the latter provision. The relevant cases fully support the conclusion that FDA may decide not to do so.

The Supreme Court's decision in *Heckler v. Chaney*[140] upheld FDA's enforcement discretion and suggested that a decision not to enforce will not be judicially reviewable in the absence of a clear statutory standard for review. In *Chaney*, prison inmates challenged FDA's decision not to take enforcement action against the unapproved use of certain drugs for administration of the death penalty by lethal injection. The D.C. Circuit held that FDA's decision was reviewable and overturned FDA's decision as arbitrary and capricious.[141]

In a unanimous decision, the Supreme Court reversed the D.C. Circuit. It held that an action is committed to agency discretion where "no judicially manageable standards are available for judging how and when an agency should exercise its discretion."[142] Section 706 of the Administrative Procedure Act limits a court's ability to set aside an agency action to situations where the action was "arbitrary, capricious, and abuse of discretion, or otherwise not in accordance with law."[143] In all other cases, however, Section 701(a)(2) precludes a court's review over matters "committed to agency discretion by law."[144] The Court explained that there can be no judicial review if the exercise of discretion is such that "a court would have no meaningful standard against which to judge the agency's exercise of discretion."[145]

An enforcement decision is a prototypical example of a decision committed to an agency's absolute discretion. In these cases, courts' "recognition of the existence of discretion is attributable in no small part to the general unsuitability for judicial review of agency decisions to refuse enforcement."[146] Such decisions are unsuitable for judicial review because a court is ill-equipped to second-guess the factors that led to the agency's decision, which may be peculiarly within the agency's expertise.[147] An agency's nonenforcement decision is essentially equivalent to a prosecutor's decision not to indict a suspect.[148] The latter class of decisions has "long been regarded as the special province of the Executive Branch, inasmuch as it is the Executive who is charged by the Constitution to 'take Care that the Laws be faithfully executed.'"[149] Given scarce FDA resources, a "court should not force the agency to funnel its efforts in any one direction."[150]

The Court in *Chaney* further noted that an agency's decision to refrain from enforcement is qualitatively different from the usual decision reviewed by courts, which is a decision to take some action. When an agency chooses not to act, it "generally does not exercise its *coercive* power over an individual's liberty or property rights, and thus does not infringe upon areas that courts are often called upon to protect."[151]

FDA has argued that *Chaney* gave it "wide latitude in matters of enforcement discretion."[152] In *Heterochemical Corporation v. FDA*,[153] for example, the plaintiff petitioned FDA to take regulatory action against its competitors. After investigating the matter, FDA declined to take enforcement action against the competitors, and the plaintiff brought suit. The court found that because the Agency had made extensive investigations into the matter, its refusal to take enforcement action was arguably arbitrary and capricious. The court therefore rejected FDA's argument that *Chaney* was fully dispositive on the issue. However, the court noted that *Chaney* clearly established one point. FDA is never required to investigate alleged violations of the FD&C Act and other statutes. The Supreme Court in *Chaney* "established a presumption that 'Refusals to take enforcement steps' are not reviewable."[154]

FDA Has Exercised Enforcement Discretion in Other Matters

FDA has on prior occasions chosen to exercise enforcement discretion out of concern that enforcing a statutory provision would contravene constitutional rights.

For example, recognizing First Amendment limits on its authority, FDA has issued a Compliance Policy Guide detailing when it will institute a seizure action against books that constitute misleading labeling.[155] The FD&C Act regulates printed material that promotes the use of a product and "accompanies" the product.[156] Such promotional labeling may not be false or misleading.[157] If FDA finds that it is false or misleading, the Agency's general enforcement practice is to recommend seizure of both the product and the offending labeling. Such action presents free speech concerns, however, and the burden on free speech is particularly troubling where the labeling takes the form of a book. Recognizing those concerns, FDA has announced that where the labeling is a book, rather than recommending outright seizure, the Agency will "consider filing a complaint for forfeiture against the product and an injunction to halt, after a hearing, the misuse of the book."[158]

Similarly, at times FDA has chosen not to pursue an appeal of an adverse decision on the constitutionality of a statutory provision it enforces.[159] Finally, in other contexts with less bearing on constitutional rights, FDA likewise has exercised enforcement discretion.[160] These examples illustrate FDA's past willingness to refrain from enforcing a statutory provision for constitutional or other reasons. In light of the First Amendment implications, FDA can exercise its discretion to permit conventional food manufacturers to make the same qualified disease claims that dietary supplement manufacturers may now make.

The Impact of the FDA Policies Established in December 2002

The policies announced by FDA in December 2002 represent a renewed agency commitment to comply with the mandate of the First Amendment.[161]

Qualified Disease Claims for Conventional Food

In its December 2002 Guidance, FDA reversed its initial position that it could not permit qualified disease claims for conventional food absent a court order, and determined instead that the rules developed under the *Pearson* decision will be applied to all forms of food, including conventional food and dietary supplements. Anyone who wishes to include a disease claim in food labeling must still submit a disease claim petition for review and approval by FDA. FDA will not permit use of a disease claim—whether qualified or unqualified—absent a credible body of scientific evidence. FDA will approve a claim that does not meet the higher standard of significant scientific agreement if that claim meets the lower "weight-of-the-scientific-evidence" standard as appropriately qualified. Claims that could threaten public health and safety, or that are misleading even when qualified, will continue to be banned.

One more change in FDA policy was still required, however, to bring the agency in full compliance with the *Pearson* decision and other First Amendment jurisprudence. In the *Whitaker* decision, handed down just a few days after FDA announced its December 2002 policy, the district court unequivocally determined that the FDA weight-of-the-scientific-evidence standard does not meet the requirements of the First Amendment, as articulated by the D.C. Circuit in the *Pearson* decision.[162] The court made it clear that FDA cannot constitutionally ban any disease claim for food that rests on "credible" scientific evidence unless the agency can demonstrate by empirical evidence that an appropriate disclaimer or other qualifying information would confuse consumers and fail to correct for deceptiveness. Accordingly, FDA was required to revise its December 2002 policy to replace the weight-of-the-scientific-evidence standard with the criteria established in the *Whitaker* decision.

When FDA announced its December 2002 policy, the agency established a Task Force to prepare a report on the implementation of this initiative and invited comment from interested persons. In response, a Coalition of major food trade associations submitted to FDA a proposed FDA regulation to establish a premarket notification program for qualified health claims for food labeling. The Coalition urged FDA to establish a premarket notification program rather than a premarket approval system. The proposed regulation established procedures and a definition of the type of "credible evidence" that would justify a qualified health claim. Of major importance, the regulation proposed by the Coalition emphasized that the focus of any system must be on the relationship of the scientific evidence to the proposed qualified disease claim, and not to the underlying substance-disease relationship.

In July 2003, FDA released the report of the Task Force and two final interim guidance documents on (1) an Interim Evidence-Based Ranking System for Scientific Data and (2) the Interim Procedures for Health Claims in the Labeling of Conventional Human Food and Human Dietary Supplements.[163] The interim guidance on an evidence-based ranking system for scientific data designated four categories of scientific data (A through D) and established a six-part evidence-based rating system that would result in ranking the totality of the evidence into one of the four categories. The interim guidance establishing procedures for qualified health claims established a premarket notification program rather than a premarket approval system. The Coalition again submitted comments, strongly supporting the premarket notification program but raising substantial questions about attempting to cabin scientific evidence into four arbitrary categories.

Finally, FDA published an advance notice of proposed rulemaking in November 2003, requesting comment on a large number of issues relating to qualified disease claims for food.[164] The coalition submitted both of its prior comments, and GMA and other members of the coalition submitted additional comments as well. In the interim, FDA began to implement its December 2002 policy of recognizing qualified disease claims for conventional food as well as dietary supplements, based upon the new standard of "credible evidence" and using a premarket notification procedure.

The Reasonable Person Standard

In cases brought by FDA against what the Agency concluded was misleading labeling, the courts generally adopted FDA's position that all claims must be judged against the standard of the "ignorant, the unthinking, and the credulous" consumer.[165] In contrast, in 1984 the Federal Trade Commission (FTC) adopted a reasonable person standard in determining whether advertising is misleading.[166] Recognizing that the ignorant, unthinking, and credulous standard is not consistent with governing First Amendment case law,[167] and with the explicit intent to conform the standard between labeling and advertising, in its December 2002 policies FDA adopted the reasonable person standard. Thus, a realistic standard now exists by which to judge whether disclaimers and other qualifying or explanatory language in a food disease claim are misleading.

Conclusion

FDA's initial failure to permit qualified disease claims contravened FDA's obligation to enforce the FD&C Act in a constitutional manner. FDA has the authority to permit manufacturers to make those claims, either by exercising its enforcement discretion as it has done for the dietary supplement claims, or by reinterpreting the significant scientific agreement standard, or both. The *Pearson* opinions and the First Amendment required FDA to permit in conventional food labeling the same qualified disease claims that it permits in dietary supplement labeling. The policies established by FDA in December 2002 have now realigned the disease claims permitted for conventional food and for dietary supplements and thus promote the dissemination of useful disease claims to consumers. They also adopt a reasonable person standard for judging whether a claim is misleading in order to bring FDA regulation of claims within current First Amendment jurisprudence. In response to the *Whitaker*[168] decision, in its July and November 2003 announcements FDA also revised the weight-of-the-scientific-evidence standard that the agency adopted in its December 2002 guidance, to reflect First Amendment jurisprudence as articulated through the extensive *Pearson* litigation.

Notes

1. 67 Fed. Reg. 78002 (December 20, 2002). See "FDA Announces Initiative to Provide Better Health Information for Consumers," FDA News P02-54 (December 18, 2002).

2. 164 F.3d 650 (D.C. Cir. 1999), rehearing denied, 172 F.3d 72 (D.C. Cir. 1999) (en banc).

3. *Pearson v. Shalala*, 130 F. Supp. 2d 105 (D.D.C. 2001), on reconsideration, Pearson v. Thompson, 141 F. Supp. 2d 105 (D.D.C. 2001); *Whitaker v. Thompson*, 248 F. Supp. 2d 1(D.D.C. 2002).

4. FDA generally refers to claims authorized by section 403(r)(1)(B) of the Federal Food, Drug, and Cosmetic Act (FD&C Act) as "health" claims. We refer to them as "disease" claims to distinguish them from structure/

function claims and because section 403(r)(1)(B) defines such claims as characterizing "the relationship of any nutrient . . . to a disease . . . ."

5. 52 Stat. 1040 (1938), as amended, 21 U.S.C. 301 et seq.
6. 65 Fed. Reg. 59855 (October 6, 2000).
7. 104 Stat. 2353 (1990).
8. Section 403(r)(1)(B) of the FD&C Act, 21 U.S.C. 343(r)(1)(B).
9. Section 403(r)(3)(B) of the FD&C Act, 21 U.S.C. 343(r)(3)(B).
10. FDA has authorized twelve disease claims under the "significant scientific agreement" standard. These claims may be made in both conventional food labeling and dietary supplement labeling. 21 C.F.R. 101.72-101.83.
11. Section 201(g)(1) of the FD&C Act, 21 U.S.C. 321(g)(1).
12. Section 403(r)(5)(D) of the FD&C Act 21 U.S.C. 343(r)(5)(D).
13. The Dietary Supplement Act of 1992, 106 Stat. 4491, 4500 (1992), imposed a moratorium on implementation of NLEA with respect to dietary supplements until December 15, 1993. NLEA had directed FDA to consider ten specific disease claims. These claims were exempt from the moratorium.
14. 58 Fed. Reg. 2478 (January 6, 1993).
15. Id. at 2503-2509.
16. Id. at 2503; 21 C.F.R. § 101.14(c).
17. 58 Fed. Reg. at 2504; id. at 2506.
18. Id. at 2504.
19. Id. at 2505.
20. 56 Fed. Reg. 60537 (November 27, 1991) (first proposed rule); 58 Fed. Reg. 33700 (June 18, 1993) (second proposed rule); 59 Fed. Reg. 395 (January 4, 1994) (final rule). The first disease claims proposal pertained to dietary supplements as well as to conventional foods. After Congress passed the Dietary Supplement Act in 1992, FDA finalized the rule as to conventional foods and issued a new proposal pertaining to dietary supplements.
21. E.g., Brief for Appellees in *Pearson v. Shalala* (No. 98-5043) (D.C. Cir.) at 6 ("FDA proposed using the same standard for dietary supplements that Congress in the NLEA mandated for all other foods— i.e., the 'significant scientific agreement'" standard); id. at 8 (In 1992, "the Agency reissued proposed regulations for dietary supplement health claims, again proposing to use the same standard— significant scientific agreement . . . "); id. ("The Agency concluded that 'subject[ing] dietary supplements to the same standard that applies to foods in conventional form . . . strikes the appropriate balance.'").
22. 56 Fed. Reg. at 60540 ("FDA believes that there would be significant potential for consumer confusion when confronted with a situation in which there would be health claims for substances when they are present in supplements but not when they are present in conventional foods.").
23. 56 Fed. Reg. at 60540 ("FDA has an obligation to treat all segments of the regulated food industry with fairness. If dietary supplements were subject to different rules, whether with respect to the procedure for assessment of conformity with the scientific standard or to the manner in which claims are made, there is a possibility that supplements could be made to appear somehow superior to conventional foods that contain the same nutrient. Such an appearance would not only be untrue, it would be unfair to firms producing conventional foods.").
24. 111 Stat. 2296 (1997).
25. To date, two authoritative statement disease claims have been permitted.
26. "Guidance for Industry: Notification of a Health Claim or Nutrient Content Claim Based on an Authoritative Statement of a Scientific Body" (June 11, 1998).
27. Id. at 2.
28. Id. at 3.
29. 64 Fed. Reg. 3520 (January 21, 1999).
30 Sections 3(b)(1)(A)(vi) and (x) of the NLEA, 104 Stat. at 2361.
31. 58 Fed. Reg. 2622 (January 6, 1993) (anti-oxidants and cancer); 58 Fed. Reg. 2537 (January 6, 1993) (dietary fiber and cancer); 58 Fed. Reg. 2682 (January 6, 1993) (omega-3 fatty acids and coronary heart disease); 58 Fed. Reg. 2606 (January 6, 1993) (folic acid and neural tube defects).
32. 58 Fed. Reg. 53296 (October 14, 1993).
33. 58 Fed. Reg. 53254 (October 14, 1993).
34. 59 Fed. Reg. 395 (January 4, 1994) (dietary fiber, antioxidant vitamins, and omega-3 fatty acids); 59 Fed. Reg. 433 (January 4, 1994) (folate).
35. 61 Fed. Reg. 8750 (March 5, 1996). Footnote 65 infra discusses the relationship between folic acid and folate.

36. 5 U.S.C. 706.
37. *Pearson*, 14 F. Supp.2d 10 (D.D.C. 1998).
38. *Pearson*, 164 F.3d 650 (D.C. Cir. 1999).
39. Id. at 655.
40. 447 U.S. 557 (1980).
41. 164 F.3d at 655.
42. Id.
43. Id. at 656.
44. Id. As explained by the D.C. Circuit, under *Central Hudson* a court evaluates a government scheme to regulate potentially misleading speech by applying a three-part test. First, the court asks whether the asserted government interest is substantial. Second, the court determines whether the regulation directly advances the government interest asserted. Third, the court determines whether the fit between the government's ends and the means chosen to accomplish those ends is reasonable. 164 F.3d at 655-656.
45. Id.
46. Id. at 657.
47. Id.
48. See also *Thompson v. Western States Medical Center*, 535 U.S. 357, 376 (2002), where the Court observed that even where there is a substantial risk of patient confusion, the government must consider whether labeling can alleviate that risk before it imposes an outright ban on accurate and nonmisleading advertising.
49. *Pearson*, 164 F.3d at 661, invalidating 21 C.F.R. 101.71(a), 21 C.F.R. 101.71(c), 21 C.F.R. 101.71(e), and 21 C.F.R. 101.79(c)(2)(i)(G).
50. 164 F.3d at 660.
51. Id. at 661.
52. 172 F.3d 72 (D.C. Cir. 1999).
53. 64 Fed. Reg. 67289 (December 1, 1999).
54. 21 C.F.R. 101.14.
55. 64 Fed. Reg. at 67290.
56. 64 Fed. Reg. 71794 (December 22, 1999) (announcing availability of guidance); "Guidance for Industry: Significant Scientific Agreement in the Review of Health Claims for Conventional Foods and Dietary Supplements" (December 22, 1999).
57. Id. at 3 ("This standard applies to conventional foods health claims by statute; FDA applied the same standard to dietary supplement health claims by regulation.")
58. Letter from Melinda K. Plaisier (FDA) to the Honorable David M. McIntosh (U.S. House of Representatives) (May 16, 2000). Joseph Levitt, Director of the Center for Food Safety and Applied Nutrition, also told a reporter in October 2000 that *Pearson* and the new interim standard of proof apply only to dietary supplements. "FDA to Allow Dietary Supplement Claims Failing to Meet its 'Gold Standard' Proof," *Dietary Supplement and Food Labeling News* 1, 8 (October 11, 2000). The December 2002 announcement by FDA, note 1 supra, directly reverses these earlier positions taken by FDA.
59. FDA later determined that such claims are not permissible NLEA disease claims. Letter from Joseph A. Levitt (FDA) to Jonathan W. Emord (Docket No. 99P-3030) (May 26, 2000).
60. 65 Fed. Reg. 14219, 14221 (March 16, 2000).
61. 65 Fed. Reg. 58917 (October 3, 2000).
62. 65 Fed. Reg. 59855 (October 6, 2000); see also FDA Talk Paper T00-51 (October 11, 2000).
63. 65 Fed. Reg. at 59856.
64. Letter from Christine J. Lewis (FDA) to Jonathan W. Emord (Docket No. 91N-0098) (October 10, 2000).
65. Letter from Christine J. Lewis (FDA) to Jonathan W. Emord (October 10, 2000) (Docket No. 91N-100H). "'Folate' is the generic term for all forms of the vitamin and includes both naturally occurring 'food folate' and the synthetic form of 'folic acid' that is added to fortified food and dietary supplements." Id.
66. *Pearson*, 130 F. Supp. 2d 105 (D.D.C. 2001).
67. Id. at 112.
68. Id. at 120. The court also stated that FDA should respond within 60 days of the decision. Id. at 120.
69. *Pearson*, 141 F. Supp. 2d 105 (D.D.C. 2001).
70 Id. at 112.
71. *Pearson*, 164 F.3d at 659-660.
72. Letter from Christine J. Lewis (FDA) to Jonathan W. Emord (April 3, 2001) (Docket No. 91N-100H).
73. Letter from Christine J. Lewis (FDA) to Jonathan W. Emord (October 31, 2000) (Docket No. 91N-0103).

74. Letter from Christine J. Lewis (FDA) to Jonathan W. Emord (February 8, 2002) (Docket 91N-0103).
75. Letter from Christine J. Lewis (FDA) to Jonathan W. Emord (May 4, 2001) (Docket No. 91N-0101).
76. *Whitaker v. Thompson*, 248 F. Supp. 2d 1 (D.D.C. 2002).
77. *Western States*, 535 U.S. at 371.
78. *Whitaker*, 248 F. Supp. 2d at 10.
79. Id.
80. Id. at 11.
81. Note 1 supra.
82. *Whitaker*, 248 F. Supp. 2d at 10.
83. Id. at 13.
84. Id.
85. Letter from Christine L. Taylor (FDA) to Jonathan W. Emord (April 1, 2003) (Docket No. 91N-0101).
86. Letter from Christine J. Lewis (FDA) to Jonathan W. Emord (November 28, 2000) (Docket No. 99P-3029).
87. 65 Fed. Reg. 14219 (March 16, 2000).
88. Letter from Joseph A. Levitt (FDA) to Jonathan W. Emord (May 26, 2000) (Docket No. 99P-3030).
89. *Whitaker v. Thompson*, 239 F. Supp. 2d 43 (D.D.C. 2003).
90. *Whitaker v. Thompson*, No. 99cv 03247 (D.C. Cir. 2004).
91. Letter from Christine J. Lewis (FDA) to Jonathan W. Emord (October 10, 2000) (Docket No. 91N-0098).
92. Letter from Christine J. Lewis (FDA) to Jonathan W. Emord (April 3, 2001) (Docket No. 91N-100H).
93. Letter from Christine J. Lewis (FDA) to Jonathan W. Emord (October 31, 2000) (Docket No. 91N-1013).
94. Letter from Christine J. Taylor (FDA) to Jonathan W. Emord (February 8, 2002) (Docket 91N-0103).
95. Letter from Christine J. Lewis (FDA) to Jonathan W. Emord (November 28, 2000) (Docket No. 99P-3029).
96. CFSAN Office of Nutritional Products, Labeling, and Dietary Supplements, "Settlement Reached for Health Claim Relating B Vitamins and Vascular Disease" (May 15, 2001), available at http://www.cfsan.fda.gov/~dms/ds-hclbv.html.
97. Letter from Christine J. Lewis (FDA) to James H. Skiles (June 22, 2001) (Docket No. 91N-0098).
98. 537 U.S. 357 (2002). See note 115 infra and related text.
99. 67 Fed. Reg. 34942 (May 16, 2002).
100. *Pearson*, 164 F.3d at 655.
101. 21 C.F.R. 101.14(g).
102. In recent cases involving hybrid speech, the Court has applied a rigorous form of *Central Hudson*. E.g., *Greater New Orleans Broadcasting Association v. United States*, 527 U.S. 173 (1999); *44 Liquormart, Inc. v. Rhode Island*, 517 U.S. 484 (1996); *Rubin v. Coors Brewing Company*, 514 U.S. 476 (1995). Cf. *Bolger v. Youngs Drug Products Corp.*, 463 U.S. 60 (1983) (advertisements containing discussions of important public issues such as venereal disease and family planning); *Consolidated Edison Co. v. Public Service Commission of New York*, 447 U.S. 530 (1980) (inclusion in monthly bills of inserts discussing political issues).
103. 447 U.S. 557 (1980).
104. *Ibanez v. Florida Department of Business and Professional Regulation*, 512 U.S. 136, 142 (1994) (citing *Zauderer v. Office of Disciplinary Counsel of Supreme Court of Ohio*, 471 U.S. 626, 638 (1985)).
105. *Board of Trustees of State Univ. of New York v. Fox*, 492 U.S. 469, 480 (1989).
106. Cf. *Ibanez*, 512 U.S. at 142.
107. *Greater New Orleans Broadcasting*, 527 U.S. at 174.
108. *Ibanez*, 512 U.S. at 143; *Edenfield v. Fane*, 507 U.S. 761, 770-771 (1993); *Zauderer*, 471 U.S. at 648-649.
109. *44 Liquormart*, 517 U.S. at 503.
110. *Virginia Board of Pharmacy v. Virginia Citizens Consumer Council*, 425 U.S. 748, 770 (1976).
111. *Central Hudson*, 447 U.S. at 562.
112. *Fox*, 492 U.S. at 480.
113. Id.
114. Id. at 478.
115. 535 U.S. 357 (2002).
116. Sections 503A(a) and 503A(c) of the FD&C Act, 21 U.S.C. 353A(a) & 353A(c).
117. *Western States*, 535 U.S. at 371.
118. Id.
119. Id. at 373.
120. Id. at 374.
121. Id. (citing *Virginia Board of Pharmacy*, 425 U.S. at 769).

122. Id. at 376.
123. The FD&C Act prohibits a manufacturer from including in its labeling a disease claim that is false or misleading. Section 403(a)(1) of the FD&C Act, 21 U.S.C. 343(a)(1); see also 21 C.F.R. 101.14(d)(2)(iii).
124. FDA's interpretations must, of course, comport with the relevant statutory provisions. *Heckler v. Chaney*, 470 U.S. 821, 833 n.4 (1985). As this section demonstrates, the operative statutory language affords more than one reasonable interpretation.
125. 5 U.S.C. 706(2)(A). Arguably, FDA's decision to explicate the meaning of the "significant scientific agreement" standard in a Guidance Document violated the administrative law requirement that legislative rules be promulgated pursuant to formal rulemaking procedures, which FDA did not follow in this case.
126. FDA recognizes this ambiguity in its Guidance Document on "significant scientific agreement," where it observes that:

"Significant scientific agreement does not require a consensus or agreement based on unanimous and incontrovertible scientific opinion. However, on the continuum of scientific discovery that extends from emerging evidence to consensus, it represents an area on the continuum that lies closer to the latter than to the former."

Guidance for Industry: Significant Scientific Agreement in the Review of Health Claims for Conventional Foods and Dietary Supplements, December 22, 1999, available at http://www.cfsan.fda.gov/~dms/ssaguide.html (accessed November 15, 2002).
127. 164 F.3d 650 (D.C. Cir. 1999).
128. Id. at 660-61.
129. *Communications Workers of America v. Beck*, 487 U.S. 735, 762 (1988).
130. *Chaney*, 470 U.S. at 834; see also *United States v. Juvenile No. 1*, 118 F.3d 298 (5th Cir. 1997)(holding that the statutory phrase "substantial federal interest" does not provide a justiciable standard).
131. U.S. Constitution Article VI, Section 2.•
132. 5 U.S. (1 Cranch) 137, 177 (1803). See also *The Federalist No. 78*, at 467 (Alexander Hamilton) (Clinton Rossiter ed., 1961) ("[E]very act of a delegated authority, contrary to the tenor of the commission under which it is exercised, is void. No legislative act, therefore, contrary to the Constitution, can be valid.").
133. Id. at 180 ("a law repugnant to the constitution is void, and . . . courts, *as well as other departments*, are bound by [the Constitution]") (emphasis added).
134. Statement of James Wilson on December 1, 1787 on the Adoption of the Federal Constitution, reprinted in 2 Jonathan Elliot, *Debates on the Federal Constitution* 418 (1836).
135. Id. at 446.
136. "The executive Power shall be vested in a President." U.S. Constitution Article II, Section 1(1).
137. U.S. Constitution Article II, Section 1(8).
138. Section 403(r)(3)(B)(i)of the FD&C Act, 21 U.S.C. 343(r)(3)(B)(i).
139. Section 403(r)(3)(A) of the FD&C Act, 21 U.S.C. 343(r)(3)(A).
140. *Heckler v. Chaney*, 470 U.S. 821 (1985).
141. *Chaney v. Heckler*, 718 F.2d 1174 (D.C. Cir. 1983).
142. *Chaney*, 470 U.S. at 830.
143. 5 U.S.C. 706(2)(A).
144. 5 U.S.C. 701(a)(2).
145. *Chaney*, 470 U.S. at 830.
146. Id. at 831 (citations omitted).
147. Id.
148. *National Milk Producers Federation v. Harris*, 653 F.2d 339, 343 (8th Cir. 1981) (observing that in general, both enforcement and prosecutorial decisions by executive branch agencies are committed to agency discretion).
149. *Chaney*, 470 U.S. at 831 (quoting U.S. Constitution, Article II, Section 3).
150. *Robbins v. Reagan*, 780 F.2d 37, 47 (D.C. Cir. 1985).
151. *Chaney*, 470 U.S. at 832.
152. 65 Fed. Reg. 59855, 59857 (October 6, 2000).
153. 644 F. Supp. 271, 273 (E.D.N.Y. 1986).
154. Id. (quoting *Chaney*, 470 U.S. at 831).
155. CPG 140.100 (revised 8/31/89).
156. Section 201(m)(2) of the FD&C Act, 21 U.S.C. 321(m)(2).
157. Section 403(a) of the FD&C Act, 21 U.S.C. 343(a).
158. Id. In another constitutional context, the D.C. Circuit restricted FDA's authority to seize literature in conjunction with an unapproved device. In *Founding Church of Scientology v. United States*, 409 F.2d 1146 (D.C.

Cir. 1969), FDA seized several electrical instruments and a large quantity of literature owned by The Founding Church of Scientology of Washington, D.C. Because the appellants had made out a prima facie case that Scientology is a religion, the D.C. Circuit held that FDA could not seize general literature which merely sketched out the doctrinal theory of Scientology, even though it also discussed the unapproved electrical instruments.

159. FDA's decision not to appeal the holding in *Milnot Co. v. Richardson*, 350 F. Supp. 221 (S.D. Ill. 1972), illustrates this point. An Illinois District Court held the Filled Milk Act to be unconstitutional on due process grounds, ignoring earlier Supreme Court precedent upholding the Act. Although Congress never repealed the Act, FDA chose not to pursue its appeal and instead exercised its discretion to cease enforcing the Act. 38 Fed. Reg. 20,748 (August 2, 1973).

160. E.g., *Extra-Label Policy Based on "Enforcement Discretion," FDA Says*, Food Chemical News, January 26, 1987, at 9 (FDA allowed veterinarians to use animal drugs in an extra-label fashion); Guidance for Industry on Levothyroxine Sodium Products — Enforcement of August 14, 2001, Compliance Date and Submission of New Applications; Availability, 66 Fed. Reg. 36,794 (July 13, 2001) (FDA allowed transition period during which unapproved product could be sold); CDRH Interim Policy Regarding Parents' Access to Tests for Drugs of Abuse, available at http://www.fda.gov/cdrh/dsma/113.html (FDA announced it would not take enforcement action against distributors or unapproved home drug test collection systems).

161. Note 1 supra.

162. Note 76 supra and accompanying text.

163. 68 Fed. Reg. 41387 (July 11, 2003).

164. 68 Fed. Reg. 66040 (November 25, 2003).

165. E.g., *United States v. El-O-Pathic Pharmacy*, 192 F.2d 62, 75 (9th Cir. 1951); *United States v. An Article of Food . . . "Manischewitz . . . Diet Thins*," 377 F. Supp. 746, 749 (E.D.N.Y. 1974).

166. FTC, *Deception Policy Statement*, 103 F.T.C. 110, 174, 177 (1984).

167. *Bolger v. Youngs Drug Products Corp.*, 463 U.S. 60, 73-74 (1983) ("the government may not 'reduce the adult population . . . to reading only what is fit for children.'").

168. Note 76 supra and accompanying text.

9 Dietary Supplements and Drug Constituents: The *Pharmanex v. Shalala* Case and Implications for the Pharmaceutical and Dietary Supplement Industries

Daniel A. Kracov, Paul D. Rubin, and Lisa M. Dwyer[1]

Introduction

On March 30, 2001, in *Pharmanex, Inc. v. Shalala* ("*Pharmanex III*"),[2] a federal district court in Utah found that Cholestin®, a red yeast rice product that had been marketed by Pharmanex, Inc. ("Pharmanex") as a "dietary supplement," was actually an unapproved "new drug." This decision turned on the interpretation of the exclusionary clause in the "dietary supplement" definition in the Federal Food, Drug, and Cosmetic Act ("FFDCA").[3]

As explained below, this decision established an important boundary between drug products, which require Food and Drug Administration ("FDA") preapproval, and "dietary supplement" products. In response to this holding, pharmaceutical companies should review whether the protections of the exclusionary clause are applicable prior to embarking upon a substantial investigational program for drug products under development; if not protected by the exclusionary clause, "dietary supplement" companies may be able to market similar products without engaging in the extensive research required to obtain drug approval. Similarly, "dietary supplement" companies should evaluate whether the exclusionary clause could prohibit the marketing of certain ingredients in "dietary supplement" products. Such marketing may result in FDA determining that a dietary ingredient is actually a drug ingredient that is not permissible for supplement marketing.

The exclusionary clause provides in pertinent part that the term "dietary supplement" does **not** include:

> an <u>article</u> that is approved as a new drug under Section 505 [21 U.S.C. § 355] [or certified as an antibiotic or licensed as a biologic] . . . or an article authorized for investigation as a new drug, antibiotic, or biological for which substantial clinical investigations have been instituted and for which the existence of such investigations has been made public, which was not **before such approval**, certification, licensing, or authorization **marketed** as a dietary supplement or as a food . . . .[4]

Simply put, a product is not a "dietary supplement" if: (1) it contains an "article" approved as a new or investigational drug, and (2) the "article" was not marketed as a food or "dietary supplement" prior to the drug approval.

Pharmanex III was the last of three court decisions in the Cholestin case that collectively resolved two issues of first impression concerning the appropriate interpretation of the exclusionary clause.[5]

First, according to the Tenth Circuit in *Pharmanex II*, the term "article" in the phrase "article approved as a new drug" can refer to **either** a finished drug product or any of the drug product's components.[6] Second, on remanding *Pharmanex III*, the Utah district court held that an article is not previously "marketed" as a food or "dietary supplement" prior to approval, under the "prior market clause" (that is, the portion of the exclusionary clause that requires "marketing" as a food or "dietary supplement" prior to drug approval or investigation authorization), without evidence suggesting more than the mere presence of the article in the food supply.

For a better understanding of the Cholestin case, this chapter summarizes the legal framework for "dietary supplements," the FDA's decision in the Cholestin matter and the three subsequent court cases, as well as the potential implications of the Cholestin decisions.

Legal Framework

The Regulatory Framework for "Dietary Supplements"

This case arose under the FFDCA, 21 U.S.C. § 321 et seq. (Supp. 2002), as amended by the Dietary Supplement Health and Education Act of 1994 ("DSHEA").[7] DSHEA, which was signed into law on October 25, 1994, altered the regulatory landscape with respect to "dietary supplements" marketed in the United States. Congress enacted DSHEA to resolve issues as to how to regulate "dietary supplements," to respond to public demand for accurate information about supplements, and to empower consumers to make choices about preventative health care programs based on available scientific evidence about "dietary supplements."[8]

DSHEA expressly provided in its findings that "although the Federal Government should take swift action against products that are unsafe or adulterated, it should not take any actions to impose unreasonable regulatory barriers limiting or slowing the flow of safe products and accurate information to consumers . . . ."[9] To protect consumer access to safe products, DSHEA designated "dietary supplements" as a distinct regulatory subcategory of "food," different from conventional foods, food additives, and drugs, and imposed a less restrictive regulatory regime on "dietary supplements" than drugs.

Under DSHEA, manufacturers of "dietary supplements" may make statements of nutritional support, including statements about the product's effect on the structure or function of the body, without the product being considered a "drug."[10] Moreover, under DSHEA, "dietary supplements," unlike "drugs," need not receive premarket approval.[11] However, manufacturers must notify the FDA before marketing a "dietary supplement" that contains a new dietary ingredient.[12] In addition, DSHEA places the burden of proof on the FDA in order to claim that a "dietary supplement" is unsafe and must be removed from the market.[13]

The "Dietary Supplement" Definition

The centerpiece of DSHEA is the definition of the term "dietary supplement."[14] Section 321(ff) of the FFDCA, which was added by DSHEA, provides in pertinent part that the term "dietary supplement":

(1) means a product (other than tobacco) intended to supplement the diet that bears or contains one or more of the following dietary ingredients:
  (A) a vitamin;
  (B) a mineral;
  (C) an herb or other botanical;
  (D) an amino acid;
  (E) a dietary substance for use by man to supplement the diet by increasing the total dietary intake; or
  (F) a concentrate, metabolite, constituent, extract, or combination of any ingredient described in clause (A), (B), (C), (D), or (E);
(2) means a product that
  (A) (i) is intended for ingestion in [tablet, capsule, powder, softgel, gelcap, or liquid] form described in section 411(c)(1)(B)(i); or
      (ii) [if not intended for ingestion in such form, is not represented as conventional food and is not represented as a sole item of a meal or of the diet];
  (B) is not represented for use as a conventional food or as a sole item of a meal or the diet; and
  (C) is labeled as a dietary supplement; and
(3) does . . .
  (B) not include
      (i) an article that is approved as a new drug under section 505, certified as an antibiotic under section 507, or licensed as a biologic under section 351 of the Public Health Act (42 U.S.C. 262), or
      (ii) an article authorized for investigation as a new drug, antibiotic, or biological for which substantial clinical investigations have been instituted and for which the existence of such investigations has been made public,
      which was not before such approval, certification, licensing, or authorization marketed as a dietary supplement or food unless the Secretary, in the Secretary's discretion, has issued a regulation after notice and comment, finding that the article would be lawful under this act.[15]

According to FDA's decision in the Cholestin case, the objective of the "exclusionary clause" was to prevent the "dietary supplement" definition from undermining the "'incentive of pharmaceutical manufacturers to develop and bring new drugs to market.'"[16]

Based on this definition, to be a "dietary supplement," a product must: (1) contain a dietary ingredient listed in Section 321(ff)(1), (2) conform with the form and labeling requirements in Section 321(ff)(2), and (3) not be excluded by Section 321(ff)(3)(B). Accordingly, even if a product that is labeled as a "dietary supplement" meets the criteria in Sections 321(ff)(1) and (2), it is not a "dietary supplement" if: (1) it contains an "article" approved as a new or investigational drug, and (2) the "article" does not meet the "prior market clause" (that is, the article was not marketed as a food or "dietary supplement" prior to the drug approval).

Cholestin Case Background

From November 1996, until the decision was issued in *Pharmanex III*, Pharmanex marketed the initial formulation of Cholestin, which was a capsule that contained milled red yeast rice. Milled red yeast rice is a traditional food that has been eaten and valued for its health benefits in China and elsewhere in East Asia for centuries, and in the United States for decades.

The initial formulation of Cholestin was intended to help maintain healthful cholesterol levels in connection with diet and exercise.[17] The formulation was the result of a university research program in China that sought to develop two separate but related products from the red yeast rice: (1) a traditional Chinese medicinal product in the form of a red yeast rice extract (called Xuezhikang), and (2) a standardized traditional Chinese red yeast rice food product (Zhi-Tai). Both Xuezhikang and Cholestin contained a range of HMG-CoA reductase inhibitors and unsaturated fatty acids.[18]

One of at least ten HMG-CoA reductase inhibitor constituents in the initial formulation of Cholestin was "mevinolin." In this case, FDA took the position that mevinolin, a naturally occurring substance, is chemically indistinguishable from synthetic lovastatin, the active ingredient in the prescription drug Mevacor.[19] Mevacor, which is marketed by Merck & Co., Inc. ("Merck"), was approved by FDA in 1987 for the treatment of high cholesterol.[20]

Based on the similarity between mevinolin and lovastatin, FDA advised Pharmanex on April 7, 1997, that it considered Cholestin to be a "new drug," which could not be marketed without FDA approval, rather than a "dietary supplement," as Pharmanex believed. Shortly thereafter, while discussions between the parties were ongoing, the FDA issued a Notice of Detention and Hearing that prevented importation of a shipment of red yeast rice for use in Cholestin.[21]

Agency and Court Proceedings

FDA's Decision

On May 20, 1998, after substantial interactions between FDA and Pharmanex, FDA issued a decision holding that Cholestin did not meet the definition of a "dietary supplement" because of the exclusionary clause in 21 U.S.C. § 321(ff)(B). Thus, according to FDA, Cholestin was subject to regulation as a "new drug."[22]

Under the exclusionary clause, a product is not a "dietary supplement" if: (1) it contains an "article" approved as a new or investigational drug, and (2) the "article" does not meet the "prior market clause" (i.e., the article was not sold as a food or "dietary supplement" prior to the approval). In reaching the conclusion that Cholestin was not a "dietary supplement," FDA determined that the first criterion of the exclusionary clause was met because the relevant "article" was lovastatin, which was approved as a "new drug" in 1997. According to FDA, the term "article" in the phrase "article approved as a new drug under section 505 [21 U.S.C. § 355]" in the exclusionary clause can refer either to a finished drug product or any of that drug product's individual components. Whether an "article" is a finished drug product or a component depends upon the circumstances surrounding the manufacturing and marketing practices of the product at issue.[23] If a company's manufacturing and marketing practices emphasize a component of the product rather than the product as a whole, the relevant "article" would be the component.

Applying that rule, FDA concluded that the relevant "article" in this case was lovastatin rather than the finished product itself (red yeast rice).[24] According to FDA, Pharmanex was marketing, or "touting," the product for the properties of lovastatin, and manufacturing the product to heighten the concentration of lovastatin in the red yeast rice.[25]

Contending that Pharmanex was actually marketing or "touting" lovastatin content rather than red yeast rice, FDA pointed to: (1) promotions directed to pharmacists stating that Cholestin contained the same active ingredients as those currently available in a prescription drug for treating high cholesterol, (2) promotions linking Cholestin to a Chinese pharmaceutical product that is expressly designed to lower cholesterol, and (3) academic articles used in marketing that allegedly linked Cholestin to lovastatin.[26] Alleging that Pharmanex was actually manufacturing lovastatin rather than red yeast rice, FDA: (1) noted that several studies demonstrating that traditional red yeast rice did not contain significant amounts of lovastatin, whereas Cholestin did, and (2) noted that Pharmanex's manufacturing process, unlike the traditional manufacturing process, was designed to maximize and standardize the levels of lovastatin.[27] Notably, however, as the district court observed on appeal, FDA "never justified its determination to treat lovastatin as the relevant "article" approved as a "new drug," as opposed to Mevacor itself or any of Mevacor's other ingredients."[28] FDA simply concluded that lovastatin was the relevant "article" "based on lovastatin's status as the active ingredient in Mevacor."[29]

FDA also concluded that the "prior market clause" did not prevent Cholestin from being deemed a "new drug" rather than a "dietary supplement."[30] According to FDA, in order for a food component to be considered "marketed as a dietary supplement or as a food" prior to the component's approval as a "new drug," the manufacturer of the food or "dietary supplement" must have promoted the product based upon the properties of the component at issue, or in some way increased the concentration of that component, prior to the component's approval as a "new drug." In other words, the mere presence of the relevant "article" in a previously marketed food would not satisfy the "prior market clause." Thus, FDA determined that lovastatin was not marketed as a "dietary supplement" or as a food, even though lovastatin, or mevinolin, has been present naturally in food products that have been sold for years, such as red yeast rice and oyster mushrooms.[31]

The District Court's Initial Decision: Pharmanex I

Following FDA's decision, Pharmanex filed an action seeking declaratory and injunctive relief with the U.S. District Court for Utah and requested that the Court hold unlawful and set aside the FDA's decision, pursuant to the Administrative Procedure Act ("APA").[32] The district court granted the preliminary injunction in favor of Pharmanex and entered a final order on February 16, 1999, setting aside the FDA decision and holding that Cholestin is a "dietary supplement."[33]

The district court's decision turned on FDA's interpretation of the phrase in the exclusionary clause "an article that is approved as a new drug." A court's review of an agency's interpretation of a statute is governed by the two-step process set forth in *Chevron U.S.A. v. NRDC*.[34] Under *Chevron*, the court asks: (1) whether Congress has unambiguously expressed its intent with regard to the issue, and if not, (2) whether the agency's interpretation is a permissible construction of the statute. Judicial deference to an agency interpretation is appropriate only if Congressional intent is unclear and if the agency's interpretation is permissible.[35]

Using the *Chevron* framework, FDA argued that the term "article" is unambiguous and can include either a finished product, such as Mevacor, or any of its components, such as lovastatin. As support, FDA pointed to the use of that term to refer to either a finished product or its components in the definition of "food" in the FFDCA, among other things.

Section 321(f) of the FFDCA provides that "food" means "articles used for food or drink" and "articles used for components of any such article."[36] According to FDA, because Congress' use of the term "article" was unambiguous, there was no need for the court to proceed to the second step of the *Chevron* test.[37]

Unlike FDA, Pharmanex focused on the entire exclusionary clause "an article that is approved as a new drug," rather than the term "article" itself. Pharmanex argued that the entire phrase is unambiguous and that only a finished drug product can be approved as a "new drug." As support for that proposition, Pharmanex cited various FDA regulations, court decisions, an FDA determination, and a different provision in the FFDCA. Based on Pharmanex's interpretation, the exclusionary clause could not apply to Cholestin because it had not been approved as a finished drug product.[38]

In response, FDA contended that if Congress had intended for the term "article" to apply only to finished drug products, it would have used the term "product" rather than "article."[39] The court, however, rejected that argument, reasoning that a Senate report prior to passage of DSHEA suggested that the exclusionary clause was added to the definition of "dietary supplement" to prevent manufacturers and importers of drugs from avoiding the drug approval process by selling their drugs as "dietary supplements."[40] According to the court, in addressing this issue, ". . . Congress was not writing on a blank slate," and its use of the phrase "article approved as a new drug" was a reference to "previously defined subject matter, namely, drugs," which uses the term "article" in its definition.[41]

Agreeing with Pharmanex, the district court found that the phrase "an article that is approved as a new drug" can apply only to finished products and not to individual components. Accordingly, the district court concluded that only Cholestin could be the relevant "article" for the purposes of the exclusionary clause. Given that Cholestin had never been approved as a "new drug," as required by the first criterion of the exclusionary clause, the district court set aside FDA's decision.[42]

The Tenth Circuit's Decision: Pharmanex II

The FDA appealed the district court's decision to the U.S. Court of Appeals for the Tenth Circuit. On July 21, 2000, the Tenth Circuit reversed the lower court's decision, holding that the phrase, "an article that is approved as a new drug," could apply either to a finished product or to a component.[43] In reaching that decision, the Tenth Circuit rejected Pharmanex's contention that the term "article" as used in the phrase "an article that is approved as a new drug" is unambiguous. The court decided that the term was ambiguous because other provisions of the FFDCA used the term "article" to refer to either finished drug products or to their individual components. Moreover, Congress could have used the term "product" or "active ingredient" to clarify its intent—but did not do so.

The Tenth Circuit also criticized the district court's reliance on the Senate report in finding that Congress clearly intended the exclusionary clause to apply only to finished drug products. The Tenth Circuit noted that the Senate report is irrelevant because the Statement of Agreement for DSHEA[44] explicitly disclaimed all other DSHEA legislative history as a source of legislative intent, and was itself silent regarding the application of the exclusionary clause.[45]

In addition, the Tenth Circuit rejected the argument that the policy underlying DSHEA—"to alleviate the regulatory burdens on the dietary supplement industry, allowing consumers greater access to safe dietary supplements in order to promote greater wellness among the American population"[46]—supports the argument that the term "article" in the exclusionary clause refers only to finished drug products. The court reasoned that the exclusionary clause actually operates as a "limiting principle" to that policy, in accordance with the clause's purpose of excluding products from the definition of "dietary supplement."[47] The court also opined that permitting a "dietary supplement" company to market a naturally occurring substance that has the same pharmacological effect as the active ingredient in a drug as a "dietary supplement" would: (1) obviate the practical application of the exclusionary clause, and (2) permit companies to market products that are essentially drug products without first proving that the products are safe and effective for their intended use.[48]

Having determined that the use of the term "article" in the exclusionary clause is ambiguous, the court went on to conclude that FDA's interpretation of that term is permissible, thus warranting *Chevron* deference.[49] The Tenth Circuit rejected Pharmanex's argument that FDA's interpretation would produce absurd results by requiring all traditional food products that contain a small amount of an active drug ingredient to be regulated as drugs rather than foods.[50] The court reasoned that even under FDA's interpretation, food products containing active ingredients of "new drugs" would be protected from drug regulation if they had a prior history of marketing as a "dietary supplement" or food. Accordingly, the Tenth Circuit held that the term "article," as used in the exclusionary clause, could be interpreted by FDA as including "components" as well as "finished drug products."[51]

The Tenth Circuit remanded the case to the district court to determine whether, in light of this definition, Cholestin is a "dietary supplement" or a "new drug."[52] In doing so, the court noted that the district court had not reached the following issues in *Pharmanex I*: (1) whether Pharmanex, in manufacturing and marketing Cholestin, was actually manufacturing and marketing lovastatin, (2) whether FDA's finding that lovastatin had not been "marketed as a dietary supplement or as a food" prior to its approval as a "new drug" is arbitrary and capricious, (3) whether the manufacturing practices for a supplement are relevant to Section 321(ff)(3), and (4) whether FDA adequately explained its departure from its prior interpretation that approval of a "new drug" is an approval only of a product, not of an active ingredient.[53]

The District Court's Decision on Remand: Pharmanex III

On remand, the district court reversed its earlier determination that Cholestin is a "dietary supplement" product. The court focused on two issues: (1) whether FDA was arbitrary and capricious in finding that Pharmanex, in manufacturing and marketing Cholestin, was actually manufacturing and marketing lovastatin, and (2) whether FDA was arbitrary and capricious in finding that lovastatin had "not been marketed as a dietary supplement or as a food," within the meaning of the "prior market clause," prior to its approval as a "new drug."[54]

As it argued in its initial submission to FDA and in *Pharmanex I*, Pharmanex contended that the relevant "article" in Cholestin for the purpose of the exclusionary clause was red yeast rice, not lovastatin, and that, as such, marketing Cholestin as a "dietary supplement" was lawful and consistent with DSHEA. The court noted that Pharmanex, in making that argument, "skillfully raised a variety of arguments as to why FDA's determination that Pharmanex [was actually] manufacturing and marketing lovastatin was not supported by the record."[55] However, FDA cited other evidence in the record that in the agency's view suggested that: (1) Pharmanex was in fact manufacturing red yeast rice to maximize the lovastatin content (for example, Pharmanex's use of a strain of red yeast rice that contains lovastatin, Pharmanex's use of temperature controls, and Pharmanex's tracking of HMG-CoA inhibitor levels), and (2) Pharmanex, in marketing Cholestin, was actually marketing lovastatin.[56] Ultimately, the court agreed with FDA, reasoning that "there [was] evidence in the record that a reasonable mind might accept as adequate to support FDA's conclusion."[57]

Having deferred to FDA's determination that lovastatin is the relevant "article" for the purposes of the exclusionary clause, the court turned to the issue of whether FDA's decision that lovastatin did not meet the "prior market clause" was in error.[58] Pharmanex and FDA disagreed on two critical points. Pharmanex disagreed with FDA's assertion that a food constituent must have been actively heightened, or actively touted, prior to the date the constituent was approved as a "new drug" for the "prior market clause" to have been met. According to Pharmanex, the plain language of the "prior market clause" merely requires that the article at issue be "marketed as"—that is, "sold as" or present in—a food or "dietary supplement" prior to the date the article was approved. Moreover, Pharmanex noted that FDA's past interpretations of the term "marketed as" were inconsistent with its interpretation of that term in this case.[59]

Pharmanex also disagreed with FDA as to whether lovastatin was actually present in food prior to 1987—the date that lovastatin was approved as a "new drug." Pharmanex asserted that lovastatin was present in red yeast rice and oyster mushrooms, as well as other foods that were marketed in the United States, prior to 1987. Therefore, according to Pharmanex, FDA's finding that lovastatin was not present in foods prior to 1987 was arbitrary and capricious. Notably, FDA's finding was based solely on the fact that certain tests in the record indicated that: (1) traditional red yeast rice sold currently, and sold prior to 1987, did not contain lovastatin, and (2) Cholestin did not contain traditional red yeast rice.[60]

The court ultimately found in FDA's favor and noted that the term "marketed as" is ambiguous and FDA's interpretation of that term to require more than mere presence of a constituent in a food "is rational and consistent with the statute."[61]

The court also rejected Pharmanex's factual argument regarding the presence of lovastatin in certain food products prior to 1987, because FDA's determinations did find support in the administrative record.[62] Therefore, the court affirmed FDA's conclusion that lovastatin did not meet the "prior market clause," and ultimately held that Cholestin was a "new drug," rather than a "dietary supplement."[63]

Cholestin Case Implications

The decisions in the Cholestin case provide important insights for the "dietary supplement" and pharmaceutical industries in developing and investigating new ingredients. The decisions clarify that under the exclusionary clause, even if a product is labeled as a "dietary supplement" and meets all of the other "dietary supplement" criteria, it is not a "dietary supplement" if: (1) it contains an "article" approved as a new or investigational drug, and (2) the "article" does not meet the "prior market clause" (*i.e.*, the article was not marketed as a food or "dietary supplement" prior to the approval or investigation). As explained in detail above, the Cholestin case resolved two issues of first impression regarding the appropriate interpretations of the term "article" and of the "prior market clause." The implications of these interpretations are the true legacy of the Cholestin case.

Interpretation of the Term "Article"

Under the first prong of the exclusionary clause, a product might not be a "dietary supplement" if it contains an "article" approved as a new or investigational drug.[64] After the Cholestin case, it has become clear that:

- (1) Either an entire product, or individual components of a product, may be deemed "an article that is approved as a new drug" or an article "authorized for investigation as a new drug" within the meaning of the exclusionary clause.

  In the Cholestin case, it was ultimately found that the lovastatin component of Cholestin, rather than the entire Cholestin product, was the relevant "article" for purposes of the exclusionary clause.

- (2) If a manufacturer of a "dietary supplement" wants the product rather than an individual constituent or component to be considered the relevant "article" for the purposes of the exclusionary clause, the manufacturer can increase the likelihood of this outcome by not using manufacturing processes that are designed to heighten the concentration of an individual ingredient. Furthermore, the manufacturer should not promote the "dietary supplement" with reference to a particular component or the properties thereof.

  In the Cholestin case, FDA found that lovastatin, rather than red yeast rice, was the relevant "article" for the purposes of the exclusionary clause because it concluded that Pharmanex was marketing, or touting, the product for the properties of lovastatin, and manufacturing the product to heighten the concentration of lovastatin in the red yeast rice.[65] If a company did not highlight or manufacture a product to maximize the content of a specific constituent, the company may, in light of these decisions, have a stronger argument that the relevant "article" is the product itself and not an individual constituent or component.

Interpretation of the "Prior Market Clause"

Under the exclusionary clause, even if the "article" at issue is approved as a "new drug" or has been studied as a "new drug" (that is, has gained recognition in the marketplace as a "new drug" or an "investigational new drug"), it is still appropriately marketed as a "dietary supplement" if it complies with the "prior market clause." The Utah district court, in *Pharmanex III*, deferred to FDA's interpretation of the "prior market clause" as requiring more than the mere presence of lovastatin in red yeast rice, oyster mushrooms, or other foods prior to the approval of lovastatin as a "new drug."

Accordingly, based upon the Cholestin case, it is now clear that FDA currently takes the position that:

- (1) The mere presence of a food component in the food supply prior to the approval of that component as a "new drug" or an "investigational new drug" is not enough to bring it within the scope of the "prior market clause."
- (2) In order for a food component to be considered "marketed as a dietary supplement or as a food," prior to the component's approval as a "new drug" or an "investigational

new drug," the manufacturer of the food must have promoted the food based upon the properties of the food component at issue, or in some way increased the concentration of the component at issue, prior to the component's approval as a "new drug."

The interpretation of the "prior market clause" in the Cholestin case may also have an effect on subsequent interpretations of the definition of the term, "new dietary ingredient" in Section 413 of the FFDCA.[66] "New dietary ingredient" is defined as a "dietary ingredient that was not marketed in the United States before October 15, 1994 and does not include any dietary ingredient which was marketed in the United States before October 15, 1994."[67] Given its position in the Cholestin case, FDA is likely to interpret the term "marketed" as requiring a dietary ingredient to have been somehow emphasized in the marketing or manufacturing of a product prior to October 15, 1994, to avoid designation as a "new dietary ingredient." Furthermore, some "new dietary ingredient" submissions have resulted in FDA taking the position that the ingredients were not permissible for marketing in "dietary supplement" products based upon the exclusionary clause of DSHEA (that is, the ingredients were already under investigation as "investigational new drugs" or were already approved as "new drugs" prior to the "new dietary ingredient" submission).

Conclusion

In resolving issues of first impression regarding the appropriate interpretation of the exclusionary clause in the "dietary supplement" definition in the FFDCA, the Cholestin case teaches that manufacturers of "dietary supplements" and research-based pharmaceutical companies alike should conduct reconnaissance before developing, manufacturing, or marketing drug or "dietary supplement" products.

The best way for a "dietary supplement" manufacturer to avoid having a product fall within the exclusionary clause of the "dietary supplement" definition is to ensure that all constituents present in the product have not previously been approved as "new drugs" by the FDA or been subject to drug investigations. In the event that a natural constituent of a product has been approved as a "new drug" or for investigational use by the FDA, and the "dietary supplement" company intends to argue that the relevant "article" is the product, and not that constituent, the company should not: (1) use manufacturing processes that are designed to heighten the concentration of that constituent, or (2) promote the "dietary supplement" with reference to the particular constituent or the properties thereof. Finally, under the safe harbor of the "prior market" clause, if the "dietary supplement" company can prove that the applicable constituent was "marketed" as a "dietary supplement" or food prior to drug approval or drug investigation, the product should still be characterized by the FDA as a supplement.

For research-based pharmaceutical companies that are seeking to develop drug products and initiate clinical trials, it is imperative to first check whether the drug ingredient was previously "marketed" as a "dietary supplement" or food ingredient. If a pharmaceutical company can demonstrate that its active ingredient was not in a food or "dietary supplement" that was manufactured or touted in a way that emphasized that ingredient prior to the date its ingredient was recognized as a new drug or "investigational new drug," the pharmaceutical company may be able to block "dietary supplement" companies from emerging as potential competitors. One should recall, however, that under FDA's interpretation, the mere presence of an ingredient in the food supply is not sufficient to constitute "marketing." Rather, according to FDA, the ingredient must have been promoted in some way for food-related use. If such promotion or marketing occurred, however, a pharmaceutical company may want to reconsider its drug development efforts, because a "dietary supplement" company could include the ingredient in a supplement product—and would not need to invest the millions of dollars necessary to obtain drug approval.

Notes

1. Mr. Kracov and Mr. Rubin are partners, and Ms. Dwyer an associate, in the FDA Practice Group in the Washington, D.C., office of Patton Boggs LLP. They may be reached at dkracov@pattonboggs.com, prubin@pattonboggs.com, and ldwyer@pattonboggs.com. Patton Boggs LLP represented Pharmanex throughout the proceedings described in this chapter.
2. Pharmanex, Inc. v. Shalala, 2001 WL 741419 (D. Utah March 30, 2001) ("Pharmanex III").
3. 21 U.S.C. § 321 et seq. (Supp. 2002).
4. Id. § 321(ff)(3)(B)(Supp. 2002)(emphasis added).
5. See Pharmanex, Inc. v. Shalala, ("Pharmanex I"), 35 F. Supp.2d 1341 (D. Utah, 1999); Pharmanex, Inc. v. Shalala, ("Pharmanex II"), 221 F. 3d 1151, 1159–1160 (10th Cir. 2000).
6. Pharmanex II, 221 F.3d at 1159–1160.
7. Pub. L. No. 103-417, 108 Stat. 4325 (1994).
8. See Pub. L. 103-417, § 2, 108 Stat. 4325–26 (Oct. 25, 1994); see also Stuart M. Pape, Daniel A. Kracov, Paul D. Rubin, Dietary Supplements and Functional Foods: A Practical Guide to FDA Regulation, Thompson Publishing Group (2001).
9. DSHEA, Pub. L. 103-417, § 2(13), 108 Stat. 4325–4326 (Oct. 25, 1994), codified at 21 U.S.C. § 321, note.
10. The FFDCA defines "drug" as "(A) articles recognized in the official United States Pharmacopoeia, official Homeopathic Pharmacopoeia of the United States, or official National Formulary, or any supplement of them; and (B) articles intended for use in the diagnosis, cure, mitigation, treatment, or prevention of diseases in man or other animals; and (C) articles (other than food) intended to affect the structure or any function of the body of man or other animals; and (D) articles intended for use as a component of an article specified in clauses (A), (B), or (C) of this paragraph. A food or dietary supplement for which [a health claim permitted under the FFDCA is made] is not a drug solely because the label or labeling contains such a claim . . . ." 21 U.S.C. § 321(g)(1) (Supp. 2002).
11. See id. § 355(a) (requiring pre-market approval for "new drugs").
12. See id. § 350b.
13. See id. § 342(f).
14. DSHEA, Pub. L. 103-417, § 3, 108 Stat. 4327 (Oct. 25, 1994), codified at 21 U.S.C. § 321(ff) (Supp. 2002).
15. 21 U.S.C. § 321(ff) (Supp. 2002).
16. FDA's Letter Determination for Cholestin ("FDA Letter Determination") (quoting Letter from Orrin Hatch, United States Senator, to Dr. Michael A. Friedman, M.D., Lead Deputy Commissioner, FDA (Dec. 22, 1997)).
17. Pharmanex II, 221 F.3d at 1151.
18. Pharmanex I, 35 F.Supp.2d at 1341.
19. FDA Letter Determination, at 7–10.
20. Pharmanex II, 221 F.3d at 1151, 1153.
21. FDA Letter Determination, at 1–3.
22. See generally, FDA Letter Determination.
23. See id. at 5–8, 10–12.
24. See id. at 5–8.
25. See id. at 10–22.
26. See id. at 10–13.
27. See id.
28. Pharmanex I, 35 F. Supp.2d at 1346.
29. See id.
30. See FDA Letter Determination, at 3–4, 22–27.
31. See id. at 21 (conceding that at least three samples of red yeast rice tested by FDA contained lovastatin).
32. 5 U.S.C. § 706(2)(A), (C) (Supp. 2002); see Pharmanex I, 35 F. Supp.2d at 1342.
33. See Pharmanex I, 35 F. Supp. 2d at 1341.

34. Chevron U.S.A. v. NRDC, 467 U.S. 837 (1984).
35. See id.
36. 21 U.S.C. § 321(f) (Supp. 2002).
37. Pharmanex I, 35 F. Supp.2d at 1346.
38. See id. at 1346–47.
39. See id. at 1347.
40. See id. at 1348 (citing S. Rep. No. 103-410, at 20 (1994)).
41. See id.
42. See Pharmanex I, 35 F.Supp.2d at 1349.
43. See Pharmanex II, 221 F.3d at 1159–60.
44. Statement of Agreement, 140 Cong. Rec. S14801 (Oct. 7, 1994), reprinted in 1994 U.S.S.C.A.N. 3523 ("It is the intent of the chief sponsors of the bill . . . that no other reports or statements be considered as legislative history for the bill.").
45. See Pharmanex II, 221 F.3d at 1158.
46. Id. at 1158–59.
47. See id.
48. See id. at 1159.
49. See id. at 1160.
50. See id.
51. See id. at 1159–60.
52. See id. at 1153, 1160.
53. See id.
54. Pharmanex III, 2001 WL 741419, *1 (D. Utah March 30, 2001).
55. See id. at *3.
56. See id.
57. See id. (citing Olenhouse v. Commodity Credit Corp., 42 F.3d 1560, 1575 (10th Cir. 1994)).
58. See id. at *4.
59. See id.
60. See id. at *4.
61. See id. (footnote 5).
62. See id.
63. See id. at *5.
64. See id.
65. See at FDA Letter Determination, at 10–22.
66. 21 U.S.C. § 350b (Supp. 2002).
67. Id.

10 The Role of the Federal Trade Commission in the Marketing of "Functional Foods"

Lesley Fair[1]

Introduction

Marketers of food, over-the-counter drugs, and dietary supplements are usually well acquainted with the regulatory framework of the Food and Drug Administration (FDA). They may be less familiar with the law enforcement mission of the Federal Trade Commission (FTC) and how the truth-in-advertising laws enforced by the FTC apply to the marketing of functional foods[2] and other products advertised to provide health benefits to consumers.

The FTC's mission is to ensure that the nation's markets function vigorously and competitively. Established in 1914, the FTC is an independent agency whose five members—no more than three of whom may be members of the same political party—are appointed by the President for staggered seven-year terms. The FTC's two primary law enforcement divisions perform complementary functions to fulfill this mission. The Bureau of Competition enforces antitrust statutes outlawing anticompetitive behavior, such as price fixing and monopoly. The Bureau of Consumer Protection protects the buying public from fraud and deception by enforcing federal truth-in-advertising laws. A third operating division, the Bureau of Economics, evaluates the economic impact of proposed Commission actions.

From the FTC's perspective, truthful advertising benefits consumers by encouraging a healthy rivalry among competitors. By ensuring that marketers may truthfully advertise any aspect of their product that they wish—superior health benefits, preferred ingredients, lower prices, and so on—the FTC's regulatory framework fosters innovation and maximizes consumer choice. However, when marketers harm consumers and competitors by advertising deceptively, the FTC's role is to protect consumers and restore fair competition through vigorous law enforcement.

Section 5 of the FTC Act outlaws "unfair or deceptive acts or practices in or affecting commerce."[3] Section 12 specifically prohibits false advertisement likely to induce the purchase of food, drugs, devices, or cosmetics.[4] The statute defines an "unfair" practice as one that causes or is likely to cause "substantial injury to consumers which is not reasonably avoidable by consumers themselves and not outweighed by countervailing benefits to consumers or to competition."[5] Although the Commission has brought a number of cases over the years pursuant to its unfairness jurisdiction, it allocates the majority of its law enforcement resources to challenging deceptive practices, especially in the marketing of health-related products such as food, drugs, and dietary supplements. Therefore, marketers of functional foods need to pay careful attention to: (1) how the FTC coordinates its law enforcement efforts with the FDA; (2) how it evaluates claims in advertising; (3) the FTC's requirement that marketers substantiate all objective product representations, including health claims for functional foods, that reasonable consumers take from their ads; and (4) the potential consequences of a violation of Section 5.

FTC-FDA Law Enforcement Framework

The FTC and the FDA share jurisdiction over the marketing of food, over-the-counter (OTC) drugs, medical devices, and cosmetics. The two agencies work closely to ensure that their enforcement efforts are, to the fullest extent feasible, consistent and efficient. Traditionally, the FTC has relied heavily on the FDA's scientific expertise, and the FDA has looked to the FTC's know-how in marketing and consumer behavior.[6] To coordinate their efforts, a 1971 liaison agreement governs the division of responsibilities between the two agencies.[7] Pursuant to the agreement—which reflects a voluntary allocation of resources, as opposed to a division of legal jurisdiction—the FDA has primary responsibility for the labeling of food, OTC drugs, devices, and cosmetics, while the FTC has primary responsibility for the advertising of those products. The one exception to this general division of labor is prescription drugs, over which the FDA exercises primary authority for labeling and advertising.[8]

For marketers accustomed to the FDA's regulatory approach, the FTC's law enforcement framework may seem curious indeed. A review of the Code of Federal Regulations yields no specific FTC rules on the advertising of health-related products. Unlike the FDA, the FTC does not approve products or preclear advertising claims. Generally speaking, whether a product is classified by the FDA as a food, an OTC drug, or a dietary supplement is immaterial to the FTC's legal analysis. From the FTC's perspective, the primary concern is that advertising claims are substantiated by competent and reliable scientific evidence, regardless of the FDA's legal classification of the product or the nature of the representation.

That is not to say that FDA's expertise is irrelevant to the FTC. On the contrary, the Commission accords substantial weight to the FDA's scientific determinations and has specifically held that FDA final monographs may form a reasonable basis for health-related advertising claims.[9] In appropriate circumstances, the FTC has approved "safe harbor" provisions expressly permitting companies to make claims in advertising that the FDA has approved under a final monograph or a new drug application. For example, in settling a number of cases challenging deceptive health claims, Commission orders expressly provide that "Nothing in this order shall prohibit respondents from making any representation for any drug that is permitted in labeling for such drug under any tentative final or final standard promulgated by the Food and Drug Administration, or under any new drug application approved by the Food and Drug Administration."[10] Similarly, the FTC has made clear that order provisions shall not be construed to prohibit advertisers from "making any representation for any product that is specifically permitted in labeling for such product by regulations promulgated by the Food and Drug Administration pursuant to the Nutrition Labeling and Education Act of 1990."[11] The FTC has also modified earlier orders based on more recent monographs.[12] Conversely, marketers who disseminate representations expressly disallowed by an FDA monograph can expect their ads to face careful FTC scrutiny.

Recent years have seen substantial change in the FDA's approach to health claims made for food, most notably through the passage of the Nutrition Labeling and Education Act of 1990 (NLEA).[13] Soon after the FDA issued regulations implementing the NLEA, the FTC promulgated its Enforcement Policy Statement on Food Advertising ("Food Policy Statement").[14] An overarching theme of the Food Policy Statement is the FTC's recognition of the importance of consistent treatment of health claims in advertising and on labels and its goal of harmonizing its approach with the FDA's to the extent feasible under the FTC Act. This policy serves two important interests: (1) making it easier for consumers to make informed purchase decisions by applying one consistent set of descriptors to labels and ads; and (2) simplifying compliance for food marketers who must follow both FDA labeling standards and FTC advertising standards.

Since the promulgation of the Food Policy Statement, the FTC has been consistent in applying to advertising the definitions of nutrient content descriptors promulgated by the FDA for labeling. For example, in The Isaly Klondike Company,[15] ads for Klondike Lite ice cream bars stated, "If you don't believe that something lite can taste delicious, then try new Klondike Lite. It's 93% fat-free. Low in cholesterol." The FTC challenged four misleading health-related claims. It alleged that the "93% fat-free" claim was deceptive because the product actually contained at least 14 percent fat by weight when the entire bar, including the chocolate coating, was considered. It challenged the low-fat claim because the bar had ten grams of fat. According to the FTC's complaint, the claims that the Klondike Lite bar has significantly less fat and fewer calories than the regular Klondike bar was deceptive because the Lite was smaller than the regular bar and on a per-weight basis did not have significantly less fat or fewer calories. The FTC also challenged the claim that the Klondike Lite bar will cause little or no increase in cholesterol levels, because the bar contained a substantial amount of saturated fat and would in many cases cause a substantial increase in cholesterol. Central to each of the allegations was the fact that the marketer had used in its advertising nutrient descriptors that would have violated FDA regulations had they appeared on the product label.[16]

The Food Policy Statement left open the possibility that in certain limited instances, the FTC may allow in advertising a carefully qualified claim that has not been authorized by the FDA for labeling. However, the Commission has made clear that such a claim must be supported by strong scientific evidence and that advertisers must take the utmost care in the wording and presentation of such a claim.

The FDA experienced another major change to its enforcement regimen with the passage of the Dietary Supplement Health and Education Act of 1994 (DSHEA).[17] DSHEA modified the FDA's role in regulating dietary supplements, establishing separate frameworks for "structure/function" claims, which are permitted without prior FDA approval, and "health" claims, which must be submitted for FDA approval. Some marketers mistakenly assumed that the passage of DSHEA created a free-for-all in how dietary supplements could be advertised. Others reached the erroneous conclusion that DSHEA's "structure/function" versus health claim distinction modified the kind of substantiation necessary to support claims under Section 5 of the FTC Act. However, DSHEA in no way altered the FTC's approach to truth in advertising, a conclusion emphasized in the Commission's 1998 staff publication, *Dietary Supplements: An Advertising Guide for Industry* ("*Dietary Supplement Guide*").[18]

As the *Dietary Supplement Guide* makes clear, the marketers of dietary supplements remain free—as do the marketers of food and functional foods—to highlight in their advertising virtually any truthful and nondeceptive health claim that they wish as long as they possess what the FTC considers to be "competent and reliable scientific evidence" to support their representations. Therefore, aside from any requirements imposed by FDA laws or regulations, companies contemplating the introduction of a functional food into the marketplace must ask themselves two critical questions necessary for compliance with Section 5 of the FTC Act: (1) what claims—express **and** implied—does the ad convey to reasonable consumers?; and (2) does the advertiser have the requisite level of substantiation necessary to support each claim?

When Is an Advertisement Deceptive under the FTC Act?

Over the years, legal decisions from the Commission and federal courts have added flesh to the FTC Act's prohibition of "unfair or deceptive acts or practices." The FTC's 1983 Deception Policy Statement is the seminal document that forms the foundation of the agency's approach to advertising law.[19] According to the Deception Policy Statement, in determining whether a claim is deceptive under Section 5, the Commission applies a three-part test. **First**, there must be a representation or an omission of information that is likely to mislead the consumer. **Second**, the Commission evaluates the advertisement from the point of view of a consumer acting reasonably in the circumstances. **Third**, the representation or omission must be "material," meaning that it would likely affect the consumer's conduct or decision toward a product. The Commission uses this three-part test to evaluate claims for all products and applies the test regardless of the media in which advertisers disseminate their claims. Thus, communicating to consumers in virtually any fashion, including disseminating statements on Internet Web sites, triggers the provisions of the FTC Act.[20]

The first step in a deception analysis is to identify the representations made by an advertisement. To determine the claims an ad conveys, the FTC looks at an ad from the point of view of a reasonable consumer. The Commission examines "the entire mosaic, rather than each tile separately" to determine the net impression consumers take from the ad.[21] Although evidence of the claims that an advertiser *intended* to convey may be relevant to a consideration of materiality, it is irrelevant to the issue of consumer perception. What matters are the claims consumers took from the ad, regardless of whether it was the advertiser's intention to convey that information.

Furthermore, advertisers are required to substantiate not only all express claims that reasonable consumers take from their ads, but all implied claims as well. An express claim makes a representation directly. The identification of an implied claim requires an examination of the entire transaction in context, including the advertising copy, the product name, all depictions and imagery, the disclosure of any qualifying information, and the juxtaposition of phrases and graphics.[22] As one federal appellate court explained the distinction, it would be an *express* representation for an advertiser to claim in an ad that a certain car gets 30 miles per gallon. It would be an *implied* representation to call the car "The Miser" and include in its commercials a depiction of the car passing gas stations and a claim that it is inexpensive to operate.[23] Although the second ad never specifically mentioned anything about gas mileage, the court held that it would be reasonable for a consumer to conclude from the statements and images that the car gets good mileage. Therefore, even though the first representation is made expressly and the second by inference, both would convey to consumers a representation that the car gets good gas mileage. If, in fact, the car gets only ten miles a gallon, the ad would violate the FTC Act regardless of whether the claim was conveyed expressly or by implication.

Applying these principles in the context of functional foods, suppose that an advertisement claims that 80 percent of cardiologists regularly drink redfruit juice. Certainly the advertiser must possess a methodologically sound survey demonstrating that express claim.[24] However, it is also likely that the FTC will construe the claim to imply that redfruit juice has some proven health benefit for the heart. To comply with Section 5, the advertiser would have to substantiate both the express and the implied representation.[25]

The FTC may rely on its own administrative expertise in determining the express claims that reasonable consumers take from an ad. For implied claims, the Commission may also rely on its own reasoned analysis to determine the claims that are conveyed to consumers, as long as those claims are reasonably clear on the ad's face.[26] However, if a review of the ad is insufficient to allow the FTC to conclude with confidence that a particular claim is conveyed, it may look to extrinsic evidence, including copy testing, focus groups, consumer surveys, and expert testimony, for assistance in making that determination. For such extrinsic evidence to be relevant, it must use methodologically sound procedures accepted by experts in the field to produce accurate and reproducible results.[27]

The Commission's approach takes into account that reasonable consumers can take multiple meanings from the same ad. A consumer's interpretation may still be reasonable if it was not what the advertiser meant to convey or even if the interpretation is not shared by a majority of consumers.[28] If an ad is subject to multiple reasonable interpretations, one of which is deceptive and the rest of which are nondeceptive, the advertiser may still be held liable for conveying the deceptive claim.[29] Similarly, under FTC caselaw, even if the wording of an ad is literally truthful, the company may nonetheless have violated Section 5 if the net impression that consumers take from the ad is deceptive.[30]

Marketers accustomed to the FDA's distinction between "structure/function" claims and health claims in the marketing of dietary supplements should be particularly careful in how they phrase advertising representations. An advertiser's intended claim that a product has an effect on the structure or function of the body may convey to consumers that the product is beneficial for the treatment of a disease. If reasonable consumers take such a claim from an ad, the advertiser must be able to substantiate the implied disease claim even if the ad contains no express reference to disease.

For example, suppose that the marketer of XYZ Redfruit Juice advertises that its product boosts the immune system to help maintain a healthy nose and throat during the winter season. The ad features images of one consumer sneezing and coughing and a healthy-looking consumer drinking a glass of juice. The advertiser may argue that its intent was to convey a structure/function claim; however, the FTC is likely to conclude that the various elements of the ad—the depiction of a cold sufferer sneezing and coughing and the reference to nose and throat health during the winter season—convey to consumers the implied representation that the product can prevent colds. Therefore, the advertiser must be able to substantiate that claim even though the ad contains no express claim of cold prevention.[31]

To be actionable, a misleading claim must be material. A misrepresentation is material if it involves information that is important to consumers and is therefore likely to affect their decision to buy or use a product.[32] Proof of actual consumer injury, however, is not required.[33] The FTC considers certain categories of information to be presumptively material, including claims that the advertiser intended to make. In addition, materiality may be presumed when the claims at issue involve health, safety, or other central characteristics with which reasonable consumers would be concerned, such as the purpose, efficacy, or cost of a product or service.[34]

The omission of material information may also be deceptive under the FTC Act. The Commission has made clear that an ad may be deceptive if it fails to disclose information that is necessary to prevent a claim from misleading consumers. For example, in *North American Philips Corporation*, the FTC challenged advertising claims that the Norelco

Clean Water Machine made tap water "clean" or "cleaner" by removing chemicals from the tap water because the company failed to disclose that the product also added to the water a small amount of methylene chloride, a chemical that public health authorities classified as a possible human carcinogen.[35] In *Campbell Soup Company*, the FTC charged that an ad linking the low-fat and low-cholesterol content of Campbell's soups to a reduced risk of heart disease was deceptive because the ad failed to disclose that the soups were high in sodium and that diets high in sodium may increase the risk of heart disease.[36] The fact that the sodium content of the product was accurately disclosed on the product label was not considered a defense to charges of deceptive advertising. However, "[n]ot all omissions are deceptive, even if providing the information would benefit consumers."[37] As with any other advertisement, the legal test for whether an omission is actionable is whether the net impression conveyed to consumers is deceptive.

The FTC's primary purpose in bringing law enforcement actions is not to punish advertisers who violate the law, but rather to protect consumers by deterring companies from making deceptive claims in the future. Therefore, to prevail in a prosecution filed pursuant to Section 5, the FTC need not prove that an advertiser *intended* to deceive consumers, only that the ad was likely to mislead consumers acting reasonably under the circumstances. Although evidence that an advertiser knowingly violated the law may justify a stronger sanction, the ultimate goal is to encourage advertisers to set up internal checks to ensure that questionable claims are red-flagged before they are disseminated. Therefore, proof of scienter—knowledge that an act or practice violated Section 5—is not a necessary element in an FTC prosecution.

Another important factor in determining the claims an ad conveys is the identity of the target market. In general, the FTC reviews ads from the point of view of the reasonable consumer, often defined as the "average listener" or the "general populace"[38] and opposed to the two opposite ends of the spectrum—the particularly savvy and sophisticated consumer or "the ignorant, the unthinking and the credulous."[39] However, when an ad is directed to a specific audience, the FTC will take into account the characteristics of the targeted group. For example, if a company markets a product to vulnerable groups that may require special protection, for example, children, the elderly, or people with serious illnesses, the Commission will consider the extent to which members of groups may be more susceptible to deceptive practices.

If a particular claim is relevant only in certain circumstances or to certain consumers, advertisers must make extra efforts to qualify the claim carefully.[40] Marketers should not underestimate how difficult it can be to convey appropriately qualified information to consumers. According to its Food Policy Statement, the FTC will be "especially vigilant in examining whether qualified claims are presented in a manner that ensures that consumers understand both the extent of the support for the claim and the existence of any significant contrary view within the scientific community. In the absence of adequate qualification, the Commission will find such claims deceptive." Vague qualifying terms—for example, that the product "may" have the claimed benefit or "helps" achieve the claimed benefit—are unlikely to be adequate under Section 5.

For example, suppose that the advertisers of XYZ Redfruit Juice possess substantiation that redfruit juice has been proven effective in treating a certain mineral deficiency that results in fatigue. However, less than 2 percent of the general population suffers from this deficiency. In that circumstance, it would be deceptive for the company to tout the product's antifatigue benefits without adequately disclosing that only the small percentage of people who suffer from this particular deficiency are likely to experience any reduction in fatigue from their use of the product. Nor would a representation that redfruit juice "may" relieve fatigue or will "help" relieve fatigue be sufficient to qualify the claim adequately under Section 5.[41]

Similarly, ads that make either an express or an implied claim of safety should include information about any significant safety risks.[42] In *Panda Herbal International, Inc.*[43] and *ForMor Inc.*,[44] the marketers of dietary supplements containing St. John's Wort claimed that their products had been proven efficacious in the treatment of HIV and AIDS and had no known side effects or contraindications. According to the Commission, the companies not only lacked the substantiation to prove the underlying efficacy claim but also failed to disclose to consumers that the use of St. John's Wort can reduce the effectiveness of many prescription medications, including drugs used to treat HIV and AIDS. The companies agreed to place the following disclosure on future advertising and labels that contain any reference to the product's efficacy, performance, or safety:

> **WARNING:** St. John's Wort can have potentially dangerous interactions with some prescription drugs. Consult your physician before taking St. John's Wort if you are currently taking anticoagulants, oral contraceptives, antidepressants, antiseizure medications, drugs to treat HIV or prevent transplant rejection, or any other prescription drug. This product is not recommended for use if you are or could be pregnant unless a qualified health care provider tells you to use it. The product may not be safe for your developing baby.

However, even in the absence of an express or implied safety representation, the risks posed by some products may be so great as to trigger an affirmative duty to disclose safety concerns about the use of the product.[45] For example, in *FTC v. Western Botanicals*[46] and *FTC v. Christopher Enterprises, Inc.*,[47] marketers of dietary supplements containing comfrey agreed to disclose the following information on all advertising and all labeling:

> **WARNING:** External Use Only. Consuming this product can cause serious liver damage. This product contains comfrey. Comfrey contains pyrrolizidine alkaloids, which may cause serious illness or death. This product should not be taken orally, used as a suppository, or applied to broken skin. For further information contact the Food and Drug Administration: http//vm.cfsan.fda.gov.

Another major mistake made by advertisers is to assume that a fine-print footnote or superscript will be sufficient to qualify the general claim. However, as Commission caselaw has established, if the disclosure of information is necessary to prevent an ad from being deceptive, then the disclosure must be "clear and conspicuous"—meaning prominent enough to be noticed, read, and understood by reasonable consumers.[48] A fine-print disclosure at the bottom of a print ad, a disclaimer buried in a dense block of text, a fleeting superscript on a television, or a vaguely labeled hyperlink on an Internet Web site are not likely to be considered adequate. To ensure that disclosures are effective, marketers should use clear language, avoid small type, place any qualifying information close to the claim being qualified, and avoid making inconsistent statements or adding distracting elements that could undercut, contradict, or divert consumer attention away from the disclosure. Because consumers are likely to be confused by ads that include inconsistent or contradictory information, disclosures need to be both direct and unambiguous to be effective.[49] For television ads, the FTC has given substantial weight to consumer research suggesting that dual modality disclosures—disclosures that simultaneously appear on the screen and are read in voice-over—are most effective for conveying information to consumers.[50]

In the past decade, the FTC has waged a substantial effort against ineffective disclosures in advertising, including fine-print footnotes in ads extolling the health benefits and nutrient content of foods and dietary supplements. For example, in *Häagen-Dazs Company*, the headline of a print ad prominently featuring a picture of Jackie Gleason and Audrey Meadows from the 1950s television series "The Honeymooners" stated, "Why is Häagen-Dazs Frozen Yogurt better than your first true love? Häagen-Dazs is still 98% fat free*."[51] The asterisk referred to a fine-print footnote at the bottom of the ad stating, "*frozen yogurt and sorbet combinations." The Commission's complaint alleged that of the entire multiproduct line of Häagen-Dazs Frozen Yogurt, most contained a substantial amount of fat, vastly more than the "98% fat free claim" in the headline. Only the frozen yogurt and sorbet combinations were 98 percent fat free. The FTC argued that the fine print footnote was insufficient to qualify the net impression that the entire line was 98 percent fat-free.

The Commission reached a similar result in considering the use of a fine-print footnote in *Stouffer Foods Corp*.[52] In that case, the FTC challenged allegedly low sodium or reduced sodium claims for the company's Lean Cuisine line of entrees. The ad stated, "Of all the things we at Stouffer's pack into our 34 Lean Cuisine entrees—the freshest ingredients, the ripest vegetables and the perfect blend of herbs and spices—there are some things we skimp on: Calories. Fat. Sodium. With less than 300 calories, controlled fat and always less than 1 gram of sodium* per entree, we make good sense taste good." In fact, according to the Commission's complaint, the Lean Cuisine entree line contained an average of 850 milligrams of sodium per serving, more than one-third of the recommended maximum daily intake and substantially more than the limits imposed by the FDA for low-sodium or reduced-sodium products. One of the company's defenses was that the ad included a footnote that read "*All Lean Cuisine entrees have been reformulated to contain less than 1 gram (1000 mg.) of sodium." An administrative law judge found, and the Commission agreed, that the footnote was insufficient to qualify the reduced sodium claim conveyed to consumers through the text of the ad.

Substantiating Health Claims under the FTC Act

The touchstone for federal truth-in-advertising enforcement is the requirement of substantiation. According to the FTC's Policy Statement Regarding Advertising Substantiation,[53] advertisers must have substantiation for all objective advertising claims before disseminating claims to consumers. This requirement is based on the premise that an objective claim about a product carries with it an expressed or implied representation that the advertiser possesses a "reasonable basis" for the claim.

The kind of substantiation required under the law will depend on the nature of the claim, not on the nature of the product. Certainly, if an advertiser makes an express representation about the level of substantiation it possesses—for example, "Scientific testing at a major university medical school proves that two eight-ounce servings a day of XYZ Redfruit Juice will reduce cholesterol by an average of 30 points"—the advertiser must, in fact, possess the specified level of evidence. A reference to a scientific study also likely conveys to consumers the implied claim that the study is methodologically sound and that the results are statistically significant.[54] However, a taste claim about the same product—"Start your day off with a smile by enjoying delicious XYZ Redfruit Juice"—would likely be considered puffery, subjective representations for which consumers would not expect the advertiser to possess substantiation.[55]

If the ad makes no express claim about the level of underlying scientific support, the FTC will evaluate the level of requisite substantiation by applying what have come to be known as the "*Pfizer* factors."[56] The FTC's standard for evaluating substantiation is sufficiently flexible to ensure that consumers have access to information about emerging areas of science. At the same time, it is rigorous enough to ensure that consumers can have confidence in the accuracy of health claims made in advertising.

The factors that the Commission will consider include:

1. The type of product: Products related to consumer health or safety will generally require a high level of substantiation.
2. The type of claim: "Credence claims," representations that may be difficult for consumers to assess themselves, will be held to a more exacting standard. Examples of credence claims for a functional food would include representations about cardiovascular health or cancer risk reduction, as opposed to subjective claims about the same product's pleasing taste or convenient packaging.
3. The benefit of a truthful representation and the feasibility of developing substantiation for the claim: These claims are often weighed together to ensure that valuable product information is not withheld from consumers because of the difficulty of developing accurate testing. This does not mean, however, that advertisers are exempt from the legal requirement of substantiation simply because the requisite level of research would be expensive to conduct.
4. The consequences of a false claim: This factor includes a consideration to possible physical injury to consumers by relying on an unsubstantiated claim and foregoing a proven treatment and possible economic injury by spending money on a product that does not provide the advertising therapeutic benefit.
5. The kind of substantiation that experts in the field believe is reasonable: In making this determination, the Commission consults with experts in the field to determine the accepted norms in the relevant field of research. Where there is an existing standard for substantiation developed by a government agency or authoritative body, the FTC accords substantial deference to that standard.

For health-related advertising claims—regardless of whether the product in question is a food, functional food, OTC drug, dietary supplement, cosmetics, or device—the required level of substantiation is typically "competent and reliable scientific evidence," defined in caselaw to mean "tests, analyses, research, studies, or other evidence based on the expertise of professionals in the relevant area, that has been conducted and evaluated in an objective manner by persons qualified to do so, using procedures generally accepted in the profession to yield accurate and reliable results."[57] Given the broad range of potential advertising claims, there is no way for the FTC to promulgate a universal testing methodology that will be sufficient to substantiate all health-related representations. However, the Commission's *Dietary Supplement Guide* outlines a number of considerations to assist an advertiser in assessing the adequacy of the scientific support for a specific advertising claim. These include the amount and type of evidence, the quality of evidence, the totality of evidence, and the relevance of the evidence to the specific representation.

The Amount and Type of Evidence

If the ad includes no specific reference to the amount or type of evidence the advertiser possesses, the kind of proof required by the FTC will depend on the nature of the claim. In recent years the FTC has given great weight to the final *Pfizer* factor—the kind of substantiation that experts in the relevant field believe would be necessary to support the representation—and has applied the accepted principle that well-controlled human clinical studies are the most reliable method of substantiating health claims.

For certain health claims, the FTC has defined "competent and reliable scientific evidence" to require "at least two adequate and well-controlled, double-blinded clinical studies that conform to acceptable designs and protocols and are conducted by different persons, independently of each other."[58] The Commission's rationale behind the requirement of multiple studies is that the replication of research results adds to the weight of the evidence. In other cases, depending on the nature of the claim and the opinion of experts in the relevant field, the Commission has allowed the somewhat more flexible standard of "adequate and well-controlled clinical testing."[59] Depending on the circumstances, the FTC may also consider results of animal and *in vitro* studies, particularly where they are widely considered to be acceptable substitutes for human research or where human research is infeasible. In addition, epidemiologic evidence may be relevant, especially when supported by other evidence, such as research explaining the biological mechanism underlying the claimed effect.[60]

The FTC and courts have made clear, however, that anecdotal evidence about the individual experience of consumers is not sufficient to substantiate a health claim.[61] As the United States Court of Appeals for the Ninth Circuit held in *Simeon Management Corp. v. FTC*, "Anecdotal evidence, such as testimonials by satisfied patients or statements by doctors that, based on their experience, they 'believe' a drug is effective do not constitute adequate and well-controlled investigations and cannot, therefore, provide substantial evidence of effectiveness." The Ninth Circuit later held in *FTC v. Pantron I Corporation*[62] that consumer satisfaction surveys are insufficient to meet the "competent and reliable scientific evidence" standard.[63]

For example, suppose that the marketer of XYZ Redfruit Juice uses in its advertising a testimonial from a consumer stating, "Since adding XYZ Redfruit Juice to my diet, my cholesterol is down 23 points." Even assuming that the testimonial accurately reflects the consumer's experience, the representation would violate Section 5 if the marketer did not possess competent and reliable scientific evidence to support the cholesterol reduction claim.

The FTC's *Guides Concerning Use of Endorsements and Testimonials in Advertising*[64] offer marketers substantial counsel in the use of endorsements by consumers, celebrities, professional organizations, and experts. The most important lesson for advertisers is that they may not make a health claim through the use of an endorsement that they cannot otherwise substantiate with the requisite level of competent and reliable scientific evidence. It is not enough that the testimonial represents the honest opinion of the endorser. Under FTC law, advertisers must also have appropriate scientific evidence to back up the underlying claim.

The Quality of Evidence

In addition to the amount and type of evidence, the FTC will also examine the internal validity of each piece of evidence relied upon as substantiation. Study design, implementation, and methodology and the statistical significance of test results are all important in assessing the adequacy of the substantiation. Although advertisers have substantial flexibility in how they choose to substantiate their claims, the FTC has issued guidance on principles generally accepted in the scientific community to enhance the validity of test results. For example, the FTC has made clear that well-controlled studies with blinding of subjects and researchers are likely to yield more reliable results. In addition, the FTC gives greater weight to studies of longer duration and larger size. Evidence of a dose-response relationship or a documented biological or chemical mechanism to explain the effect are also factors that add weight to the findings.[65]

The statistical significance of results is another important consideration. A study that fails to show a statistically significant difference between the test group and the control group may indicate that any perceived benefit is merely the result of the placebo effect or chance. The results should also translate into a meaningful benefit for consumers. Some results that are statistically significant may be so small that they would translate to a *de minimis* benefit to consumer health. The nature and quality of the report of the research may also be relevant. Although the FTC will consider unpublished proprietary research, a peer-reviewed study published in a reputable journal carries a substantial amount of weight. The FTC will also consider studies conducted in foreign countries as long as the design and implementation of the study are scientifically sound.

The Totality of Evidence

Under Section 5, marketers must consider all evidence relevant to the advertised claim, and not just the research that supports the representations they intend to disseminate. Generally speaking, even if a well-designed clinical study supports the claim, the ad may still violate FTC law if other methodologically sound studies show no effect. If anything, inconsistent or conflicting results should raise serious questions in an advertiser's mind about the adequacy of its substantiation. The surrounding body of evidence will have a significant impact both on what type, amount, and quality of evidence is required to substantiate a claim and on how that claim is presented—that is, how carefully the claim is qualified to reflect accurately the strength of the evidence. If a stronger body of surrounding evidence runs contrary to a claimed effect, even a qualified claim is likely to be deceptive.[66]

The Relevance of the Evidence to the Advertised Claim

One mistake that marketers make in the dissemination of health-related representations is failing to ensure that the science "fits" the advertising claim. In some FTC cases, companies have violated Section 5 by relying on consumer testimonials, popular press articles, or other nonscientific material to substantiate health-related claims. However, the more common error is for companies to take a nugget of information from a clinical study of limited application and use it to substantiate a broad, unqualified advertising claim.

Advertisers must take care to ensure that the composition, dosage, and form of administration to the advertised product is the same or substantially similar to the tested product. For example, suppose that sound clinical studies exist to demonstrate that the topical administration of 50 units of a highly concentrated extract of redfruit juice will treat psoriasis. It would likely violate Section 5 for the marketer of XYZ Redfruit Juice to claim that drinking eight ounces of redfruit juice a day will have a beneficial effect for psoriasis sufferers, because the composition, dosage, and form of administration of the tested product is different from the advertised product.

Marketers should also consider whether the study population reflects the characteristics of consumers likely to use the product, if there are any differences between the research conditions and how consumers will actually use the product, and if the advertised product contains additional ingredients that might alter the effect of the ingredient in the study. For example, if a well-controlled study demonstrates that redfruit juice has a beneficial effect on the cognitive processes of healthy adults, it would likely violate Section 5 for a company to run an ad claiming that XYZ Redfruit Juice would treat memory loss due to Alzheimer's disease. Conversely, a study of redfruit juice demonstrating a cognitive benefit for people with Alzheimer's disease would not necessarily support a representation aimed at the general public.[67]

The FTC's settlement in *Schering-Plough Healthcare Products* demonstrates the principle that the substantiation must reflect how consumers will actually use the product. In that case, the FTC challenged the claim that Coppertone Kids sunscreen would offer children sunscreen protection for six hours even during sustained activity in and out of the water.[68] According to the FTC's complaint, the claim was unsubstantiated because none of the company's tests evaluated a single application of the product under the advertised conditions. Similarly, in *Honeywell, Inc.*,[69] the Commission alleged that the company's efficacy claims for an air filtration device were deceptive because the substantiation reflected performance under laboratory conditions rather than in normal household use.

Companies have also run afoul of Section 5 by overstating the health benefits that consumers can expect to receive by using a product. In *Interstate Bakeries Corporation*,[70] ads for Wonder Bread stated, "Moms know it helps build strong bodies. But did you know it helps build strong minds, too? Neurons in your brain need calcium to transmit signals. Without it, they can be, well, a little slow." The ad contained a rather fanciful before-and-after depiction of listless neurons becoming energized after a child eats a slice of Wonder Bread. The FTC's complaint alleged that the ad conveyed the claim that Wonder Bread could help children's minds work better and help them remember things. Although the product did indeed contain calcium, and calcium is an essential nutrient for proper brain function, the Commission charged that the company lacked substantiation for the specific cognition representations conveyed in the ad.

The FTC's settlements in *QVC, Inc.*[71] and *Quigley Corporation*[72] illustrate the importance of ensuring that the substantiation "fits" the advertising claim. In that case, some clinical evidence existed to suggest that zinc taken at the first sign of cold symptoms could reduce the severity and duration of cold symptoms. However, the claim disseminated in the advertising was "If you take these on a preventative basis, you may never get a cold." Obviously, a study demonstrating a reduction in the severity and duration of symptoms cannot be used to support a prevention claim.

As more and more foods are formulated to include dietary supplements, herbs, or other ingredients advertised to provide a health benefit to consumers, marketers must take care to ensure that their claims comport with Section 5. In *Unither Pharma, Inc.*,[73] the FTC challenged claims for food bars and powders enriched with L-Arginine, vitamins, and minerals and marketed under the brand name HeartBar. According to the complaint, Unither failed to have the appropriate level of substantiation to support its claim that HeartBar reverses damage to the heart caused by high cholesterol, smoking, diabetes, or estrogen deficiency; prevents age-related vascular problems, including "hardening of the arteries" and plaque formation; reduces the risk of developing cardiovascular disease; reduces or eliminates the need for surgery or medication in patients with cardiovascular disease; and improves endurance and energy for the general population. The FTC alleged that several of Unither's representations were not supported by any clinical studies on humans. Others were based on studies that suffer from various flaws, including the failure to account for the placebo effect and extremely small sample sizes.

The Consequences of Deceptive Advertising under the FTC Act

When an advertiser is found to have engaged in unfair or deceptive acts or practices, the FTC Act, as interpreted by more than eighty years of precedent, gives the Commission the right to seek a broad range of remedies to provide restitution to aggrieved consumers and to deter future violations. At minimum, the FTC will insist on a cease and desist order that requires the company to stop disseminating the deceptive claims and imposes civil penalties of $11,000 per day per ad if the company is found to have violated the order in the future.[74] However, it is important for advertisers to understand that almost all FTC orders will include some measure of "fencing in" relief—meaning that companies will be held liable for civil penalties not only if they engage in the same form of deception that was the subject of the original law enforcement action, but also if they commit any of a broad range of infractions. As the Supreme Court held in *FTC v. Ruberoid Company*,[75] "If the Commission is to attain the objectives Congress envisioned, it cannot be required to confine its road block to the narrow lane the transgressor has traveled; it must be allowed effectively to close all roads to the prohibited goal, so that its order may not be by-passed with impunity."

In the most serious cases, the FTC imposes an "all products" or "all claims" order. In those instances, any future violation of Section 5 will trigger the civil penalties provision. In other cases, the Commission may limit its order to include only future health-related claims or only claims made for food, drugs, dietary supplements, cosmetics, or devices. Among the factors the FTC will consider in determining the appropriate remedy include the seriousness of the violation, the violator's past record with respect to deceptive practices, and the potential transferability of the illegal practice to other products.[76]

An essential feature of any cease and desist order is a requirement that for the next twenty years, the company must file periodic compliance reports with the FTC about its marketing activities and the steps it has taken to prevent future violations of the FTC Act. It may also require corporate defendants to distribute a copy of the order to current and future employees or require an individual defendant to disclose the existence of the order to any future employers.

In appropriate cases, the Commission may seek full financial restitution directly to consumers. In the past decade, cases involving deceptive health claims have resulted in multimillion dollar redress orders against advertisers.[77] In the most serious cases, courts have required violators to post substantial bonds before continuing to market products to consumers, or have banned violators outright from particular industries.[78]

The Commission has been particularly aggressive in fighting deceptive health claims by mandating informational remedies, including corrective advertising and mandatory disclosures in future advertising and on product labeling. The FTC first won the right to seek corrective advertising in the landmark case of *FTC v. Warner-Lambert*.[79] For decades the marketer of Listerine had advertised the product as effective in preventing the common cold, a claim that the FTC alleged—and the Court agreed—was deceptive. In fashioning a remedy in that case, the Commission was concerned that simply ordering Warner-Lambert not to disseminate the deceptive claim in the future would do nothing to correct the misimpression left with millions of consumers over the decades-long campaign that the product was effective for that purpose. Establishing the use of corrective advertising as a remedy in deceptive advertising cases, the Court held, "If a deceptive advertisement has played a substantial role in creating or reinforcing in the public's mind a false and material belief which lives on after the false advertising ceases, there is clear and continuing injury to competition and to the consuming public as consumers continue to make purchasing decisions based on the false belief. Since this injury cannot be averted by merely requiring respondent to cease disseminating the advertisement, we may appropriately order respondent to take affirmative action designed to terminate the otherwise continuing ill effects of the advertisement."

In *Novartis v. FTC*,[80] the Commission challenged the long-standing advertising campaign that Doan's pills contained an ingredient that was superior to competing products for relieving back pain. Although the ingredient in Doan's was recognized by the FDA as an effective analgesic, the company could not substantiate its representation that it was superior to other analgesics for the treatment of back pain. Upholding the propriety of corrective advertising in the case, the court held that the FTC had carried its burden under *Warner-Lambert* of proving that the ads played a substantial role in reinforcing in the public's mind a false belief about the product that would linger on even after the false advertising ceased.

Informational remedies have been particularly important to the FTC to correct consumer misimpressions about the safety or correct use of health-related products. When the advertiser is able to identify the consumers who purchased the product, the FTC has required the company to contact consumers directly to warn them of the health or safety problem posed by the product.[81] In other cases, the Commission has required disclosures in future ads and on labels to alert consumers to possible risks.[82] For example, the FTC challenged claims made by two dietary supplement companies—MetRx and AST—that its muscle-building products were safe alternatives to anabolic steroids.[83] According to the Commission's complaint, both products contained androstenedione, a steroid hormone that posed safety risks to consumers. To settle the case, the companies agreed to disclose clearly and conspicuously in future advertising and on labeling the following information:

> **WARNING:** This product contains steroid hormones that may cause breast enlargement, testicle shrinkage and infertility in males, and increased facial and body hair, voice deepening, and clitoral enlargement in females. Higher doses may increase these risks. <u>If you are at risk for prostate or breast cancer, you should not use this product.</u>

In a similar fashion, the FTC alleged that Eggland's Best[84] made unsubstantiated cholesterol reduction claims for its eggs. The Commission's settlement required the company to include for one year on all packaging a corrective notice regarding the product's effect on serum cholesterol levels.

The majority of FTC law enforcement actions target the company primarily responsible for advertising and marketing the product. Those companies may be held strictly liable for violations of the FTC Act, meaning that the Commission may find them legally responsible regardless of whether they intended to violate the law.[85] Advertisers should be aware, however, that the reach of the FTC Act is wide. In certain instances, the FTC may take action against not only the corporate entity but also corporate officers in their individual capacity.[86] For others who participate in the dissemination of misleading claims, most notably advertising agencies, the Commission may take action if they "knew or should have known" that the acts in question were illegal.[87] Over the years, the Commissioner has broadened the scope of liability by taking action against infomercial producers,[88] shop-at-home channels,[89] catalog companies,[90] and other marketers who play a role in the dissemination of deceptive representations.

In ads using what the FTC considers to be deceptive expert endorsements, the agency has taken action against both the marketer and the expert endorser. According to the FTC's *Guides Concerning Use of Endorsements and Testimonials in Advertising,* endorsers represented directly or by implication to be experts must have qualifications sufficient to give them the represented expertise.[91] Both the company and the endorser are responsible for ensuring that the endorser conducts an examination or testing of the product that would be generally recognized in the relevant field as sufficient to support the endorsement. Endorsements by groups that consumers would perceive to be expert organizations must meet the same high level of substantiation and must also be reached by a process sufficient to ensure that the endorsement fairly reflects the collective judgment of the organization.[92]

For example, suppose that an ad for XYZ Redfruit Juice contained an endorsement of the product's cholesterol-reducing properties by a person wearing a white coat and referred to as "Dr. Mary Smith." It is likely to be a violation of Section 5 if, in fact, "Dr. Smith" was not a physician but rather possessed a Ph.D. in physical therapy. But even if Dr. Smith were a duly licensed cardiologist, it would likely be a violation of Section 5 if she did not base her recommendation on evidence that cardiologists would generally accept as necessary for supporting a cholesterol reduction claim.[93] If the ad states that XYZ Redfruit Juice carries the seal of approval of the "National Academy of Heart Surgeons," the group must, in fact, be a bona fide association of experts in the field and the seal must be awarded based on independent professional testing or evaluation that reflects the consensus of the group.[94]

In one of the FTC's first infomercial cases, the agency alleged that the expert endorsements of a baldness product by two dermatologists, Patricia Wexler, M.D.,[95] and Steven Victor, M.D.,[96] were not based on the requisite level of competent and reliable scientific evidence. In addition to holding the infomercial company and its corporate officers liable for the deceptive expert endorsements,[97] the FTC sued the dermatologists in their individual capacities. Given the high level of credence consumers give to the opinion of physicians and other scientific experts, the Commission has insisted that markets and professionals exercise appropriate care when using expert endorsements in advertising.[98]

Whenever an expert or consumer endorser is used, the advertiser should disclose any material connection between the endorser and the advertiser of the product.[99] A material connection is one that would affect the weight or credibility of the endorsement—for example, a personal, financial, or similar connection that consumers would not reasonably expect. For example, in *Numex Corporation*[100] the FTC alleged that it was a violation of Section 5 for a company advertising a pain relief device to feature the endorsement of a physician and a former professional athlete without disclosing that they were both officers of the corporation selling the product.

Conclusion

Functional foods offer substantial promise to consumers in the form of health promotion and disease prevention. But those benefits will be realized only if marketers are careful to substantiate their claims in keeping with the FTC's long-standing truth-in-advertising standards. To comply with the law and encourage consumer confidence in this new product market, companies selling functional foods must: (1) carefully draft their advertising claims with particular attention to how claims are qualified and what express and implied messages are actually conveyed to consumers; and (2) carefully evaluate their substantiation for those claims to ensure that it is scientifically sound and relevant to the specific product and claim advertised.

Notes

1. The author is a senior attorney with the Federal Trade Commission's Division of Advertising Practices in Washington, D.C. The opinions expressed herein are solely those of the author and do not necessarily reflect the opinions of the Federal Trade Commission or any of its bureaus, divisions, or offices. In addition, the citation of Federal Trade Commission settlements—called consent orders—is for illustrative purposes only. A company's decision to settle FTC charges against it does not constitute an admission of a law violation.

2. The term "functional food" appears nowhere in FTC statutes or regulations. For the purposes of this chapter, the term is used to describe foods advertised directly or by implication to provide health benefits to consumers beyond general nutritional needs.

3. 15 U.S.C. § 45.

4. Section 15 of the FTC Act defines the terms "food," "drug," "device," and "cosmetic." 15 U.S.C. ' 55. According to Section 15(b), "The term 'food' means (1) articles used for food or drink for man or other animals, (2) chewing gum, and (3) articles used for components of any such article." Section 15(c) defines "drug" to mean "(1) articles recognized in the official United States Pharmacopoeia, official Homoeopathic Pharmacopoeia of the United States, or official National Formulary, or any supplement to any of them; and (2) articles intended for use in the diagnosis, cure, mitigation, treatment, or prevention of disease in man or other animals; and (3) articles (other than food) intended to affect the structure or any function of the body of man or other animals; and (4) articles intended for use as a component of any article specified in clause (1), (2), or (3); but does not include devices or their components, parts, or accessories."

5. 15 U.S.C. § 45(n).

6. An agency of attorneys and economists, the FTC does not employ an in-house scientific staff. Rather it relies on the expertise of the FDA, the National Institutes of Health, the Centers for Disease Control and Prevention, and other federal and state agencies, as well as experts in academia and the private sector.

7. Working Agreement Between FTC and FDA, 4 Trade Red. Rep. (CCH) & 9.850.01 (1971) (updating and replacing prior agreement of 1954 and 1958).

8. Even in the area of prescription drug advertising, the FTC shares its expertise in consumer perception with the FDA. See In the Matter of Request for Comments on Consumer-Directed Promotion, Docket No. 2003N-0344, Comments of the Staff of the Bureau of Consumer Protection, the Bureau of Economics, and the Office of Policy Planning of the Federal Trade Commission (Dec. 1, 2003).

9. See Thompson Medical Co., Inc.,104 F.T.C. 648, 826 (1984), aff'd, 791 F.2d 189 (D.C. Cir. 1986), cert. denied, 479 U.S. 1086 (1987).

10. See, e.g., Weider Nutrition International, Inc., C-3983 (Nov. 17, 2000) (consent order) (challenging weight loss claims for PhenCal, a dietary supplement marketed as a safe alternative to prescription drug combination Phen-Fen); SmartScience Laboratories, Inc., C-3980 (Nov. 7, 2000) (consent order) (challenging pain relief

claims for Joint Flex, a topically applied cream containing glucosamine and chondroitin sulfate); Efamol Nutraceuticals, Inc., C-3958 (May 23, 2000) (consent order)(challenging efficacy claims for dietary supplements marketed to treat Attention Deficit Disorder and hyperactivity).

11. Interstate Bakeries Corp., C-4042 (Apr. 19, 2002) (consent order) (challenging claims that Wonder Bread containing added calcium could improve children's brain function and memory).

12. See Chesebrough Ponds, Inc., 105 F.T.C. 567 (1985) (modifying 1963 order against marketer of the Vaseline petroleum jelly based on the FDA's 1983 monograph on skin protectants).

13. Pub. L. No. 101-535, 104 Stat. 2353 (1990).

14. 59 Fed. Reg. 28388 (June 1, 1994).

15. 116 F.T.C. 74 (1993) (consent order).

16. See also Pizzeria Uno 123 F.T.C. 1038 (1997) (consent order) (challenging misleading low-fat claims for "Thinzettas" line of pizzas); Mrs. Fields Cookies, 121 F.T.C. 599 (1996) (consent order) (challenging misleading low-fat claims for cookies).

17. Pub. L. No. 103-417, 108 Stat. 4325 (1994).

18. Federal Trade Commission, Dietary Supplements: An Advertising Guide for Industry (1998), available at www.ftc.gov/bcp/conline/pubs/buspubs/dietsupp.pdf. Although the Dietary Supplement Guide is a plain-language business education piece aimed at the marketers of dietary supplements, the principles are equally applicable to the marketers of functional foods. Many of the examples cited herein are adapted from the Commission's Dietary Supplement Guide.

19. Deception Policy Statement, appended to Cliffdale Associates, Inc., 103 F.T.C. 110, 174 (1984). Numerous federal courts have cited the Deception Policy Statement with approval in evaluating deceptive advertising claims. See, e.g., Novartis Corp. v. FTC, 223 F.3d 783 (D.C. Cir. 2000); FTC v. Pantron I Corp., 33 F.3d 1088 (9th Cir. 19994); Rosa v. Gaynor, 784 F. Supp. 1 (D. Conn. 1989).

20. It is also important for marketers to appreciate that the FTC's definition of "advertising" extends to a wide variety of communications between a company and a consumer. Therefore, Section 5 applies not only to statements disseminated via print media, such as newspapers and magazines, and broadcast media, including television commercials and radio spots, but also to direct mail promotions, telemarketing scripts, infomercials, shop-at-home television channels, billboard advertising, point-of-purchase displays, statements made by call center personnel, statements made by retail sales staff, and virtually any representation made on a company's Internet Web site or through pop-up ads, banner ads, or e-mail. Although the FTC-FDA Liaison Agreement gives primary authority over labeling to FDA, the FTC has also challenged deceptive claims made on product labels.

21. FTC v. Sterling Drug, 317 F.2d 669, 674 (2d Cir. 1964).

22. Deception Statement, 103 F.T.C. at 176.

23. Kraft, Inc., 970 F.2d 311, n.4 (7th Cir. 1992), cert. denied, 507 U.S. 909 (1993).

24. See Gerber Products Co., 123 F.T.C. 1365 (1997) (consent order) (challenging baby food marketer's survey methodology supporting its claim that "4 out of 5 pediatricians recommend Gerber.").

25. See Dietary Supplement Guide at p. 4, Example 2.

26. Id.

27. Although there is no legal requirement that companies copy test their ads before dissemination, they are required to substantiate all express and implied claims that consumers take from an ad, regardless of whether the advertiser intended to convey those claims. Therefore, copy testing before the dissemination of a new ad campaign serves two important purposes: (1) it can protect companies in the event of a subsequent FTC inquiry; and (2) it can help ensure that consumers are taking the messages that the advertiser intends to convey.

28. Heinz W. Kirchner, 63 F.T.C. 1282 (1963).

29. Sears, Roebuck & Co., 95 F.T.C. 406, 511 (1980), aff'd, 676 F.2d 385 (9th Cir. 1982).

30. Grolier, Inc., 91 F.T.C. 315 (1978) ("The literal truth employed in a particular context may be used to deceive and deception, moreover, may be accomplished by innuendo as well as by outright false statements.")

31. Dietary Supplement Guide at p. 4, Example 3.

32. Deception Statement, 103 F.T.C. at n.45.

33. See Cliffdale Associates, Inc., 103 F.T.C. at 166 n.11

34. Deception Statement, 103 F.T.C. at 182 and n.51, citing American Home Products Corp., 98 F.T.C. at 368–69 ("The very fact that AHP sought to distinguish its products from aspirin strongly implies that knowledge of the true ingredients of those products would be material to purchasers."), aff'd, 695 F.2d 681 (3d Cir. 1982).

35. 111 F.T.C. 150 (1988) (Commission Decision).

36. 115 F.T.C. 788 (1992) (consent order).

37. Deception Statement, 103 F.T.C. at 175 n.4.

38. Warner-Lambert Co., 86 F.T.C. 1398, 1415 n.4 (1975), aff'd, 562 F.2d 749 (D.C. Cir. 1977), cert. denied, 435 U.S. 950 (1978).

39. FTC v. Balme, 23 F.2d 615 (2d Cir. 1928), quoting Florence Manufacturing Co. v. J.C. Dowd & Co., 178 F. 73 (2d Cir. 1910).

40. See infra at 160.

41. See Dietary Supplement Guide at p. 7.

42. See Dietary Supplement Guide at p. 6, Example 7.

43. C-4018 (Aug. 3, 2001) (consent order).

44. C-4012 (Aug. 3, 2001) (consent order).

45. See Dietary Supplement Guide at p. 6, Example 8.

46. No. CIV.S-01-1332 DFL GGH (E.D. Cal. July 11, 2001) (stipulated final order).

47. No. 2:01 CV-0505 ST (D. Utah Nov. 29, 2001) (stipulated final order).

48. See, e.g., United States v. Bayer Corp., No. CV 00-132 (NHP) (D.N.J. Jan. 11, 2000) (consent decree); Palm, Inc., C-4044 (Apr. 19, 2002) (consent order); Gateway Corp., C-4015 (June 24, 2001) (consent order); Stouffer Food Corp., 118 F.T.C. 746 (1994).

49. FTC staff has suggested a shorthand checklist for advertisers to consider in reviewing the use of disclosures in ads: Prominence—Is the information big enough for consumers to notice and read? Presentation—Is the wording clear enough for consumers to understand? Placement—Is the information located in a place where consumers will look? and Proximity—Is the information located close to the claim it qualifies? In 2001, the FTC sponsored with the National Advertising Division of the Council of Better Business Bureaus, Inc., a workshop on the use of disclosures in advertising, the proceedings of which are available at www.ftc.gov/bcp/workshops/disclosures/index.html. The FTC has also issued guidance to businesses on the most effective means of disclosing information to consumers on the Internet. See Federal Trade Commission, Dot.Com Disclosures (1999), available at www.ftc.gov/bcp/conline/pubs/buspubs/dotcom/index.pdf.

50. See Joint FTC-FCC Policy Statement on the Advertising of Dial-Around and Other Long-Distance Services to Consumers (March 1, 2000) ("Furthermore, research suggests that disclosures that are made simultaneously in both visual and audio modes generally are more effectively communicated than disclosures made in either mode alone.")

51. 119 F.T.C. 762 (1995) (consent order).

52. 118 F.T.C. 746 (1994).

53. Appended to Thompson Medical Co., 104 F.T.C. 648, 839 (1984), aff'd, 791 F.2d 189 (D.C. Cir. 1986), cert. denied, 479 U.S. 1086 (1987).

54. See Dietary Supplement Guide at p. 4, Example 1.

55. Depending on the nature of the representation, even subjective taste claims must be supported by a reasonable basis. For example, if an ad states, "In a blind taste test, seven out of ten ABC Redfruit Juice drinkers preferred XYZ Redfruit Juice," the advertiser must support that claim with methodologically sound taste testing.

56. Pfizer, Inc., 81 F.T.C. 23 (1972).

57. See, e.g., Interstate Bakeries Corp., C-4042 (Apr. 19, 2002) (challenging memory and brain function claims for Wonder Bread) (consent order).

58. See Schering Corp., 118 F.T.C. 1030 (1994) (requiring two well-controlled double-blinded studies to substantiate weight-loss claims for fiber supplement); Thompson Medical Co., 104 F.T.C. 648 (1984), aff'd, 791 F.2d 189 (D.C. Cir. 1986), cert. denied, 479 U.S. 1086 (1987) (requiring two well-controlled clinical studies to substantiate certain analgesic claims).

59. Removatron International Corp., 111 F.T.C. 206 (1988), aff'd, 884 F.2d 1489 (1st Cir. 1989) (requiring "adequate and well-controlled clinical testing" to substantiate claims for hair removal device).

60. See Dietary Supplement Guide at p. 10.

61. 579 F.2d 1137 (9th Cir. 1978).

62. 33 F.3d 1088 (9th Cir. 1994), cert. denied, 514 U.S. 1083 (1995).

63. See Dietary Supplement Guide at p. 18.

64. 16 C.F.R. ' 255.

65. See Dietary Supplement Guide at p. 12.

66. See Dietary Supplement Guide at p. 14.

67. See Dietary Supplement Guide at p. 16.

68. 123 F.T.C. 1301 (1997) (consent order).

69. 126 F.T.C. 202 (1998) (consent order).

70. C-4042 (Apr. 19, 2002) (consent order).

71. C-3955 (June 16, 2000) (consent order).
72. C-3926 (Feb. 10, 2000) (consent order).
73. C-4089 (July 29, 2003) (consent order).
74. Each dissemination of a deceptive claim is considered to be a separate violation of the law. Therefore, civil penalties can often total in the millions of dollars. See, e.g., United States v. Nu Skin International, Inc., No. 97-CV-0626G (D. Utah Aug. 6, 1997) (stipulated permanent injunction) ($1.5 million civil penalty against seller of weight loss products for violating FTC order barring deceptive claims); In re Dahlberg, No. 4-94-CV-165 (D. Minn. Nov. 21, 1995) (stipulated permanent injunction) ($2.75 million civil penalty against hearing aid manufacturer for violating FTC order barring false or unsubstantiated performance claims); United States v. General Nutrition Corp., No. 94-686 (W.D. Pa. Apr. 28, 1994) (stipulated permanent injunction) ($2.4 million civil penalty for violating FTC order requiring substantiation for disease, weight loss, and muscle building claims). Furthermore, the fact that a company has already discontinued an ad campaign or pledges not to disseminate the same deceptive claims in the future is not a defense to an FTC action.
75. 343 U.S. 470, 473 (1952). See also FTC v. National Lead Co., 352 U.S. 419 (1957) ("Those caught violating the [FTC] Act must expect some fencing in.").
76. Sears, Roebuck & Co. v. FTC, 676 F.2d 385, 391 (9th Cir. 1982).
77. Whenever the Commission can identify consumers who purchased the product, the agency's policy is to return money directly to them. If that is not feasible, any financial restitution won by the agency escheats to the U.S. Treasury. No portion of judgments goes to the FTC's operating budget. See, e.g., FTC v. Rexall Sundown, Inc., Civ. No. 00-706-CIV (S.D. Fla. March 11, 2003) (stipulated final order) (between $8–12 million in consumer redress for deceptive efficacy representations for Cellasene, a purported anti-cellulite dietary supplement); FTC v. Blue Stuff, Inc., No. Civ-02-1631W (W.D. Okla. Nov. 18, 2002) (stipulated final order) ($3 million in consumer redress for allegedly deceptive claims for Blue Stuff pain reliever and two dietary supplements advertised to reduce cholesterol and reverse bone loss); FTC v. Enforma Natural Products, Inc., No. 04376JSL(CWx) (C.D. Cal. April 26, 2000) (stipulated final order) (ordering $10 million in consumer redress from marketer of purported weight loss product); FTC v. American Urological Corp., No. 98-CVC-2199-JOD (N.D. Ga. April 29, 1999) (permanent injunction) ($18.5 million judgment against marketers of "Väegra," a dietary supplement purporting to treat impotence); FTC v. SlimAmerica, Inc., No. 97-6072-Civ (S.D. Fla. 1999) (permanent injunction) (ordering $8.3 million in consumer redress from marketer of purported weight loss product).
78. See Synchronal Corp., 116 F.T.C. 1189 (1993) (consent order) (requiring corporate officer who sold baldness, anti-cellulite, and skin care products to establish $500,000 escrow account before marketing certain products in the future to fund consumer redress, if necessary); FTC v. American Urological Corp., No. 98-CVC-2199-JOD (N.D. Ga. April 29, 1999) (final order for permanent injunction) (imposing $6 million bond on marketer of "Väegra," dietary supplement purporting to treat impotence); FTC v. Sloniker, No. CIV '02 1256 PHX RCB (D. Az. Feb. 6, 2003) (stipulated final judgment) (banning principals for life from any future telemarketing activities).
79. 86 F.T.C. 1398, 1415 n.4 (1975), aff'd, 562 F.2d 749 (D.C. Cir. 1977), cert. denied, 435 U.S. 950 (1978).
80. 223 F.3d 783 (D.C. Cir. 2000).
81. See PhaseOut of America, Inc., 123 F.T.C. 395 (1997) (consent order) (requiring marketer of device advertised to reduce health risks of smoking to notify purchasers that the product has not been proven to reduce the risk of smoking-related diseases); Consumer Direct, Inc., 113 F.T.C. 923 (1990) (consent order) (requiring marketer of Gut Buster exercise device to mail warnings to purchasers regarding serious safety hazard of product).
82. See United States v. Bayer Corp., No. CV 00-132 (NHP) (D.N.J. Jan. 11, 2000) (consent decree) (requiring disclosure in ads that "Aspirin is not appropriate for everyone, so be sure to talk with your doctor before beginning an aspirin regimen."); California SunCare, Inc., 123 F.T.C. 332 (1997) (consent order) (requiring prominent cautionary statement in future advertising for suntanning products about hazards of sun exposure).
83. No. SAC V-99-1407 (D. Colo. Nov. 15, 1999), and No. 99-WI-2197 (C.D. Cal. Nov. 15, 1999) (stipulated final orders).
84. 118 F.T.C. 340 (1994) (consent order).
85. See Chrysler Corp. v. FTC, 561 F.2d 357, 363 & n.5 (D.C. Cir. 1977).
86. See, e.g., FTC v. Affordable Media, 179 F.3d 1228 (9th Cir. 1999).
87. See Campbell Mithun, L.L.C., C-4043 (Apr. 19, 2002) (consent order) (challenging agency's role in advertisements claiming that Wonder Bread with added calcium could improve children's brain function and memory); Jordan McGrath Case & Taylor, 122 F.T.C. 152 (1996) (consent order) (challenging ad agency's role in advertisements making deceptive claims for pain relief superiority of Doan's pills); NW Ayer & Son, Inc.,121 F.T.C. 656 (1996) (consent order) (challenging agency's role in advertisements containing deceptive claims regarding the

effect of Eggland's Best eggs on blood cholesterol); BBDO Worldwide, Inc., 121 F.T.C. 33 (1996) (consent order) (challenging agency's role in advertisements containing deceptive claims for nutritional content of Häagen-Dazs frozen yogurt).

88. See, e.g., World Media TV, Inc., C-5717 (Feb. 28, 1997) (consent order) (challenging role of infomercial producer for role in disseminating claims for a pain relief device).

89. See, e.g., Home Shopping Network, Inc., 122 F.T.C. 227 (1996) (consent order) (holding home shopping company liable for deceptive claims for vitamin and stop-smoking sprays); United States v. Home Shopping Network, Inc., No. 99-897-CIV-T-25C (M.D. Fla. April 15, 1999) (consent decree) ($1.1 million civil penalty for violating previous FTC order barring false and unsubstantiated claims for skin care, weight-loss, and PMS/menopause products); QVC, Inc., C-3955 (June 16, 2000) (consent order) (holding home shopping company liable for deceptive cold prevention claims for zinc supplement) (consent order); ValueVision International, Inc., C-4022 (Aug. 24, 2001) (holding home shopping company liable for deceptive claims for weight loss, cellulite, and baldness products).

90. See, e.g., Sharper Image, 116 F.T.C. 606 (1993) (consent order) (holding cataloger liable for unsubstantiated claims for exercise device and dietary supplement) (consent order); Lifestyle Fascination, Inc., 118 F.T.C. 171 (1994) (consent order) (holding cataloger liable for deceptive claims for pain relief device and "Brain Tuner" for addiction and depression).

91. See 16 C.F.R. ' 255.3(a); FTC v. Kendall, No. 00-09358-AHM (AIJx) (C.D. Cal. Aug. 31, 2000) (challenging false representation that person who appeared on an infomercial touting a weight loss product was a nutritionist) (stipulated final order).

92. 16 C.F.R. ' 255.4.

93. See Dietary Supplement Guide at p. 19–20.

94. See Black & Decker (U.S.) Inc., 113 F.T.C. 63 (1990) (consent order) (challenging as deceptive the claim that Black & Decker Automatic Shut-Off Iron received the endorsement of the National Fire Safety Council because the group did not have expertise to evaluate appliance safety and did not base its endorsement on appropriate professional standards).

95. 115 F.T.C. 849 (1992) (consent order).

96. 116 F.T.C. 1189 (1993) (consent order).

97. Synchronal Corp., 116 F.T.C. 1189 (1993) (consent order).

98. See, e.g., Robert M. Currier, D.O., C-4067 (Dec. 20, 2002) (consent order) (challenging deceptive representations made by eye doctor for a purported anti-snoring treatment); William S. Gandee, 123 F.T.C. 698 (1997) (consent order) (challenging chiropractor's role in the marketing of a purported pain relief device); William E. Shell, M.D., 123 F.T.C. 1519 (1997) (consent order) (challenging physician's role in the marketing of a purported weight loss pill).

99. 16 C.F.R. ' 255.5.

100. Numex Corp., 116 F.T.C. 1078 (1993); and James McElhaney, M.D., 116 F.T.C. 1137 (1993) (consent orders).

11 Functional Foods: Regulatory and Marketing Developments in the United States

Ilene Ringel Heller

Introduction

The expanding market for so-called "functional foods" in the United States has raised a multitude of complex regulatory and public policy issues. This paper will review U.S. regulatory requirements, marketplace implications, and consumer impact. It will then conclude by stating that the Food and Drug Administration (FDA) should ensure that functional foods are regulated in a manner that maximizes health benefits and minimizes health risks.

Regulatory Requirements

Although the term "functional foods" has entered the popular lexicon, its definition remains amorphous. The U.S. General Accounting Office (GAO) defines functional foods as foods "that claim to have health benefits beyond basic nutrition."[1] This definition includes foods that have something added to them for a particular health benefit, such as a margarine substitute containing a derivative from pine trees. It may also include foods that may simply contain a natural ingredient that has become associated with a particular health benefit (for example, tomato sauce made with lycopene-rich tomatoes).[2]

This definition, however, does not address the fact that in some sense all foods are functional foods. For example, in recent years, we have learned that a low-fat diet rich in fruits and vegetables can help reduce the risk of heart disease and cancer, and thus, fruits and vegetables can be considered "functional." In addition, certain foods traditionally have been used for their "functional" properties: prunes to aid regularity and coffee to combat fatigue are some examples. But the unprecedented emergence in recent years of a multitude of foods[3] promoted for almost everything from fighting depression to warding off the common cold calls for an examination of whether functional foods, no matter how they are defined, are being regulated effectively.

The United States does not have any specific regulations pertaining to functional foods. Functional foods are regulated as "foods," "dietary supplements," "drugs," "medical foods," or "food for special dietary use," depending upon the claims that are made as well as the ingredients that are used.[4] FDA states that it will apply those existing rules to functional foods on a case-by-case basis.[5] Some FDA officials, however, concede that the agency lacks a coherent policy. As one retired FDA official stated: "The market is moving faster than we can sit down and think things through."[6]

Whether a manufacturer successfully positions a particular functional food as a conventional food, a dietary supplement, or a product fitting into another regulatory category is central to determining how FDA will regulate both the use of a functional ingredient and any accompanying labeling claims.

Regulation of Ingredient Safety

Under the Federal Food, Drug, and Cosmetic Act (FDCA),[7] manufacturers must obtain premarket approval for food additives or demonstrate that such ingredients are "generally recognized as safe" (GRAS).[8] If a manufacturer does not consider the ingredient to be GRAS,[9] or its GRAS determination is challenged by FDA, the company must submit a food additive petition establishing the safety of the ingredient under the conditions of its expected use before FDA will grant approval.[10] Because the filing of a food additive petition can be costly and take several years, both the food and dietary supplement industries have pressed Congress to deregulate the process.

The supplement industry has been more successful in that effort than has the food industry. After a grassroots consumer letter-writing campaign, largely organized by the industry, Congress enacted the Dietary Supplement Health and Education Act of 1994 (DSHEA).[11] That law prevents dietary supplements from being subjected to the approval requirements applied to food additives. The definition of a dietary supplement includes substances that may be regulated as drugs in other countries.[12] New dietary ingredients may be marketed 75 days after the manufacturer provides notice and substantiation to FDA that the ingredient can "reasonably be expected to be safe."[13] FDA approval, however, is not required. Ingredients marketed prior to October 15, 1994, are exempt from even this minimal requirement.

DSHEA also makes it more difficult for FDA to remove an unsafe, or potentially unsafe, product from the market. Prior to DSHEA, FDA used its authority to regulate food additives as an expedited means of declaring a supplement ingredient unsafe or inadequately tested. If FDA charged that the dietary ingredient was unsafe under the food additive provisions of the FDCA, a manufacturer had the burden of proving that the affected ingredient was safe. By eliminating FDA's authority to regulate dietary supplements as food additives, DSHEA shifted the burden of proof to FDA.[14] To take a product off the market, DSHEA requires FDA to establish that the product "presents a significant or unreasonable risk of illness or injury"[15] or poses an "imminent hazard to public health or safety."[16] As a practical matter, this means that FDA must first build a convincing case of substantial harm to public health and prevail in court before it can act.

This process can take years. For example, ephedra has been associated with more adverse effects, including numerous deaths, than any other supplement. Although FDA issued a Medical Bulletin in September 1994,[17] the Agency was unable to ban the sale of ephedra as a dietary supplement until April 12, 2004.[18] Some members of Congress have begun to question whether DSHEA prevents effective enforcement.[19]

In light of the difficulty of issuing binding regulations limiting the use of hazardous substances, FDA has resorted to issuing public warnings.[20] In March 2002, the agency issued a consumer advisory regarding the potential risk of severe liver injury associated with the use of kava-containing dietary supplements.[21] The issuance of the Consumer Advisory followed reports of more than 25 adverse events in other countries. Switzerland, France, Canada, and Singapore, in contrast to FDA, have taken steps to remove kava-containing products from the market.[22] Germany has banned the sale of kava except in homeopathic products containing only minute amounts.[23]

To escape strict FDA regulation of the safety of functional substances, it is tempting for some manufacturers to market a functional food as a dietary supplement rather than as a conventional food with added ingredients. By marketing functional foods as dietary supplements, companies hope to escape the legal obligation to prove that added ingredients are GRAS, and when questions arise, such products are allowed to stay on the market until FDA proves in court that they may be harmful.

Regulation of Product Claims

Claims for functional foods are subject to a variety of regulations depending on the regulatory category into which the product is placed—for example, food, dietary supplement, medical food, or drug.

Health Claims for Foods and Dietary Supplements Positioned as Functional Foods

Health claims for foods and dietary supplements are subject to the Nutrition Labeling and Education Act of 1990 (NLEA),[24] which requires FDA premarket approval or authorization.[25] Under the NLEA, FDA must determine, based on "the totality of publicly available scientific evidence," that the claim is supported by "significant scientific agreement, among experts qualified by scientific training and experience."[26] If FDA determines that such a claim may be made, it issues a regulation that allows any qualifying product to bear the claim.

The United States was the first to enact a law allowing health claims for nutrients naturally contained in conventional foods and dietary supplements.[27] These statutory provisions are now being relied on by companies wishing to develop and market functional foods in the United States.

Government policies in the United States regarding the use of health claims have a rather checkered history. In 1987, FDA began experimenting with allowing health claims for conventional foods without requiring premarket approval by the agency.[28] The result was a few truthful claims followed by a flood of products claiming to cure almost every ailment under the sun. After several years of such experimentation, the cover of *Business Week* magazine proclaimed "Health Claims for Foods Are Becoming Ridiculous."[29]

Congress then passed the NLEA, which permitted certain health claims approved by FDA. The claims had to be based on the presence or absence of a nutrient that is linked to a disease or health-related condition.[30] For example, certain foods low in sodium can carry a claim that diets low in sodium can help reduce the risk of high blood pressure. Adding less salt to a processed food can, in a sense, turn a conventional product into a functional food.

After NLEA's enactment, from 1990–1994, misleading claims practically disappeared, and an orderly system for approving health claims on qualifying food products prevailed. Although the food industry objected to the new system, numerous companies, including Kellogg, Quaker Oats, Campbell Soup, and Pillsbury, began utilizing approved claims. Beginning in 1994, however, Congress, and then the courts, began rolling back the requirements of the law.[31] As a result, misleading claims again are appearing.

Under amendments to the law resulting from passage of the Food and Drug Administration Modernization Act of 1997 (FDAMA),[32] manufacturers may now make health claims for foods and dietary supplements based on authoritative statements published by a scientific body of the U.S. government about "the relationship between a nutrient and a disease or health-related condition to which the claim refers."[33]

Such claims do not require FDA approval but still must be authorized by the agency.

Manufacturers must: (1) notify FDA at least 120 days in advance of marketing a product with the prospective health claim;[34] (2) demonstrate that the claim is based on an "authoritative statement" of a scientific body "with official responsibility for public health protection or research directly relating to human nutrition";[35] and (3) submit a balanced representation of the scientific literature on which the claim is based.[36] After a manufacturer takes those steps, the claim can be made unless FDA, within the 120-day period, issues an interim final regulation prohibiting the claim or successfully brings a lawsuit against the company.

Public health groups have voiced considerable anxiety over the effect that FDAMA would have on FDA's ability to prohibit poorly substantiated and misleading claims before they appear on labels. Actions by FDA, however, along with industry's reluctance to use the procedure, have allayed some of those fears. First, FDA announced that the 1997 amendments to the law did not change the "significant scientific agreement" standard, and health claims based on authoritative statements still have to meet that requirement, based on the totality of publicly available scientific evidence.[37]

Second, FDA denied the first series of attempts to make health claims based on authoritative statements. Within days of the effective date of FDAMA, Weider Nutrition International, Inc., a dietary supplement marketer, submitted nine prospective health claims to FDA based on what it considered "authoritative statements."[38] FDA rejected all nine of the claims within the 120-day period established by law; the agency issued interim final regulations prohibiting their use.[39]

A month later, however, FDA permitted a health claim based on an authoritative statement to go into effect. General Mills requested the approval of a claim linking the consumption of diets high in whole grains, including cereal, to a decreased occurrence of coronary heart disease and certain cancers.[40] Most recently, it has permitted a health claim linking consumption of foods that are good sources of potassium and low in sodium with a reduced risk of high blood pressure and stroke.[41]

FDAMA also attempted to expedite the issuance of health claims by permitting FDA to implement proposed rules effective upon publication, with a comment period to follow only after the fact. FDA may take such steps if it determines that: (1) the rule would "enable consumers to develop and maintain healthy dietary practices"; (2) that the rule would "enable consumers to be informed promptly and effectively of important new knowledge regarding nutritional and health benefits of food"; or (3) the rule will "ensure that scientifically sound nutritional and health information is provided to consumers as soon as possible."[42] To date, this procedure has been used three times with respect to health claims for plant stanol esters and the reduction of coronary heart disease (CHD),[43] beta glucan soluble fiber from whole oats sources and reduced risk of CHD[44] and the sugar D-tagatose, and the reduction of tooth decay.[45] Consumer groups oppose the use of such a procedure because issuing interim final rules without prior notice and an opportunity for comment limits the ability of FDA to ensure that it has considered information from all relevant sources.

In any event, whether a manufacturer submits a petition for a new health claim under the NLEA or a notification of a new health claim based on an authoritative statement of another government agency pursuant to FDAMA, health claims for functional foods—as a general matter—may not be used on products that FDA has determined contain excessive levels of fat, saturated fat, cholesterol, sodium, or other substances specified in FDA regulations.[46] Similarly, health claims are not permitted for products that do not contain, prior to any nutrient addition, at least 10 percent of the Reference Daily Intake or Reference Daily Value of vitamin A, vitamin C, calcium, protein, or fiber per reference amount customarily consumed.[47] Thus, health claims for functional soda pops, chewing gum, bottled waters, and other foods of low nutritional value would not be permitted under what has been nicknamed the "Jelly Bean Rule" unless there is a specific exemption as there is for products containing ingredients that do not promote tooth decay.[48]

FDA's refusal to allow health claims based on nutrient fortification of the products is rooted in a fortification policy that it has followed for many years.[49] The policy, however, is not legally binding and thus may have a limited effect on the development of functional foods.[50] Health claims also are prohibited for both foods and dietary supplements if the claim relates to a substance that does not contribute taste, aroma, or nutritive value, or does not perform a technological function on the food itself.[51]

FDA's ability to deny health claims for foods and dietary supplements on the basis of inadequate scientific support may be limited by a U.S. Court of Appeals decision, *Pearson v. Shalala.*[52] The court concluded that FDA should approve claims even where there was conflicting or insufficient scientific evidence, as long as claims could be rendered "not misleading" by the inclusion of a disclosure such as "the scientific evidence supporting this claim is inconclusive."[53] The court determined that FDA's blanket refusal to permit preliminary health claims with such disclosures violates the commercial free speech protections afforded by the First Amendment.[54] The court also directed FDA to issue a clarification as to how the agency defines the term "significant scientific agreement."[55] Although a number of First Amendment challenges have been brought against the NLEA, this is the first decision to limit FDA's authority over health claims.

In response to the *Pearson* decision, FDA has issued a clarification of the term "significant scientific agreement."[56] It also has decided to exercise its "enforcement discretion" with respect to claims for dietary supplements that do not meet the significant scientific agreement standard as long as they are qualified by disclosures about the strength of the scientific evidence. Qualified claims must also meet the following requirements:

1. The claim is the subject of a health claim that meets the requirements of section 101.70.
2. The scientific evidence in support of the claim outweighs the scientific evidence against the claim, the claim is qualified appropriately, and all statements in the claim are consistent with the weight of the scientific evidence.
3. Consumer health and safety are not threatened.
4. The claim meets the general requirements for health claims in section 101.14, except for the requirement that the evidence supporting the claim meet the significant scientific agreement standard and the requirement that the claim be made in accordance with an authorizing regulation.[57]

Although FDA initially took the position that the *Pearson* decision applied only to dietary supplements, it announced on December 18, 2002, that it would apply the same criteria for exercising its enforcement discretion to both supplements and foods. In addition, it announced that it would now be applying the same standard as the Federal Trade Commission (FTC) for determining whether a health claim is inherently misleading: the "reasonable consumer" test. Traditionally, the agency has evaluated claims based on its role of protecting "the ignorant, the unthinking and the credulous."[58]

In July 2003, FDA established an interim policy for reviewing qualified claims as well as a ranking system for characterizing the level of scientific support for particular claims. Claims would be ranked from "A," for those that meet the significant scientific agreement standard, to "D," for those with little scientific support.[59]

FDA's move is highly controversial particularly because, earlier in 2002, the National Academy of Science's Institute of Medicine (IOM) cautioned against permitting claims based on preliminary scientific studies. The IOM stated that "IOM [c]laims about nutrient-disease relationships are more easily made than scientifically supported. Because the implications for public health are so important, caution is urged prior to accepting such claims without supportive evidence from appropriately designed clinical trials."[60] The IOM cautioned that further study of an "appealing hypothesis" may result in a finding that the nutrient actually causes harm. For example, although preliminary evidence suggested that beta-carotene supplements could reduce the risk of lung cancer, clinical intervention trials later demonstrated that beta-carotene supplements actually increased the risk of lung cancer in smokers.[61]

Consumer groups expressed disappointment that the FDA had, once again, bowed to industry pressure to extend the *Pearson* decision to foods. In a lawsuit challenging the policy, consumer groups argued that FDA violated the NLEA by permitting the use of claims that do not meet the significant scientific agreement standard set forth in the Act. Consumer groups also maintained that the policy violates NLEA procedural requirements because qualified health claims would not be subject to the notice and comment rulemaking proceedings specified by Congress. The lawsuit was dismissed on procedural grounds and an appeal has been filed (*CSPI v. FDA,* No. 03-1962 D.D.C. July 30, 2004).

FDA has authorized the following qualified claims linking: (1) folic acid and the reduction of neural tube defects; (2) omega-3 fatty acids and the reduction of coronary heart disease (CHD); (3) B vitamins and the reduction of CHD; (4) selenium and the reduction of cancer risk; and (6) phosphatidylserine and a decreased risk of cognitive dysfunction and dementia. FDA authorized the first qualified health claims for conventional foods in July 2003. Those claims link the consumption of walnuts, almonds, hazelnuts, pecans, some pine nuts, and pistachio nuts with a reduction of CHD.[62]

Health Claims for Foods for Special Dietary Use and Medical Foods

U.S. law also recognizes two other categories of foods that are often thought to have a bearing on how FDA can regulate health claims for functional foods. They are "foods for special dietary use"[63] and "medical foods."[64]

"Foods for special dietary use" encompass certain products that: (1) are used for supplying particular dietary needs that exist by reason of age, including but not limited to infancy and childhood; (2) are used for supplementing or fortifying the ordinary or usual diet with any vitamin, mineral, or other dietary property; (3) contain artificial sweeteners.[65] Those products include hypoallergenic foods, certain infant foods, and products useful in reducing or maintaining weight.[66]

Health claims for foods for special dietary use are covered specifically by the NLEA's premarket approval requirements. If FDA, however, already has issued a food for special dietary use regulation that authorizes a particular health claim, a manufacturer need not reapply under the NLEA for approval.

Companies seeking to market functional foods could, theoretically, petition FDA to issue a regulation for a food for special dietary use instead of petitioning the agency to permit a health claim under the NLEA. FDA, however, has not recently made use of this authority and has instead withdrawn regulations for some foods for special dietary use.[67]

Some companies have relied on another provision of the FDCA concerning the regulation of "medical foods" as a mechanism for marketing functional foods without FDA approval. The FDCA defines a medical food as a:

> food which is formulated to be consumed or administered internally under the *supervision of a physician* and which is intended for the *specific dietary management* of a disease or condition for which *distinctive nutritional requirements*, based on recognized scientific principles, are established by medical evaluation.[68]

Many medical foods are intended for institutional use in hospitals and nursing homes. For example, TraumaCal Liquid is used to feed burn patients and Travasorb Hepatic Powder is for patients suffering from liver failure. As a matter of practice, traditional medical food suppliers generally require written orders from a physician before filling orders.[69] Neither label claims nor formulations have to be approved by FDA prior to marketing.[70] FDA instead depends on the judgment of individual physicians who traditionally authorize the use of such foods for patients.[71]

Increasingly, however, manufacturers have begun to call foods sold to the general public and used by consumers without the supervision of a physician "medical foods" in order to escape FDA approval requirements for health claims.[72] FDA has recognized the irony of permitting medical foods, which are intended for consumption by individuals with serious health conditions, to receive less regulatory scrutiny than conventional foods marketed with health claims to the average consumer.[73] The agency had been attempting to close the gap by issuing new regulations. In April 2003, however, FDA proposed the withdrawal of its pending rulemaking proceeding.[74]

Structure/Function Claims for Foods and Dietary Supplements

Perhaps the biggest loophole in the U.S. regulatory scheme utilized by marketers of functional foods is a series of statutory amendments enacted in 1994 permitting supplement companies to make structure/function claims.[75] Such statements explain how a substance affects the structure of the human body or the normal functioning of an organ or system; an example is "vitamin A is essential for normal vision."[76] The impact of these amendments has been worsened by FDA's regulatory and enforcement policies for structure/function claims that were developed to implement the 1994 amendments. As a result, some companies are deciding to forgo attempts to gain FDA approval for health claims and instead are marketing functional foods on the basis of structure/function claims, which can be made without approval. Supplement manufacturers that wish to make structure/function claims must notify FDA 30 days after initial marketing.[77] Although they must have substantiation for their claims, they need not show such information to FDA.[78] Instead, supplement labels must include the following disclaimer:

> This statement has not been evaluated by the Food and Drug Administration. This product is not intended to diagnose, treat, cure, or prevent any disease.[79]

Food products that bear structure/function claims do not have to meet such requirements, although a report published by the GAO recommends that foods making structure/function claims also should be subject to the notification and disclaimer requirements.[80]

Much controversy has arisen over the fine line separating structure/function claims, which do not require FDA premarket approval, from health claims, which do require such approval. A structure/function claim "describes the role of a nutrient or dietary ingredient intended to affect the structure or function in humans, [and] characterizes the documented mechanism by which a nutrient or dietary ingredient acts to maintain such structure or function . . . ."[81] A health claim "characterizes the relationship of any nutrient to a disease or health related-condition."[82] In its implementing regulations, however, FDA substituted the word "substance" for the word "nutrient,"[83] and defined "substance" as any food or food component, including "vitamins, minerals, herbs, or other similar nutritional substances."[84] FDA's definition of the word "substance" is problematic because many herbs are not considered to be nutritional substances, that is, they do not have "nutritive value."[85]

Those ambiguous definitions have triggered a game of statutory semantics with regard to the labeling of foods and supplements that contain functional ingredients. A final rule for dietary supplements that many companies believe is applicable to foods is intended to clarify the line between permissible structure/function claims and impermissible disease claims. The rule, however, draws hairline distinctions that will, no doubt, be lost on most consumers. For example, FDA states that a manufacturer can claim that a supplement "helps to maintain cholesterol levels that are already within normal range."[86] But a manufacturer is prohibited from claiming that a product "lowers cholesterol" or "maintains healthy cholesterol" (unless it obtains FDA approval for a health claim). The agency said it will prohibit the use of brand names such as CarpalCare that make an implied health claim (that the product will help treat carpal-tunnel syndrome), but will permit brand names like "Heart Tabs" if the product claims "to maintain healthy circulation."[87]

Furthermore, FDA has failed to propose how much scientific evidence a company must have to make a permissible structure/function claim. Traditionally, consumer organizations felt that structure/function claims should be statements of undisputed fact such as "calcium is necessary for bone growth and development." Recently, however, some companies have been making structure/function claims based on preliminary scientific studies.[88] FDA's failure to articulate a substantiation standard for structure/function claims has contributed to the prevalence of misleading claims.

The root of the problem is that FDA has been forced by Congress to make sense out of a series of inconsistent and overlapping statutory amendments that reflect political pressures instead of sound public policy. Unfortunately, until the law is further amended, it will be nearly impossible for FDA to develop sensible regulatory policies.

Marketplace Implications and Consumer Impact

Many companies, both large and small, are selling or developing products that could be considered to be functional foods. Some of the biggest names in the industry are involved, including Kellogg, Campbell Soup, ConAgra, Quaker Oats, and Mars.[89]

Some companies sell such products with FDA-approved health claims. For example, in 1996, Kellogg petitioned FDA for a health claim that soluble fiber from psyllium may reduce the risk of heart disease when eaten as part of a diet low in saturated fat and cholesterol.[90] The company also demonstrated to FDA that psyllium, which it adds to its Bran Buds cereal, is GRAS.[91] Recently, however, companies have begun to market functional foods on the basis of structure/function claims that do not require FDA approval. For example, Kellogg has decided to make structure/function claims in cases for which FDA regulations prohibit health claims. Other companies are marketing functional foods as dietary supplements or medical foods to avoid FDA premarket approval requirements.[92] These practices raise a number of troubling questions and serious legal issues.

Some Foods and Supplements Are Sold with FDA-Approved Health Claims

The U.S. government has authorized 22 health claims (see Table 11.1), most of which apply only to foods.[93] Some of the claims, however, may be used on both foods and supplement products, and six apply only to supplements.[94]

Some companies have used such FDA-approved claims to reposition conventional food products as functional foods. For example, Kellogg markets its Product 19 cereal with the approved FDA claim linking insufficient consumption of folic acid in women of childbearing age to neural tube birth defects.

Initially, FDA-approved health claims encompassed general diet/disease relationships, for example, diets low in saturated fat and cholesterol could reduce the risk of heart disease.[95] Major food companies utilized these claims. For example, Campbell's Pork and Beans has used an FDA-approved health claim regarding diets low in saturated fat and cholesterol and the risk of heart disease (the product contains minimal amounts of fat),[96] and Pillsbury's Green Giant brand frozen broccoli spears has used an FDA-approved claim that links diets low in fat and high in vegetables containing certain nutrients to a reduced risk of cancer.

Several years later, in response to a petition from the Quaker Oats Company, FDA approved a new health claim linking soluble fiber specifically from oatmeal and oat bran to a reduction in the risk of heart disease.[97] Product labels may state: "Diets low in saturated fat and cholesterol that include soluble fiber from whole oats may reduce the risk of heart disease."[98] The approved health claim allows Quaker, which dominates the oatmeal market, to specifically highlight the benefits of eating oatmeal, even though certain fruits, vegetables, and other grain products can provide the same benefit.

Kellogg followed suit with a petition, approved by FDA in 1998, to make a claim linking soluble fiber from psyllium to a reduced risk of heart disease.[99] At the time of Kellogg's petition, only one nationally distributed brand-name cereal, Kellogg's Bran Buds, contained psyllium. Then, under the authoritative statements provisions of FDAMA, General Mills was permitted to make health claims linking products containing whole grain foods to a lower risk of heart disease and certain cancers.[100]

More recently, manufacturers of products containing plant sterol or stanol esters were given permission to make claims linking the substances to a reduced risk of coronary heart disease.[101] Then Tropicana Products was permitted to make a health claim, based on an authoritative statement, linking foods that are good sources of potassium and low in sodium to a reduced risk of high blood pressure and stroke.[102] FDA's authority to approve health claims concerning the relationship of a nutrient to a disease has thus been used by some companies as a mechanism to gain approval for what could be characterized as claims for functional foods.

Table 11.1. Health claims authorized in the United States

Authorized Diet and Disease Relationship	Model Claims	Criteria
Calcium and osteoporosis*	Regular exercise and a healthful diet with enough calcium helps teens and young adult white and Asian women maintain good bone health and may reduce their risk of osteoporosis.	High in calcium Assimilable (Bioavailable) Supplement must disintegrate and dissolve Phosphorus content cannot exceed calcium content
Dietary fat and cancer	Development of cancer depends on many factors. A diet low in total fat may reduce the risk of some cancers.	Low fat (Fish and game meats: "Extra Lean")
Sodium and hypertension	Diets low in sodium may reduce the risk of high blood pressure, a disease associated with many factors.	Low sodium
Dietary saturated fat and cholesterol and coronary heart disease	While many factors affect heart disease, diets low in saturated fat and cholesterol may reduce the risk of this disease.	Low saturated fat Low cholesterol Low fat (Fish and game meats: "Extra Lean")
Fiber containing grain products, fruits, and vegetables and cancer	Low-fat diets rich in fiber containing grain products, fruits, and vegetables may reduce the risk of some types of cancer, a disease associated with many factors.	A grain product, fruit, or vegetable that contains dietary fiber Low fat Good source of dietary fiber (without fortification)
Fruits, vegetables, and grain products that contain fiber, particularly soluble fiber and coronary heart disease	Diets low in saturated fat and cholesterol and rich in fruits, vegetables, and grain products that contain some types of dietary fiber, particularly soluble fiber, may reduce the risk of heart disease, a disease associated with many factors.	A fruit, vegetable, or grain product that contains dietary fiber Low saturated fat Low cholesterol Low fat At least 0.6 g of soluble fiber per reference amount (without fortification) Soluble fiber content provided on product label
Fruits and vegetables and cancer	Low-fat diets rich in fruits and vegetables (foods that are low in fat and may contain dietary fiber, vitamin A or vitamin C may reduce the risk of some types of cancer, a disease associated with many factors (e.g., broccoli is high in vitamins A and C and is a good source of dietary fiber).	Fruit or vegetable Low fat Good source (without fortification) of at least one of the following: vitamin A, vitamin C, or dietary fiber
Folate and neural tube birth defects*	Healthful diets with an adequate daily folate intake may reduce a woman's risk of having a child with a brain or spinal cord birth defect.	Good source of folate

(continued)

Table 11.1. Health claims authorized in the United States (*Cont.*)

Authorized Diet and Disease Relationship	Model Claims	Criteria
Sugar alcohols and dental caries	Frequent eating of foods high in sugars and starches as between-meal snacks can promote tooth decay. The sugar alcohol (name optional) used to sweeten this food may reduce the risk of dental caries. For D-tagatose: Tagatose, the sugar used to sweeten this food, unlike other sugars, may reduce the risk of dental caries.	Sugar (mono and disaccharides) free Contain xylitol, maltitol, isomalt, lactitol, hydrogenated starch hydrolysates, hydrogenated glucose syrup, or a combination of these or the sugar D-tagatose If fermentable carbohydrates are present, the food shall not lower plaque pH below 5.7 by bacterial fermentation during consumption or for up to 30 minutes after consumption
Foods that contain soluble fiber from whole-oat or psyllium products and coronary heart disease**	Diets low in saturated fat and cholesterol that include soluble fiber from whole oats or psyllium may reduce the risk of heart disease.	At least 0.75 g per reference amount of soluble fiber from whole oats; 1.7g per reference amount of soluble fiber from psyllium husks; or 0.75 of beta-glucan soluble fiber per reference amount Soluble fiber content provided on the label
Soy protein and risk of coronary heart disease**	Diets low in saturated fat and cholesterol that include 25 g of soy protein a day may reduce the risk of heart disease. One serving of [name of food] provides __ g of soy protein.	At least 6.25 g of soy protein per reference amount must be low in saturated fat, unless food consists of or is derived from whole soybeans and it contains no fat in addition to that naturally present in whole soybeans Must be low cholesterol
Whole grain foods and coronary heart disease and certain cancers[1]	Diets rich in whole grain foods and other plant foods and low in total fat, saturated fat, and cholesterol may help reduce the risk of coronary heart disease and certain cancers.	Must contain 51% or more whole grain ingredient(s) by weight by reference amount
Plant sterol/stanol esters* and risk of coronary heart disease[2]	Plant Sterol Esters: Foods containing at least 0.65 g per serving of plant sterol esters, eaten twice a day with meals for a total intake of at least 1.3 g, as part of a diet low in saturated fat and cholesterol, may reduce the risk of heart disease. A serving of [name the food] supplies __ g of vegetable oil sterol esters.	0.65 g per serving plant sterol esters 1.7 g per serving plant stanol esters

(*continued*)

Table 11.1. Health claims authorized in the United States (*Cont.*)

Authorized Diet and Disease Relationship	Model Claims	Criteria
	Plant Stanol Esters: Foods containing at least 1.7 g per serving of plant stanol esters, eaten twice a day with meals for a total intake of at least 3.4 g, as part of a diet low in saturated fat and cholesterol, may reduce the risk of heart disease. A serving of [name of the food] supplies __ g of vegetable oil stanol esters.	
Potassium-containing foods and reduced risk of high blood pressure and stroke[1]	Diets containing foods that are good sources of potassium and are low in sodium may reduce the risk of high blood pressure and stroke.	At least 350 mg of potassium per reference amount Less than 140 mg of sodium per reference amount
Folic acid and neural tube defects[3]	0.8 mg folic acid in a dietary supplement is more effective in reducing the risk of neural tube defects than a lower amount in foods in common form. FDA does not endorse this claim. Public health authorities recommend that women consume 0.4 mg folic acid daily from fortified foods or dietary supplements or both to reduce the risk of neural tube defects.	Good source of folate
Omega-3 fatty acids and coronary heart disease[3]	Consumption of omega-3 fatty acids may reduce the risk of coronary heart disease. FDA evaluated the data and determined that, although there is scientific evidence supporting the claim, the evidence is not conclusive.	Should not exceed daily intake of 2 g per day Manufacturers are encouraged to set a daily intake of 1 g or below
Folic acid, vitamin B6, and vitamin B12 and vascular disease[3]	As part of a well-balanced diet that is low in saturated fat and cholesterol, folic acid, vitamin B6, and vitamin B12 may reduce the risk of vascular disease. FDA evaluated the above claim and found, that while it is known that diets low in saturated fat and cholesterol reduce the risk of heart disease and other vascular diseases, the evidence in support of the above claim is inconclusive.	Good source of folate

(*continued*)

Table 11.1. Health claims authorized in the United States (*Cont.*)

Authorized Diet and Disease Relationship	Model Claims	Criteria
Antioxidant vitamins and cancer[3]	Some scientific evidence suggests that consumption of antioxidant vitamins may reduce the risk of certain forms of cancer. However, FDA has determined that this evidence is limited and not conclusive.	Dietary supplements containing vitamin E and/or vitamin C
Selenium and cancer[3]	Selenium may reduce the risk of certain cancers. Some scientific evidence suggests that consumption of selenium may reduce the risk of certain forms of cancer. However, FDA has determined that this evidence is limited and not conclusive.	Dietary supplements containing selenium
Phosphatidyl serine and cognitive dysfunction and dementia[3]	Consumption of phosphatidyl serine may reduce the risk of cognitive dysfunction or dementia in the elderly. Very limited and preliminary scientific research suggests that phosphatidyl serine may reduce the risk of cognitive dysfunction or dementia in the elderly. FDA concludes that there is little scientific evidence supporting this claim.	Dietary supplements containing soy-derived phosphatidyl serine
Nuts and coronary heart disease[3]	Scientific evidence suggests but does not prove that eating 1.5 ounces per day of most nuts (such as *name of specific nut*) as part of a diet low in saturated fat and cholesterol may reduce the risk of heart disease. (See nutrition fat content if claim is made for whole or chopped nuts.)	Whole or chopped almonds, hazelnuts, peanuts, pecans, some pine nuts, pistachio nuts, and walnuts Nuts must not exceed disqualifying level for saturated fat (4g saturated fat per 50 g nuts) Nut-containing products other than whole or chopped nuts that contain at least 11 grams of one or more of nuts Must comply with definition of low saturated fat food and low cholesterol food Must comply with all disqualifying levels for total fat, saturated fat, cholesterol and sodium Food must contain a minimum of 10 percent of the Daily Value per Reference Amount Customarily Consumed (RACC) of vitamin A, vitamin C, iron, calcium, protein, or dietary fiber prior to any nutrient addition

(*continued*)

Table 11.1. Health claims authorized in the United States (*Cont.*)

Authorized Diet and Disease Relationship	Model Claims	Criteria
Walnuts and heart disease[3]	Supportive but not conclusive research shows that eating 1.5 ounces per day of walnuts as part of a diet low in saturated fat and cholesterol may reduce the risk of heart disease. See nutrition information for fat content.	Whole or chopped walnuts
Whole grain foods and coronary heart disease[1]	Diets rich in whole grain foods and other plant foods, and low in saturated fat and cholesterol, may reduce the risk of heart disease.	Must contain a minimum of 51% whole grain (using dietary fiber as a marker) Must meet the regulatory definitions for "low saturated fat" and "low cholesterol" Must bear quantitative *trans* fat labeling Must contain less than 6.5 grams total fat and 0.5 gram or less *trans* fat per RACC

* Indicates that dietary supplements are explicitly authorized to make this claim.
** Indicates that although this health claim does not mention supplements explicitly, FDA believes that supplements may make this claim. Ultimately, other supplements may be eligible for such claims. Telephone conversation with James E. Hoadley, Acting Team Leader, Nutrition Labeling and Program Team, Division of the Office of Nutritional Products, Labeling, and Dietary Supplements, FDA.

[1] This health claim is based on an authoritative statement from the National Academy of Sciences.
[2] Health claims are permitted expressly for dietary supplements containing stanol esters but not sterol esters. It is likely that health claims will also be permitted for dietary supplements containing stanol esters.
[3] This "qualified health claim" is authored pursuant to an FDA decision to exercise its enforcement discretion. These claims do not meet the significant scientific agreement standard set forth in the Nutrition Labeling and Education Act and must be accompanied by specific language specified by FDA that indicates the level of scientific support for the statement.

Functional Foods Are Often Sold on the Basis of Structure/Function Claims to Avoid FDA Approval Requirements That Apply to Health Claims

Although companies use the health claims originally approved by FDA, and have persuaded the agency to approve new claims, the industry generally remains dissatisfied with the length of the FDA approval process, which often can take several years. To avoid FDA approval requirements, some companies have begun making structure/function claims in lieu of health claims.

For example, Kellogg briefly marketed a new line of functional foods (called Ensemble) that contained psyllium. The product line consisted of 22 items including breads, pasta, frozen entrees such as lasagna, snack chips, cakes, and cookies.[103] Although all the products contained either psyllium or oat bran, both of which have been recognized by FDA as being useful in reducing cholesterol and lowering the risk of heart disease,[104] some items contained too much fat to qualify for an FDA-approved health claim concerning psyllium;

others provided inadequate nutritional value, and under FDA's Jelly Bean Rule were not entitled to make any health claims at all. To get around these restrictions, Kellogg made structure/function claims on products not meeting the criteria for health claims. For example, the Kellogg's Ensemble carrot cake label stated: "made with a natural soluble fiber [psyllium] that actively works to promote heart health." This statement most certainly implies that the product can help reduce the risk of heart disease, and should be considered by the Agency as an implied health claim requiring premarket approval even though the claim does not specifically mention heart disease.

It is possible that FDA would consider the statement to be a structure/function claim. In that case, the statement should have, nonetheless, been prohibited under the agency's general authority to stop misleading labeling claims because the carrot cake contains too much fat to be considered healthy for the heart.

Kellogg has also used structure/function claims to circumvent two previous FDA denials of approval for a health claim linking adequate intakes of folic acid and vitamins B6 and B12 to a reduced risk of heart disease. FDA had prohibited use of such a claim because it concluded that it was not supported by significant scientific agreement[105] or an authoritative statement.[106] Nonetheless, on April 4, 2000, Kellogg announced that a variety of its cereals would claim that "adequate intakes of folic acid, vitamin B6, and B12 may promote a healthy vascular system."[107] Kellogg boasted that "it is pleased to launch the first major application of the Food and Drug Administration's new regulations concerning structure/function claims on packaged foods,"[108] despite the fact that the rules apply only to dietary supplements and not packaged foods. Moreover, when seen in the context of its packaging and the Web site to which it refers, the claim is in actuality an unapproved health claim. FDA sent Kellogg a letter in October 2000 expressing its belief that "the relationship correlating folic acid, vitamins B6 and B12, and homocysteine levels to 'heart health' is [not] clearly established."[109]

Subsequently, as required by the *Pearson* decision, FDA approved a qualified health claim for dietary supplements on the link between folic acid, vitamins B6 and B12, and a reduced risk of heart disease.[110] Even if the approval, which applies only to supplements, were applicable to food, Kellogg's claim would have been in violation of the law because it did not use the qualifying language specified by FDA.[111]

Similarly, the Campbell Soup Company, which produces V8 Juice, makes a structure/function claim on the label that states: "Research . . . suggests that antioxidants [in V8 juice] may play an important role in slowing changes that occur with normal aging." The product contains too much sodium to make an FDA-approved health claim concerning benefits from consuming fruits and vegetables.[112] The company, however, can evade FDA premarket approval requirements by making a structure/function claim instead of a health claim. In this case the structure/function claim on the label is especially misleading because diets high in sodium are linked to high blood pressure, a condition that is associated with aging.[113] Moreover, the evidence linking the consumption of antioxidants to aging is preliminary. Nonetheless, the company continues to market V8 as a functional food that may slow changes that occur with aging.

Some Companies Are Selling Functional Foods As Dietary Supplements to Avoid FDA Regulation

Other companies are attempting to avoid FDA regulation by positioning functional food products as dietary supplements. Since 1994, food and drug companies increasingly have envied the rights of dietary supplement manufacturers to market products with ingredients that have not been approved by FDA. As sales of dietary supplements grew after the passage of DSHEA, more and more companies began to consider how they could avoid complying with FDA approval processes by marketing functional foods as dietary supplements.

In the most heavily publicized case,[114] McNeil Consumer Health Care, a subsidiary of Johnson & Johnson, attempted to market Benecol, a margarine substitute promoted on the basis of a cholesterol-lowering functional ingredient as a dietary supplement.[115] Had the company been permitted to do so, it could have avoided compliance with the U.S. law that requires that ingredients be approved as food additives or recognized as GRAS.[116] Although the product has been sold in Finland since 1995, in a precedent-setting action, FDA told the company that it would be illegal to sell the product in the United States. FDA stated that Benecol could not legitimately be marketed as a dietary supplement, and that the plant stanol esters that it contains would be considered unapproved food additives or perhaps even drugs, both of which require FDA premarket approval.[117] In response to FDA pressure, the company decided to market the product as a conventional food, after first demonstrating to the agency that plant stanol esters are GRAS.[118] It also revised the product's original claim ("promotes healthier lower cholesterol levels") to "benefits cholesterol levels." Take Control, a similar margarine-type product also began marketing after demonstrating to FDA that the sterol esters in its product are GRAS.[119] Ultimately, however, the manufacturers of Benecol and Take Control filed petitions for health claim approval and were granted permission to make a claim expressly linking the plant stanol/sterol esters to a reduced risk of coronary heart disease.[120]

The Benecol and Take Control spreads also illustrate the significance of the nutritive value requirement. Dietary supplements can, unlike foods, make structure/function claims without meeting the requirement that the substance, which is the subject of a claim, provide nutritive value.[121] Because FDA had concluded that the products were foods and not dietary supplements, no structure/function claims could have been permitted unless the plant stanol/sterol esters contained nutritive value—FDA concluded that they do. The "nutritive value" requirement is, thus, an important safeguard for preventing drugs and medicinal herbs from being added to the food supply.

As previously mentioned, dietary supplement ingredients do not have to be approved by FDA or be recognized as GRAS.[122] It is thus not surprising that many food companies are increasingly trying to market functional foods as dietary supplements.

For example, The Hain Food Group, Inc. began selling a line of canned soups containing medicinal herbs. The products were labeled as supplements and were sold as a part of a new product line that was boldly called Kitchen Prescription.[123] The labels for those products contained structure/function claims normally found on dietary supplement labels. Hain's Chicken Broth and Noodles with echinacea made the structure/function claim for "support of the immune system." The label of Hain's Split Pea Soup containing St. John's wort claimed to "give your mood a natural lift."[124] By labeling the canned soups as a "supplement," the company apparently hoped to escape its obligation to obtain an FDA food additive approval or demonstrate to FDA that echinacea and St. John's wort are GRAS when used as food ingredients. Also, by labeling the soups as a "supplement," the company wanted to be able to make structure/function claims even though the herbs did not provide taste, aroma, or nutritive value.[125]

The Hain Food Group, however, abandoned its marketing of these soups after FDA sent a courtesy letter concluding that the products were foods that contained unapproved additives, that they were making therapeutic claims that must be redrafted as structure/function claims, and that any claims to affect the structure or function of the body must be achieved through nutritive value.

Another example of this trend is illustrated by Amurol Confections' marketing of Stay Alert Caffeine Supplement Gum. The company labeled the product as a dietary supplement, even though the name of the product includes the name of a conventional food, "gum," and the product looks like and is marketed in stores next to other chewing gums. Significantly, caffeine is approved for food use only in cola-type beverages[126] and is otherwise considered a drug covered by FDA regulations for nonprescription stimulants.[127] FDA, however, has taken no enforcement action.

In an interesting variation on the theme of marketing a food as a dietary supplement, The Dannon Company continues to market Actimel, a breakfast drink with three active yogurt cultures, as a dietary supplement, even though FDA sent a courtesy letter informing the company that the product is a food.[128] FDA also had concerns that the claim that the product is "clinically proven to help fortify your body's natural defenses" is not worded as a structure/function claim.[129] It is unclear why Dannon wants to maintain its dietary supplement status for the product, even though FDA stated that it expects the ingredient to be GRAS for use in food.

An even more interesting case involves Cholestin, a product promoted to reduce cholesterol levels, that was available in both capsule and candy bar form.[130] From May 1998 until March 30, 2001, FDA and Pharmanex, Inc. litigated whether the red rice yeast in Cholestin was a dietary supplement or a drug.[131] The court ultimately ruled that the active ingredient in the product was the drug ingredient Lovastatin, not red rice yeast. Because Lovastatin was not marketed as a dietary supplement or food prior to its approval as a new drug, Cholestin cannot be marketed as a dietary supplement.[132]

It is not known the extent to which FDA will enforce the law and prevent additional food companies from marketing functional foods as dietary supplements to avoid FDA requirements intended to ensure the safety of food ingredients and the validity of claims on food labels. Although FDA has taken action in the case of Benecol, there is no indication that the agency is willing to devote the resources necessary to take enforcement action in cases that raise similar issues. Initially, FDA issued only a small number of warning letters to manufacturers of functional food products containing herbal ingredients that it does not believe to be GRAS for use in food.[133] These herbs include some of the most popular on the market, such as kava, ginkgo, echinacea, and St. John's wort. FDA also warned these companies that their claims are false or misleading.[134]

In response to the "significant growth in the marketplace of conventional food products that contain novel ingredients, such as added botanical ingredients,"[135] FDA ultimately issued a Letter to Industry to remind manufacturers of longstanding legal requirements governing conventional food products that some companies are not following.[136]

The agency reminded manufacturers that:

- Ingredients that are neither approved food additives nor generally recognized as safe for use in food may not be added to products.
- Health claims and nutrient content claims may not be used unless the claims have been authorized by regulation or by statute under the authoritative statement notification procedure.
- Structure/function claims may not claim to diagnose, treat, cure, or prevent disease.

- Structure/function claims may not be false or misleading.
- The claimed effect in a structure/function claim must be achieved through nutritive value.[137]

FDA has done only limited follow up on this Industry Letter by sending individual warning letters to manufacturers of functional foods that are not in compliance with these requirements.[138] In a revealing article published in a leading food industry trade publication, a senior FDA official admitted that the agency would not be prosecuting companies whose products contained only minimal amounts of ingredients not considered to be GRAS. This policy places the consumer in a Catch 22 situation by encouraging companies to seek GRAS status by selling products that contain misleadingly small amounts of the purported functional ingredient. In such situations, consumers are assured that a product is safe only because it lacks any significant amount of the functional ingredient.[139]

FDA also has signaled its intent to ensure that the Internet is not used as a means of circumventing FDA requirements for health claims. In a January 19, 2001, letter to Ocean Spray Cranberries, Inc.,[140] FDA determined that references to the company's Web site on its product labels caused the Web site to be "labeling." As a result, any claims made on the label are to be in accordance with FDA requirements. FDA determined that Ocean Spray's Web site violated the law by: (1) making a number of unauthorized health claims; (2) not including all the required elements when it used authorized health claims; and (3) making claims to treat, cure, or mitigate disease that are "beyond the scope of the types of claims that are permitted on foods."[141]

FDA was particularly concerned that statements regarding grapefruit juice and cancer care posed a serious safety threat by overstating any possible benefits of drinking that beverage and understating the real risks of interactions between certain drugs and grapefruit juice. For example, Ocean Spray claims that "grapefruit juice can lower cholesterol in people with high blood cholesterol levels." FDA pointed out, however, that the labels of such cholesterol-lowering drugs, such as Simvastatin, warn consumers about the risk of grapefruit juice interaction.

FDA's warning letter to Ocean Spray prompted an outcry from the Washington Legal Foundation (WLF) that FDA had overreached its authority and was in violation of the First Amendment. WLF petitioned FDA to reverse this opinion and to issue a policy statement clarifying that material on the Internet constitutes advertising and that FDA would defer to the FTC.[142] FDA rejected this petition, determining that it must have the flexibility to consider such issues on a case-by-case basis. Because of the fast-changing nature of Internet technology, the agency did not want to be bound by a policy that could quickly become obsolete.[143]

FDA action in this case is significant especially because Internet Web sites have been viewed by the FTC as advertising, not labeling.[144] FTC has applied weaker substantiation requirements to advertising than FDA has applied to labeling. It is not clear how the agencies will share jurisdiction in this area.

Meanwhile, some states are considering taking action against manufacturers of functional food products that contain illegal ingredients and/or make false or misleading claims, and consumer groups continue to urge FDA to halt the sale of these products.[145]

The National Advertising Division (NAD) of the Council of Better Business Bureaus, a self-regulating industry group, also has asked FDA to review and take "possible law enforcement action" against the manufacturer of Arizona Rx Memory Mind Elixir.[146] NAD determined that the elixir's label created the impression that the product could improve or enhance memory, but there was no evidence in the record to support the message. When the manufacturer refused to comply with NAD's recommendations, NAD took the highly unusual step of referring the matter to FDA. Normally, NAD limits itself to advertising issues.[147]

Some Functional Foods Are Sold As Medical Foods to Avoid FDA Regulation

Medical foods are intended for consumption under the supervision of a physician. The failure of FDA to complete its rulemaking on medical foods and establish requirements for approval of medical food claims, however, has inspired some manufacturers to market functional foods (which are not consumed under the supervision of a physician) as medical foods. These manufacturers pay lip service to the requirement that the products be used under medical supervision by simply stating on the product label, "use under the supervision of a physician."[148] But, given the fact that these products are marketed directly to consumers, the products may be used without a physician's knowledge.

One industry observer predicts that medical foods are positioned for rapid growth in the next five years and "will be the biggest branch of the developing food market."[149] A brochure advertising a March 1997 conference sponsored by the Pharmaceutical Division of the Institute for International Research boasts that "[m]edical food regulations provide a pathway for health claims for products as aids in the dietary management of diseases or conditions."[150] During one discussion session, a product called Cardia Salt (a salt alternative)[151] was to be featured as a case study of how functional foods can escape FDA approval requirements for health claims by being positioned as medical foods.[152]

Other prominent examples of functional foods that have been marketed as medical foods include NiteBite[153] and Zbars,[154] snack bars designed to reduce diabetics' risk of developing hypoglycemia. They bear little resemblance to bona fide medical foods, such as TraumaCal Liquid or Travasorb Hepatic Powder, and often are not consumed under the supervision of a physician. Rather, these products appear to be similar to "foods for special dietary use." Health claims for those products, however, must be pre-approved by FDA.[155] Not surprisingly, the manufacturers are attempting to market the products in a manner that does not require FDA approval of label claims. A similar example involves a product called HeartBar that is being promoted as the first medical food developed for the management of coronary artery and vascular disease.[156] The product contains arginine, a crystalline basic amino acid that its manufacturer claims will enhance the body's production of nitric oxide. The manufacturer states that raising the levels of nitric oxide will increase blood flow and pain-free exercise.[157] Like NiteBite and Zbars, however, HeartBar is sold directly to consumers and appears to be labeled as a medical food in order to avoid FDA approval requirements that might apply if the product were labeled correctly.

The potential for harm to consumers is readily apparent. Contrary to FDA intentions, the latest medical foods to enter the marketplace are not used by consumers under the supervision of a physician. They are sold with health claims that are neither evaluated by a physician nor approved by FDA.[158] Thus, there is no assurance that such claims are valid.

Companies Limit Health Claims to Advertising to Avoid FDA Scrutiny

In the United States, the responsibility for stopping deceptive claims in food advertising is shared by FDA and FTC.[159] During the last several decades, FTC has eschewed industry-wide regulation in favor of case-by-case enforcement and employed enforcement standards that were often more lenient than FDA's. In some situations, FTC has permitted companies to make health claims in advertising that are not permitted on labeling.[160] Not surprisingly, the food industry has favored FTC enforcement policies over those of FDA. Since 1954, FTC and FDA have operated under a Memorandum of Understanding under which FTC has assumed primary responsibility for food advertising and FDA has assumed primary responsibility for product labeling.[161] At the time of the agreement, FDA's reliance on FTC to regulate food advertising may have made sense. Now that many food advertisements are filled with claims involving complex scientific matters, however, FDA's reliance on FTC (an agency with no scientific expertise) to police deceptive claims may no longer be appropriate.

In an attempt to harmonize its enforcement policies with those of FDA, FTC issued an *Enforcement Policy Statement on Food Advertising* following the implementation of NLEA.[162] Although FTC's enforcement policy states that the agency will defer to FDA's scientific expertise, FTC, as a matter of practice, fails to take enforcement action against claims in advertising, even though such claims have not been approved by FDA for labeling.

This inconsistency in the U.S. regulatory scheme is sometimes exploited by food and dietary supplement companies. For example, Welch Foods, Inc. advertised that "[g]rowing evidence suggests that diets rich in antioxidants may reduce the risk of some cancers . . ." and noted that ". . . Welch's purple grape juice has more than three times the naturally-occurring antioxidants of other popular juices."[163] FDA, however, rejected the health claim that diets rich in antioxidants may reduce the risk of some cancers because there was not "significant scientific agreement" to support the claim.[164] Nonetheless, FTC took no action to stop the claim in advertisements.

Heinz advertised that lycopene in its Ketchup "may help reduce the risk of prostate and cervical cancer."[165] Heinz has not petitioned FDA for permission to make such a claim on labels, and unapproved health claims on labeling are prohibited expressly by law. The case for advertising, however, is much different. Heinz does not need FTC permission to make such claims in advertisements. Indeed, the company had run such claims in major U.S. publications since January 1999.[166] FTC wrote to Heinz on December 15, 1999, expressing its concern that the claims may have violated the Commission's advertising substantiation standards. Nevertheless, FTC closed its investigation because Heinz had ceased running the advertisement and stated that it would not resume using that specific advertisement in the future. (Heinz, however, continues to make similar claims in promotional materials distributed to the media.) FTC could have ordered the company to cease and desist after the fact if it determined that the cancer prevention claims were not supported by competent and reliable scientific evidence or were otherwise false or misleading. That process, however, can take years and the Commission apparently did not want to invest the resources.[167]

Meanwhile, Campbell ran similar ads for its Tomato Soup.[168] Although the NAD found evidence for such a claim to be competent and reliable, it concluded that the company had not adequately communicated that the studies show only a causal relationship between the ingestion of tomato products and the reduction of cancer risk. The studies "do not prove lycopene's role as a chemo-preventive agent."[169] The FTC has taken little enforcement action in such cases despite complaints from consumer organizations.[170]

To date, the FTC has undertaken only two enforcement actions in such cases, against

Wonder Bread and HeartBar. Wonder Bread claimed that, as a good source of calcium, the product helps children's minds work better and helps them remember things. The FTC concluded that the claims were unsubstantiated and entered into a consent agreement obligating the manufacturer of Wonder Bread to refrain from making such claims until the company has "competent and reliable scientific evidence that substantiates the representation."[171]

The FTC recently entered into a consent agreement with HeartBar's manufacturer, Unither Pharma, Inc., under which the company agreed not to make advertising claims unless they are supported by scientific evidence. Among the unsubstantiated advertising claims cited by FTC were claims to reduce the risk of developing heart disease, reverse damage to the heart, and reduce the need for heart medication in patients with diagnosed heart conditions.[172]

Conclusions

The marketing of functional foods raises fundamental questions concerning the nature of foods and drugs. The central issue for FDA, and for society at large, to deliberate is whether—and if so, where—the line between foods and drugs should be drawn. Traditionally, FDA has drawn clear distinctions between foods and drugs, but that distinction is becoming fuzzy. For example, psyllium is both an over-the-counter (OTC) laxative and a food ingredient (with an FDA-approved health claim).[173] Caffeine is sold as an OTC stimulant and is used in soft drinks as a flavor additive. Recently, however, manufacturers have begun adding caffeine to chewing gum and promoting the gum on the basis of its stimulant properties. The development of dozens of other functional foods containing additives with drug-like effects ultimately will force FDA to determine how such products should be regulated.

Traditionally, the addition of ingredients not found normally in food transforms a food into a drug. In accordance with that view, FDA has adhered to a policy that food additives must have a "physical, nutritive, or other technical effect in food"[174] (that is, they must provide nutrients, act as a preservative, color additive, flavoring agent, and so on). In addition, FDA regulations provide that a substance that is the subject of a health claim must "contribute taste, aroma, or nutritive value, or any other technical effect . . . to the food."[175] Similar requirements apply to substances in foods that are the subject of structure/function claims.[176]

FDA reaffirmed the importance of nutritive value in the position it took on Benecol and Take Control. If FDA abandons its traditional position, companies might be able to add with impunity a variety of drug-like ingredients to food products. The industry, however, argues that FDA's current requirements inappropriately prohibit companies from adding beneficial substances to foods and thereby improve the public's health.

The agency should encourage debate on this issue within the academic, public health, and consumer communities. While the public debate continues, FDA and FTC should act to ensure that functional foods are regulated in a manner that maximizes health benefits and minimizes health risks. The agencies should ensure that only safe and effective products are marketed and that claims on labels and in advertisements are supported by adequate scientific evidence and are not otherwise misleading.

The steps outlined in the following sections should be taken to accomplish those objectives.

FDA Should Prohibit Companies from Marketing Functional Foods As Dietary Supplements

DSHEA provides a low level of protection from potentially hazardous ingredients and misleading claims. When companies attempt to label a food as a dietary supplement, FDA should evaluate the product comprehensively to determine in which regulatory category it should be placed. If, for example, a product replacing another food in the diet is packaged like a food, is sold in grocery stores next to foods, and/or is marketed, in part, on the basis of flavor or taste, then FDA should determine that the product is a food and not a dietary supplement.[177] In such cases, the agency should take prompt enforcement action.

FDA Should Close the Loophole That Allows Unapproved Health Claims on the Labels of Medical Foods

FDA should require that health claims for medical foods be approved prior to marketing, as is required for conventional foods. Presently, companies are taking advantage of the medical foods exemption from NLEA and are selling functional foods as medical foods to escape FDA's health claims approval requirements. Furthermore, companies should be prohibited from calling functional foods sold in grocery and health food stores "medical foods" unless they are designed to be used under the supervision of a physician.

FDA should promptly end those practices by completing its rulemaking proceeding for medical foods. In the interim, the agency should take case-by-case enforcement actions.

Functional Ingredients in Foods Should Be GRAS or Regulated As Food Additives

Functional ingredients added to foods to provide a physiological effect should be regulated no less strictly than are ingredients added to preserve, flavor, or color a food. Manufacturers should be required to submit a petition to FDA demonstrating that the ingredients pose a reasonable certainty of no harm (the safety standard applicable to food additives) or demonstrate that they are GRAS. FDA should increase its number of enforcement actions and ensure that all products are in compliance with the agency's regulations.

In deciding whether to permit foods to have a functional ingredient with drug-like properties, FDA must make safety its top concern. The ingredient would have to present a wide margin of safety so that it could be eaten by virtually all consumers in large amounts without any adverse effects. The substance would have to be safe for all consumers including children, restaurant patrons, or individuals with compromised immune systems, who might unintentionally or unknowingly consume the product. Furthermore, the substance would have to be promoted for health conditions that could be evaluated by the consumer and that do not require the intervention of a health professional.

In certain cases, a post-marketing surveillance system should be established to monitor adverse effects from functional ingredients in foods. For example, manufacturers of products containing Olestra, a fat substitute approved by FDA as a food additive, are required to forward reports of adverse effects to FDA. Such reporting requirements also should be required for manufacturers of functional foods to track problems if particular ingredients are suspected of causing problems.

Health Claims for Functional Foods Should Be Approved Pursuant to the NLEA

FDA should make it clear that the NLEA applies to health claims for functional foods as well as conventional foods, and should require companies to comply. Companies should be made to demonstrate that claims are supported by significant scientific agreement based on the totality of the publicly available scientific evidence.

FDA Should Issue Regulations Controlling Structure/Function Claims for Foods

Some companies are trying to evade FDA's strict health claim requirements by taking advantage of a legal loophole permitting them to make structure/function claims that do not require the agency's approval. Many of these claims are unsubstantiated or otherwise misleading. FDA should end the practice by requiring that structure/function claims be based on universally recognized factual statements concerning known and substantively significant relationships regarding the effect of a substance on the structure or function of the body. For example, claims such as "calcium is necessary for bone growth and development" or "vitamin A is necessary for good vision" could be permitted, whereas claims that "antioxidants may play an important role in slowing changes that occur with normal aging" or that "lycopene may help ensure normal functioning of the prostate gland" should be prohibited until those relationships have been established.

Also, FDA should require that foods making structure/function claims, similar to those making health claims, meet specified nutrient levels. Functional foods, whether they make health claims or structure/function claims, or simply promote the presence of an ingredient, should not be permitted to have unhealthful levels of fat, saturated fat, cholesterol, or sodium.[178] Furthermore, as are conventional foods that make health claims, functional foods should be required to contain, prior to fortification, at least 10 percent of the Reference Daily Intake of one or more of the following nutrients: vitamin A, vitamin C, iron, calcium, protein, or fiber.[179] These requirements would prohibit functional foods such as fatty cakes and cookies that contain herbs.

FTC Should Update Its Enforcement Policy Statement on Food Advertising to Clarify That Health Claims for Functional Foods Will Be Considered Nondeceptive Only if They Meet Requirements Set Forth by FDA

Advertisements that hype functional ingredients and make unapproved health claims, or that make misleading or inadequately substantiated structure/function claims, undermine FDA's efforts to protect consumers. FTC's *Enforcement Policy Statement on Food Advertising* should be updated to address specifically the marketing of functional foods and should be fully harmonized with FDA's regulatory policies. Furthermore, FTC should work with FDA to develop a consistent policy for the regulation of structure/function claims.

These steps will protect consumers, establish a level competitive playing field for industry, and help to ensure that the full potential of functional foods is realized.

Notes

1. General Accounting Office, Food Safety Improvements Needed in Overseeing the Safety of Dietary Supplements and "Functional Foods," (GAO/RCED-00-156) July 2000, at 3.

2. Id. at 6.

3. Functional food sales increased from $11.3 billion to an estimated $16.2 billion from 1995–1999. Projected sales are expected to reach $49 billion by 2010. See id. at 6–7.

4. FDCA §§ 201 (f) (food), 201 (ff) (dietary supplements), 201 (g) (drugs), 411 (c) (foods for special dietary use); Orphan Drug Act § 5(b) (medical foods); 21 C.F.R. §§ 321(f), (ff), and (g), 350 (c), 360ee(b)(3).

5. FDA has issued a handful of letters on functional foods containing ingredients that it does not recognize to be "generally recognized as safe" for use in foods. See Foret Letters, infra note 134.

6. FDA Labeling Policy "Established Through Enforcement": Campbell, Food Reg. Wkly., Jan. 4, 1999, at 4 (reporting on speech by Betty Campbell, acting director of the Office of Food Labeling [retired], Center for Food Safety and Applied Nutrition (CFSAN), Food and Drug Administration [FDA], to the FDLI annual meeting [December 1998]).

7. Pub. L. No. 75-717, 52 Stat. 1040 (1938) (codified as amended 21 U.S.C. §§ 301 et seq. (2002)).

8. General recognition of safety must be based on the views of experts throughout the scientific community who have common knowledge, based on accepted scientific procedure, that a substance is safe. In the case of substances used in food prior to 1958, however, general recognition of safety can be based simply on common experience gained from use of the substance in food within or outside the United States. 21 U.S.C. § 321 (FDCA § 201(s)).

9. Pesticides, pesticide residues, color additives, "prior-sanctioned" ingredients, new animal drugs, and dietary supplement ingredients are exempted from the food additive classification. Prior sanctions exist for specific uses of particular substances that were explicitly approved by FDA on an informal basis prior to the passage of the 1958 Food Additives Amendment. 21 U.S.C. § 321(s) (FDCA § 201(s)). FDA has published a list of prior-sanctioned substances, and manufacturers may petition for the recognition of additional prior sanctions. 21 C.F.R. pt 181.

10. 21 U.S.C. § 348.

11. Pub. L. No. 103-417, 108 Stat. 4325 (codified at 21 U.S.C. § 301 note (2002)).

12. "The term 'dietary supplement':
  1. means a product (other than tobacco) intended to supplement the diet that bears or contains one or more of the following dietary ingredients:
  (A) a vitamin; (B) a mineral; (C) an herb or other botanical; (D) an amino acid; (E) a dietary substance for use by man to supplement the diet by increasing the total dietary intake; or (F) a concentrate, metabolite, constituent, extract, or combination of any ingredient described in clause (A), (B), (C), (D), or (E);
  2. means a product that . . . (B) is not represented for use as a conventional food or as a sole item of a meal or the diet; and (C) is labeled as a dietary supplement . . .
  3. . . . except for purposes of section 201(g) [definition of a drug] a dietary supplement shall be deemed a food within the meaning of this Act." 21 U.S.C. § 321 (ff).

13. Id. at § 350b.

14. I. Scott Bass & Anthony L. Young, Dietary Supplement Health and Education Act: A Legislative History and Analysis 43–44 (FDLI 1996).

15. 21 U.S.C. § 342(f)(1)(A)&(B).

16. Id. at § 342(f)(1)(C).

17. FDA, Adverse Events with Ephedra and Other Botanical Dietary Supplements, FDA Medical Bulletin, Sept. 1994, available at www.cfsan.fda.gov/~dms/ds-ephe2.htm.

18. 69 Fed. Reg. 6788 (Feb. 11, 2004).

19. See, e.g., Press Release, Sen. Dick Durbin, Durbin Renews Call on Bush Administration to Suspend Sales of Diet Supplement Ephedra, Aug. 16, 2002, available at http://durbin.senate.gov/durbinnew200/press2002/08/200294B58.ht. ("The only law now on the books to regulate dietary supplements was written to cover products like vitamins, minerals and herbal food products such as garlic . . . . The existing laws simply do not address the current dietary supplement situation and our government shouldn't just stand by as more and more American consumers unwittingly risk their health and even their lives.") Id.

20. FDA Warns Consumers Against Consuming Dietary Supplements Containing Tiractricol (Nov. 21, 2000); FDA Concerned About Botanical Product, including Dietary Supplements, Containing Aristolochic Acid (Apr.

11, 2001); FDA Public Health Advisory: Risk of Drug Interactions With St. John's Wort and Indinavire and Other Drugs (Feb. 10, 2000); FDA Warns Against Consuming Metabolic Accelerator (Nov. 11, 1999) available at (last visited June 1, 2001) vm.cfsan.fda.gov/~dms/ds-warn.html.

21. FDA, Consumer Advisory, Kava-Containing Dietary Supplements May be Associated with Severe Liver Injury, Mar. 25, 2002, available at http://www.cfsan.fda.gov/~dms/addskava.htm.

22. Id., FDA Warns of Kava Link to Serious Liver Damage, Newsday, Mar. 26, 2002 at A40. Canada, Singapore Ban Sales of Kava, Whole Foods, Oct. 2002 at 16.

23. Germans warned against kava products on Internet, Reuters, Oct. 10, 2002.

24. Pub. L. No. 101-535, 104 Stat. 2353 (codified at 21 U.S.C. §§ 301 note, 321, 337, 343, 343 notes, 343-1, note, 345, 371 (2002)).

25. 21 U.S.C. §§ 343(r)(1)(B), (r)(3)(C) (FDCA §§ 403(r)(1)(B), (r)(3)(C)).

26. Id. at § 343(r)(3)(B)(i).

27. U.S. law further permits claims about how foods and dietary supplement ingredients can affect the structure of the body or the functioning of bodily systems (structure/function claims). 21 U.S.C. §§ 201(g)(1)(C), 343(r)(6) (FDCA §§ 201(g)(1)(C), 403(r)(6)).

28. 52 Fed. Reg. 28,843 (Aug. 4, 1987).

29. Zachary Schiller et al., Can Cornflakes Cure Cancer?, Business Week, Oct. 9, 1989, at 114.

30. 21 U.S.C. §§ 343(r)(1)(B), (r)(5)(D) (FDCA §§ 403(r)(1)(B), (r)(5)(D)).

31. See, e.g., Passage of the Dietary Supplement Health and Education Act, supra note 11. Pearson v. Shalala, 164 F.3d 650 (D.C. Cir. 1999). See infra note 53 and accompanying text.

32. Pub. L. No. 105-115, 111 Stat. 2296 (1997) (amending 21 U.S.C. §§ 301 et seq. (2002)).

33. 21 U.S.C. § 343(r)(3)(C).

34. Id. at § 343 (r)(3)(C)(ii).

35. Id. at § 343(r)(3)(C)(i).

36. Id.

37. Office of Food Labeling, Center for Food Safety and Applied Nutrition (CFSAN), FDA, Guidance for Industry, Notification of a Health Claim or Nutrient Content Claim Based on an Authoritative Statement of a Scientific Body (June 1, 1998).

38. 63 Fed. Reg. 34,084-34,117 (June 22, 1998).

39. FDA concluded that the statements on which Weider relied were not authoritative because they were not based on a "deliberative review of all relevant scientific evidence." Id. at 34,095-96. FDA explained that the reports from government contractors, press releases, and other informational material on which the company relied did not constitute "authoritative statements" within the meaning of the law. Id. at 34,108-09. Other claims were disqualified because FDA already had issued an applicable health claim, and the proposed claim was inconsistent with that claim, or the claim itself was not considered to be a health claim. Id. at 34,094-102.

40. FDA, Health Claim Notification for Whole Grain Foods (July, 1999) available at (last visited Dec. 5, 2000) vm.cfsan.fda.gov/~dms/flgrains.html. Congressional attempts to modify the health claims provisions continue. For example, Rep. Frank Pallone (D-NJ) introduced legislation, The Nutraceutical Research and Education Act, H.R. 3001, 106th Cong. (1999), that would permit manufacturers who successfully have petitioned FDA to approve a health claim based on at least one clinical study to have the exclusive right to use that health claim for a period of 10 years. The bill is important because it reflects an increasingly popular view that research on the safety and efficacy of dietary supplements, functional foods, and medical foods must be done, but that manufacturers do not have the incentive to conduct such research unless they receive some sort of competitive advantage, such as exclusivity. Id.

41. FDA, Health Claim Notification for Potassium-Containing Foods (Oct. 31, 2000) (last visited Nov. 17, 2000) available at vm.cfsan.fda.gov/~dms/hclm-k.html.

42. FDCA § 403 (r)(7), 21 U.S.C. § 343 (r)(7).

43. 65 Fed. Reg. 54,686 (Sept. 8, 2000).

44. 67 Fed. Reg. 61,773 (Oct. 2, 2002).

45. 67 Fed. Reg. 71,461 (Dec. 2, 2002).

46. 21 U.S.C. § 343(r)(3)(A)(ii), 21 C.F.R. §§ 101.14(e)(3) and (4). One exception is health claim approvals for margarine substitutes containing plant stanol/sterol esters. Those products, which are high in fat, may claim to reduce cholesterol and the risk of heart disease. 21 C.F.R. § 101.83.

47. 21 C.F.R. § 101.14(e)(6).

48. See, 21 C.F.R. § 101.80(c).

49. Id. at § 104.20(a). FDA explains that:
"The fundamental objective of this [policy] is to establish a uniform set of principles that will serve as a model for the rational addition of nutrients to foods. The achievement and maintenance of a desirable level of nutritional quality in the nation's food supply is an important public health objective. The addition of nutrients to specific foods can be an effective way of maintaining and improving the overall nutritional quality of the food supply. However, random fortification of foods could result in over- or under-fortification in consumer diets and create nutrient imbalances in the food supply. It could also result in deceptive or misleading claims for certain foods. The Food and Drug Administration does not encourage indiscriminate addition of nutrients to foods, nor does it consider it appropriate to fortify fresh produce, meat, poultry, or fish products, sugars, or snack foods such as candies and carbonated beverages. To preserve a balance of nutrients in the diet, manufacturers who elect to fortify foods are urged to utilize these principles when adding nutrients to food."

50. FDA's ability to regulate products containing vitamins and minerals was curtailed severely by what is commonly known as the Proxmire Amendment. Under the amendments, FDA may not set maximum limits of potency, may not regulate a dietary supplement as a drug because its potency is greater than what FDA determines to be nutritionally rational, and may not limit the combination or number of vitamins or minerals in a product. Health Research and Health Services Amendments, Pub. L. No. 94-278, 90 Stat. 410 (Apr. 23, 1976), adding FDCA § 411 (21 U.S.C. § 350 (as amended)). In fact, more and more products are appearing on the market as candy or sodas filled with vitamins and minerals and herbal remedies implicitly promoted for a multitude of symptoms. Even a popular children's candy, Gummi Bears, is being marketed as a functional food called Yummi Bears. The product contains concentrates of a variety of vegetables and boasts that two bears contain 50 milligrams of vegetables. But that translates into less than one hundredth of a standard serving of vegetables, and Yummi Bears costs about 10 times as much as the regular gummies. See also Karen Springen, Candy and Coffee Are the New Health Foods? We Wish, Newsweek, May 25, 1998, at 14. The product sends a misleading message to children and adults alike, that is, that junk foods with added ingredients are healthful.

51. 21 C.F.R. § 101.14(b)(3)(i).

52. 164 F.3d 650 (D.C. Cir.) reh'g en banc denied, 1999 U.S. App. Lexis 5954 (1999).

53. Id. at 659.

54. U.S. Constitution, amend. I.

55. Pearson, 164 F.3d, at 660.

56. 64 Fed. Reg. 71,794 (Dec. 22, 1999).

57. 65 Fed. Reg. 59,855-57 (Oct. 6, 2000).

58. FDA, Press Release, FDA Announces Initiative to Provide Better Health Information for Consumers, (Dec. 18, 2002).

59. Press Release, FDA, FDA to Encourage Science-Based Labeling and Competition for Healthier Dietary Choices (July 10, 2003).

60. National Academy of Sciences, Institute of Medicine, Evolution of Evidence for Selected Nutrient and Disease Relationships (2002) at 58.

61. Id.

62. Letters from Christine Taylor, Dir., ONPLDS to D.J. Soetaert, Pres., International Tree Nut Council and Sarah Taylor, Attorney, California Walnut Council (July 14, 2003), available at http://www.cfsan.fda.gov/qhcnut2.html.

63. 21 U.S.C. § 350(c)(3) (FDCA § 411(c)(3)).

64. Id. at § 360ee(b)(3).

65. 21 C.F.R. §§ 105.62, 105.65, 105.66.

66. FDA regulations require such foods to make specific label statements concerning the use of the product. Such statements are not considered to be health claims. Any additional statements relating to the relationship between a nutrient and a disease or health-related condition, however, are considered to be health claims and are subject to FDA approval requirements. 58 Fed. Reg. 2478, 2482 (Jan. 6, 1993).

67. Stephen H. McNamara, So You Want to Market a Food and to Make Health-Related Claims? How Far Can You Go? What Rules of Law Will Govern the Claims You Want to Make?, 53 Food & Drug Law Journal, 421, 434 (1998).

68. 21 U.S.C. § 360ee(b)(3) (emphasis added).

69. Telephone interview with Robert Moore, Division of Special Nutritionals, CFSAN, FDA (Jan. 1999).

70. Id.

71. Id.

72. Lisa Benavides, NiteBite Snack Bar Gives Diabetics a Sweet Solution, Better Business Journal, Aug. 30, 1996, at 3; Martha Groves, Preventive Measures; Search for Therapeutic Foods Picks Up Steam, L.A. Times, Nov. 7, 1996, at D2. The labeling for ZBar specifically stated that it is a new medical food; Conference Notice, Promoting Functional Foods, Medical Foods and Nutritionals, Nutrimarket '97, Pharmaceutical Division of the Institute for International Research (Mar. 24–25, 1997), at 5.

73. 61 Fed. Reg. 60,661, 60,663-64 (Nov. 29, 1996).

74. 68 Fed. Reg. 19,766 (Apr. 22, 2003).

75. In addition, dietary supplements are permitted to make statements of general well-being without being classified as drugs. 21 U.S.C. § 343(r)(6). As with any statement on a food label, such claims must not be "false or misleading in any particular." 21 U.S.C. § 343(a) (FDCA § 403(a)).

76. The structure/function provision originated in 1938 when Congress added it to the definition of a "drug" in section 201(g)(1)(C) of the FDCA. Prior to this revision, a product was considered a drug if it was recognized as such in the U.S. Pharmacopoeia or was intended to have a therapeutic effect, that is, it was used in the "diagnosis, cure, mitigation, treatment or prevention of disease." Charles Wesley Dunn, Federal Food, Drug, and Cosmetic Act, A Statement of its Legislative Record 1053 (1938) (G.E. Stechert & Co. 1987). Congress was troubled that the drug definition did not encompass products that had only physiologic effects, such as anti-fat products. Congress, therefore, broadened the reach of the drug definition so that FDA could appropriately regulate products having only a physiologic effect, that is, those affecting a structure or function of the body—as opposed to a therapeutic one. At the same time, however, Congress specifically exempted foods that make structure/function claims from the definition of a drug. Id. at 239, 1053, 1126, 1247. With the passage of DSHEA, dietary supplements can also avoid regulation as a drug as long as the claims made fall within the parameters of a structure/function claim or claim of general well-being, and the products follow other labeling requirements applicable to dietary supplements.

77. 21 U.S.C. § 343(r)(6) (FDCA § 403(r)(6)).

78. 21 C.F.R. § 101.93(a).

79. 21 U.S.C. § 343(r)(6)(C).

80. See U.S. General Accounting Office, supra note 1, at 11, 26.

81. 21 U.S.C. § 343(r)(6)(A).

82. Id. at § 343(r)(1)(B). In its regulations implementing the NLEA, FDA defines a "health claim" as "any claim made on the label or in labeling of a food, including a dietary supplement, that expressly or by implication, including "third party" references, written statements (e.g., a brand name including a term such as 'heart'), symbols (e.g., a heart symbol), or vignettes, characterizes the relationship of any substance to a disease or health-related condition . . ." 21 C.F.R. § 101.14(a)(1).

83. Id.

84. Id. at § 101.14(a)(2).

85. Id. at § 101.14(a)(2), (b)(3)(i). "The substance must, regardless of whether the food is a conventional food or a dietary supplement, contribute taste, aroma, or nutritive value, or any other technical effect . . . to the food and must retain that attribute when consumed at levels that are necessary to justify a claim." Id. at § 101.14(b)(3)(i).

86. FDA Regulations on Statements Made for Dietary Supplements Concerning the Effect of the Product on the Structure or Function of the Body; Final Rule, 65 Fed. Reg. 999, 1019 (Jan. 6, 2000) (codified at 21 C.F.R. § 101.93).

87. Id. at 1022.

88. For example, some Kellogg's products, such as Special K, claim that "scientific studies from around the world suggest that an adequate intake of folic acid (a B-vitamin), B6, and B12 helps maintain a healthy cardiovascular system." Prior to the time Kellogg attempted to make this structure/function claim, FDA had denied requests to approve a "health claim" version of this statement. FDA stated it could not conclude that "based on the totality of publicly available scientific evidence, there is significant scientific agreement among experts qualified by training and experience to evaluate such evidence that a relationship between folic acid, vitamin B6, and vitamin B12 dietary supplements and risk of vascular disease is supported by the available evidence." Letter from Elizabeth Yetley, Director, Office of Special Nutritionals, FDA, to Jonathan W. Emord, attorney for Julian Whitaker, M.D., Durk Pearson, and Sandy Shaw, American Preventive Medical Association, and Pure Encapsulations, Inc. (Nov. 30, 1999) (on file with author). FDA ultimately authorized a health claim that must state that "the evidence in support of the claim is inconclusive."

89. Carole Sugarman, Magic Bullets; Pumped-Up Foods Promise to Make Us Happier and Healthier. We'll See, Wash. Post, Oct. 21, 1998, at E1; Vanessa O'Connell, Food for Thought: How Campbell Saw a Breakthrough Menu Turn into Leftovers, Wall Street Journal, Oct. 6, 1998, A1.; Kellogg Will Ask FDA to Approve New Health Claims for Functional Foods Line, Food Labeling & Nutrition News, Mar, 4, 1998, at 1.

90. 62 Fed. Reg. 28,234, 28,235 (May 22, 1997).

91. Id. at 28,235.

92. CSPI, Functional Foods: Public Health Boon or 21st Century Quackery (Mar. 1999) (last visited June 5, 2001) available at www.cspinet.org/reports/functional_foods/index.html.

93. See Health Claims Authorized in the United States, infra Appendix. These include claims linking diets low in saturated fat and cholesterol to a reduced risk of heart disease; low-fat diets rich in fruits and vegetables are linked to a reduced risk of cancer, sugar alcohols, and a reduced risk of dental cavities. Authorized health claims like these are estimated to appear on several hundred products found in a typical supermarket. Data from FDA's 1995 Food Labeling and Packaging Survey (FLAPS) indicates that 24 out of 1,030 products surveyed included a health claim stating the relationship between a component of the product and a disease. Although these findings are based on nonsales weighted raw data, they indicate that a typical supermarket stocked with 15,000 food items may contain as many as 345 food items that bear health claims. FDA, 1995 FLAPS (Oct. 25, 1995), computations performed by CSPI based on data provided by FDA.

94. See supra notes 53–60 and accompanying text.

95. 21 C.F.R. § 101.75.

96. The Campbell Soup Company experimented with a line of mail-order frozen meals and snacks designed to reverse such medical conditions as high blood pressure and elevated cholesterol and blood sugar levels. The line, called Intelligent Quisine, was in compliance with FDA health claim requirements, but was discontinued after poor results. See Sugarman, supra note 91, at E1.

97. 62 Fed. Reg. 3584 (Jan. 23, 1997) (codified at 21 C.F.R. § 101.81).

98. Id.

99. 63 Fed. Reg. 8103 (Feb. 18, 1998.)

100. FDA, Health Claim Notification for Whole Grain Foods (July 1999) available at (last visited Dec. 5, 2000) vm.cfsan.fda.gov/~dms/flgrains.html.

101. 65 Fed. Reg. 64,686 (Sept. 8, 2000). This marked the first time FDA approved a health claim as "an interim final rule" and permitted public comment only after the issuance of the final rule. Previously, public comment was invited on proposed rules, and the comments were taken into account in revising final regulations. This health claim is also controversial because for the first time it exempted a category of foods from the 13-gram total fat disqualifying level that has been established for other substances making health claims.

102. FDA, Health Claim Notification for Potassium-Containing Foods, (Oct. 31, 2000) available at (last visited Dec. 5, 2000) vm.cfsan.fda.gov/-dms/hclm-k.html.

103. Kellogg has started a $65 million nutraceutical/functional food division, the W.K. Kellogg Institute for Food and Nutritional Research. Nancy Millman, Campbell Cooks Up Health Food for the Masses; Firm Will Be First Giant to Serve Up Meal Plans to Lower Blood Pressure, Blood Sugar and Cholesterol Levels, Chicago Tribune, Dec. 29, 1996, at Cl.

104. Sixteen of the products made FDA-approved health claims. Six other items in the product line, such as cakes and potato crisps, however, made structure/function claims instead of health claims. The similarity between these two types of claims, and the fact that both claims would appear under the Ensemble product-line logo, could have misled consumers into believing that all the products in the line were equally healthful.

105. See Yetley Letter, supra note 89.

106. 63 Fed. Reg. 34,097 (June 22, 1998).

107. Kellogg Press Release, Using New FDA Regulations, Kellogg Takes the Lead in Promoting Heart-Health Benefits of Folic Acid, April 5, 2000 (on file with author).

108. Id.

109. Letter from Christine L. Lewis, Ph.D. FDA to Guy Johnson, Ph.D., Vice President, Nutrition Kellogg Company (Oct. 30, 2000). Kellogg continues to make claims linking adequate intake of folic acid, B6, and B12 to a healthy heart. See www.kellogg.com/nutrition/folic_acid/heart_disease/heart_role.html. (visited Dec. 4, 2002).

110. See supra note 53 and accompanying text.

111. See Table 11.1.

112. Products carrying health claims must contain no more than 480 milligrams of sodium per reference amount customarily consumed. 21 C.F.R. § 101.14(a)(4). An 8-ounce serving of V8 contains 620 mg.

113. 21 C.F.R. § 101.74(a)(2).

114. See, for example, Nancy Hellmich, Supplement or Additive? Issue Is on FDA's Plate, USA Today, Nov. 4, 1998, at 1D ("The spread is one of the highest-profile entries into an exploding area of new products known in nutrition circles as 'functional foods. . .'").

115. Letter from Joseph A. Levitt, Director, CFSAN, FDA to Brian D. Perkins, President, McNeil Consumer Products Co. (Oct. 28, 1998) (on file with author).

116. 21 U.S.C. §§ 321(s), 348 (FDCA §§ 201(s), 409).

117. See Levitt Letter, supra note 116.

118. Agreement Reached on Regulatory Approach for Benecol, Food Labeling & Nutrition News, Feb. 3, 1999, at 3. Johnson & Johnson also is releasing Benecol snack bars containing the recommended serving of the stanol esters. McNeil Benecol Broadening With Snack Bars, Tub, F-D-C Rep. ("The Tan Sheet"), Oct. 25, 1999, at 9.

119. R.M., The Margarine Wars, "Benecol," "Take Control" Hit the Shelves, Nutraceuticals World, July/Aug. 1999, at 92.

120. See supra note 102 and accompanying text.

121. 62 Fed. Reg. 49,859, 49,861 (Sept. 23, 1997).

122. 21 U.S.C. §§ 350(b), 321(s)(6) (FDCA §§ 413(a), 201(s)(6)).

123. Letter from John B. Foret, Director, Division of Programs and Enforcement Policy, Office of Food Labeling, FDA, to Myron Cooper, Hain Food Group (June 21, 1999) (on file with author).

124. Id. The label of another product, "Think! Interactive Bar," (manufactured by Personal Health Development) contains ginkgo biloba and choline and states that it will promote "concentration, calmness, and stamina." The product, however, resembles a candy bar. Thus, the added ingredients may be considered by FDA to be unapproved food additives or drugs.

125. See Foret Letter, supra note 124.

126. 21 C.F.R. § 182.1180.

127. Id. § 340.10.

128. Letter from Lynn A. Larsen, Director, Division of Programs and Enforcement Policy, Office of Special Nutritionals, FDA, to The Dannon Company, Inc. (Aug. 26, 1999) (on file with author). See also (last visited Nov. 7, 2002) www.dannon.com.

129. Id.

130. Pharmanex Inc. Press Release, Pharmanex, Inc. Introduces the Cholestin Bar Natural Health-Promoting Benefits of Cholestin Now Available in a Tasty, Ready-to-Eat Bar (Oct. 25, 1999), PRN Newswire (on file with author).

131. Pharmanex, Inc. v. Shalala, 35 F. Supp. 1341 (C.D. Utah 1999), rev'd and remanded, 221 F.3d 1151 (10th Cit. 2000); Pharmanex, Inc. v. Shalala, Memorandum Decision and Order (unpublished) Case No: 2:97CV262(K) (Mar. 30, 2000) available at (last visited June 5, 2001) www.findlaw.com (District Court of Utah file).

132. Id.

133. Warning Letters from John B. Foret, Director, Division of Compliance and Enforcement, Office of Nutritional Products, Labeling, and Dietary Supplements, to Vitamin Classic's, Inc. (Feb. 21, 2001); South Beach Beverage Co. (Feb. 1, 2000); Robert's American Gourmet (Jan. 27, 2000); Langer Juice Co. (Sept. 28, 1999) available at (last visited May 11, 2001) www.fda.gov/scripts/wlcfm/ indexissuer.cfm. See also Courtesy Letter from John B. Foret, Office of Nutritional Products, Labeling, and Dietary Supplements, to Hain Food Group (June 21, 1999) (on file with author).

134. Id.

135. Letter from Christine J. Lewis, Director, Office of Nutritional Products, Labeling, and Dietary Supplements, FDA, to manufacturers of functional foods (Jan. 30, 2000), at 1 (on file with author).

136. Id.

137. Id. at 1–2.

138. Warning Letters from John B. Foret, Director of Compliance and Enforcement, CFSAN, FDA, available at (last visited Jun. 28, 2001) www.fda.gov/scripts/wlcfm/indexissuer.cfm to Julia Sabin, President Smucker Quality Beverages, Inc. (June 8, 2001 ONPLDS 14-01); Cynthia Davis, Executive Vice President, US Mills Inc. (June 5, 2001 ONPLDS 13-01); Doug Levin, CEO, Fresh Samantha, Inc. (June 4, 2001 ONPLDS 11-01); Rodney C. Stacks, Chairman, Hansen Beverage Co. (June 4, 2001 ONPLDS 12-01).

139. FDA Continues to Review Use of Supplements in Food, Food Chem. News, Aug. 5, 2002, at 1, 9–10.

140. Warning Letter NWE-08-O1W from Gail T. Costello, District Director, New England District Office, FDA, to Robert Hawthorne, President, Ocean Spray Cranberries, Inc. (Jan. 19, 2001) available at (last visited May 11, 2001) www.fda.gov/scripts/wlcfm/indexissuer.cfm.

141. Id.

142. Washington Legal Foundation Citizen Petition to Exempt Internet Information from FDA Labeling Requirements, Docket 01P-018 (Apr. 13, 2001).

143. FDA, FDA Letter on Labeling Food Products Presented or Available on the Internet to Daniel J. Popeo and Paul Kamenar, Washington Legal Foundation (Nov. 1, 2001) available at http://www.cfsan.fda.gov~dms/lab-www.htm.

144. FTC, Advertising and Marketing on the Internet: The Rules of the Road (Apr. 1998).

145. Statement of Connecticut Attorney General Richard Blumenthal at Press Conference (July 18, 2000) ("I'm recommending that state attorney generals make functional foods a top consumer protection initiative.") CSPI Press Release, FDA Urged to Halt Sale of Functional Foods Containing Illegal Ingredients, July 18, 2000.

146. NAD Press Release, NAD Recommends That Arizona Iced Tea Discontinue its Rx Memory Label, Apr. 2, 2001.

147. Id.

148. HeartBar is an example of a product sold to consumers directly that states, "Use under the supervision of a physician."

149. See Benavides, supra note 73, at 3.

150. See Conference Notice, supra note 73.

151. Elizabeth Seay, New Salt Substitute Seeks Same Status on Shelf as Aspirin, Wall Street Journal, Apr. 1, 1996, at A7D.

152. See Conference Notice, supra note 73.

153. See Benavides, supra note 73, at 3.

154. Martha Groves, Preventative Measures: Search for Therapeutic Foods Picks up Steam, L.A. Times, Nov. 7, 1992, at D2. The labeling for Zbar states specifically that "Zbar is a new medical food."

155. There is no specific exemption from the NLEA for health claims made on these products. Some but not all of the foods for special dietary use regulations have been amended in light of the NLEA. FDA has indicated that any label statements used in accordance with FDA regulations for foods for special dietary use will not be regulated as health claims. 58 Fed. Reg. 2578, 2482 (Jan. 6, 1993).

156. Food Labeling & Nutrition News, Oct. 28, 1998, at 2.

157. Id.

158. 61 Fed. Reg. 60,661, 60,663 (Nov. 29, 1996).

159. The United States is one of a minority of countries in the developed world that separates responsibility for regulating food-labeling claims from responsibility for regulating food-advertising claims and applies different standards to each. FTC is responsible for regulating most advertising in the United States. FDA and FTC attempt to coordinate standards for food-labeling and food-advertising claims, and FTC usually defers to FDA's scientific expertise. FTC, however, has reserved the right to permit health claims in advertisements that are not permitted on labels when it believes such claims comply with the FTC Act. FTC, Enforcement Policy Statement on Food Advertising, 59 Fed. Reg. 28,338 (June 1, 1994).

160. FTC, Enforcement Policy Statement on Food Advertising (May 1994) available at (last visited May 11, 2001) www.ftc.gov ("The Commission recognizes that there may be certain limited instances in which carefully qualified health claims may be permitted under section 5 [of the FTC Act] although not yet authorized by the FDA.") (emphasis added).

161. Working Agreement between FTC and FDA, 3 Trade Reg. Rep. (CCH) ¶ 9,851 (1971).

162. 59 Fed. Reg. at 28,388.

163. Newsweek, Sept. 27, 1999.

164. 58 Fed. Reg. 2,622, 2,630–31 (Jan. 6, 1993).

165. New York Times Magazine, Jan. 10, 1999, at 55.

166. Letter from C. Lee Peeler, Associate Director, Division of Advertising Practices, FTC, to Edward P. Henneberry, Counsel for H.J. Heinz Co. (Dec. 15, 1999) (on file with author).

167. Id.

168. Campbell's Classics: Woman in Kitchen With Tomato Wallpaper, Storyboard 001-05646, (NBC television broadcast, Jan. 20, 2000) available from Video Monitoring Services of America (New York, NY).

169. National Advertising Division News Release, Campbell Soup Participates in NAD Self-Regulatory Process, Apr. 23, 2002.

170. See, for example, Letter from C. Lee Peeler, Associate Director, Director of Advertising Practices, FTC, to Bruce Silverglade, Director of Legal Affairs, CSPI (Mar. 21, 2001) (denial of petition to prohibit health claims

about the relationship of almond consumption to the reduced risk of heart disease by the Almond Board of California).

171. Interstate Bakeries Corp. FTC. Decision and Order, Docket No. C-4042, Apr. 16, 2002.

172. Press Release, FTC, FTC Alleges Maryland Companies Lack Support for Claims that HeartBar is Effective Against Cardiovascular Disease (June 12, 2003).

173. Psyllium is the active ingredient in such over-the-counter laxatives as Metamucil and also is the subject of an approved health claim. 21 C.F.R. § 101.81.

174. Id. at § 172.5(a)(1). The leading definition of food has been set forth in Nutrilab, Inc. v. Schweiker, 713 F.2d 335 (7th Cit. 1983).

175. 21 C.F.R. § 101.14(b)(3)(i).

176. 62 Fed. Reg. at 49,860–61.

177. FDA stated in its Oct. 28, 1998 letter to McNeil:

> The label for Benecol spread, through statements that the product replaces butter or margarine, vignettes picturing the product in common butter or margarine uses, statements promoting the flavor and texture of the product, and statements such as "...help(s) you manage your cholesterol naturally through the foods you eat," represents this product for use as a conventional food.

178. FDA also should set disqualifying levels for added sweeteners.

179. 21 C.F.R. § 101.14(e)(6). FDA should consider adding additional nutrients such as folate to this list.

12 The Nutraceutical Health Sector: A Point of View

Stephen L. DeFelice

Voltaire, largely influenced by Aristotle, once said, "If you wish to converse with me, you must define your terms." Global laws and regulations regarding the categories of foods and dietary supplements have not defined or properly characterized these two entities. Therein lies the major problem confronting the research, development and commercialization of these products. It not only causes confusion in the market place but also deprives both the healthy and the sick from receiving their much-needed benefits.

Regarding definitions, let's start off with the term, "Nutraceutical," that I coined in 1989 while enjoying a wonderful meal and great wine on the Piazza Navona in Rome. (**Nutraceutical**, by the way, is now in the *Webster's Collegiate Dictionary*.) I coined the term to create a clear, specific category on which the Congress or any international governmental body could enact legislation that would create a proprietary, high-margin, research-driven, ethical nutraceutical industry in place of the current market-driven, commodity or generic, low-margin one.

A nutraceutical is a food or part of a food that has a medical or health benefit, including the prevention and treatment of disease. As with a pharmaceutical, it can exist in a variety of dosage forms. For example, an orange (food), orange juice (fortified food or functional food), and a capsule containing vitamin C (dietary supplement) can all be used either to prevent or treat scurvy. The dosage form is merely a nutraceutical delivery system. When we consider legislation, therefore, as with pharmaceuticals, the dosage form should not be the issue of primary concern. Rather, of primary concern should be the issue of encouraging nutraceutical clinical research to demonstrate whether a specific product is both effective and acceptably safe.

Another core problem with most nutraceutical laws and regulations is the attempt to separate claims into categories of "health" and "disease" or "medical" ones, which even our greatest living epistemologic sophist would find impossible to do.

Regarding definitions, more than twenty years ago, I asked myself the following questions:

- What is health?
- What is disease?
- What is a health claim?
- What is a medical or disease claim?

After exhaustively combing the world of dictionaries, textbooks, and the minds of my colleagues, the following definitions represent the majority consensus of my inquiry:

- Health is the absence of disease or any abnormal condition that may generally require medical management.

- Disease is a condition that impairs health and usually benefits from medical management.
- A health claim deals with a substance that has a beneficial clinical effect on a disease or abnormal condition, be it prevention or treatment.
- Similarly, a medical-disease claim deals with a substance that has a beneficial clinical effect on a disease or abnormal condition, be it prevention or treatment.

In conclusion, little difference exists between a health or medical-disease claim, both of which generally deal with either preventing or treating disease. It follows, therefore, that any law or regulation that attempts to divide the two into different categories is faulty.

The Dietary Supplement Health and Education Act of 1994 (DSHEA), and the regulations based on that legislation, represent a classic example of faulty definitions that not only harm the consumer but also discourage the birth of a thriving ethical OTC nutraceutical industry.

DSHEA permits "statements of nutritional support," or structure/function claims, for a dietary supplement, without approval of the "health claim." But the language, "this product is not intended to diagnose, treat, cure or prevent any disease," must be placed on the label.

The structure/function claims permitted are the following: (1) a statement that "claims a benefit related to a classic nutrient deficiency disease and discloses the prevalence of such disease in the United States"; (2) a statement that "describes the role of a nutrient or dietary ingredient intended to affect the structure or function in humans"; (3) a statement that "characterizes the documented mechanism by which a nutrient or dietary ingredient acts to maintain such structure or function"; and (4) a statement that "describes general well-being from consumption of a nutrient or dietary ingredient."

FDA realized that all four categories are not only "health" claims but also "medical-disease" claims, which, as previously mentioned, are usually one and the same.

To fulfill the spirit of DSHEA, which generally prohibits medical or disease claims, FDA has attempted, by the use of legerdemain, to artificially separate the two. Examples are as follows:

- Permissible: "Helps to maintain cholesterol levels that are already within the normal range."
- Impermissible: "Lowers cholesterol."
- Permissible: "Maintains healthy lung function."
- Impermissible: "Maintains healthy lungs in smokers."
- Permissible: "For the relief of occasional sleeplessness."
- Impermissible: "Helps to reduce difficulty falling asleep."

The absurdity of these attempts to distinguish a health claim from a medical-disease claim is self-evident and does not require detailed explanation. Thus, if a company sponsors clinical studies on a dietary supplement that proves to clear up the lungs of smokers and enables them to breathe more easily, the company is prohibited from making the claim.

The tragic conclusion is that DSHEA prohibits companies from telling the truth about their products to consumers and, by doing so, offers a powerful incentive for such companies *not* to invest in costly clinical studies.

The laws and regulations also have greatly stimulated a market-driven, commodity, and low profit-margin nutraceutical industry that, despite the law, has not blocked the growth of unsubstantiated and misleading disease and health claims. Of the thousands of nutraceutical products sold, be they food, functional food, or dietary supplement, I estimate that less than one percent of one percent of the *specific* products sold have been clinically evaluated for effectiveness and safety.

All this reflects the interesting and puzzling fact that, as we shall see, the current nutraceutical industry is not listening to its customer—the consumer!

A few years ago, FIM, the Foundation for Innovation in Medicine, conducted a survey on consumer attitudes toward dietary supplements. The majority of consumers had significant doubts regarding the safety and effectiveness of the products they were taking. Also, a significant majority expressed a strong desire that dietary supplements be clinically evaluated for effectiveness and safety before being consumed.

Prevention magazine also conducted a survey regarding the rapidly growing movement of self-care. The survey revealed that 72 percent of Americans take vitamins, 50–77 percent use herbs, 15–20 percent use homeopathic remedies, and 20 percent use aromatherapy.

Nearly every other person interviewed by *Prevention* believed that eating naturally reduces the risk of disease, including major conditions such as heart disease and cancer. The survey showed that 54 percent of shoppers (representing 54.3 million households) had recently purchased a food to achieve a health benefit. The survey concluded that what consumers urgently need is accurate, consistent advice on using all types of nutraceutical dosage forms, both to prevent and treat disease.

HealthFocus, Inc., in its own survey, found that more than 70 percent of those surveyed saw a connection between nutrition and health. A whopping 79 percent believed that foods could prevent cancer and heart disease.

The overwhelming conclusion of these and other surveys is that people consume nutraceuticals to prevent and treat disease—to be healthy!

Though the nutraceutical health sector is presently stalled, it is reasonable to assume that prevailing forces will inevitably lead to change.

One possible scenario that was repeatedly discussed at past FIM conferences has already begun. Because of high visibility, increased consumer nutraceutical consumption, and pervasive unsubstantiated and misleading claims, it was inevitable that the government and other institutions would sponsor well-controlled clinical trials to evaluate both the effectiveness and safety of selected, highly consumed dietary supplements. Major products such as vitamin E for the prevention of heart disease, St. John's wort for the treatment of depression, and echinacea for the treatment of the common cold were all found to be ineffective under the conditions of the clinical designs.

To add to the bad news, St. John's wort was shown to speed up the metabolism of a variety of drugs, rendering them less effective or ineffective, including anti-AIDS drugs. The latter sparked intensive mass-media coverage, alerting both physicians and consumers and leading to a national concern for dietary supplements in general. As a result, we now have initiated a national medical movement to search for nutraceutical adverse effects and toxicities, and there is little doubt that they will be found for the simple fact that practically all substances have bad effects—even water! Not too long ago, it was reported that excessive water intake by infants led to water toxicity in certain parts of the country. The more you look for toxic effects, the more you will find them.

When these toxicities are discovered, they will have a major and constant presence on the national radar screen. Congress will, based on fear instead of delivering the nutraceutical promise, pass new legislation, the primary impact of which will be a reduction of the freedom, size, and growth of the nutraceutical market. The fact that this has not already occurred attests to the political clout of the consumers. They are extremely faithful customers.

Another possible scenario is a much preferred, logical, and beneficial one. It may occur after the enormity of the nutraceutical market is recognized by industry leaders.

How large, then, is the current nutraceutical market? Well, it depends on how one defines nutraceutical. As an example, let's look at the approximately $600 million functional food market. As mentioned previously, surveys report that more than 50 percent of consumers who buy such products do so to help prevent or treat disease, or "be healthy." Fifty percent of $600 million leaves us with a $300 million market. Add to that 50 percent of the multibillion dollar dietary supplement sector.

Using this simple method for rough assessment, the nutraceutical market is larger than the U.S. ethical pharmaceutical and OTC markets combined! There is presently, however, one major difference. The current nutraceutical market is a commodity, low-profit one, whereas the pharmaceutical-OTC market is a proprietary, highly profitable one.

Just imagine the size of the nutraceutical market, if it were also a proprietary and high-profit margin one. It's breathtaking!

Shortly after I coined the term "nutraceutical," I held a small meeting with a group of medical, legal, and business experts to discuss what types of legislation Congress could enact that would give nutraceuticals sufficient, proprietary protection to justify the costs to sponsor the clinical studies that are necessary to determine whether the *particular* product sold by a company is both effective and acceptably safe.

I told them the story of my personal experience with carnitine. Carnitine is a naturally occurring substance with many potential clinical uses, one being for a rare type of carnitine deficiency, which is often fatal in children. Lacking sufficient patent protection, no company would support a major New Drug Application (NDA) effort for any medical use. In 1983, however, Congress passed the Orphan Drug Act, which significantly reduced the cost of obtaining an NDA while granting the sponsor of the NDA a seven-year period of exclusivity to make claims on the specific product, based on the results of the clinical studies. This Act has been an unequivocal success, and many thousands of patients in need of new therapies have benefited from it. Carnitine itself was approved as an orphan drug and has and continues to save the lives of those who suffer from this rare disease.

I proposed this as a model for nutraceuticals. The experts unanimously agreed. Thus, the concept of the NREA, or Nutraceutical Research and Education Act, was born.

Subsequently, I met with multiple food and health food companies to gather support for the effort to persuade Congress to enact the NREA. None had interest. I was told that the nutraceutical market was booming and should not be tampered with.

Time passed. In 1998, I met Congressman Frank Pallone (D-NJ), who had a particular interest in the potential beneficial effects of nutraceuticals. Convinced of the importance of clinical studies, in 1999 he introduced the NREA in Congress (see Appendix 1). And guess what? History repeated itself. No one had interest, not even members of the scientific-nutritional community. Shortly thereafter, the nutraceutical market began and continues to lose its thrust, and toxicity issues tenaciously cling to the national radar screen.

The following is an example of how the NREA, if enacted, would work: The complications of diabetes are well known. More than 50 percent of diabetics have intracellular magnesium deficiency. With this deficiency, many pathological processes have been reported to occur, such as increased blood clotting, constriction of small arteries, and insulin resistance. All these events can eventually lead to decreased availability of oxygen to body tissues, which plays a major role in diabetic complications such as heart attacks, loss of limbs, blindness, kidney failure, neuropathy, and others. Reasonable clinical evidence exists that magnesium supplementation may reverse these three pathologic processes as well as other detrimental metabolic changes. Also, clinical studies of large populations have shown that magnesium deficiencies are associated with an increased incidence of diabetes.

Myocardial ischemia, or lack of oxygen to the heart due to blockage of the coronary arteries, is a common complication in diabetics. A number of clinical pharmacological studies, including those involving treadmill tests and atrial pacing—all of which severely stress the ischemic heart—have shown that carnitine significantly protects the distressed ischemic heart.

Insulin resistance, which is common in the adult type diabetic, is an independent cardiovascular risk factor. Chromium and alpha lipoic acid both have been reported to reduce insulin resistance.

A combination of the aforementioned substances could be clinically tested using surrogate or biomarkers, such as better performance on the treadmill, a decrease in intracellular free radical activity, or improved performance on a broad metabolic rating scale. Because of the amount of these nutraceuticals needed, it would be impractical to take the number of pills required to obtain sufficient blood levels. For this reason, one must use a food or functional food dosage form that can more easily accommodate the total amount, both to ensure absorption and to mask the taste, resulting in increased consumer compliance.

If the clinical results are positive, the FDA will then grant the company an exclusive right to make the claim for the product studied. Also, it is important to note that, for very good practical and historical reasons, physicians would be strongly inclined to recommend the particular product, and not generic competitors, simply because it has been clinically tested and, therefore, assumed to be a trustworthy product. There is little doubt that proven nutraceutical products will win out over unproven ones.

The same conceptual approach can be applied to all other conditions, such as chronic fatigue and impaired immune function, and even improve the effectiveness of pharmaceuticals. With respect to the latter, niacin and Benecol are examples of a nutraceutical working together with a pharmaceutical to lower unacceptably high levels of blood cholesterol.

And now the sad enigma: To my knowledge, this is the first time that a U.S. health industry sector has consistently ignored the compelling, well-documented desires of its customers and missed the opportunity to create a huge, profitable market catering to such desires. The leaders of this health sector continue to fail to see that the NREA or equivalent legislation is required to establish its foundation.

Over the years I have been asked more than a hundred times about why the industry fails to grasp this unique opportunity. My reply is "I really don't know." I do, however, have an opinion. It may be due to the fact that the NREA would create a scientifically and clinically driven research market, which would present an entirely new culture in which to operate for these companies. It probably would require a major change of corporate mentality and behavior, which is extremely difficult to bring about. There are ways to resolve this problem, such as alliances with pharmaceutical companies.

In conclusion, the current nutraceutical health sector has lost its dynamic growth and fails to see the future largely because it does not understand its customers. In my opinion, it is inevitable that this health sector will change in the near future either in a positive or negative way. Industry leaders can either seize the moment and strive to create a new, highly profitable, OTC nutraceutical health sector or remain passive, waiting for inevitable events beyond its control to determine its destiny.

Appendix 1
Nutraceutical Research and Education Act

106TH CONGRESS
1ST Session

H.R.3001

IN THE HOUSE OF REPRESENTATIVES

Mr. Pallone introduced the following bill; which was referred to the Committee on October 1, 1999.

A BILL

To amend the Federal Food, Drug, and Cosmetic Act to promote clinical research and development on dietary supplements and foods for their health benefits; to establish a new legal classification for dietary supplements and foods with health benefits, and for other purposes

Be it enacted by the Senate and House of Representatives of the United States of America in Congress assembled,

SECTION 1. SHORT TITLE AND REFERENCE.

(a) SHORT TITLE.—This Act may be cited as the "Nutraceutical Research and Education Act".

(b) REFERENCE.—Whenever in this Act an amendment or repeal is expressed in terms of an amendment to, or repeal of, a section or other provision, the reference shall be considered to be made to a section or other provision of the Federal Food, Drug, and Cosmetic Act.

SEC. 2. FINDINGS AND STATEMENT OF PURPOSE.

(a) Findings.—The Congress finds the following:

(1) Consumers spend annually an estimated $12,000,000,000 on dietary supplements and billions more on medical and similar foods. Nevertheless, the health benefits of these products have not, in most cases, been demonstrated by clinical testing or other means. In consequence, specific health claims may not be advanced for them. Consumers are thus left uncertain as to the value of these products in promoting health and well-being, and preventing or reducing the risk of disease, including the manage-

ment of a disease or condition. The companies that produce these products desire to provide them to consumers with specific health claims based on clinical testing. The Federal Government demands sound scientific evidence of safety and effectiveness in order to fulfill its statutory mandate to protect and promote the public health.

(2) Because dietary supplements and similar foods are natural products widely available without a strong proprietary position, a person who now finances the cost of research successfully demonstrating the health benefits of such a product receives no special economic benefit in the marketplace to repay that cost. Others, who have not contributed to those research costs, may nevertheless embrace the findings of that research to support identical claims for their own versions of the product. Without economic incentive to research and develop new products, those who would finance the cost of research are presently focusing their efforts on promotional activities to the disservice of the public interest and health.

(3) It is in the national interest to encourage clinical research into the health benefits of dietary supplements, medical foods, and other foods.

(4) Current regulatory and epistemological chaos exists with regard to health claims for foods, dietary supplement, and medical foods. It is in the national interest to provide a category of products that have recognized health benefits but are not drugs and to recognize that these products are safe when used as indicated on their labeling.

(5) It is necessary to promote research into the health benefits of dietary supplements, medical foods, and other foods and to require that these health benefits be established by the results of clinical studies.

(6) It is necessary to establish a regulatory system within the Food and Drug Administration for reviewing health claims of health benefits of such products which is less burdensome than the traditional regulatory scheme for drugs and to stimulate the industry to devote resources to proving the health claims anticipated under this Act since such claims relate to the possibility of preventing or reducing the risk of disease, including the management of a disease or health condition.

(7) It is necessary to update the present regulatory scheme to reflect the fact that such products can safely prevent disease and health conditions, manage or improve health, or reduce the risk of disease.

(b) STATEMENT OF PURPOSE.—It is the purpose of this Act to:

(1) promote research into the health benefits of dietary supplements, medical foods, and other foods.

(2) establish a simplified process within the Food and Drug Administration for reviewing, on a case by case method, health claims of health benefits of such nutraceutical products made under a petition under section 403(r)(4) of the Federal Food, Drug, and Cosmetic Act (21 U.S.C. 343(r)(3)).

(3) prescribe a period of exclusive marketing protection for a person that demonstrates the health benefits of a dietary supplement, medical food, or other food, and who markets such product in association with approved labeling that describes its contribution to human health; and

(4) confirm the health benefits of these products as determined by clinical trials, and disseminate this information to the public and the health care profession, so that the public and the health care profession may integrate this knowledge into practice.

SEC. 3. DEFINITIONS.

Section 201 (21 U.S.C. 321) is amended by adding at the end the following:

"(kk) The term 'nutraceutical' means a dietary supplement, food, or medical food, as respectively defined in paragraphs (f) and (95) and section 5(h)(3) of the Orphan Drug Act (21 U.S.C. 360ee(b)(3)), that-

"(1) possesses health benefits; and

"(2) is safe for human consumption in such quantity, and with such frequency, as require to realize such properties.

"(ll) The term 'health benefit', when used with reference to a nutraceutical, means a benefit which prevents or reduces the risk of a disease or health condition, including the management of a disease or health condition or the improvement of health.

SEC. 4. HEALTH CLAIMS.

(a) NUTRACEUTICAL HEALTH CLAIM.—Section 403(r)(5)(D) (21 U.S.C. 343(r)(5)(D)) is amended by inserting before the period the following: "except that in the ease of a claim made with respect to a nutraceutical, the regulation shall be issued by the Secretary under subparagraph (4)(D)".

(b) PETITION.—Section 403(r)(4) (21 U.S.C. 343(r)(4)) is amended by adding at the end the following:

"(D)(i) Any person may file a petition with the Secretary to issue a regulation relating to a claim for a nutraceutical described in subparagraph (5)(D).

"(ii) A petition filed under subclause (i) shall be prepared in such form, and submitted in such manner, as the Secretary may prescribe, and, with respect to the product sought to be introduced as a nutraceutical, shall containing the following:

"(I) A report of at least 1 clinical trial which has been conducted on the product which is the subject of the petition. Such clinical trial results shall address the potential health benefits of the product and its safety. The results of the clinical trial must demonstrate and characterize the beneficial relationship or the significance of the relationship of the nutraceutical in such product to a disease or its affects on a health related condition, health problem, or health status. The clinical trial must have a sufficient size to prove the benefits and may have as its endpoints either surrogate markers or clinical endpoints to support the claim. The application may also include epidemiological or preclinical studies in support of the clinical trial. The amount of evidence necessary to sustain a claim will be determined by the Secretary on a ease by ease basis.

" (II) Evidence that it is safe for human consumption in such quantity, and with such frequency, as required to provide the health benefits.

"(III) A complete description, in the ease of a processed product, as to its ingredients or chemical composition.

"(IV) Information adequate to enable the Secretary to determine, where pertinent, that the methods used in, and the facilities and controls used for, processing and packing the product are sufficient to preserve its identity, strength, quality, and purity.

"(V) Such samples of the products as the Secretary may require.

"(VI) A specimen of the labeling proposed to be used with the product, when introduced or delivered for introduction into commerce as a nutraceutical, that accurately and completely describes its health benefits under its stated conditions of use.

"(iii) Within 7 days of the receipt of a petition, the Secretary shall cause it to be published in the Federal Register to provide notice to the public that the petition has been filed. Such notice shall contain the name of the petitioner, date and time of filing, a summary and description of the proposed product, and the nature of the proposed health claim.

"(iv) When a petition is filed for a nutraceutical claim under subparagraph (5)(D), no other petition for a product which is the same as or similar to the product for which a petition has been filed and no other petition for a claim which is the same or similar to the claim for which a petition has been filed may be filled until final action has been taken on the first petition.

"(v) A person who files a petition for a claim for a nutraceutical claim under subparagraph (5) (1)) may apply to the Secretary to amend the petition when the amendment is required by a change in the product clue to new and unexpected findings in research on the product or the disease or condition for which the product is being proposed.

"(vi) The Secretary shall refer any petition filed for a nutraceutical claim under subclause (i) to the Advisory Council on Nutraceuticals established under section 7 of the Nutraceutical Research and Education Act.

"(vii) The Secretary shall take final action on a petition which-

"(I) was filed under subclause (i), and
"(II) was determined by such Advisory Council on Nutraceuticals to be worthy of review, not later than 6 months after the date the petition is filed."

SEC. 5. MARKET PROTECTION FOR NUTRACEUTICAL.

(a) IN GENERAL.—Section 403(r) is amended by adding at the end the following:

"(8) If the Secretary issues a regulation in response to a petition filed under subparagraph (4) relating to a, claim for a nutraceutical described in subparagraph (5)(D), the Secretary may not issue another regulation for an essentially identical nutraceutical claim during the 10-year period that begins on the date that the Secretary approved the original petition, except that—

"(A) if a petition is submitted for an essentially identical nutraceutical claim for a nutraceutical the intended use of which provides greater effectiveness, greater safety, or otherwise a major contribution to patent care, the Secretary may issue a regulation under subparagraph (4)(D) for such claim; or

"(B) if a petition is subsequently revoked, another petition may be submitted to the Secretary for an essentially identical nutraceutical claim.".

(b) MISBRANDING.—Section 402 (21 U.S.C. 342) is amended by adding at the end the following:

"(h) If it is a nutraceutical and it has not had a petition approved under section 403(r)(4)(D).".

SEC. 6. GOOD MANUFACTURING PRACTICES.

Section 402(g) (21 U.S.C. 342(g)) is amended by—

(1) inserting ", including a nutraceutical" after "dietary supplement" in subparagraph (1); and

(2) inserting ", including nutraceuticals" after "dietary supplements" in subparagraph (2).

SEC. 7. ADVISORY COUNCIL ON NUTRACEUTICALS.

(a) ESTABLISHMENT.—There is established within the Food and Drug Administration advisory council to be known as the "Advisory Council on Nutraceuticals".

(b) DUTIES.—The Advisory Council shall evaluate the merit of each petition filed for a nutraceutical health claim under section 403(r)(4)(D) of the Federal Food, Drug, and Cosmetic Act, including the proposed labeling of the product that is the subject of the petition, and submit its evaluation to the Secretary. The evaluation of the Advisory Council shall determine if a, petition is worthy of review by the Food and Drug Administration and whether it conflicts with any other petition.

(c) MEMBERSHIP.—

(1) IN GENERAL.—The Advisory Council shall consist of ex officio members and not more than 6 additional members appointed by the Secretary. The ex officio members shall be nonvoting members.

(2) EX OFFICIO MEMBERS.—The ex officio members of the Advisory Council shall be the Secretary, the Director of the National Institutes of Health (hereinafter in this Act referred to as the "Director of NIH"), and such additional officers or employees of the United States as the Secretary determines necessary for the Advisory Council to carry out its functions.

(3) OTHER MEMBERS.—The members of the Advisory Council who are not ex officio members shall be appointed by the Secretary from among individuals distinguished in the fields of health, nutrition, or biomedical research.

(d) COMPENSATION.—Members of the Advisory Council who are officers or employees of the United States shall serve on the Advisory Council as part of their official duties, and shall not receive additional compensation therefore. Other members of the Advisory Council shall receive, for each day (including travel time) they are engaged in the performance of Advisory Council functions, compensation at rates not to exceed the daily equivalent of the annual rate in effect for grade ES-1 (5 U.S.C. 5382). Such other members, when performing Advisory Council functions (including travel to and from Advisory Council meetings), shall be entitled to travel expenses (including per diem in lieu of subsistence) as authorized by section 5703 of title 5, United States Code, for persons in the Government service employed intermittently.

(e) TERM.—The term of office of an appointed member of the Advisory Council is 4 years, except that any member appointed to fill a vacancy for an unexpired term shall be appointed for the remainder of such term and the Secretary shall make appointments to the Advisory Council in such a manner as to ensure that the terms of the members do not all expire in the same year. A member may serve after the expiration of the member's term for 180 days after the date of such expiration. A member who has been appointed for a term of 4 years may not be reappointed to the Advisory Council before 2 years from the date of expiration of such term of office. If a, vacancy occurs in the Advisory Council among the appointed members, the Secretary shall make an appointment to fill the vacancy within 90 days from the date the vacancy occurs.

(f) CHAIR.—The Secretary shall select the chair of the Advisory Council from among the appointed members. The term of office of the chair shall be 2 years.

(g) MEETINGS AND PROCEDURES.—The Advisory Council shall meet at the call of the chair, or at the direction of the Director of the National Institutes of Health, but with sufficient frequency to ensure prompt evaluation of every nutraceutical petition referred to it by the Secretary The Advisory Council shall adopt rules governing its procedures.

(h) FEDERAL ADVISORY COMMITTEE ACT.—Meetings and proceedings of the Advisory Council shall not be subject to the Federal Advisory Committee Act (5 U.S.C. Appendix).

SEC. 8. NUTRACEUTICAL INDEX.

The Secretary shall maintain, and periodically publish in the Federal Register, an index that shall list-

(1) the name and description of each nutraceutical for which there is an approved petition, the name and address of the applicant, and the date upon which the Secretary approved the petition; and

(2) each petition pending with the Secretary, the date upon which it was filed with the Secretary, the name and address of the applicant, and a description of the nutraceutical and the claim made for the nutraceutical that is the subject of that petition.

SEC. 10. SMALL BUSINESS ANTITRUST EXEMPTION.

(a) EXEMPTION.—It shall not be unlawful under the antitrust laws for 2 or more small businesses to agree to combine their resources to meet the requirements of section 403(r) of the Federal Food, Drug, and Cosmetic Act (21 U.S.C. 353(r)) for claims of health benefits of a nutraceutical.

(b) DEFINITIONS.—

(1) ANTITRUST LAWS.—The term "antitrust laws" has the meaning given such term in subsection (a) of the first section of the Clayton Act (15 U.S.C. 12(a)), except that such term includes section 5 of the Federal Trade Commission Act (15 U.S.C. 45) to the extent quell section applies to unfair methods of competition.

(2) NUTRACEUTICAL.—The term "nutraceutical" has the meaning given such term in section 201(k)(k) of the Federal Food, Drug, and Cosmetic Act (21 U.S.C. 321(k)(k)).

(3) SMALL BUSINESS.—The term "small business" has the meaning given such term in section 736(d)(3)(A) of the Federal Food, Drug, and Cosmetic Act (21 U.S.C. 379h(d)(3)(A)).

SEC. 11. EFFECTIVE DATE.

This Act and the amendments made by this Act shall take effect 90 days after the dated of its enactment.

13 Regulatory Issues Related to Functional Foods and Natural Health Products in Canada

Kelley Fitzpatrick

The Canadian functional food and natural health product (NHP) sector has grown substantially to reflect the growing demand for nutritional products based on increasing scientific evidence linking diet to health outcomes as well as increasing consumer interest in self-care and alternative medicine. According to a recent study conducted for Agriculture and Agri-Food Canada (AAFC), up to CA$1B of farm production value is estimated to be devoted to supplying the functional foods and NHP industry (1). This estimate does not include the marine industry, which contributes to the sector through the production of omega-3 fatty acids and other marine-based products.

Canadian sales figures for functional foods and NHPs are difficult to interpret because much of it is extrapolated from U.S. sales and adjusted downward. In addition, strict regulations have forced companies to label products either as foods or drugs, thus functional food and NHP products may not be accounted for because they will overlap into the food processing or pharmaceutical industries. Estimates are that Canadians purchased approximately U.S.$4.2 billion worth of dietary supplements (defined as NHPs in Canada) and functional food products in 2001 (2). Additional data show that Canadian sales of functional foods in 2001 were valued at U.S.$2.8 billion (3). The latter study indicated that although functional food sales in the United States represent approximately 4.5 percent of total food sales, Canada's portion of total food sales is only 2.2 percent, representing a significant growth potential for the domestic industry.

In regard to dietary supplements, Canadian sales figures in 2001 were approximately U.S.$0.8 billion (3). Whereas supplements account for almost half of the global nutrition industry, within Canada, retail sales account for 21 percent of total industry sales. Lower figures are believed to be due to stricter regulations that have historically been present in Canada. Compared to the United States, Canada is generally considered to be between 12 to 18 months behind in launching new products, again believed to be due to a more restrictive regulatory climate.

Globally, Canada's participation in the industry is growing and is demonstrated through:

- Increasing agricultural crop production and development of varieties targeted at enhanced human health
- Development of new technologies that allow for the processing of supplements and ingredients that provide a health benefit
- The increasing emphasis on clinical validation of functional foods and nutraceuticals
- The upsurge in entrepreneurial activity establishing new and innovative companies throughout Canada

A report by KPMG Consultants (4) as part of a "Canadian Technological Road Mapping on Functional Foods and Nutraceuticals: MARKET" found that 215 Canadian companies are involved in functional foods and nutraceuticals and distributed geographically as shown in Table 13:

	Number of Companies	% of Companies in each Province
British Columbia	47	22
Alberta	15	7
Saskatchewan	26	12
Manitoba	7	3
Ontario	52	24
Quebec	49	23
New Brunswick	2	1
Nova Scotia	7	3
Newfoundland	6	3
Prince Edward Island	4	2
Total	**215**	**100**

The Canadian industry is not only reacting to an increasingly lucrative global market but also targeting health care concerns "at home." Lifestyle-related chronic disorders are a major component of increasing health care expenditures in Canada (5). The proportion of disease onset attributable to diet is estimated to be approximately 40 to 50 percent for cardiovascular disorders and diabetes, while 35 to 50 percent of all cancers are directly related to dietary factors. Approximately 20 percent of osteoporosis is diet related. In this report, the author presents strong arguments to support the role of functional foods and NHPs in reducing the prevalence of chronic disease in Canada and providing impressive savings in health care costs without significant overall dietary changes.

The Canadian functional food and NHP sector has also garnered a great deal of attention and enthusiasm from support/partner organizations such as governments and health care and research communities for its significant potential to provide:

- Diversification and market growth for Canadian agriculture and marine-based industries
- Increased economic development in Canada
- Reduced health care costs for all Canadians
- Improved health and well-being and disease reduction for all Canadians

Canada has the potential of being recognized as a global leader in the production and exporting of functional food ingredients and products as well as being a model to the world of a healthy nation driven by a philosophy that fosters proper nutrition. The industry, and indeed Canadian consumers, have been negatively impacted by a regulatory environment that has restricted all but the most stringent drug-type health claims on functional foods and NHPs, a situation that must be rectified if we are to take full advantage of the opportunities that product development in this area offer the economy as well as the health of the population.

The Regulatory Challenge in Canada

The Canadian Food and Drugs Act and Regulations was passed into law in 1953. The definition of food under the Food and Drugs Act includes ". . . any article manufactured, sold or represented for use as food or drink for human beings, chewing gum, and any ingredient that may be mixed with food for any purpose whatever." Drugs are defined as ". . . any substance or mixture of substances manufactured, sold or represented for use in: the diagnosis, treatment, mitigation or prevention of a disease, disorder or abnormal physical state, or its symptoms, restoring, correcting or modifying organic function."

Through its definitions of "food" and "drug," this legislation currently restricts health-related claims for foods, food ingredients and NHPs. Since 1953, these products have been considered as either foods or drugs depending on the type and concentration of the "active ingredient" and whether claims are made. For products regulated as a food, the legislation has no provisions for making claims of a "health" or "therapeutic" nature regarding the use of, or possible side effects of, the product. Thus, the consumer may be inadequately informed about the use of the product. To protect consumers, Health Canada has banned or restricted the use of certain NHPs because the agency was unsure of the product's safety as a food. According to many industry members, the standards and regulations applied to NHPs in Canada are inappropriate, leading to restricted market entry of many safe and effective products. Research conducted by the Canadian Health Food Association (1998), for example, indicated that 42 percent of the 858 NHPs sold by a typical large U.S. supplier were prohibited for sale by a Canadian supplier in Canada. U.S.-approved products such as melatonin and single amino acids are prohibited for sale in Canada.

Any supplement or functional food that carries a health claim, or levels of ingredients not permitted for conventional foods, are currently treated as drugs. This requires a Drug Identification Number (DIN), which is issued by the Therapeutic Products Directorate (TPD) of the Health Protection Branch (HPB) after it reviews the DIN application. The cost of a review will vary but requires a significant investment for smaller companies. Companies have the choice of obtaining a DIN, manufacturing under drug Good Manufacturing Practices (GMP) in many cases and marketing their NHPs, nutraceutical, or functional food product as a drug, or selling it without claims. Of interest is the fact that Tropicana chose to market its product carrying a DIN and providing dosage information on packaging as a "calcium and vitamin supplement," otherwise known as orange juice in the U.S. (6).

Further limiting the development of health claims in Canada are statements in Section 3 of the Act and Regulations that prohibit the sale or advertisement to the general public of any food, drug, cosmetic, or device that indicates a treatment, cure, or preventive role for diseases or disorders referred to in Schedule A that include heart disease, diabetes, cancer, hypertension, obesity, and arthritis. These are the most common causes of morbidity and mortality in Canada and are diseases for which functional foods and NHP have the potential to be most beneficial. Physiological effects that relate to these conditions, such as lowering of serum cholesterol or glucose, are also considered under the umbrella of Schedule A and are precluded from appearing on labeling and in advertising. For health claims to appear in Canada, changes would have to be made to allow claims about Schedule A diseases or physiological effects related to these diseases.

The regulatory environment in Canada is believed to have stifled innovation, competition, and investment in the industry. A report released in 2001 concluded that the burden of Canadian regulations has led to significant lost opportunities and sales (7). Additionally, the authors identify disadvantages to consumers who have fewer choices of functional foods and NHPs and much less information to help in their selection and use. Of significance, Health Canada has recognized that the regulatory framework does not support labeling and advertising of the potential health benefits of NHPs and functional foods to consumers and, as of 1996, began a series of regulatory reviews for these product categories. A great amount of progress has been made in the area of regulatory developments for NHPs, whereas functional food initiatives have been less than speedy.

Functional Foods

In the summer of 1996, the Food Directorate of HPB began deliberations that resulted in a Policy Options paper entitled "Recommendations for Defining and Dealing with Functional Foods" (8). This paper presented the different definitions of functional foods, described approaches to health claims in several jurisdictions, and discussed possible regulatory alternatives to permit health claims in Canada.

In 1997, a joint strategy was initiated by the Food Directorate and Therapeutic Products Programme to discuss further the possibility of allowing health claims for foods in Canada. The strategy involved consultations with major stakeholders and resulted in a Policy Options Analysis in October 1997, followed by a Final Policy paper in November 1998 (9). In these documents, various possible options for regulations were identified, with the final option chosen being:

> to permit structure/function and risk reduction claims for food products, and continue to regulate remaining health claims as drugs

The proposed framework, outlined in Table 13.1, indicates that health claims would be permitted for foods and food products but that therapeutic claims, those intended to cure, prevent, treat or mitigate a disease, would continue to be regulated as drugs. The proposal also indicated that product-specific risk-reduction claims would be considered, as long as a regulatory framework could be developed to allow such claims within the current Food and Drugs Act. The structure/function scenario would be similar to the regulatory framework for dietary supplements in the United States known as the Dietary Supplement Health and Education Act of 1994 (DSHEA). Risk-reduction claims are currently permitted for certain food and food constituents in the United States under the Nutrition Labeling and Education Act of 1990 (NLEA). One key difference between the United States and Canada is that the proposed changes could allow product specific claims whereby a company sponsoring research and development on its product would be allowed an exclusive claim. Another food with the same "active ingredient" would not be able to make the claim unless supported by research.

The Final Policy document (9) identified three major components for implementation of health claims:

1. Adoption, where applicable, of certain diet-based risk reduction claims currently approved in the U.S. under the Nutrition Labeling and Education Act
2. Development of standards of evidence for evaluating foods with new health claims and a Guidance Document which will provide more detailed information regarding the preparation of submissions for health claim review
3. Development of a regulatory framework for foods with new health claims

Table 13.1. Analytical framework for health claims for nutraceuticals/functional foods

	Structure/Function Claims	Risk Reduction Claims		Therapeutic Claims
		Product Specific	Generic	
Definition	Asserts the role of a nutrient or other dietary component intended to affect the structure or physiological function in humans	Asserts a relationship between a specific food product and reduced risk of a disease or specific health condition	Asserts a relationship between a nutrient in a diet and reduced risk of a disease	Asserts a relationship between a nutrient(s) and the cure, treatment, mitigation of a disease, disorder, or abnormal physical state
Prior presence of recognized disease state in target group	No	No	No	Yes
Target group	General population	General population and subgroups	General population and subgroups	Specific individuals with a disease (cure/treat/mitigate), General population and subgroups (prevent)

In this document, Health Canada indicated several underlying principles in relation to the development of health claims in Canada:

- Products with proven physiological benefits should be available to Canadians.
- A regulatory environment that is conducive to this aim will fairly and responsibly permit the promotion to consumers of food and drug products that have been shown by valid scientific evidence to improve health.
- It is postulated that health claims benefit consumers provided that the information is substantiated, truthful, not misleading, and not likely to lead to harm.

The proposed definition in the Policy Options paper for a nutraceutical was a product that has been isolated or purified from foods and generally sold in medicinal forms not usually associated with food and exhibiting a physiological benefit or provide protection against chronic disease (9). With the establishment of the Natural Health Products Directorate (NHPD) and a definition for NHP, the proposed category of "nutraceuticals" has been included in the regulatory framework for NHP, which is described later in this document.

In the same proposal (9), a functional food was defined as similar in appearance to a conventional food, consumed as part of the usual diet, with demonstrated physiological benefits, and/or consumed to reduce the risk of chronic disease beyond basic nutritional functions. However, in late 2001, Health Canada announced that a new regulatory definition for functional foods would not be required under the current Canadian Food and Drugs Act to permit health claims for foods. However, this term is used extensively in Canada to describe foods with health benefits beyond basic nutrition.

Adoption of Certain of the Initial 10 Diet-Based Disease Risk-Reduction Claims Approved in the United States under the NLEA

At the time of the Final Policy paper, 10 generic health claims had been approved in the United States under the Nutrition Labeling and Education Act (NLEA). Such claims apply to a food or a group of foods that have compositional characteristic(s) that contribute to a dietary pattern associated with reducing the risk of a disease or health condition. After the claim is authorized, any food that meets the specified conditions for composition and labeling may carry the claim without further assessment. For each of these 10 claims, Health Canada asked key scientists in Canada to review the literature and submit reports addressing the legitimacy of the existing U.S. claims. In 2000, it was announced by Health Canada that five of the NLEA claims were considered valid in a Canadian context. These were as follows:

- Sodium and hypertension
- Calcium and osteoporosis
- Saturated and trans fat and cholesterol and coronary heart disease
- Fruits and vegetables and cancer
- Sugar alcohols and dental caries

Each of these health claims required an individual amendment to the Foods and Drug Act. Details of the specific wording of the claims and criteria for the nutrient content of the food and other factors were published in January 2003 in *Canada Gazette* Part II, "Regulations Amending the Food and Drug Regulations (Nutrition Labeling, Nutrition Claims and Health Claims)" as part of mandatory nutrition labeling regulations (10). By 2005, food labels in Canada will be required to provide more nutrition information such as core information on calories, fat, saturated fat, trans fat, cholesterol, sodium, carbohydrate, fibre, sugar, protein, vitamin A, vitamin C, calcium, and iron.

Health Canada has estimated that these new labeling regulations with five approved claims will result in an estimated $5 billion in health-care cost reductions during the next 20 years, specifically in direct and indirect costs associated with three diagnostic categories: cancer, diabetes, and cardiovascular and stroke.

For the other five generic U.S. claims considered, Health Canada concluded that issues remained that did not allow their immediate approval. In the case of fat and cancer, there had been considerable new evidence to question the role of dietary fat in cancer risk, such that Health Canada decided not to proceed with this claim. For the following four claims, decisions are still pending:

- Folate and neural tube defects
- Fibre-containing grain products, fruits, and vegetables and cancer
- Fruits, vegetables, and grain products that contain fibre, particularly soluble fibre and risk of coronary heart disease
- Soluble fibre and risk of coronary heart disease

Development of Standards of Evidence and a Guidance Document on Data Requirements for Supporting the Validation of New Health Claims for Foods

For other health claims, the Health Products and Foods Branch, using an internal working group and in consultation with an Expert Advisory Panel, has developed a framework for the standards of scientific evidence required for health claims and a guidance document for industry. In June 2000, Health Canada published a Consultation Document entitled "Standards of Evidence for Evaluating Foods with Health Claims: A proposed framework" (11).

The principles governing the proposed standards have three important elements. The first is product safety, which Health Canada interprets to be reasonable assurance of no adverse health effects. Second is claim validity as determined by demonstration of product efficacy and effectiveness based upon establishing an etiologic link between the desired effect and consumption of the food or bioactive substance, at the recommended level of intake in the target population that will most likely benefit. Third is quality assurance, meaning that foods bearing health claims should be able to identify, measure, and maintain a consistent level of the bioactive substance to ensure efficacy without jeopardizing safety. Health Canada proposes that initially only claims with the potential for major public health benefit and for which there is sufficient acceptable scientific evidence should be given priority for evaluation.

According to Health Canada, the proposed standards are intended to reassure consumers that the products carrying the claims are safe and that the claims are valid. The standards would apply to all food or beverage products that bear direct or indirect claims about their health benefits.

As examples of "direct" and "indirect" claims, Health Canada states:

- A direct health claim is clearly stated and straightforward. Example: "Calcium reduces the risk of osteoporosis."
- An example of an indirect or implicit health claim would be "Contains calcium," thereby relying on consumer awareness of the proposed benefits of the ingredient.

Such food or beverage products would include:

- Conventional foods
- Foods to which biologically active substances have been added
- Foods that have been modified by other means, including having been derived from biotechnology

The proposed standards do not include nutraceuticals or other NHPs sold in dosage form. Health Canada states, however, that ". . . if such products are added to foods, the appropriateness of their inclusion would be assessed. If the inclusion is found to be appropriate, the modified food would be evaluated according to the proposed standards."

Development of an Appropriate Regulatory Framework to Allow Product-Specific Health Claims for Foods

In October 2001, Health Canada released a proposed approach to regulating product-specific health claims for foods (12). This authorization is proposed for specific foods having a direct, measurable metabolic effect beyond normal growth, development or health maintenance, reducing disease risk or aiding in the dietary management of a disease or condition. In the November 1998 policy options paper, Health Canada recognized that a claim concerning the effect of a food or its ingredient(s) or component(s) should not be generalized to other similar products unless acceptable supporting evidence was provided. This is the rationale for product-specific authorization whereby each product with the intended claim is evaluated on its own merit. This concept recognizes that food matrices and processing conditions could have an effect on the physiological property of foods. Therefore, an application containing product-specific evidence would be required for a similar claim on another product, unless generally accepted or specific nutritional or food science theory or knowledge would indicate otherwise.

It was proposed that a food that is manufactured, sold, or represented to have a direct, measurable effect on a body function or structure beyond normal growth and development or maintenance of good health (previously termed "structure/function claims") be required to submit detailed information to support such an effect before being advertised or offered for sale. The conditions, which must be met before a food can be authorized to carry a claim or a representation conveying such an effect, are outlined in this proposal.

An authorized claim would be identified by a Claim Identification Number (CIN), which would be displayed in product labelling. Authorization would be granted on a product-by-product basis. Adequacy of the evidence supporting a claim would be reviewed on a case-by-case basis, based on six underlying principles:

- Totality of the evidence (not only supporting evidence)
- Evidence supportive of a causal relationship between the food intake and its health effect
- Evidence relevant and generalizable to the target population
- A systematic approach used to ensure that all evidence is considered and conclusions are justified
- High level of certainty of claim validity based on best practices in science review
- Acceptable design and quality of studies based on best practices in scientific review

Product-specific authorization will require **Type 1** direct evidence consisting principally of controlled human trials of the food or biologically active compound, as well as supportive **Type 2** published literature consisting of systemic reviews of human studies, if available. If the submission were for product-specific authorization of a risk reduction type of health claim, evidence from human observational studies (prospective, retrospective cohorts, case-controls) would also be required. Generic authorization will require **Type 2** published literature consisting of human observational studies (prospective, retrospective cohorts, case-controls) and systematic reviews, and human experimental evidence if available. Data from other observational studies in humans and animals or *in vitro* experiments would be applicable to both types of evidence as supportive data only. Health Canada has released an *Interim Guidance Document on Preparing a Submission for Foods with Health Claims*, in which further information may be found regarding the proposed requirements (13).

Given the current definitions of **food** and **drug** within the Food and Drugs Act and Regulations, as well as the restrictions imposed under Schedule A, new health claims on foods still require individual amendments be made to the Act. This process is laborious and, as evidenced with the five generic claims recently approved, can take upwards of three years to complete the regulatory process. Of interest, as of November 2003, no applications for product-specific health claims had been received by Health Canada.

A potential solution to the impediments of the Act would be to allow foods with health claims to be regulated as a subset of drugs, the route that new NHP regulations are taking in Canada (14). In addition, it is encouraging that Health Canada announced in February 2003 that a Working Committee had been established to examine and make recommendations as to modifications of Schedule A (15). Both issues are discussed in detail in the following sections.

Natural Health Products

The Office of Natural Health Products was formed in March 1999 to establish and implement a new regulatory framework for NHPs. In 2001, the Office became the NHP Directorate (NHPD), a parallel and equal department to Health Canada's Food Directorate and Therapeutic Products Programme. The Directorate moved quickly to develop NHP regulation to allow health claims and increased consumer access.

Proposed regulations for NHPs were initially published in *Canada Gazette Part I*, the first step toward amendments to the Foods and Drug Act and Regulations, on December 21, 2001 (16). Final regulations were published in Canada Gazette II on Wednesday, June 18, 2003 (17). The main components of the regulations include NHP definitions, product licensing, adverse reactions reporting, site licensing, good manufacturing practices (requirements for product specifications [identity, purity, potency], premises, equipment, personnel, sanitations program, operations, quality assurance, stability, records, sterility, lot or batch samples, and recall reporting), standards of evidence for safety and health claims, and labeling and packaging.

Definition and Health Claims

Until publication of Canada Gazette II, the working definition for a *nutraceutical* in Canada has been "a product that has been isolated or purified from foods and generally sold in medicinal forms not usually associated with food. Nutraceuticals have been shown to exhibit a physiological benefit or provide protection against chronic disease." Health Canada decided that the product category of nutraceuticals would be encompassed within NHP regulations. Under NHP proposals, such products will include homeopathic preparations, substances used in traditional medicine, a mineral or trace element, a vitamin, an amino acid, an essential fatty acid, or other botanical-, animal-, or microorganism-derived substance.

Regarding health claims, NHP products will be allowed to be manufactured, sold, or represented for use in: (i) the diagnosis, treatment, mitigation, or prevention of a disease, disorder, or abnormal physical state or its symptoms in humans (that is, claims are currently allowed in Canada only for drug products); (ii) resorting or correcting organic functions in humans, or (iii) maintaining or promoting health or otherwise modifying organic functions in humans. In order for "drug type" claims to be allowed within the Foods and Drug Act and Regulations, these products will be regulated under a subsection of the Drug Regulations but will still be referred to as NHPs. The drug legislation is the vehicle to manage the process without requiring the opening of the Food and Drugs Act and Regulations or the time-consuming steps involved in establishing a new regulation for **each** new health claim. The standards of evidence for safety and a health claim will correspond to the strength of the claim. For example, traditional references will be required for a product to carry "self care" claims such as "Traditionally used to . . . ." This is in contrast to a "disease related" claim such as "Clinical trials show . . . ." Such a claim will require the submission of data from a meta-analysis of randomized controlled trials, or at least one well-designed (multicenter) randomized controlled trial.

One area of debate continues to be the definition of NHPs. Even though the proposal does not include conventional foods and is not intended to capture a product in a food medium, the rule is unclear as to whether foods bearing health claims or structure/function claims relating to a nutrient (for example, a beverage with the statement "contains calcium to help build strong bones") are outside the scope of the NHP regulation. For example, the proposed regulations state, "Although 'dosage form' is not an express part of the definition, the NHPD recognizes that NHPs are usually sold in capsule, pill, tablet or liquid form. As well, certain other forms, such as gum or bars, have come to be considered acceptable dosage forms."

In comparing the Canadian NHP regulations to that of DSHEA, there remains the possibility that NHPs available in Canada may be able to legally carry a much wider range of health claims than can their U.S. counterparts.

Natural Health Product Regulations—Canada Gazette II

Effective January 1, 2004, product licences for NHPs will be available through the NHPD. For NHPs already licenced as drugs (with a Drug Identification Number, or DIN), a Product Licence application and market authorization from the NHPD will be required by December 31, 2009. An abbreviated application process will be used to issue a market authorization in the form of a Product Licence. During the transition period, DIN products will continue to be regulated under the Food and Drug Regulations until they obtain NHP product licences.

The transition period will be four years (that is, by December 31, 2007) for products that do not hold DINs. Applications for a product licence must provide specific information about the NHP including the quantity of the medicinal ingredients it contains, the specification with which it complies, the recommended use or purpose for which the NHP is intended to be sold, and the supporting safety and efficacy data. After a licence is granted, a natural product number (NPN) will be issued. Requirements for Adverse Event Reporting (AER) come into effect as products obtain their NHP licences.

The transition period for good manufacturing practices (GMPs) and site licencing will be two years. Those companies that already manufacture, package, label, or import NHPs or hold an establishment licence under the Food and Drug Regulations must obtain a site licence from the NHPD. Adherence to GMPs is the main prerequisite that must be met before a site licence is issued. Site licence applicants must submit a report from a qualified quality assurance person; this individual can be an internal employee or a third-party auditor.

Distributors will not be required to hold a site licence, because their activities are usually limited to the handling of a finished, packaged product and do not involve manufacturing, packaging, or labeling. Because of jurisdictional issues, foreign sites will not receive site licences, but they will be required to meet the same site licencing requirements to obtain a foreign site authority number.

There is a "default licencing" provision whereby products containing a single active ingredient for which Health Canada has created a monograph will be automatically granted a licence within 60 days. This provision does not come into effect until July 1, 2004. The

NHPD is working on a combination product policy that may allow products with multiple active ingredients to use this default-licencing provision.

Most products that fall within the NHP definition and do not hold drug licences are currently allowed to stay on the market as "Products Subject to Special Measures," even though they do not specifically comply with the Food and Drug Regulations. It is expected that the NHPD will introduce a new policy to allow these products to stay on the market throughout the transition period.

A Transition Guidance Document, which outlines the requirements and time frames for DIN products to become compliant with NHP regulations and provides information on how to prepare Product Licence and Site Licence submission packages, was published in November 2003 (18).

Applicants with a DIN submission for an NHP in process at the Therapeutic Products Directorate (TPD) before October 3, 2003, may consent to having their submission transferred to the NHPD prior to January 1, 2004, for assessment under the NHP regulations. Submissions transferred through this process will be placed in a priority queue at NHPD for assessment.

After December 31, 2003, the TPD and the NHPD jointly will contact those applicants for submissions considered to be NHPs in queue at TPD who have not been issued a DIN. These applicants will be asked to identify their product as an NHP or provide a rationale as to why their product is not considered to be a natural health product, and submit this information to TPD on or before January 30th, 2004.

Of significance to the NHP industry, health claims will be considered for NHP that are not assessed through the "drug" route. Until amendments are made to the Food and Drugs Act and Regulations' Schedule A, a lengthy list of diseases and disorders (including biomarkers associated with disease), significant limitations on claims for certain disease states (for example, heart disease, arthritis, depression, and so on) will exist.

A working group was established by Health Canada in February 2003 (15) to make recommendations intended to address this regulatory impediment. This group held meetings in April and May of 2003. The Working Group agreed that "prevention" and "risk reduction" in fact did not have the same meaning—risk reduction was considered to be related to a health continuum, whereas prevention refers to an absolute term. Such a differentiation has potential significance in the allowance of risk reduction for both functional foods and NHPs.

The new regulations are expected to hold the industry to more uniform standards for the manufacturing and formulation of NHPs, and this uniformity is hoped to have a positive impact on consumer confidence. Improved labeling is assumed to help consumers in making informed choices and take more control over their health care decisions.

One unknown with regard to the implementation of these new regulations is how long the NHPD will take to evaluate product licence applications and grant licences for new products and for products already on the market that do not hold DINs. Some in the Canadian NHP industry fear that Health Canada will not assign sufficient resources to product evaluation and licencing and that backlogs will result, particularly for new ingredients and for products with more than one active ingredient. Such delays may hamper the ability of manufacturers to introduce innovative products.

Another concern expressed is the potential impact of the new regulations on the small and medium-sized businesses that dominate the industry. Responding to this concern, Health Canada has indicated that a sustained effort will be undertaken to provide working tools and processes to support these enterprises, including initiatives such as the public education program that will be launched in late fall of 2003.

Natural Health Products Research Program

In its series of recommendations, the Standing Committee on Health identified the need for more focused research in NHPs. In response, the NHPD held a series of consultations with stakeholders to identify research priorities and develop appropriate strategies (19).

In September 2003, the NHPD announced the Natural Health Products Research Program (NHPRP). Over the next five years, the NHPD will invest $5 million in supporting NHP research, with $2 million dollars allocated to a partnership with the Canadian Institutes of Health Research. The priorities of the NHPRP include building research capacity, committing to conduct research of the highest quality, developing community infrastructure and partnerships, and enhancing knowledge transfer and information retrieval.

The scope of the NHPRP is to encourage research interest in natural health product research and direct funding to research and related activities that:

- Are relevant to the regulatory function of the NHPD
- Enable consumers to make informed choices about natural health products
- Are sensitive to the needs of the communities that produce and utilize these products

The objectives of the NHPRP are to the following:

- Contribute to improved knowledge of natural health products, including their safe and effective use, by supporting research in various forms including:
  –biomedical
  –product quality
  –clinical
  –health services
  –population health
- Make information accessible to stakeholders and consumers

The program will support research and knowledge-based development through various means, both individually and in partnerships with other funding agencies, most notably the Canadian Institutes of Health Research (CIHR).

The Canadian Institutes of Health Research is the major federal agency responsible for funding health research in Canada. The objective of the CIHR is to excel, according to internationally accepted standards of scientific excellence, in the creation of new knowledge and its translation into improved health for Canadians, more effective health services and products, and a strengthened Canadian health care system.

Conclusion

It is hoped that the proposed changes outlined in this chapter will have a positive impact on the domestic market. A more open Canadian regulatory system would allow companies to develop products and initially test market and commercialize them in Canada before exporting them. This is the normal route to a successful export strategy. Growth in the international market will help the industry develop core competencies, which will enhance the competitiveness of Canadian companies.

The growing demand for NHPs and functional foods to meet consumers' desire to lead healthier lifestyles presents significant opportunities for the Canadian industry. Canada faces the ongoing challenge of having an abundance of natural resources but a fragmented, regionalized industry spread across a vast border with the United States. Having the United States serve as our largest trading partner, and we as its, is both a strength and weakness when trying to develop more value-added opportunities and an economically viable industry under free trade. The relatively high growth rate of various segments of the nutritional market is attracting pharmaceutical, chemical, and food-processing companies, which increasingly are requiring good sources of raw materials and ingredients. Global companies are interested in locating innovative products. Developing and supplying such products represents a significant opportunity for Canadian companies.

The Canadian environment is viewed globally as clean and pristine, which is appealing to an industry built upon the concepts of **pure** and **natural**. Canadian companies have developed specialized expertise in production and processing. Canadian producers are innovative and very receptive to diversification opportunities. The country has valuable expertise in several aspects of functional food and NHP research, which provides a foundation to building an industry.

As described in this chapter, very significant and positive progress has been made in a relatively short period of time in Canada with regard to the regulation of NHPs. The Canadian population and the Canadian industry deserve the same commitment to progress in the area of functional foods.

References

1. Scott Wolfe Management. Potential benefits of functional foods and nutraceuticals to the Agri-Food industry in Canada. Report submitted to Agriculture and Agri-Food Canada, Food Bureau. Ottawa, ON. March 2002.

2. Ferrier, G. *Nutrition Business Journal.* San Diego, CA. 2002. Vol. 3, No. 9.

3. Health Strategy Consultants. *The Canadian Nutrition Industry Overview.* Providence, R.I.: September 2002.

4. KPMG Consultants. 2002. *Canadian Technological Road Mapping on Functional Foods and Nutraceuticals: Market.* Report commissioned by the National Research Council of Canada, Ottawa, ON and University of Laval, QC.

5. Holub, B. Potential benefits of functional foods and nutraceuticals to reduce the risk and costs of diseases in Canada. Report submitted to Agriculture and Agri-Food Canada, Food Bureau Ottawa, ON: June 2002.

6. Singer, Z. Fortified juice must be sold as a drug in Canada. *Ottawa Citizen.* August 24, 1999.

7. Smith, B.L., Marcotte, M., and Harrison, G. 1996. A Comparative Analysis of the Regulatory Framework Affecting Functional Food Development and Commercialization in Canada, Japan, the European Union and the United States of America. Commissioned by Agriculture and Agri-Food Canada to Intersect Alliance. Ottawa, ON.

8. Health Canada. 1997. Discussion Document. Functional Foods and Nutraceuticals. Internet: http://www.hc sc.gc.ca/main/drugs/zfiles/english/ffn/ffdscdoc_e.html (accessed December 12, 2001).

9. Health Canada. Final Policy Paper on Nutraceuticals/Functional Foods and Health Claims on Foods. November 1998. Internet: http://www.hc-sc.gc.ca/food-aliment/ns-sc/ne-en/health_claims-allegations_sante/ e_nutra-funct_foods.html (accessed October 24, 2003).

10. Government of Canada. Regulations Amending the Food and Drug Regulations (Nutrition Labeling, Nutrition Claims and Health Claims). *Canada Gazette Part II.* January 2003. Internet: http://canadagazette.gc.ca/ partII/2003/20030101/pdf/g2-13701.pdf (accessed January 4, 2003).

11. Health Canada. Consultation Document: Standards of Evidence for Evaluating Foods with Health Claims: A Proposed Framework. June 2000. Internet: http://www.hc-sc.gc.ca/food-aliment/english/subjects/health_ claims/standards_of_evidence (accessed July 13, 2000).

12. Health Canada. Product-Specific Authorization of Health Claims for Foods: A Proposed Regulatory Framework. October 2001. Internet: http://www.hc-sc.gc.ca/food-aliment/ns-sc/ne-en/health_claims-allegations_sante/e_finalproposal01.html (accessed November 11, 2003).

13. Health Canada. Interim Guidance Document: Preparing a Submission for Foods with Health Claims Incorporating Standards of Evidence for Evaluating Foods with Health Claims. Undated. Internet: http://www.hc-sc.gc.ca/food-aliment/ns-sc/ne-en/health_claims-allegations_sante/pdf/e_guidance_doc_interim.pdf (accessed November 9, 2003).

14. Stephen, A.M., Liston, A.J., Anthony, S.P., Munro, I.A., and Anderson, G.H. (2002). Regulation of foods with health claims: A proposal. *Can. J. Public Health.* 93:328–331.

15. Health Canada. Health Canada Reviews Schedule A to the Food and Drugs Act. February 2003. Internet: http://www.hc-sc.gc.ca/hpfb-dgpsa/sched_a_review_e.html (accessed June 23, 2003).

16. Government of Canada. Natural Health Products Regulations. *Canada Gazette Part 1*. December 2001. Internet: http://canadagazette.gc.ca/partI/2001/20011222/pdf/g1-13551.pdf (accessed December 22, 2001).

17. Health Canada. Natural Health Products Regulations. Canada Gazette Part 2. June 2003. Internet: http://canadagazette.gc.ca/partII/2003/20030618/html/sor196-e.html (accessed June 18, 2003).

18. Health Canada. Natural Health Products: Guidance Documents. November 2003. Internet: http://www.hc-sc.gc.ca/hpfb-dgpsa/nhpd-dpsn/index_e.html (accessed November 15, 2003).

19. Health Canada. Natural Health Products Research Program. September 2003. Internet: http://www.hc-sc.gc.ca/hpfb-dgpsa/nhpd-dpsn/research_links_e.html (accessed November 15, 2003).

14 The Regulation of Functional Foods and Nutraceuticals in the European Union

Peter Berry Ottaway

On March 25, 1957, the Treaty of Rome was signed, bringing into existence the European Economic Community (EEC), with the six signatories of the Treaty: Belgium, France, Germany, Italy, Luxembourg, and the Netherlands becoming the founder members of the Community.

Over the following 35 years, the Community has been expanded by an additional nine European countries to bring the total to 15 members. During the later stages of development, the Community changed its name to the European Union (EU). Beginning in 2004, a further period of enlargement will take place that will bring the total membership to 25 with a population exceeding 470 million.

In the early days of the EEC the emphasis was on the reduction and eventual elimination of barriers to trade between the member countries. One of the main trade sectors identified as requiring almost total harmonization was the food industry. By 2002, well over 90 percent of the food law in any member state was based on European law. However, some areas of food legislation still had not been harmonized and were controlled at a national level. The national laws are allowed to continue until superseded by European law.

As of early 2004, functional foods, many forms of dietary supplements, and nutraceuticals were falling into the complex problem of being regulated by both European law and the national legislation of each member state in which they are intended to be sold. This means that a large number of products still cannot be sold freely across the whole of the EU, and this situation will continue for a number of years, probably well past 2006.

The History of European Food Law

The harmonization of EU food law has been in progress for forty years. From the early 1960s until 1985, most of the focus was on vertical legislation, with "recipe" or compositional criteria being developed for product categories such as cocoa and chocolate products, fruit juices and fruit jams, and jellies and marmalades. Of the original program of more than thirty "recipe" directives, only nine were agreed to and adopted in the first twenty years of the European Community. In 1985 a White Paper from the European Commission to the European Council recommended that a complete reevaluation of the approach to food legislation was required and that the original concept of restrictive vertical legislation should be replaced by a system of horizontal legislation.

The recommendations in the White Paper were accepted and a new concept of European food law based on five horizontal framework directives was proposed and eventually adopted by the European Council and Parliament. The framework directives were developed to comply with identified requirements of public health and safety, consumer information, and general food control measures. The five framework directives that were considered to be crucial to a harmonized system of food law were:

- Food labeling and presentation
- Food additives
- Materials and articles in contact with food
- Official control of foodstuffs
- Foods for particular nutritional uses (PARNUTS)

These framework directives were required to lay down the general principles for control in the relevant areas. Specific technical directives were to be developed as adjuncts to the framework directives where necessary. However, even with the more simplified and streamlined system introduced with the 1985 initiative, progress was slow and much of the proposed legislation was not in place by the introduction of the Single Market in January 1993.[1]

In January 2000 the European Commission published its White Paper on Food Safety, which included details and a timetable for 84 items of legislation that were still outstanding.[2] Some of this outstanding legislation was of direct relevance to the nutraceutical, functional food, and supplement industries, and important legislation was still not completed by the end of 2003.

As already mentioned, European food legislation is based around five framework directives. All products sold under food law must comply with the requirements of these directives, together with any specific directives and regulations in force. This requirement embraces functional foods, food supplements, and foods for particular nutritional uses (PARNUTS).

Food Labeling

In the EU, all functional foods and products containing nutraceuticals sold under food law, as opposed to medicines law, have to comply with the general requirements of the food labeling directive 2000/13/EC[3] and with specific requirements of product category directives such as those for food supplements and PARNUTS. In addition, many member states have national labeling requirements for products in which certain ingredients are present.

The core labeling requirements for almost all prepacked foods are given in Table 14.1. Some of these requirements are very detailed. For example, all ingredients and additives must be declared in descending order by weight of input according to prescribed rules. If specific reference is made to an ingredient in label or promotional copy, the percentage of the ingredient in the product must be stated in the ingredients list. Food additives, colors, and sweeteners must be listed and preceded by their additive category description (for example, "sweetener: aspartame"). Additives can be listed either in the form of their "E" number or by their official name as given in the relevant directive. Thus, aspartame can also be listed as E951.

Historically, European food labeling law has contained a concession whereby compounded ingredients that were present at an amount less than 25 percent of the total weight of the food product did not have to have their subcomponents declared. As a result of the need to introduce the requirement to highlight potential allergens on the label, the concession for compounded ingredients is being removed and the legislation to enact this is to come fully into effect in November 2005. The result will be a significant increase in the length of the ingredient listing, particularly when coated or microencapsulated ingredients such as vitamins are used.

Table 14.1. Essential requirements for the labeling of food products in the European Union

1. The generic (full descriptive) name under which the product is sold
2. A list of ingredients
3. In the case of prepackaged foodstuffs, the net quantity
4. The date of minimum durability (or, for highly perishable foodstuffs, the 'use by' date)
5. Any special storage conditions or conditions of use
6. The name and address of the manufacturer, or the packager, or of a seller established within the EU
7. Particulars of place of origin or provenance where failure to give them may mislead the consumer
8. Instructions for use
9. For beverages containing more than 1.2% by volume of alcohol, the alcoholic strength by volume
10. Lot or batch number (preceded by letter L if not obvious, for example, L123A45)
11. Statements or warnings (for example, GMOs, sweeteners and so on)

The product label must also contain a generic name or description of the product if a brand or fantasy name is used. In some instances, the generic name is prescribed by law, as in the case of food supplements.

An open expiration date is required on each pack. For the longer-shelf-life products, this must be in the form of "Best before *end: month/year.*" There are no official rules as to how the expiration date is determined. For products containing active ingredients that can deteriorate on storage, such as vitamins, the expiration date should be derived from properly conducted stability studies. The use of such studies is particularly important for some groups of nutraceuticals, such as the carotenoids that can deteriorate or interact with other ingredients.

In addition to the open expiration date, the pack must also contain a batch or lot number, and there are rules as to how this information must be provided.

Over the past few years an increasing number of mandatory label statements or warnings have been introduced into EU food law, particularly in relation to sweeteners and ingredients derived from genetically modified organisms. Although such statements are compulsory in all member states of the EU, some countries also have national requirements, such as the inclusion of a warning for pregnant women on products containing vitamin A as retinol; the British and Italian authorities also require label warnings on certain products containing flavonoids.

EU law requires that the label contain a contact name and address of a manufacturer, packer, or seller established within the EU. In the case of imported products intended for sale, the name and address of an EU importer or marketing company must be included on the label.

Nutrition Labeling

Unless required by specific product category legislation, or unless a nutrient claim is made on the label or promotional material for a product (for example, "Rich in calcium"), nutrition labeling is not mandatory. However, if nutrition labeling is used, the layout and content of the information must comply with very strict rules as laid down in the directive on nutrition labeling (90/496/EEC).[4]

For the purpose of the law, "nutrition labeling" means any information appearing on a label relating to the energy value of a food and the following nutrients:

- Protein
- Carbohydrate
- Fat
- Fiber
- Sodium
- Vitamins and minerals listed in the annex to the directive (12 vitamins and six minerals)

The directive makes it clear that a reference to the quantity or quality of a nutrient as required by the legislation does not constitute a nutrition claim. However, any representation or advertising message stating, suggesting, or implying that a food has particular nutritional properties due to its energy or nutrient content is regarded as a nutrition claim. Article 3 of the directive states that the only permissible nutrition claims are those relating to energy and to the nutrients listed above. This means that until the list of vitamins and minerals is amended, claims cannot be made for a number of trace minerals such as selenium, copper, manganese, and molybdenum.

For the purpose of nutrition labeling, definitions are given for many of the nutrients, some of which differ from those in the United States of America and other non-EU areas. Protein content is calculated as total Kjeldahl nitrogen × 6.25. **Carbohydrate** means any carbohydrate that is metabolized and includes polyols. **Sugars** includes all mono- and disaccharides present in foods but excludes polyols. **Fat** refers to total lipids and includes phospholipids, saturated fatty acids, and mono and polyunsaturates.

Nutrition labeling may include the quantities of starch, polyols, mono-unsaturates, polyunsaturates, cholesterol, and the quantities of any of the vitamins and minerals listed in the annex provided that they are present in significant amounts. For the micronutrients, a significant amount is considered to be 15 percent of the recommended amount of the nutrient supplied by 100g or 100ml, or per package if the pack contains only a single portion. The recommended daily allowance for each listed vitamin and mineral is given in the annex to the directive.

The directive also gives the conversion factors to be used for the calculation of the energy value (Table 14.2) and lays down in detail the formats in which the declarations must be made together with the prescribed units of measurement and the symbols to be used. The latter is very specific in that energy must be given first in kilojoules as kJ, followed by kilocalories shown as kcal. Both values must be given.

Vitamins and minerals must be declared in either milligrams or micrograms as required by law with the symbol for micrograms being given as μg and not mcg. The declared con-

Table 14.2. Conversion factors to be used in calculation of energy values for foods in the European Union*

Carbohydrates (except polyols)	4 kcal/g–17 kJ/g
Polyols	2.4 kcal/g–10 kJ/g
Protein[+]	4 kcal/g–17 kJ/g
Fat	9 kcal/g–37 kJ/g
Alcohol (ethanol)	7 kcal/g–29 kJ/g
Organic acid	3 kcal/g–13 kJ/g

* Directive 90/496/EEC
+ Protein = N x 6.25

tent of vitamins and minerals must also be given as a percentage of the Recommended Daily Allowance.

With the exception of food supplements, the nutrient declarations must be based on the content per 100g or 100ml. They can also be given per serving or per portion as quantified on the label, provided that the number of portions contained in the package is stated on the label.

Where there is a marketing or nutritional requirement to specify a particular carbohydrate or fat form, a complete declaration is required as follows:

for carbohydrates:
    Carbohydrate   g
    of which
        -sugars   g
        -polyols   g
        -starch   g
for fats:
    Fat   g
    of which
        -saturates   g
        -monounsaturates   g
        -polyunsaturates   g
        -cholesterol   mg

For food supplements, the directive requires only the vitamins and minerals to be listed. Previously there has been an inconsistency across the EU, with some member states requiring full nutritional labeling for supplements.

From August 1, 2005, all food supplements are required to declare the active levels of nutrients or substances with a physiological effect on the label in numerical forms. The units to be used for the vitamins and minerals are given in Annex I of the directive. The declaration of the active components must be as the recommended daily intake, not per individual dose where the daily intake is split between two or more doses. The information on the vitamins and minerals must also be expressed as a percentage of the reference values (RDAs) given in Directive 90/496/EEC on nutrition labeling.

The directive on nutrition labeling is undergoing a revision by the European Commission. One of the objectives is to increase the list of RDAs to include all the vitamins and minerals approved for use in food supplements. The indications are that some of the existing values for micronutrients will be revised, in some cases with major changes to the amounts.

Claims

European food law prohibits the attribution to a food the property of preventing, treating, or curing a human disease, or any reference to such properties. These claims are considered to be medicinal and are illegal when made for foods sold under food law.

The European Commission has been attempting to introduce legislation on food claims for more than 20 years; the first proposal for a directive was circulated in 1980. Agreement could not be reached on the proposal and it was eventually dropped, only to be resurrected in 1992 with a revised proposal. The new proposal was also contentious and did not reach agreement even though it was considerably revised during the discussion; it was withdrawn by the Commission in 1995.

In mid-2001, more than five years later, the Commission circulated a discussion paper covering only nutrient and nutrient function claims. The important area of health claims was avoided in the paper even though a number of member states had already introduced internal procedures for the review and approval of such claims.

Finally, in June 2002, the Commission published a draft proposal for a regulation on nutrition, functional, and health claims made for foods. This proposal was intended to cover all claims made for nutrient content and included, in addition to the generally recognized nutrients, substances with a physiological effect. This proposal was withdrawn for reworking, and almost a year elapsed before a considerably revised version was available. This version again required amending, and the revised document was released in July 2003.[5]

The July 2003 proposal contained a number of very controversial points, the main ones being:

- The category of functional claims was removed, leaving only nutrition and health claims.
- The concept of "good" and "bad" foods was introduced, with claims being permitted only for foods that fall into acceptable nutrient profiles. The profiles are to be established with particular reference to the fat, saturated fat, and trans-fatty acid, sugar, and salt/sodium content of the food.
- All claims that make reference to general, nonspecific benefits of the nutrient or food for overall good health and well-being would be prohibited.
- Claims that make reference to slimming or weight control, or to the rate or amount of weight loss that may result from their use, or to a reduction in the sense of hunger or an increase in the sense of satiety, or to the reduction of the available energy from the diet would be prohibited.
- Claims that make reference to the advice of doctors or other health professionals, or their professional associations, or charities, or suggest that health could be affected by not consuming the food, would also be prohibited.
- Only those claims that appear in an official register would be allowed.
- Health claims not on the register would be allowed only after a critical review of the scientific substantiation of the claim.
- No claims would be allowed on products containing more than 1.2 percent by volume of alcohol.

Through the second half of 2003 the draft was severely criticized by both the European food industry and a number of member state governments. An opinion from the European Parliament's Committee on Legal Affairs and the Internal Market recommended 67 amendments in November 2003.

National Initiatives on Health Claims

In the absence of positive direction from the European Commission on health claims, a number of member states have undertaken internal initiatives and introduced national procedures for the approval of claims. The main ones are described in the following sections.

Belgium

Work on a Code of Conduct on Health Claims was undertaken by Fevia, the food industry federation of Belgium, and completed in 1999.[6] The code covers all health claims, including disease risk-reduction claims, and goes into considerable detail as to how claims can be made. Although all the normal legal controls apply, the code makes some concessions provided that strict conditions are applied. These concessions allow certain words or descriptions to be used to explain the particular health claim more precisely. It is an essential requirement of the code that any claim be substantiated scientifically and all the data kept in a dossier and retained by the person responsible for making the claim.

France

During a plenary session held in September 1997, the French *Conseil National de l'Alimentation* (CNA—National Dietary Council) undertook the task of considering claims linking diet and health, including functional claims.[7]

The deliberations of the Council have resulted in an opinion following a wide consultation that included consumer associations, the food and drink federations, and food scientists. The Council has also heard evidence from people involved in advertising and from market researchers studying dietary behavior.

The opinion of the Council was that the prohibition on therapeutic claims must remain but, provided that certain principles were maintained, health claims would be acceptable. Claims that could be permitted with suitable controls were as follows:

- A claim for the food product as aiding a reduction of a risk of a disease
- A claim for a positive contribution to health when presented as having an influence on the modification of a physiological state or biological parameter
- A functional nutritional claim that describes the positive role of the nutrient in the normal functioning of the body.

The Council felt that such claims would require prior authorization before marketing unless they appeared on an officially approved list.

For new claims, and possibly for all claims, the validation of the scientific data that forms the basis of the claims should be undertaken by independent organizations such as the Human Nutrition Research Centers.

Netherlands

After a long period of discussion, a Code of Practice for assessing the scientific evidence for health benefits from food and health claims made for foods and drinks has been adopted in the Netherlands. This code was drawn up by the *Voedingscentrum* (the Netherlands Nutrition Center) and contains a procedure for scientific assessment of the data supporting health claims. The code has the official support of the Dutch government.[8]

The main requirement of the code is that health benefit claims are assessed by an independent panel of experts appointed by the Netherlands Nutrition Center. There is a requirement that the evidence must be based on relevant data from human subjects and it must be demonstrated that any effectiveness of the substance determined by research is not reduced when included in a commercial product as presented to the consumer.

A procedure for appeal against the panel's decision is built into the code.

Spain

An agreement on health claims for foods was signed on March 20, 1998, between the *Ministerio de Sonidad y Consumo* (Spanish Ministry of Health) and the *Federación de Industrias de Alimentación y Bebidas* (FIAB—the Spanish Federation of Food and Drinks Manufacturers).[9] This agreement was voluntary and clarifies the situation relating to health claims within the legislation covering the labeling and advertising of foods. Within the agreement, health claims are defined as any claim relating to:

- The function of one or more nutrients in the human body
- The effects of the consumption of one or more food products on human health
- The consumption of certain foods as part of a healthy diet

The agreement is intended to apply to the labeling and advertising of all foods and drinks with the exception of products that come under the classification of PARNUTS and mineral waters. Nutrition content claims or claims such as "contains vitamin A" or "contains iron and calcium" are also excluded from the scope of the agreement because they are already controlled by specific legislation.

There is a general requirement that all health claims must be truthful and able to be clearly substantiated by scientific evidence.

There is also a requirement that whenever a health claim is made in the labeling or in promotional materials for a food, it should be accompanied by a statement about the importance of maintaining a healthy and balanced diet.

Sweden

Sweden was the first of the six EU countries developing guidelines on health claims to agree upon and adopt a self-regulating procedure. The first program to regulate health claims was agreed upon by the food industry and came into effect in August 1990. This was monitored for three years until July 1993, and revised self-regulating rules were agreed to in August 1996 and came into effect on January 1, 1997. This revised program was developed and agreed to by representatives of all sectors of the Swedish food industry including the Federation of Swedish Food Industries, the Swedish Food Retail Association, the Federation of Swedish Farmers, and the Grocery Manufacturers of Sweden.[10]

In the explanatory document, a health claim is defined as "an assessment of the positive health effects of a foodstuff, i.e. a claim that the nutritional composition of the product can be connected with prophylactic effects or the reduced risk of a diet related disease." This definition is qualified by the requirement that the health claim must be based on the importance of the product in a balanced diet and must be in line with the official Swedish dietary recommendations.

A health claim must consist of two parts. The first must provide information on the diet and the health relationship of the food. The second part must contain information on the composition of the product. Two examples given as acceptable health claims are as follows:

Part 1. Iron deficiency is common among women but can be prevented by good dietary habits.
Part 2. Product X is an important source of the type of iron that is readily absorbed by the body.

Part 1. Omega-3 fatty acids have a positive effect on blood lipid and can therefore help protect against cardiovascular disease.
Part 2. Fish product X is rich in omega-3 fatty acids.

Six other examples of approved health claims are given in the document. With each approved claim there is an example of an unapproved statement or reference regarding the product composition and effect. For example, it is permissible for the second part of a claim relating to cholesterol and cardiovascular disease to say "Brand X contains a low amount of saturated fat and total fat" but it is not permissible to say "Brand X provides excellent protection against cardiovascular disease through its low content of saturated fatty acids" or "Brand X will help you reduce your blood cholesterol levels."

United Kingdom

In December 1996 the Food Advisory Committee (FAC) of the Ministry of Agriculture, Fisheries and Food completed its review of the British market for functional foods and the control of health claims. As part of this review, the FAC published draft guidelines on health claims for foodstuffs. These guidelines were intended to set out the conditions that food manufacturers or retailers were required to follow when making health claims for their products. The scope of these guidelines was that they applied to all foods and drinks, including food supplements but excepting those products controlled under the EC directive on PARNUTS.

A health claim is defined in the draft as any statement, suggestion, or implication in food labeling or advertising that a food is beneficial to health. Nutrient content claims and medical claims are excluded from this definition. Health claims could be subdivided into three categories:

- Claims that refer to possible disease risk factors (for example, can help lower blood cholesterol)
- Nutrient function claims (for example, calcium is needed to build strong bones and teeth)
- Recommended dietary practice (for example, eat more oily fish for a healthful lifestyle)

The draft guidelines devoted a section to the principles underlying health claims and one to the scientific substantiation of claims.

Following a period of consultation that met with a generally positive response, the Joint Health Claims Initiative (JHCI) was established in June 1997. This was a joint venture between consumer organizations, enforcement authorities, and industry bodies with the objective of establishing a code of practice for the use of health claims for foods.

The draft code of practice, which was developed by the JHCI, further separates health claims into generic claims and new claims. Generic health claims are defined as those based on well-established and generally accepted knowledge, evidence in scientific literature and/or recommendations from national or international public health bodies. A regularly reviewed and updated list of generic health claims was to be maintained as part of the administration of the code.[11]

A new health claim must be based on scientific evidence as applied to either existing or new foods. The scientific evidence must be substantiated in accordance with detailed rules laid down in the code.

A number of criteria must be met when using a health claim. It is essential to demonstrate that the food, or its components, will cause or contribute to a significant and positive physiological benefit when consumed by the target population as part of its normal diet. The effect must be achieved by the consumption of a reasonable amount of food on a regular basis or by the food's making a reasonable contribution to the diet. The claimed effect must be maintained over a reasonable period of time and should not be a short-term response to which the body adjusts, unless the claim is relevant for only a short- or medium-term benefit. An example of the latter is the use of folic acid just before and during the early stages of pregnancy in the case of neural tube defects.

The scientific data needed to support a claim is described in some detail in the code. The claim must be based on a systematic review of all the available evidence, including published scientific literature, relating to the validity of the health claim. The conclusions drawn from this review should be based on the totality of the evidence and not just on the data that supports the claim.

The conclusions should be based on available human studies that are the most methodologically sound or on other human evidence and not just on biochemical, cellular, or animal studies. Although the conclusions ideally should be based on experimental studies in humans, observational studies may be acceptable in some circumstances.

Companies wishing to make a health claim must submit the details of the claim and the substantiating scientific evidence to the Code Administration Body (CAB), which is administered by the Leatherhead Food Research Association. The CAB will seek the opinion of a panel of experts, which is described in the code as the Expert Authority, and provide liaison between the company and the experts.

Substantiation of Health Claims

Criteria to be used for the scientific substantiation of health claims in the EU have been developed as part of an EU Concerted Action Project. The project developed a "Process for the Assessment of Scientific Support for Claims on Foods" (PASSCLAIM).[12] The objectives of PASSCLAIM were to:

- Produce a generic tool with principles for assessing the scientific support for health-related claims for foods and food components
- Evaluate critically the existing schemes that assess the scientific substantiation of claims
- Select common criteria for how markers should be identified, validated, and used in well-designed studies to explore links between diet and health

The outcome of this exercise has been the establishment of the main principles for use in the assessment of scientific data submitted in support of the claim. One of the key points is that the totality of the evidence must be assessed and that there should be no selectivity in the presentation of the data.

European Regulation on Novel Foods and Novel Food Ingredients

In January 1997, Regulation (EC) N 258/97 on Novel Foods and Novel Food Ingredients was adopted and came fully into force in all member states of the EU on May 15, 1997.[13]

This regulation requires that all foods, ingredients, and some processes that had not been used for human consumption within the European Community to a significant degree before May 1997 must be subject to official review and approval before being placed on the market in any EU country. Since its introduction, this law has had a very serious impact on product innovation in Europe and has particularly affected functional foods and nutraceuticals.

The scope of the regulation is such that it encompasses almost all categories of foods and ingredients that are likely to come into the market. Included are foods and ingredients with a new or intentionally modified primary molecular structure; those consisting of or isolated from microorganisms, fungi, and algae; those consisting of or isolated from plants or animals (except for those obtained by traditional propagating or breeding practices).

Also included are new or novel processes that give rise to significant changes to the composition or structure of the foods or ingredients that affect their nutritional value, effect on metabolism, or level of undesirable substances.

The regulation also covers all foods and ingredients containing or consisting of genetically modified organisms (GMOs) or those produced from, but not containing, GMOs.

The requirements for the authorization of a novel food or ingredient given in the regulation are that a full dossier on the substance or process must be reviewed by a competent authority in the EU country of intended first sale. The dossier must have the contents and format as set forth by the Commission, with the main focus being on safety. The initial assessment should be completed in 90 days. Summaries of the application must be sent to the European Commission, which is required to notify the other 14 member states. The other member states have a 60-day period in which to object to the assessment, make comments, or request clarification from the applicant. In theory, the total time should be no longer than 150 days. In practice, however, it has been found that applications are, in effect, being reviewed in detail by the competent authorities in all or most of the member states, resulting in up to 15 reviews of the same data which increases the time-scale. In addition, many of the applications have been referred by the European Commission to other EU committees, such as the Scientific Committee on Food (SCF). The SCF is not bound by law to a time limit in which it must give its opinion, and delays by the SCF have added considerably to the length of time taken to process applications, in many cases more than two years.

The costs of obtaining the data to support a novel food application, most of which relate to the required safety studies, have been found to be substantial and are considered to be prohibitive by many of the small to medium-sized innovative companies. In the first five years after the regulation came into effect there were only 37 applications, of which 11 were for genetically modified plants such as maize and rape (canola oil) intended to be used as sources of food ingredients.

As a result of this legislation there has, in reality, been only one innovative nutraceutical ingredient, a DHA-rich oil from a micro alga, introduced into the European market since early 1997; the small number of applications approved have almost no commercial benefit for companies marketing functional foods.

In June 2002 the European Commission published a discussion paper that considered some of the major issues that had emerged during the first five years of the operation of the regulation. A number of options to amend the regulation were given in the paper and comments elicited.

One of the suggestions was to remove the categories covering GMOs to specific legislation controlling new sources of GMOs. In October 2003 new legislation on the authorization of GMOs for use as food and food ingredients has removed the GMOs from the novel foods regulations to the new regulations.

Foods for Particular Nutritional Uses (PARNUTS)

In European law, PARNUTS are defined in Directive 89/398/EEC[14] as:

> foodstuffs which, owing to their special composition or manufacturing processes, are clearly distinguishable from foodstuffs for normal consumption, which are suitable for their claimed nutritional purposes and which are marketed in such a way as to indicate such suitability.

To qualify as having a particular nutritional use, the foodstuff must be able to fulfill the specific nutritional requirements of certain categories of people whose digestive processes or metabolism are disturbed (for example, diabetics or coeliacs) or people who are in a special physiological condition and who are therefore able to benefit from controlled consumption of certain substances in foodstuffs (for example, people who are controlling their weight). The third category of products is very specific and covers foodstuffs for infants and young children in good health.

A prerequisite of the directive is that the nature or composition of the product must be appropriate for the particular nutritional use for which it is intended; also required is that the product complies with any mandatory provisions applicable to foodstuffs for normal consumption except for any changes necessary to ensure their conformity to the required use. This means that the products must conform with all applicable food legislation for normal foods except where specifically exempted or controlled within that legislation or within specific directives developed to control a PARNUTS category.

The directive also provides for products that fall into PARNUTS categories to be characterized as "dietetic" or "dietary" products and prohibits the use of these terms in the labeling, presentation, and advertising of foodstuffs for normal consumption that is also suitable for a particular nutritional use to indicate such suitability.

The directive, when originally adopted, required that detailed directives were to be developed for nine categories of dietetic foods. These categories included foods for infants and young children, gluten-free foods, foods for diabetics, slimming foods, sports nutrition products, and foods for special medical purposes.

For various reasons only four of the original nine categories had been adopted by the middle of 2002, and two of these are specific to infant formulas and baby foods. The one with the greatest relevance to the nutraceutical industry is that of meal replacement products for weight control. This directive (96/8/EC), which came into force on April 1, 1999, covers two categories of meal replacement products designed for weight control.[15] These are as follows:

1. Products presented as a replacement for the whole of the daily diet
2. Products presented as a replacement for one or more meals of the daily diet

A number of detailed compositional requirements exist for both categories of product. The energy content of products in the first category must not be less than 800kcal (3360kJ) and not more than 1200kcal (5040kJ). Meal replacement products in the second category must fall within the range of 200kcal (840kJ) to 400kcal (1680kJ) per meal. Products in both categories must provide not less than 25 percent and not more than 50 percent of the total energy as protein. The quality of the protein is specified, as is the addition of isolated amino acids to improve the nutritional value of the protein if required.

The fat content must not exceed 30 percent of the total available energy of the product. For products in category 1 (replacement of the whole diet) the linoleic acid content, in the form of glycerides, must not be less than 4.5g. For meal replacement products the linoleic acid must not be less than 1g per meal. This requirement is difficult to achieve in practice.

For category 1 products the dietary fiber content must be between 10g and 30g per daily intake of product.

As an annex to the directive, there is a table listing 12 vitamins and 11 minerals and trace minerals together with amounts for each micronutrient. Products in category 1 must contain at least 100 percent of the amounts given, and those in category 2 at least 30 percent of each amount with the exception of the amount of potassium, which must be at least 500mg per meal.

The directive contains a number of labeling requirements specific to the products. It also contains a prohibition on the labeling, advertising, and presentation of the product containing any reference to the rate or amount of weight loss that may result from the use of the product, or to a reduction in the sense of hunger or an increase in the sense of satiety.

The fourth of the specific category directives to be adopted was that of foods for special medical purposes (FSMP), which after many years of discussion was finally adopted as Commission Directive 99/21/EC.[16] Foods that fall into this category are those that are specially processed or formulated and intended for the dietary management of patients and to be used under medical supervision. The foods are intended for the exclusive or partial feeding of patients with a limited, impaired, or disturbed capacity to take, digest, absorb, metabolize, or excrete ordinary foodstuffs or certain nutrients contained therein or metabolites or with other medically determined nutrient requirements, whose dietary management cannot be achieved only by the modification of the normal diet, by other foods for particular nutritional uses, or by a combination of the two.

This lengthy definition is necessary to encompass the wide range of products of different composition that may differ substantially from each other depending on the specific disease or medical condition of the patients for whom they are intended. In the preparation of the directive, the Commission accepted that it was not possible to lay down detailed compositional rules because of the wide diversity of these foods and rapidly evolving scientific knowledge.

The directive on FSMP contains a number of requirements, particularly with regard to micronutrient content, labeling, and notification of the product. For such products, notification of intended sale must be made to the relevant authorities in each EU country in which the product is to be sold.

The directive on sports nutrition products is still in the early stages of development, and the European Commission target of making the first draft available by December 2001 was not met. An early draft became available no sooner than the end of 2003. This draft limited the controls to:

- Carbohydrate-rich food products
- Carbohydrate-electrolyte drinks
- Protein concentrates
- Protein-enriched foods

Also included were provisions to allow the prohibition on the addition of certain substances to sports products.

In February 2001, Directive 2001/15/EC was adopted.[17] This directive contains a positive list of nutrient sources to supply vitamins, minerals, and amino acids. The directive applies as of April 1, 2004, and by that time all PARNUTS products can be formulated using only nutrient sources on the permitted list. Although a procedure exists for considering applications for addition to the list, recent experience has shown that the procedure is very costly and applications are taking a number of years to be considered by the authorities.

Due to delays, a number of applications had not had their assessments completed at the end of 2003. Thus, the European Commission had to produce legislation allowing the temporary continuation of use of these substances until an official decision is made or until 2006.

The PARNUTS directive (89/398/EEC) also contains a procedure for products that fall into the definition of a PARNUTS product and thus into the scope of the directive, but which do not fall into any of the prescribed categories (that is, weight control, FSMP, and so on).

This procedure is set forth in Article 9 of the directive and requires that when a product that does not belong to one of the groups in Annex I of the framework directive is placed on the market for the first time, or where a product is manufactured in a third state (that is, a non-EU state), the manufacturer or importer has to notify the competent authority in the Member State in which the product is being marketed.

This notification is initiated by the forwarding of the label, or draft of the label, used for the product. When the same product is subsequently placed on the market in another Member State, the manufacturer or importer is required to provide the competent authority of that Member State with the same information, together with details of the first notification. A list of the competent authorities in each Member State has been promulgated by the Commission.

The competent authority in each State is empowered to require the manufacturer or importer to produce the scientific evidence to prove the product's compliance with the definition of a food for particular nutritional use as given in Article 1 (2) of the directive. Evidence of the particular composition or manufacturing process that gives the product its particular nutritional characteristics may also be required. If the scientific evidence for the product is contained in a readily available publication, the manufacturer or importer need only give a reference to the publication.

Food Additives

European food law contains very detailed requirements governing the use of food additives, colors, sweeteners, extraction solvents, and flavors.[18, 19, 20, 21, 22]

Technological additives such as preservatives, antioxidants, and so on are controlled not only by the categories of food in which they are permitted but also by upper limits on their input that are specific to each permitted category. In some cases, the additives are permitted for a number of food categories and in others it may be for only one or two. If the food category does not appear in the list for a particular additive, the additive cannot be used in the product. Because functional foods cannot always be defined into the categories listed in the legislation, various complications can arise during the formulation of a product. Since the legislation came into force in March 1997, a number of instances have occurred in which the technological needs of a product could not be met by strict observance of the legislation.

The laws for colors and sweeteners follow the same format as that for the technological additives (that is, defined permitted categories with maximum levels of use).

The directive on colors can be particularly confusing as there are a number of tables which include: those for products that are not permitted to be colored; those that can contain only certain colors; and colors that can only be used in specified foods. Some groups of colors are restricted to a cumulative maximum level when used in combination.

Similar controls exist on the use of intense sweeteners and polyols (for example, sorbitol, xylitol, and so on). If an extraction solvent is used in the production of a food ingredient, both the solvent and its residue levels must comply with European legislation.

Contaminants

Over the past few years an increasing focus has been placed on contaminants in food. A number of European laws control the maximum levels of many contaminants, including pesticides, heavy metals, mycotoxins, 3-MCPD, and dioxins. Each ingredient must be checked for compliance. For example, since July 1, 2002, there has been a legal limit on dioxins in fish oils and vegetable oils.

In the first instance, the law relates only to dioxins and furans (PCDD/F) and has a limit of 2pg WHO-TEQ/g fish oil. This limit has been hard to achieve, particularly for fish liver oils, and a substantial amount of product was found not to be in compliance.

A second stage of this legislation, which will probably take effect in the latter half of early to mid-2005, is to add limits for dioxin-like polychlorinated biphenyls (PCBs).

There have also been many problems in Europe with nonauthorized pesticides on botanicals imported from outside the EU and from countries working to different pesticide regulations.

An area of concern since the second half of 2002 has been the presence of relatively high levels of Polycyclic Aromatic Hydrocarbons (PAHs) in vegetable oils, herbal extracts, and some other plant derivatives. This resulted in a number of products, particularly supplements, being ordered off the market by the national authorities in some Member States. As a consequence of the concerns the European Commission issued proposals at the end of 2003 to bring PAHs into European-wide legislation.

A number of botanicals, including some specified herbs, must comply with stringent legislation on mycotoxins.

Complex laws exist concerning flavorings that control the source materials.[22] In addition, flavorings produced from natural botanical sources must comply with requirements that impose low limits on biologically active principles (BAPs), which occur naturally in plants. BAPs subject to strict control include coumarin, safrole, and berberine. There are proposals to amend this legislation not only to increase the number of substances (BAPs) but also to reduce some of the limits.

The contaminants regulations are being very strictly enforced in many EU countries.

Irradiation

The European legislation on irradiation permits its use for a limited and specified number of ingredients. All food irradiation must be carried out in EU registered facilities under strict control.[23]

A nationwide surveillance carried out by the United Kingdom Food Standards Agency in early 2002 found that 42 percent of food supplements containing herbs had been illegally irradiated. These findings caused the European Commission to instruct the authorities in all 15 Member States to check for illegal irradiation of supplements in their own territories. There have subsequently been a number of reports from member states of illegally irradiated supplements and herbal products.

Vitamins and Minerals

With the exception of the EU directive on food supplements and specific directives on PAR-NUTS products, the addition of vitamins and minerals to foods is still regulated at a national level, and functional foods containing vitamins and minerals are subject to a diversity of national laws. Historically, many European countries have considered products containing high levels of vitamins and minerals to be drugs, and some of these countries have had national laws with strict upper limits of addition. For example, in Ireland and Spain, products must not contain more than 1 x Recommended Daily Allowance (RDA) of each vitamin and mineral per recommended daily intake. In Germany, vitamins A (as retinol) and D are not permitted in supplements. Other vitamins may be added up to 3 x RDA per day, whereas some other minerals are permitted only at less than half of the RDA. Italy allows most micronutrients at up to 1.5 x RDA, and Belgium allows some at 3 x RDA; others are restricted to 1 or 2 x RDA.

These local rules have resulted in very serious barriers to trade between EU countries because no two Member States have exactly the same controls. Manufacturers have found it very difficult to develop products with vitamins and minerals above 1 x RDA that can be sold widely across the EU.

In 2000 the European Commission published the first draft of a proposal to regulate the addition of vitamins and minerals to foods on a similar basis to that of food supplements. However, the progress on this directive was slow, and a considerably revised proposal was published at the end of 2003.

This proposal supports the concept of control of micronutrient levels in foods to maximum levels derived from a scientific safety assessment, with each vitamin or mineral being assigned its maximum level per recommended daily intake of the product.

Also included in the draft is a proposal for the control of substances, other than vitamins and minerals, which could be added to foods. These proposals are for "prohibited," "controlled," and "under scrutiny" lists. The under scrutiny section relates to the control of substances for which the European Food Safety Authority has concerns about the safety of the substance but scientific uncertainty exists. Under such circumstances, the substance is placed on the scrutiny list and the companies wishing to use it have up to four years to supply convincing scientific data attesting to its safety in foods.

Herbs and Botanicals

The marketing of products containing physiologically active herbs and botanicals in Europe is extremely complex, and considerable confusion and misunderstanding exist about the status of herbal products in the EU. At present it is impossible to produce a herbal product that can be sold as a food in a pan-European market. Companies hoping to launch an herbal product are faced with 15 different sets of laws, ranging from the relatively liberal approach in the United Kingdom to the very strict requirements of its near neighbor, Ireland, where any product containing a herbal compound is likely to have to be registered as a medicine.

French law contains a positive list of 34 herbs that can be sold as foods. Products containing herbs not on this list require registration. Most are on the list for culinary purposes and, unfortunately, very few are of commercial importance to the nutraceutical and functional food industries. A similar situation prevails in Spain.

In Germany, physiologically active herbs are normally regarded as medicinal, requiring registration of the product. Italy has recently modified its approach to products containing some active herbs and allows them to be notified under PARNUTS legislation, but there is an additional labeling requirement for such products.

Belgium has a relatively extensive list of herbs that can be added to supplements and foods provided that the authorities are notified with product details.

Even in the United Kingdom, which is regarded as being one of the more tolerant regimes in the EU with regard to nutraceuticals, the use of herbs in products is not straightforward. A large number of herbs that are considered by the British authorities to be medicinal herbs cannot be used in foods or supplements. This list has been rapidly increasing, which has made it difficult for manufacturers to predict the future legality of a product.

An EU directive on traditional herbal medicinal products, which is likely to become fully effective by 2011, would treat many herbal products as medicines under the control of medicinal law, requiring production in European authorized medicines facilities. These moves are likely to make it more difficult to market products containing herbs as functional foods or food supplements.

Food Supplements

Food supplements have been defined in European legislation as:

> Foodstuffs the purpose of which is to supplement the normal diet and which are concentrated sources of nutrients or other substances with a nutritional or physiological effect, alone or in combination, marketed in dose form, namely forms such as capsules, pastilles, tablets, pills and other similar forms, sachets of powder, ampoules of liquids, drop dispensing bottles and other similar forms of liquids and powders designed to be taken in measured small unit quantities.

The long-awaited EU Directive on Food Supplements finally came into force on July 12, 2002, and will come fully into effect on August 1, 2005.[24] In the first instance the directive is concerned only with supplements containing vitamins and minerals, but it is intended that in the future it will be extended to cover other active substances commonly used in supplements such as amino acids, fatty acids, and plant extracts.

The directive specifies the sources of vitamins and minerals that can be used in supplements and will also give maximum and minimum levels of addition for the micronutrients.

Mandatory label statements are included that are in addition to the general labeling requirements for all foods.

The directive does not exempt supplements from compliance with all relevant aspects of general food law (for example, contaminants, additives, novel ingredients, and so on).

Genetically Modified Organisms

At the end of 2003 new legislation came into force requiring the authorization, traceability, and labeling of all foods and ingredients derived from genetic modification (GM).[25, 26] This legislation requires the labeling of all ingredients derived from GMOs, including many produced from fermentation processes through which the microorganism has been genetically modified.

Companies undertaking product development will have to carry out a detailed scrutiny of ingredients to determine whether there has been any GM technology in the production of the ingredient and, if so, whether the ingredient has been officially registered in Europe.

Other Legislation

A plethora of food-related legislation must be considered before launching a functional food. For example, detailed requirements exist for packaging materials coming into contact with the food, particularly plastic materials, and also packaging waste legislation that restricts the size of packs and has stringent controls on the heavy metal content of packaging material and waste recovery requirements.

Conclusions

The European Union legislation has been in a state of flux for many years and is unlikely to settle down before 2010–2012.

This fact does not make life easy for companies wishing to market functional foods and nutraceutical products in Europe. Companies need to be aware that it has become almost impossible to directly transfer products from markets such as the U.S. or Japan into the EU and that, in most cases, significant modifications are required.

References

1. Berry Ottaway, P. (1995) Harmonization of European Food Legislation. Financial Times Management Report Series, London: Pearson Professional Ltd.
2. European Commission (2000) White Paper on Food Safety. SANCO/3578/99-2000, Brussels.
3. European Parliament and Council Directive 2000/13/EC. O.J. of E.C. L109/29 of 6 May 2000.
4. European Council Directive 90/496/EC. O.J. of E.C. L276/40 of 6 October 1990.
5. European Commission (2003) Proposal for a European Parliament and Council Regulation on Nutrition and Health Claims made on Foods. COM(2003)424-C5-0329/2003, Brussels.
6. Federatie Voedingsindustrie/Fédération de l'Industrie Alimentaire (Belgium). Health Claims Code of Conduct, draft, 21 October 1998.
7. Conseil National de l'Alimentation (France), Allégations faisant un lien entre alimentation et santé. Avis No. 21, 30 June 1998, Paris.
8. Voedingscentrum (Netherlands). Code of Practice assessing the scientific evidence for health benefits stated in health claims on food and drink products. April 1998.
9. Joint Ministerio de Sonidad y Consumo (Ministry of Health, Spain) and Federación y Bebidas (Spanish Federation of Food and Drink Manufacturers) agreement on health claims on foods of 20 March 1998.
10. Federation of Swedish Food Industries et al. Health Claims in the Labeling and Marketing of Food Products. Food Industry's Rules (Self-Regulating Programme). Revised programme of 28 August 1996.

11. Joint Health Claims Initiative (United Kingdom). Code of Practice on Health Claims on Foods. Final text, 9 November 1998.

12. Richardson, D. P. et al. (2003) PASSCLAIM—Synthesis and review of existing processes. *Eur. J. Nut.* 42 (Suppl 1), 1/96–1/111.

13. European Parliament and Council Regulation (EC) No. 258/97. *O.J. of E.C.* L43/1 of 14 February 1997.

14. European Council Directive 89/398/EEC. *O.J. of E.C.* L186/27 of 30 June 1989.

15. European Commission Directive 96/8/EC. *O.J. of E.C.* L55/22 of 6 March 1996.

16. European Commission Directive 99/21/EC. *O.J. of E.C.* L91/29 of 7 April 1999.

17. European Commission Directive 2001/15/EC. *O.J. of E.C.* L52/19 of 22 February 2001.

18. European Parliament and Council Directive 95/2/EC. *O.J. of E.C.* L61/1 of 18 March 1995 and amendments.

19. European Parliament and Council Directive 94/36/EC. *O.J. of E.C.* L237/13 of 10 September 1994.

20. European Parliament and Council Directive 94/35/EC. *O.J. of E.C.* L237/3 of 10 September 1994 and amendments.

21. European Council Directive 88/344/EEC. *O.J. of E.C.* L157/28 of 24 June 1988 and amendments.

22. European Council Directive 88/388/EEC. *O.J. of E.C.* L184/61 of 15 July 1988.

23. European Parliament and Council Directive 1999/2/EC. *O.J. of E.C.* L66/16 of 13 March 1999.

24. European Parliament and Council Directive 2002/46/EC. *O.J. of E.C.* L183/51 of 12 July 2002.

25. European Parliament and Council Regulation (EC) No. 1829/2003. *O.J. of E.U.* L268/1 of 18 October 2003.

26. European Parliament and Council Regulation (EC) No. 1830/2003. *O.J. of E.U.* L268/24 of 18 October 2003.

Note: European directives and regulations can be accessed via the Official Journal website at: http://europa.eu.int/eur-lex/en/oj/index.html.

15 Functional Foods in Japan: FOSHU ("Foods for Specified Health Uses") and "Foods with Nutrient Function Claims"

Ron Bailey

Introduction

The Japanese government recognized more than fifteen years ago that the unique demographic trends in Japan, if continued, would eventually lead to a health care crisis in that country.

The Japanese population is aging more rapidly than that of any other country in the world, according to the Japanese Ministry of Health, Labor, and Welfare (MHLW). It is estimated that by the year 2050 more than 35 percent of the population will be 65 years of age or older. For perspective, in the United States currently less than 15 percent of the population is age 65 or older, compared to approximately 17 percent in Japan. The *Nikkei Weekly* newspaper reported in August 2002 that according to MHLW data, Japan set new records for longevity in 2001, an average of 81.5 years. Both Japanese women (85 years) and men (78 years) currently lead the world in longevity.

In addition, the birth rate in Japan is at an all-time low and has been declining for many years. The *Nikkei Weekly* reported in June 2002 that the number of babies born in Japan the previous year hit a record post-war low of 1.17 million according to MHLW. The fertility rate—the average number of children women bear in their lifetime—dropped to 1.33, the lowest level ever. This compares with a U.S. fertility rate of 2.13 in 2000.

At the same time, the rate of immigration is also very low, mostly related to government policy. The net result is a rapidly aging population that is actually expected to decline significantly over the next several decades if current trends continue. Some estimates indicate that the current Japanese population of 127 million will decline to well under 100 million people in the next 50 to 75 years. The economic implications of such a dramatic decline are considerable.

Although the Japanese population has the highest longevity in the world and is considered to be relatively healthy by global standards, the increasing costs of health care are a burden on the entire population. Data from MHLW reported in the *Nikkei Weekly* in September 2002 indicated that nearly 50 percent of all medical expenses are currently for people aged 65 or older. Medical spending in Japan is up nearly 50 percent since 1989, which has been described as a "chronic rise in medical spending." Given the projected demographics trends, it is easy to understand the concerns in Japan regarding an impending health care financial crisis.

Leading causes of death in Japan are quite similar to those in the western countries—cancer, cardiac failure, and cerebrovascular failure are the top three—even though the overall mortality rate is lower than for the major industrialized countries. There are major concerns, however, about the increased incidence of diseases such as adult-onset diabetes and allergies and asthma, particularly among the younger Japanese population. More than seven million Japanese have been diagnosed with adult-onset diabetes, and another nine million are believed to be at risk of contracting the disease. On a population-adjusted basis, the diabetes problem is comparable to that in the United States.

An excellent source of up-to-date information on the nutritional status of the Japanese population is the annual nutrition survey report prepared by the National Institute of Health and Nutrition entitled *The National Nutrition Survey in Japan*. Although the publication is in Japanese only, it contains data on the vitamin and mineral intakes of the population, the food sources of key nutrients in the diet, the cholesterol and blood pressure status, and the prevalence of obesity, for example. Data are presented separately for men, women, and children by age. Trend data for the past 25 years are included. Data on adult obesity (defined in Japan as a Body Mass Index of 25 or greater) are presented from the year 2000 survey, for example, which suggest one possible reason for the increased incidence of diabetes seen in recent years in Japan. The intake data reported on calcium are watched very carefully each year, because calcium is the only mineral considered to be deficient in the diet of the Japanese, on average. The findings from the annual survey provide support for the health policy initiatives of the Japanese government, including those related to functional foods.

It was decided several years ago in Japan that it might be possible to mitigate the effects of some of the negative trends by allowing food products with demonstrated health benefits to actually make health claims on the food label. The primary objective, of course, was to keep the aging population as healthy as possible for as long as possible, to avoid overburdening the health care system. From this start in the late 1980s, the current FOSHU (Foods for Specified Health Uses) regulations evolved.

Overview of Japanese Food Regulations

Functional foods in Japan are subject to the umbrella food regulations administered by the MHLW, specifically the Food Sanitation Law that was established in 1947.

The Food Sanitation Law regulates all foods in Japan, whether produced domestically or imported. Key features of the law are the positive lists of food additives that are allowed to be used in foods. If the food additive does not appear on a positive list, it is not allowed. An excellent English-language source of detailed information on food additives is the January 2000 publication from JETRO (Japan External Trade Organization) entitled *Specifications and Standards for Foods, Food Additives, etc. under The Food Sanitation Law*. The publication includes a list of the approved chemically synthesized food additives and their standards of use where applicable, as well as a list of the nonchemically synthesized food additives (read "natural") that can be used in foods. When questions arise regarding the regulatory status of food ingredients, an excellent source of information is the nongovernment Japan Food Additives Association.

Historically, it has been very difficult to gain approval for adding new food additives to the positive lists, even those food additives widely used outside Japan. Many additives that are commonly used in pharmaceuticals in Japan have not been allowed in foods, for example. Consumer groups in Japan have been particularly vocal, often arguing that Japanese consumers are being subjected to serious health risks when less-regulated (they claim) imported foods and beverages are consumed, because of the possible presence of unapproved food additives and agricultural chemical residues.

Recently, however, this situation appears to be changing as pressure is applied from outside Japan (**gaiatsu** is the term used) on the Japanese government to include in the Japanese food additives positive list ingredients that are widely used in foods by other World Trade Organization/Codex member countries. The Food Industry Center in Japan has been active recently in requesting that the MHLW add many new food additives to the official positive lists, primarily on the basis that the more restrictive lists put the Japanese food industry at a global disadvantage competitively. In 2002 there were several recalls in Japan of foods that contained very minor amounts of nonapproved food additives in their flavor systems, even though those same additives are regularly in use in foods in North America and the EU. So far, the MHLW is being responsive to the requests, and this policy is likely to continue. The recent approvals are not included in the JETRO publication cited previously, however, and a formal English-language update has not yet been made available.

Recent problems in Japan with pesticide residues in imported frozen vegetables from China have resulted in more detailed investigations of agricultural chemicals in imported foods. The JETRO publication cited previously also has a list of Maximum Residue Limits for Agricultural Chemicals in Food, as a guide. This is consistent with the food safety focus of the Food Sanitation Law and is clearly in the best interests of concerned Japanese consumers. The Ministry of Agriculture, Forestry, and Fisheries (MAFF) recently announced that the amount of imported certified organic agricultural produce was five times that of domestic production (*Japanscan Food Industry Bulletin*, December 2002). The growth of the organic food market in Japan is another reason for the renewed focus on agricultural chemical residues in the food supply.

Food ingredients that have been genetically engineered are covered under separate regulations issued by the MAFF. In general, the Japanese regulations regarding "GMO" foods and food ingredients are much closer to the more restrictive EU regulations than is the more open U.S. Food and Drug Administration (FDA) approach. The Japanese government is continuing to study the experience of other countries in this regard, both formally and informally, and has developed special labeling requirements and other restrictions, as well as increased analysis of imported foods and food ingredients. This rapidly changing regulatory environment needs to be considered in the context of functional ingredients and functional foods imports in particular.

FOSHU (Food for Specified Health Uses)

Emergence of the FOSHU Concept

The initial indications in the mid 1980s regarding the need for a regulatory category of foods that provided special health benefits emerged from a study conducted by the Ministry of Education, Science, and Culture (*Reports of Systematic Analysis and Development of Food Functionalities*, March 1988). That study included an analysis of what were called the primary function of foods (nutrition), the secondary function of foods (sensory), and the tertiary function of foods (physiological regulation). It was from this early effort that the concept of functional foods was born, specifically targeted at the tertiary function opportunities. The MHLW (it was actually called the Ministry of Health and Welfare at the time) was assigned the primary responsibility for developing the regulatory framework for the new food category, clearly separating the functional foods from the drugs already under MHW control. For a brief period of time in the late 1980s the category was in fact informally called "Functional Foods," but the name was changed to "Food for Specified Health Uses" (FOSHU) in 1991.

It was decided that the new regulatory category could be included under the existing Nutrition Improvement Law, the law that already regulated Enriched Foods and Foods for Special Dietary Uses. The regulatory category of Foods for Special Dietary Uses included Foods for the Sick, Formulated Milk Powder for Infants, and Food for Aged Persons. The new Food for Specified Health Uses was simply added as another category without the need for a new law.

It was also decided that the rather complicated FOSHU application process would be administered by the Japan Health Food Association, the large private organization that had been responsible for developing with members from the health food industry science-based monographs for each of the popular health food ingredients. The name of the organization was changed to Japan Health Food and Nutrition Food Association (JHNFA) and a FOSHU division was added. JHNFA has become an excellent source of accurate and up-to-date Japanese- and English-language information regarding the FOSHU products and functional ingredients, as well as periodic market analyses.

FOSHU Category Overview

The FOSHU system that emerged is regulated by the MHLW. It allows approved food products to be sold with on-label health claims and a special FOSHU logo indicating that the product has been accepted by MHLW. Products are approved on a product-by-product basis after a thorough review of safety and efficacy data. Each approved FOSHU product has at least one "functional ingredient" that provides the special health benefit(s) being claimed.

The FOSHU category has grown rapidly from the first two approvals in 1993 to nearly 422 products as of June 2004. Category sales for all the approved products are now well over $5 billion U.S. dollars at retail, with more growth expected in the future. As recently as only three years ago the FOSHU category was not yet considered to be an unqualified success, but the subsequent conversions of well-known foods and beverages from non-FOSHU to FOSHU status changed the thinking in Japan. More and more, manufacturers of new ingredients and new products targeted at the growing health-oriented food market are considering FOSHU status at the initial research stages of the product development cycle.

FOSHU Definitions

An early, informal definition taken from a translation of an MHW brochure explaining the functional food concept to the Japanese food industry (Anonymous, 1989) was as follows:

> Functional food is food that is designed and prepared so as to express fully its properties having to do with regulatory functions that relate to protection of the body, regulation of bodily rhythms, prevention of disease, and recovery from illness.

In addition, it was stated that "1) functional foods must be made from ordinary materials and ingredients, and they can be consumed in a usual form and way 2) functional foods must be such that they can be eaten or drunk every day 3) the mechanism of the effects in regulatory functions must be known." This definition reflected the thinking at that time that all FOSHU foods must clearly look like foods and could not be in typical dietary supplement tablet or capsule form. This food form requirement was abandoned a few years ago, and currently there are no restrictions.

In a presentation at the "Marketing Nutraceuticals & Functional Foods" conference in Singapore in January 2002, the JHNFA representative provided an updated and more practical English-language definition of FOSHU:

> Foods for Specified Health Uses (FOSHU) are foods that are composed of functional ingredients that affect the structure/function of the body. These foods are used to maintain or regulate specific health conditions, such as gastro-intestinal conditions, blood pressure, and blood cholesterol level. (Nakajima, 2002)

This revised definition clearly reflects the FOSHU experience in Japan; it also uses elements of the proposed functional food definitions from other countries around the world in explaining the FOSHU concept.

FOSHU Approval Process

Each proposed FOSHU product requires a separate application, and each product must be approved on a product-by-product basis. The original application process required the equivalent of a "pre-approval" of the functional ingredient itself outside the food product matrix, but this extra requirement was eliminated a few years ago. The functional ingredient supplier is an important source of technical information for the FOSHU applicant, but does not necessarily have any ownership of the formal FOSHU approval.

The FOSHU application process is quite complicated in that it involves submission of information and samples to several government and nongovernment organizations. The process starts with a submission to the appropriate Public Health Center near the Japanese manufacturer's head office location. After that agency's review, the application will be forwarded to the MHLW Office of Health Policy on Newly Developed Foods. Overseas applicants who plan to export their FOSHU product to Japan file an application directly with MHLW. The extensive detailed application requirements are available from JHNFA and other official sources in Japanese, and unofficially in English from private sources.

A critical step in the FOSHU approval process is the review by one of the "Committees of Experienced Specialists," whose members generally are chosen from the academic community in Japan. Depending on the health benefit being claimed for the product, the application is assigned by JHNFA to the appropriate committee for review and comment. The committee then decides whether more data are needed, or if the application can be forwarded to MHLW with a recommendation for approval. JHNFA is involved in this review process and is able to provide additional support on a fee basis if requested by the applicant. In general, it is recommended that the FOSHU application plan be exposed to the JHNFA experts for guidance well in advance of the actual application. They are in a position to help make the approval process as efficient as possible and have the experience to advise the applicant regarding the level of product safety and efficacy support likely to be required for the claims category of interest. Even when the approval process proceeds smoothly, it is still expected that six months will be required from the date of application and submission of data to the formal FOSHU approval.

Generally, Japanese companies claim that the required supporting safety data for the functional ingredient and the FOSHU product are not much different from the data that a responsible company in Japan would develop internally before marketing the functional ingredient or the food product. There are, however, significant additional requirements for the FOSHU product applications:

- At least one clinical study showing efficacy in support of the requested health-related claim
- Evidence showing a history of safe food use for the functional ingredient
- Publication of the supporting information in a Japanese scientific journal, such as the quarterly scientific "*Journal of Nutritional Food*"

A JHNFA English-language publication from 1998 entitled *Foods for Specified Health Uses (FOSHU)—A Guideline* expressed some of the important requirements a little differently, but officially as:

- Documentation that shows clinical and nutritional proof of the product's function for the maintenance of health
- Documentation that shows clinical and nutritional proof of the intake amount of the product
- Documentation concerning the safety of the product
- Documentation concerning the stability of the product
- Documentation of physiochemical properties and the test methods for the product's functional ingredients

Information must also be provided on the processing, formulation, analytical methods, and chemical and physical analyses, as well as other specifics. Product samples and proposed labels with proposed claims must accompany the application.

The amount of supporting safety and efficacy data required is not specified. It is expected that the clinical study information will demonstrate a statistically significant end result in support of the claim, including an April 2003 requirement that at least some of the clinical data are generated in Japan on Japanese subjects. It is up to the applicant, however, to negotiate the claim(s) with the appropriate authorities. Actual food use experience is helpful, which is one reason many companies first introduce the product in Japan as a non-FOSHU food without health claims. Information generated outside Japan on the functional ingredient and the FOSHU product itself when applicable is also very useful, although all information submitted must be in the Japanese language.

FOSHU Market Development Overview

As indicated previously, the FOSHU product category has grown to nearly 422 approved products as of June 2004, with retail sales well in excess of $5 billion U.S. dollars. JHNFA (Nakajima, 2002) has indicated, however, that only 20 to 30 of the products are considered truly successful by Japanese standards, and those products account for the majority of sales. JHNFA occasionally publishes a full-color, Japanese-language FOSHU product brochure with photos of the approved FOSHU products categorized according to the primary health benefit target, and maintains a full FOSHU product display case in its head-

Table 15.1. FOSHU product approvals and sales trends

Date	Number of Products Approved	Retail Annual Sales Estimates, USD($)
June 1993	2	0
May 1995	69	?
June 1997	80	500 Million
June 1998	100	500 Million
December 1998	126	800 Million
July 1999	153	1.5 Billion
June 2000	192	3.0 Billion
September 2001	271	3.5 Billion
December 2003	398	5.1 Billion

Source: Japan Health Food and Nutrition Food Association

Official full year 1999 ($2.2 billion USD), 2001 ($3.5 billion USD) and 2003 ($5.1 Billion USD) retail sales from JHNFA

quarters office. It also occasionally has the full range of approved FOSHU products and literature on display at trade shows in Japan.

A summary of the historical product approval and sales trends is shown in Table 15.1, beginning with the first two products approved in June 1993.

It can be seen from the data that the major growth in FOSHU retail sales occurred after 1998, not in terms of product approvals but rather in terms of sales. This was directly due to the conversion of successful non-FOSHU products to FOSHU status, in particular the extensive line of lactic acid bacteria beverages and yogurts from Yakult Honsha, beginning in late 1998 and extending through the end of 2000.

According to JHNFA, only about one-half of the approved FOSHU products are currently being sold. In addition, the manufacturers of many additional approved products chose not to renew their registrations and therefore the products are no longer included in the official JHNFA product list. JHNFA maintains an up-to-date Japanese-language summary of approved FOSHU products, and occasionally publishes an English-language list that is available for purchase as well. The *Japanscan Food Industry Bulletin* published monthly in the U.K. is an excellent source of timely English-language information on new FOSHU product approvals and regulatory changes in the category.

FOSHU Product Sales by Health Benefit Claimed

A review of the targeted health benefits for the FOSHU products being sold is particularly instructive in indicating the areas of health that are of interest to Japanese consumers. The data are summarized in Table 15.2.

The success of the gastrointestinal health products is in large part related to the historical Japanese interest in maintaining intestinal health by consuming foods and beverages containing live bacteria. One reason for the consumer interest is that overprescribing of antibiotics is very common in Japan and has been for many years, and the Japanese consumers are well aware of the need to reestablish "friendly" bacteria in the gastrointestinal tract on a regular basis. The "Yakult" lactic acid beverage with active *Lactobacillus casei* has been sold in Japan for more than 65 years, mostly in small, single-serving bottles.

Table 15.2. FOSHU sales by specified health use (2001 sales)

1. Gastro-intestinal—81% (77% based on *Lactobacilli*)
2. Dental Caries—5%
3. Blood Glucose—4%
4. Neutral Fat—4%
5. Mineral Absorption—3%
6. Hypertension—2%
7. Cholesterol—1%

Source: JHNFA Singapore presentation, January 2002

More recently, yogurts with other *Lactobacillus* and *Bifidobacterium* strains have become very popular as well. Morinaga Milk Industries and Meiji Dairies converted their most popular yoghurt brands to FOSHU status in late 1996, adding considerable consumer awareness and credibility to the FOSHU category.

Examples of acceptable health-benefit claims were summarized in English in the JHNFA presentation at the Singapore conference, as follows:

- Maintenance or improvement of a health condition that is easily diagnosed, such as "helps maintain normal blood pressure"
- Maintenance or improvement of the body's physiological and structural function, such as "effective in promotion of fat metabolism"
- Improvement of a health condition that can be detected individually and that is temporary and not chronic, such as "suitable for people who feel fatigued"

Although the original intent was to avoid disease-related health claims, the claims allowed for FOSHU products have been getting stronger as new clinical efficacy support is developed. In general, the initial claims allowed for a given benefit category were quite conservative; ". . . keeps the tummy in good condition . . ." is one example for a breakfast cereal containing wheat bran as the functional ingredient. More recently, the claims have been expanded to include ". . . wheat bran very rich in dietary fiber, which regulates your intestines and maintains satisfactory bowel movement," for example. In the cholesterol-lowering area, the initial claims such as ". . . helps maintain a desirable cholesterol level . . ." have for certain products been strengthened to, for example, ". . . soy protein, which acts to lower serum cholesterol . . . and improve the diet for people beginning to be concerned about cholesterol." The claim approved early last year for the Johnson & Johnson Benecol Fat Spread is "this product contains phytostanol ester, which suppresses the absorption of cholesterol and lowers blood cholesterol, in particular LDL-cholesterol (bad cholesterol). It is recommended for people with a tendency towards high cholesterol." An approval late last year for *natto* fermented food from soybeans is a more recent example of the trend: "this product containing Vitamin K2 is an active means to promote the formation of calcium-binding bone protein (osteocalcin)."

The positive experience with the early, more conservative claims has allowed the claims review committees (and MHLW) to approve stronger claims more recently, with the confidence that they are not likely to be controversial with Japanese consumers in terms of overpromising a health-related benefit.

The JHNFA published list of FOSHU product approvals includes a summary of the permitted claim for each of the products. Some products have now been allowed to make more than one health-benefit claim, and several functional ingredients are being used in different approved FOSHU products with different claims. An example of the latter is the use of the functional ingredient indigestible dextrin ("this product contains dietary fiber . . . to compensate for the dietary deficiency of fiber and to regulate the intestines" as well as a separate claim for other products that "this product contains indigestible dextrin dietary fiber that moderates absorption of sugar . . . suitable for people beginning to be concerned about blood sugar levels"). The incentive exists for the functional ingredient manufacturers to develop FOSHU-quality support data for a range of possible health benefits, and not to limit the use of their ingredient to a single FOSHU product or claim.

FOSHU Functional Ingredients

For the first few years of the FOSHU experience it was first necessary to obtain an informal "pre-approval" of the functional ingredient before the ingredient was allowed to be considered for use in a FOSHU product. The functional ingredients were expected to come from one of the ten specific categories established in the late 1980s by MHW, as indicated in Table 15.3.

Working groups from industry were requested by MHW to help support the development of the FOSHU regulations by identifying likely functional ingredients in each category, as well as by providing data on the methods of analysis, results of animal trials, and the results of clinical trials where such information was available. This was the beginning of the process.

It was decided, however, that the pre-approval procedure for the functional ingredients was inefficient and unnecessary, because the FOSHU application procedures require supporting safety and nonclinical efficacy data for the functional ingredient in any event. In addition, dropping the pre-approval requirement allows for a measure of competitive marketing advantage because it is now possible for the company with the FOSHU approval to surprise the competition when new functional ingredients are used.

The list of functional ingredients that are allowed to be used in approved FOSHU products is not particularly long, although it is growing as new ingredients are developed. At

Table 15.3. FOSHU health benefit ingredient categories

1. Dietary Fiber
2. Oligosaccharides
3. Sugar Alcohols
4. Unsaturated Fatty Acids
5. Peptides and Proteins
6. Glycosides, Isoprenoids, and Vitamins
7. Alcohols and Phenols
8. Choline (complexed with lipids)
9. Lactic Acid Bacilli
10. Minerals
11. Others

Source: Ministry of Health, Labor, and Welfare FOSHU Guidelines

Table 15.4. FOSHU functional ingredients allowed (2003)

1. Dietary Fiber Sources—indigestible dextrin, polydextrose, partially hydrolyzed guar gum, low molecular weight sodium alginate, chitosan, psyllium seed coat, wheat bran, beer yeast fiber, agar, and corn fiber
2. Oligosaccharides—xylo-oligosaccharide, fructo-oligosaccharide, soybean oligosaccharide, isomalto-oligosaccharide, lactosucrose, lactulose, galacto-oligosaccharide, raffinose
3. Sugar Alcohols—palatinose, maltitol, xylitol, reduced palatinose
4. Unsaturated Fatty Acids
5. Peptides and Proteins—soy protein, casein phosphopeptide, casein dodecapeptide, lacto-tripeptide, sardine peptide containing valyl tyrosine, bonito oligopeptide, wheat albumin, soy peptide bound with phospholipids, casein phosphopeptide-amorphous calcium phosphate compound, casein dipeptide, milk basic protein, globin hydrolysate, soy protein isolate, fermented whey substances
6. Glycosides, Isoprenoids, and Vitamins—glycoside from Eucommia leaves (geniposidic acid), Vitamin K2, soy isoflavone, oolong tea polyphenols
7. Alcohols and Phenols—green tea polyphenol, beta-sitosterol, guava leaf polyphenol
8. Choline
9. Lactic Acid Bacilli—*Lactobacillus GG, Bifidobacterium longum* BB536, *Lactobacillus delbrueckii* 2038 and *Streptococcus salivarius* 1131, *Lactobacillus acidophilus* SBT-2062 and *Bifidobacterium longum* SBT-2928, *Yakult bacillus (L.casei, Shirota strain), Bifidobacterium breve, Natto bacillus, Bifidobacterium lactis* FK 120, *Bifidobacterium lactis* LKM512, *L. acidophilus* CK92 and *L. helveticus* CK60, *gasseri* bacteria SP strain
10. Minerals—calcium citrate malate (CCM), heme iron
11. Other—diacylglycerol, mid-chain triglyceride

Sources: Translations of JHNFA FOSHU list by JHNFA and JapanScan Food Industry Bulletin, through 2002

the start of 2003, the list of functional ingredients included the 11 categories outlined in Table 15.4.

It has become relatively common in Japan that ingredient manufacturers will consider potential FOSHU applications for their food ingredients at the time that the ingredient is in the early development stages. In at least one instance, an ingredient manufacturer has gone as far as obtaining formal FOSHU approval for a product using its ingredient as a means of demonstrating the safety and efficacy for the ingredient, even though the company has no interest in actually marketing the FOSHU product.

Some of the functional ingredients used in FOSHU products are now being sold outside Japan as well. The data generated in order to obtain FOSHU approval, including the safety and efficacy data on the functional ingredient, are often sufficient to allow the use of the ingredient in the food/beverage and/or dietary supplement market in the U.S. and other countries. The popular Econa cooking oil with its diacylglycerol functional ingredient from Kao Corporation has now been introduced into the U.S. market as Enova through an ADM Kao joint venture, and the diacylglycerol functional ingredient was granted Generally Recognized as Safe (GRAS) status required by the U.S. FDA for food product applications.

FOSHU Product "Pre-Marketing"

Another recent development being allowed by the food and beverage regulators is the use of the targeted future FOSHU health-benefit claim(s) in off-label advertising at the time of the market introduction of the product as a regular non-FOSHU food or beverage. An announcement is made that a FOSHU application is planned for the product, with the target claims identified. The product is then sold through the regular sales channels and experience is gained with the product and the functional ingredient. Because a history of safe food use is an important element of the FOSHU approval procedures, this allows the sales experience to be gained while the FOSHU application is in process. The FOSHU approval process usually takes six months from the submission of data, assuming that all data are in place and eventually accepted, so the advance sale is in effect a test market of the product acceptance in advance of the FOSHU approval. When the FOSHU approval is in hand, the FOSHU logo and on-label claims can be added.

FOSHU Access by Non-Japanese Companies

The FOSHU regulatory procedures allow the use of non-Japanese functional ingredients and applications for FOSHU status directly by overseas companies. Examples of branded functional ingredients from overseas include calcium citrate malate (CCM) for enhanced calcium absorption, from Procter & Gamble, and polydextrose as a convenient fiber source, from Pfizer, even though those two companies are not named in the FOSHU-approved products list. The data submitted in support of the FOSHU applications are considered confidential, but it can be assumed that supporting efficacy and safety data from the two companies were included in the application process through the Japanese company applicants.

Several of the FOSHU product approvals are also in the name of the Japanese subsidiary of, or joint venture with, an overseas parent, including All Bran and Bran Flakes cereals from Kellogg (Japan) K.K., AGF Vitahot powdered soft drinks from Ajinomoto General Foods, Inc., Recaldent chewing gums from Warner Lambert Inc., Rama ProActive margarine from Nippon Lever Co., and Milo milk and malt drinks and Karada Shien soft drinks and a soy protein drink from Nestle Japan. Although the FOSHU system allows an application directly from overseas companies, the detailed requirements of the application process, including the requirement that all supporting information must be in Japanese, has led to the coordination of most of the overseas developments through Japanese subsidiaries.

Two of the FOSHU products, however, the Johnson & Johnson Benecol fat spread and a Recaldent tablet candy from Pfizer Consumer, Inc., are actually designated as FOSHU products formally "approved" by the MHLW, rather than "accepted" by MHLW, which is the usual designation for the other FOSHU products. The FOSHU logo for the two products makes this distinction as well, apparently based on the primary source of the information provided in support of the two products. It is a distinction that is likely to be of little importance to Japanese consumers, however.

Non-FOSHU Foods with FOSHU Potential

Most of the growth of the FOSHU retail market has been as a result of conversion of non-FOSHU foods and beverages to FOSHU status, as mentioned previously. The Yakult experience in converting nearly their entire product line to FOSHU status attracted a lot of attention in Japan and has been reported as a positive move in terms of increased sales for Yakult. Two other examples of large-volume non-FOSHU beverages that are well-known by Japanese consumers to provide health benefits include green tea and health drinks, and these have become interesting FOSHU conversion candidates.

Green teas have been investigated in Japan for many years for anticancer, anticaries, and cholesterol-lowering effects. Although it is not likely that MHLW would allow a FOSHU food or beverage to make a cancer-related claim even if the data were in place, it can be expected that green tea extract suppliers are actively researching health-benefit applications that are more likely to be approved through the FOSHU process. FOSHU chocolate and chewing gum products with green tea polyphenols have already been approved for anticaries benefits, and new FOSHU green tea drinks with added functional ingredients such as indigestible dextrin have been approved. The leading brands in the much larger ready-to-drink green tea beverage market are not yet represented, however, and remain an interesting possibility.

Health drinks are also a very large market in Japan, with excellent consumer acceptance over many years. Popular drinks such as Oranamin C from Ohtsuka Pharmaceutical (well over 600 million bottles sold per year) and the Lipovitan series from Taisho Pharma (more than 800 million bottles sold per year) are just two of the large non-FOSHU brands in this category that could be considered for potential FOSHU status.

Foods with Nutrient Function Claims

For some time there had also been pressure on the Japanese government agencies to allow more generic health claims to be made, similar to the structure/function claims allowed for dietary supplements in the United States. The MHLW decided to recognize this interest by creating a new food category called Foods With Nutrient Function Claims, which allows foods (including dietary supplements, which are foods in Japan) and beverages to make specific claims for 12 vitamins and two minerals provided that the product meets the minimum and maximum dosage restrictions per serving. It is expected that additional regulatory initiatives will follow, depending on the success of the new category.

As of April 2001, the new category of foods was formally established by MHLW to allow generic structure/function claims to be made on-label for the first time. As long as the food or beverage includes one or more of 12 vitamins and two minerals within specified intake ranges, the generic claim(s) can be used without a formal approval from MHLW and without the need for an individual product application.

Food With Nutrient Function Claims Requirements

The 12 vitamins and two minerals, along with the permitted minimum and maximum intakes per serving, are indicated in Table 15.5.

It does not matter whether the vitamins and minerals are naturally occurring or are the result of fortification, as long as they are present in the required amounts.

In addition to the content requirement, the regulations require that specific additional information be included on the food or beverage label:

- Disclaimer that the food is not a Food for Specified Health Uses (FOSHU) and therefore not evaluated individually by MHLW
- Indication that the food is a "Food with health claim (food with nutrient function claim)"
- Appropriate warnings regarding avoiding excess intake

Table 15.5. Foods for nutrient function intake limits

Vitamin or Mineral	Intake Limits: Maximum/Minimum per Serving
Vitamin A	2000IU/600IU
Vitamin D	200 IU/35IU
Vitamin E	150mg/3mg
Vitamin B1	25 mg/0.3mg
Vitamin B2	12 mg/0.4mg
Niacin	15 mg/5mg
Vitamin B6	10mg/0.5 mg
Folic Acid	200mcg/70mcg
Vitamin B12	60mcg/0.8mcg
Biotin	500mcg/10mcg
Pantothenic Acid	30mg/2mg
Vitamin C	1000mg/35mg
Calcium	600mg/250mg
Iron	10mg/4mg

Source: Ministry of Health, Labor, and Welfare

Food with Nutrient Function Claims Marketing

Although it is relatively easy to convert a food or beverage to a Food with Nutrient Function Claim, the incentive for doing so is not yet clear to most manufacturers. It is relatively easy to find the new products in stores in Japan, but it is believed that the product sales are quite limited. There is no public evidence that conversion to Food with Nutrient Function Claims status has actually increased the sales of a food or beverage product, unlike the more positive FOSHU conversion experience for certain brands. Most of the concern seems to be with the generic claims that are allowed, because in general they do not seem to provide much new information to knowledgeable Japanese consumers. The implied approval from the government, however, is regarded as a positive for the category. A list of the allowed claims appears in Table 15.6.

Additional claims can be expected to be allowed for the initial 12 vitamins and two minerals as experience is gained with the Food with Nutrient Function Claims category. More important, it can also be expected that the permitted generic claims will be expanded to include more vitamins and minerals as well as additional nutritional ingredients that are of more interest to consumers. If the FOSHU experience is any indication, MHLW will proceed cautiously with this new category development as well.

Future Regulatory Initiatives

It is important to realize that all the positive FOSHU market development has been achieved in Japan during a period in which the overall economy has been in serious decline. In addition, the FOSHU category in particular has emerged from the recent food safety scandals in Japan with even stronger credibility. This would suggest that the Japanese FOSHU model might represent an appropriately conservative consumer/industry/government partnership for consideration by other countries in the future.

There is pressure in Japan, however, for revisions to the basic Food Sanitation Law that

Table 15.6. Foods for nutrient function authorized claims

Vitamin A—Vitamin A is a nutrient that helps maintain vision at night. Vitamin A is a nutrient that helps maintain healthy skin and mucosa.
Vitamin D—Vitamin D is a nutrient that promotes absorption of calcium and aids in the development of bone.
Vitamin E—Vitamin E is a nutrient that helps protect fat in the body from being oxidized and helps maintain healthy cells.
Vitamin B1—Vitamin B1 is a nutrient that helps produce energy from carbohydrate and helps maintain healthy skin and mucosa.
Vitamin B2—Vitamin B2 is a nutrient that helps maintain healthy skin and mucosa.
Niacin—Niacin is a nutrient that helps maintain healthy skin and mucosa.
Biotin—Biotin is a nutrient that helps maintain healthy skin and mucosa.
Pantothenic Acid—Pantothenic acid is a nutrient that helps maintain healthy skin and mucosa.
Vitamin B6—Vitamin B6 is a nutrient that helps produce energy from protein and helps maintain healthy skin and mucosa.
Folic Acid—Folic acid is a nutrient that aids in red blood cell formation. Folic acid is a nutrient that contributes to the normal growth of a fetus.
Vitamin B12—Vitamin B12 is a nutrient that aids in red blood cell formation.
Vitamin C—Vitamin C is a nutrient that helps to maintain healthy skin and mucosa and has anti-oxidizing effect.
Calcium—Calcium is a nutrient that is necessary in the development of bone and teeth.
Iron—Iron is a nutrient that is necessary for red blood cell formation.

Source: Ministry of Health, Labor, and Welfare, March 2001

would enable it to respond more effectively to the serious food safety issues that have emerged in the past three years. Protection of consumer health is expected to be the new focus of a reorganized MHLW as it relates to the food industry. The discovery of domestic cows with bovine spongiform encephalopathy in 2001 shocked Japanese consumers, coming soon after the Snow Brands Milk food poisoning crisis in 2000 that crippled that respected company. So far, most of the early focus has been placed on ensuring that food imports are in compliance with the existing food regulations, primarily by additional enforcement activities at the time that the products enter the country.

Although the future of the functional foods market in Japan is not yet clear, there is cause for optimism as long as the consumer credibility of the growing FOSHU category can be maintained. The FOSHU logo on the product, with the endorsement of the MHLW, represents an important "guarantee of quality and efficacy" in the minds of many Japanese consumers.

References

Japanscan Food Industry Bulletin. 2001-2003. Unofficial translations of Japan Health Food & Nutrition Food Association new FOSHU approval announcements. Stratford-on-Avon, U.K. www.japanscan.com

Japanscan Food Industry Bulletin. December 2002. Organic agricultural produce. Page 38. Stratford-on-Avon, U.K. www.japanscan.com

JETRO. 2000. Specifications and Standards for Foods, Food Additives, etc. under The Food Sanitation Law. January 2000. Japan External Trade Organization. Tokyo, Japan. www.jetro.go.jp/books/eng/

JHNFA. 1998. Foods For Specified Health Uses (FOSHU)—A Guideline. Tokyo Japan: Japan Health Food & Nutrition Food Association.

JHNFA. 2002. "Functional Food" in Japan: Regulatory Update and a Profile of the Japanese Market. Kaori Nakajima. Marketing Nutraceuticals & Functional Foods. Singapore, 2002.

MHLW. 2003. The National Nutrition Survey in Japan, 2001. National Institute of Health and Nutrition. Tokyo, Japan.

MOE. 1988. Reports of Systematic Analysis and Development of Food Functionalities. Ministry of Education, Science, and Culture of Japan. Tokyo, Japan.

Nikkei Weekly. 2002. Life Expectancy. August 12, 2002. Tokyo, Japan.

Nikkei Weekly. 2002. Newborns. June 24, 2002. Tokyo, Japan.

Nikkei Weekly. 2002. Medical Expenditures. September 16, 2002. Tokyo, Japan.

USATO. 1992. The Japanese Market for Health Foods and Beverages. January. U.S. Agricultural Trade Office. Tokyo, Japan.

16 Chinese Health (Functional) Food Regulations

Guangwei Huang and Karen Lapsley

Introduction

Functional foods are called health foods in China, and the Chinese have a long tradition and strong passion for using tonics and specific foods to regulate body functions. There are many traditional Chinese medicine (TCM) books about the physiological functions of different foods because it is commonly believed that foods and medicine are of the same origin. Early health foods made in China were primarily from medicinal food recipes. As the Chinese economy advanced in the late 1980s, health foods emerged as an important part of the food industry in China and reached more than 10 percent of the food industry annual revenue by 1996 (1). There were many kinds of health foods making numerous types of health claims. In order to strengthen the control and inspection of health foods and assure health food quality, the Ministry of Health (MOH) issued regulations in 1996 under the relevant provisions of the "Food Hygiene Law of the People's Republic of China" (2).

The "Administrative Regulations for Health Foods" (3) provide administrative guidelines for the supervision of the health foods in the Chinese market. These regulations focus more on the evaluation and approval process of the health food products than production of the products even though there are sections dealing with the production and marketing of health foods. The regulations have been amended with several additional provisions since its promulgation and continue to be defined. However, all these regulations and amendments are in Chinese, which creates difficulties for foreign functional foods researchers and manufacturers. A few English overviews of Chinese health (functional) foods exist but they are not complete (4–7). The China office of the Foreign Agricultural Service (FAS) of the United States Department of Agriculture (USDA) translated the administrative regulations and the hygiene standards for health foods into English. The translations are posted on its Web site (8,9). The regulations keep evolving and the statuary authority of the regulations switched from the MOH to the newly established State Food and Drug Administration (SFDA) in October 2003. More than 4,000 health foods have been approved since the promulgation of the regulations in 1996, and the imported health foods account for about 10 percent of these (10). This leads to difficulty for the supervision of the health foods in China. Several professional organizations, such as the Chinese Institute of Food Science and Technology, the International Life Sciences Institute China Focal Point, and the Chinese Nutrition Society, have organized symposiums to address the issues related to the regulations (1,10–14). This chapter focuses on the Chinese health food regulation system as related to imported health foods, and reviews the scientific requirements behind the evaluation and approval process for the health foods in China.

Evolution of the Regulations (1996–Present)

The MOH under the State Council, the Peoples' Republic of China, was the statutory authority responsible for implementation of the regulations. The National Center for Health Inspection and Supervision, a subsidiary agency of the MOH, handled the approvals of health foods. The respective provincial health authorities monitored compliance of such regulations. As of October 2003, the SFDA is the agency that handles the application and approval process for health (functional) foods. Therefore, the MOH, as the statutory authority stated in all regulatory documents or proceedings, shall be substituted by the SFDA (15, 16). A similar application and approval procedure is still being used and followed since the structural change, although some modifications for the process are expected to take place soon.

The Regulations define health (functional) foods as food products proven to have specific health functions; namely, those with physiological function regulation, designed for a specific population to consume, but which are not for the purpose of disease treatment. The legal basis of the regulations is from Articles 22 and 23 of the "Food Hygiene Laws." The laws plainly state: "A food claiming to have a specific health function, together with its product manual, should be submitted to the health administrative authority under the State Council for evaluation and approval. The relevant hygienic standards and administrative regulations for production and handling will be established by the health administrative authority. . . . A food claiming to have a specific health function must not be harmful to the human body[;] the product manual content must be accurate and truthful. The function and composition of the product must be consistent with what are claimed in the manual, and they can not be falsified."

The categories of the permitted function claims changed from the original 12 in 1996 to 24 in 1997, dropped to 22 in 1999, and went up to 27 in 2003. In 1997, two function claims, "improving sexual function" and "assisting in cancer prevention," were removed from the permitted function list. The descriptors of function claims have been modified as well (1,13,17,18). Now the descriptors used for function claims are more specific and less generic. For example, "cosmetic enhancement" was further classified as "alleviating acne," "eliminating skin pigmentation," "improving skin moisture," and "improving skin oil content"; and "improving gastrointestinal function" was replaced by "regulating gastrointestinal flora," "facilitating digestion," "alleviating constipation," and "assisting in protection against gastric mucosa injury." The new claims are more accurate as well, as reflected in such changes as "regulating blood lipids" to "assisting blood lipid reduction" and "reducing blood pressure" to "assisting in blood pressure reduction." Table 16.1 lists all English translations of the currently accepted function claims and respective Chinese function names (17).

Since the regulations' first promulgation in 1996, many supplemental procedures, notifications, directives, and guidelines have been added to complete the regulations, and a total revision is expected in the near future (20–34). These regulations are aimed at providing administrative directives on the evaluation and approval, production, marketing, and monitoring for health (functional) foods. The regulations and their subsequent additional directives have focused more on the evaluation and approval process, with very specific provisions for imported functional foods as well.

All foods claiming to have health-promotion effects must undergo the assessment and approval process set forth by the SFDA. Researchers and manufacturers shall submit an application to local provincial health administration authorities. Upon passing preliminary examination, the application shall be submitted to the SFDA for evaluation and approval. After being approved, the SFDA shall issue an "Approval Certificate of Health (Func-

Table 16.1. Function claims for China health (functional) foods (17)

English Translation of Accepted Functions	Chinese Name
Enhancing immune function	增强免疫力
Assisting in blood lipids reduction	辅助降血脂
Assisting in blood sugar reduction	辅助降血糖
Anti-oxidation	抗氧化
Assisting in memory improvement	辅助改善记忆
Reducing eye fatigue	缓解视疲劳
Facilitating lead excretion	促进排铅
Thinning throat mucus	清咽
Alleviating hypertension	辅助降血压
Enhancing sleep	改善睡眠
Facilitating lactation	促进泌乳
Alleviating physical fatigue	缓解体力疲劳
Enhancing anoxia endurance	提高缺氧耐受力
Assisting protection against irradiation hazard	对辐射危害有辅助保护
Weight reduction	减肥
Enhancing child growth and development	改善生长发育
Increasing bone density	增加骨密度
Alleviating nutritional anemia	改善营养性贫血
Assisting in protection against liver chemical injury	对化学性肝损伤有辅助保护
Alleviating acne	祛痤疮
Eliminating skin pigmentation	祛黄褐斑
Improving skin moisture	改善皮肤水份
Improving skin oil content	改善皮肤油份
Regulating gastrointestinal flora	调节肠道菌群
Facilitating digestion	促进消化
Alleviating constipation	通便
Assisting in protection against gastric mucosa injury	对胃粘膜损伤有辅助保护

tional) Foods" to the qualified products. A serial number such as "Foodstuff/Health [xxxx]"—a four-digit number signifying the year when the certificate was issued.) No##" shall be issued together with the approval certificate.

A food product granted with an "Approval Certificate of Health (Functional) Foods" shall be permitted to use the symbol of health (functional) food designed by the Ministry of Health, as illustrated in Figure 16.1. For imported health (functional) foods, an importer or representative needs to submit an application directly to the SFDA for approval. The SFDA will issue an "Approval Certificate for Imported Health (Functional) Food" with a certificate serial number to the qualified products. The certificate number together with the health food symbol must be marked on the package for the products granted with such an approval certificate.

Figure 16.1. Chinese health food symbol (sky blue color) (18)

Evaluation and Approval of Health (Functional) Foods

The regulations established minimum requirements for health (functional) foods as follows: (1) necessary animal and/or human trial experiments must have confirmed the effect and stability of the effect; (2) all raw materials and finished products must comply with relevant food hygiene requirements and shall not cause any acute, subchronic, or chronic harm to human body; (3) formulation and dosage must be based on scientific evidence, and there should be an identified functional ingredient, that is, an active ingredient or compound that is of physiological function from a natural plant or a health food formulation; if the functional ingredient cannot be identified under current technical conditions, the primary raw materials pertaining to the health-promotion function shall be listed; and (4) label, manual, and advertisement of a health (functional) food shall not make any claims of therapeutic effects.

The MOH issued a draft for comments for the regulations on administration of dietary supplements. Any companies who are interested in marketing dietary supplements in China should investigate this publication. In essence, minerals and vitamins are classified as dietary supplements. They need to get an "Approval Certificate for Health Food" prior to production. But there is no need to perform laboratory function and safety evaluation testing. Other dietary supplements such as dietary fibers, proteins, and amino acids are considered as common foods. The approval certificate and evaluation testing are not needed. However, these supplements cannot make health food claims. If a manufacturer intends to make such claims or market the supplements as health foods, the supplements need to undergo the whole process as designed for health foods (11).

Safety Control of Raw Materials and Finished Products

To further tighten supervision of the raw ingredients for health foods, the MOH issued a directive in February of 2000 to provide guidelines on raw ingredients usage. For newly discovered or imported items that were not customarily consumed as a food, they will follow the "Administrative Provisions for New Source Food Hygiene" (35). If a food additive is used in a health food, it should follow the "Administrative Provisions for Food Additive Hygiene" (37). A health food using fungus or probiotics as an ingredient should follow special guidelines, and only select fungus and probiotics are permitted as health food ingredients (24–27).

The permitted fungus species are the following: *Saccharomyces cerevisiae, Cadida atilis, Kluyveromyces lactis, Saccharomyces carlsbergensis, Paecilomyces hepiali Chen et Dai, sp. Nov, Hirsutella hepiali Chen et Shen, Ganpderma lucidum, Ganoderma sinensis, Ganoderma tsugae, Monacus anka, Monacus purpures.*

The permitted probiotics are the following: *Bifidobacterium bifidum, Bifidobacterium infantis, Bifidobacterium longum, Bifidobacterium breve, Bifidobacterium adolescentis, Lactobacillus bulgarius, Lactobacillus acidophilus, Lactobacillus Casei subsp. Casei, Streptococcus thermophilus.*

A nuclei acid type of health food (using DNA or RNA as raw materials) must follow specific requirements. The purity of DNA or RNA must be above 80 percent, and the recommended dosage for nuclei health foods should be 0.6g–1.2g per day.

Raw materials are classified into three categories: category I, raw materials considered as either foods or medicine; category II, raw materials permitted as health food ingredients; and category III, raw materials forbidden as health food ingredients. Table 16.2

Table 16.2. Classification of raw materials for Chinese health foods (7,26,43,44)

Category I. Raw Materials Considered as Food or Medicine			
Chinese Names	Mandarin Phonics	Common English Names	Pharmaceutical or Botanical Names
丁香	Dingxiang	Cloves	Flos Caryophylatae
八角茴香	Bajiaohuixiang	Star anise seed	Fructus Illicium verum
刀豆	Daodou	Sword bean	Semen Canavalia gladiata
小茴香	Xiaohuixiang	Fennel seed	Semen Foeniculi
小蓟	Xiaoji	Field thistle; Small thistle	Herba Cirsii segeti
山药	Shanyao	Chinese Yam; Dioscorea	Rhizoma Dioscoreae
山楂	Shanzha	Hawthorn fruit	Fructus Crataegi
马齿苋	Machixian	Purslane herb; Portulaca	Herba Portulacae
乌梢蛇	Wushaoshe	Black-tailed snack	Zaocys Dhumnades
乌梅	Wumei	Smoked plum; Black plum; Mume	Fructus Mume
木瓜	Mugua	Chinese quince fruit; Chaenomeles fruit	Fructus Chaenomelis
火麻仁	Huomaren	Hemp seed	Fructus Cannabis
代代花	Daidaihua		Flos Citri aurantit
玉竹	Yuzhu	Fragrant solomonseal rhizome	Rhizoma Polygonati odorati
甘草	Gancao	Licorice root	Radix Glycyrrhizae
白芷	Baizhi	Dahurian angelica root	Radix Angelicae dahuricae
白果	Baiguo	Ginkgo seed	Semen Ginkgo
白扁豆	Baibaindou	White hyacinth bean & flower	Semen Lablab album
白扁豆花	Baibiandouhua	White hyacinth bean flower	Flos Lablab album
龙眼肉（桂圆）	Longyanrou (guiyuan)	Longan aril; Arillus fruit	Arillus Longan
决明子	Juemingzi	Foetid cassia seed	Semen sennae
百合	Baihe	Lily bulb	Bulbus Lilii
肉豆蔻	Roudoukou	Nutmeg	Semen Myristicae
肉桂	Rougui	Cinnamon bark; Cassia bark	Cortex Cinnamomi
余甘子	Yuganzi	Dharty, Indian gooseberry	Fructus Phyllanthus emblicae
佛手	Foshou	Finger citron fruit	Fructus Citri sarcodactylis
杏仁（甜、苦）	Xingren (Tian, Ku)	Bitter or sweet apricot seed	Semen Armeniacae amarum
沙棘	Shaji	Seabuckthorn fruit	Fructus Hippophae
牡蛎	Muli	Oyster shell	Concha Ostreae
芡实	Qianshi	Gordon euryale seed	Semen Euryales
花椒	Huajiao	Prickly-ash peel	Pericarpium Zanthoxyli
赤小豆	Chixiaodou	Adzuki bean; Phaseolus seed	Semen Phaseoli
阿胶	Ejiao	Ass-hide glue; Donkey hide gelatin	Colla Cori asini
鸡内金	Jineijin	Chicken's gizzard skin or lining	Endothelium Corneum gigeria galli
麦芽	Maiya	Germinated barley	Fructus Hordei germinatus
昆布	Kunbu	Kelp; Kunbu sea cress; Laminaria	Thalus Laminariae seu Eckloniae
枣（大枣、酸枣、黑枣）	Zao (Dazao, Suanzao, Heizao)	Date (Chinese date, Spine date, Black date)	Fructus Jujubae, Fructus Spinosae, Black Jujubae

(continued)

Table 16.2. Classification of raw materials for Chinese health foods (7,26,43,44) (*Cont.*)

Category I. Raw Materials Considered as Food or Medicine			
Chinese Names	Mandarin Phonics	Common English Names	Pharmaceutical or Botanical Names
罗汉果	Luohanguo	Swingle	Siraitia Grosvenorii
郁李仁	Yuliren	Chinese dwarf cherry seed; Bush-cherry seed	Semen Pruni
金银花	Jinyinhua	Honeysuckle flower; Lonicera flower	Flos Lonicerae
青果	Qingguo	Chinese white olive	Fructus Canarii
鱼腥草	Yuxingcao	Heartleaf houttuynia herb	Herba Houttuyniae
姜（生姜、干姜）	Jiang (Shengjiang, ganjiang)	Ginger root (fresh, dried)	Rhizoma Zingiberis recens, R. zingiberis
枳子	Zhijuzi	Dulcis seed	Semen Hovenia dulcis
枸杞子	Gouqizi	Barbary wolfberry fruit	Fructus Lycii
栀子	Zhizi	Cape-jasmine fruit; Gardenia fruit	Fructus Gardeniae
砂仁	Sharen	Spiny amomum fruit	Fructus Amomi
胖大海	Pangdahai	Boat fruited-sterculia seed	Semen Sterculiae lychnophorae
茯苓	Fuling	Indian bread; Poria	Poria Cocos
香橼	Xiangyuan	Citron fruit	Fructus Citri
香薷	Xiangru	Elsholtzia herb; Aromatic madder	Herba Elshotziae
桃仁	Taoren	Peach seed	Semen Persicae
桑叶	Sangye	Mulberry leaf; Morus leaf	Folium Mori
桑椹	Sangshen	Mulberry fruit	Fructus Mori
桔红	Jiehong	Red tangerine peel	Exocarpium Citri rubrum
桔梗	Jiegeng	Platycodon root; Balloonflower root	Radix Platycodi
益智仁	Yizhiren	Sharpleaf galangal seed; Bitter cardamon	Semen Alpiniae oxyphyllae
荷叶	Heye	Lotus leaf	Folium Nelumbinis
莱菔子	Laifuzi	Radish seed	Semen Raphani
莲子	Lianzi	Lotus seed	Semen Nelumbinis
高良姜	Gaoliangjiang	Lesser galangal; Galanga rhizome	Rhizoma Alpiniae officinarum
淡竹叶	Danzhuye	Lophatherum herb	Herba Lophatheri
淡豆豉	Dandouchi	Fermented soybean	Semen Sojae preparatum
菊花	Juhua	Chrysanthemum flower	Flos Chrysanthemi
菊苣	Juju	Common chicory grass	Herba Cichorri intybi
黄芥子	Huangjiezi	Airpotato yam	Rhizoma Dioscoreae bulbiferae
黄精	Huangjing	Siberian solomonseal rhizome	Rhizoma Polygonati
紫苏	Zisuye	Perilla leaf	Folium Perillae
紫苏籽	Zisuzi	Perilla seed; Purple perilla fruit	Fructus Perillae
葛根	Gegen	Lobed kudzuvine root; Pureraria root	Radix Puerariae
黑芝麻	Heizhima	Black sesame seed	Semen Sesamumae indicumae

(*continued*)

Table 16.2. Classification of raw materials for Chinese health foods (7,26,43,44) (*Cont.*)

Chinese Names	Mandarin Phonics	Common English Names	Pharmaceutical or Botanical Names
Category I. Raw Materials Considered as Food or Medicine			
黑胡椒	Heihujiao	Black pepper	Semen Piper nigrum
槐米	Huaimi	Pagodatree flower-bud	Flos Sophorae immaturus
槐花	Huaihua	Pagodatree flower	Flos Sophorae
蒲公英	Pugongying	Dandelion herb	Herba Taraxaci
蜂蜜	Fengmi	Honey	Apis Melifera
榧子	Feizi	Grand torreya seed	Semen Torreya
酸枣仁	Suanzaoren	Spine date seed; Wild jujube seed	Semen Ziziphi spinosae
鲜白茅根	Xuanbaimaogen	Fresh imperata rhizome; Fresh woody grass	Rhizoma Imperatae
鲜芦根	Xuanlugen	Fresh reed root	Rhizoma Phragmitis
蝮蛇	Fushe	Pallas pit viper	Agkistrodon Halys
橘皮	Jupi	Tangerine peel	Citrus Reticulata
薄荷	Bohe	Peppermint; Mentha	Herba Menthae
薏苡仁	Yiyiren	Coix seed; Job's tears seed	Semen Coicis
薤白	Xiebai	Longstamen onion	Bulbus Allii macrostemoni
覆盆子	Fupenzi	Palmleaf raspberry fruit	Fructus Rubi
藿香	Huoxiang	Agastache; Pogostemon	Herba Agastachis
Category II. Raw Materials Permitted as Health Food Ingredients			
人参	Renshen	Ginseng root	Radix Ginseng
人参叶	Renshenye	Ginseng leaf	Folium Ginseng
人参果	Renshenguo	Ginseng fruit	Fructus Ginseng
三七	Sanqi	Notoginseng; Pseudoginseng	Radix Notoginseng
土茯苓	Tufuling	Glabrous greenbrier rhizome	Rhizoma Smilacis glabrae
大蓟	Daji	Japanese thistle	Herba seu Radix Cirsii japonici
女贞子	Nuzhenzi	Grossy privet fruit; Ligustrum seed	Fructus Ligustri lucidi
山茱萸	Shanzhuyu	Asiatic cornelian cherry fruit	Fructus Corni
川牛膝	Chuanniuqi	Cyathula root; Achyranthes root	Radix Cyathulae
川贝母	Chuanbeimu	Tendrilleaf fritillary bulb	Bulbus Fritillariae cirrhosae
川芎	Chuanxiong	Szechwan lovage rhizome	Rhizoma Chuanxiong
马鹿胎	Malutai	Deer embryo	Fetus Cervi
马鹿茸	Malurong	Deer antlers	Cornu Cervi pantotrichum
马鹿骨	Malugu	Deer bone	Fel Cervi
丹参	Dansehn	Red sage root; Salvia root	Radix Salviae militiorrhizae
五加皮	Wujiapi	Slenderstyle acanthopanax bark	Cortex Acanthopanacis
五味子	Wuweizi	Chinese magnoliavine fruit, Schhisandra fruit	Fructus Schisandrae

(*continued*)

Table 16.2. Classification of raw materials for Chinese health foods (7,26,43,44) *(Cont.)*

Category II. Raw Materials Permitted as Health Food Ingredients			
Chinese Names	Mandarin Phonics	Common English Names	Pharmaceutical or Botanical Names
升麻	Shengma	Largetrifoliolious bugbane rhizome; Cimicifuga rhizome	Rhizoma Cimicifugae
天门冬	Tianmendong	Asparagus root	Radix Asparagi
天麻	Tianma	Gastrodia tuber	Rhizoma Gastrodiae
太子参	Taizishen	Heterophylly falsestarwort root; Pseudostellaria root	Radix Pseudostellariae
巴戟天	Bajitian	Morinda root	Radix Morindae officinalis
木香	Muxiang	Aucklandia root; Costus root	Radix Aucklandiae
木贼	Muzei	Scouring rush herb; Shave grass	Herba Equiseti hiemalis
牛蒡子	Niubangzi	Burdock fruit; Arctium fruit	Fructus Arctii
牛蒡根	Niubanggeng	Burdock root	Radix Arctii
车前子	Cheqianzi	Plantain seed	Semen Plantaginis
车前草	Cheqiancao	Plantain herb	Herba Plantaginis
北沙参	Beishashen	Coastal glehia root	Radix glehniae
平贝母	Pinbeimu	Ussuri fritillary bulb	Bulbus Fritillariae ussuriensis
玄参	Xuansehn	Figwort root; Scrophularia root	Radix Scrophularlae
生地黄	Shengdihuang	Fresh rehmannia root	Radix Rehmanniae
生何首乌	Shengheshouwu	Fresh fleeceflower root	Radix Polygoni multiflori
白及	Baiji	Bletilla tuber	Rhizoma Bletillae
白术	Baizhu	Largehead atractylodes rhizome	Rhizoma Atractylodis macrocephalae
白芍	Baishao	White peony root	Radix Paeoniae alba
白豆蔻	Baidoukou	Round cardamom seed	Semen Amomi rotundus
石决明	Shijueming	Sea-ear shell; Abalon shell; Haliotis shell	Concha Haliotidis
石斛（需提供可使用证明）	Shihu	Dendrobium	Herba Dendrobii
地骨皮	Digupi	Chinese wolfberry bark	Cortex Lycii
当归	Danggui	Chinese angelica root	Radix Angelicae sinensis
竹茹	Zhuru	Bamboo shavings	Caulis Bambusae in taeniam
红花	Honghua	Safflower	Flos Carthami
红景天	Hongjintian	Integripetal rhodiola grass	Herba Rhodiolae sacrae
西洋参	Xiyangshen	American ginseng	Radix Panacis quinquefolii
吴茱萸	Wuzhuyu	Evodia fruit	Fructus Evodiae
怀牛膝	Huainiuxi	Twoteethed Achyranthes root	Radix Achyranthis bidentatae
杜仲	Duzhong	Eucommia bark	Cortex Eucommiae
杜仲叶	Duzhongye	Eucomma leaf	Folium eucommiae
沙苑子	Shayuanzi	Flatstem milkvetch seed	Radix Astragali complanati
牡丹皮	Mudanpi	Peony tree bark; Mudan bark	Cortex Moutan
芦荟	Luhui	Aloe vera; Aloes	Herba Aloe *(continued)*

Table 16.2. Classification of raw materials for Chinese health foods (7,26,43,44) (*Cont.*)

Category II. Raw Materials Permitted as Health Food Ingredients			
Chinese Names	Mandarin Phonics	Common English Names	Pharmaceutical or Botanical Names
苍术	Cangzhu	Atractylodes rhizome	Rhizoma Atractylodis
补骨脂	Buguzhi	Malaytea scurfpea fruit; Psoralea fruit	Fructus Psoraleae
诃子	Hezi	Terminalia fruit	Fructus Chebulae
赤芍	Chishao	Red Peony root	Radix Paeoniae rubra
远志	Yuanzhi	Thinileaf milkwort root; Polygala root	Radix Polygalae
麦门冬	Maimendong	Ophiopogon tuber	Radix Ophiopogonis
龟甲	Guijia	Tortoise shell	Carapax et Plastrum testudinis
佩兰	Peilan	Fortune Eupatorium herb	Herba Eupatorii
侧柏叶	Chebaiye	Chinese Arborvitae twig and leaf; Biota tops	Cacumen Biotae
制大黄	Zhidahuang	Prepared Rhubarb	Radix et Rhizoma Rhei preparata
制何首乌	Zhiheshouwu	Prepared Fleeceflower root	Radix Polygoni Multiflori preparata
刺五加	Ciwujia	Thorny Acathopanax root	Radix Acanthopanacis senticosi
刺玫果	Cimeiguo	Dahurian rose fruit	Fructus Rosae davuricae
泽兰	Zelan	Bugleweed herb	Herba Lycopi
泽泻	Zexie	Water-plantain tuber; Alismatis rhizome	Rhizoma Alismatis
玫瑰花	Meiguihua	Rose flower	Flos Rosae rugosae
玫瑰茄	Meiguique	Roselle calyx	Calyx Hibisci sabdariffae
知母	Zhimu	Anemarrhena rhizome	Rhizoma Anemarrhenae
罗布麻	Luobuma	Dogbane leaf	Folium Apocyni veniti
苦丁茶	Kudingcha	Chinese holly leaf	Folium Llicis
金荞麦	Jingqiaomai	Fagopyrum cymosum seed	Semen Cymosum
金樱子	Jinyingzi	Cheroke rose hips; Roas fruit	Fructus Rosae laevigatae
青皮	Qingpi	Green tangerine peel	Pericarpium Citri Reticulatae viride
厚朴	Houpo	Magnolia bark	Cortex Magnoliae officinalis
厚朴花	Houpohua	Magnolia flower	Flos Magnoliae
姜黄	Jianghuang	Turmeric	Rhizoma Curcumac longae
枳壳	Zhiqiao	Dried green orange fruit	Fructus Aurantii
枳实	Zhishi	Immature bitter orange fruit	Fructus Aurantii immaturus
柏子仁	Baiziren	Arborvitae seed; Biota seed	Sermen Biotae
珍珠	Zhenzhu	Pearl	Concha Margarita
绞股蓝	Jiaogunan	Miracle grass, amachazuru	Herba Gynostenma pentaphylla
胡芦巴	Huhuba	Fenugreek seed	Semen Trigonellae
茜草	Qiancao	Indian madder root; Rubia root	Radix Rubiae
荜茇	Bibo	Long pepper	Fructus Piperis longi

(*continued*)

Table 16.2. Classification of raw materials for Chinese health foods (7,26,43,44) (Cont.)

Category II. Raw Materials Permitted as Health Food Ingredients

Chinese Names	Mandarin Phonics	Common English Names	Pharmaceutical or Botanical Names
韭菜子	Jiucaizi	Chinese chive seed	Semen Allii tuberosi
首乌藤	Shouwuteng	Fleece-flower stem	Caulis Polygoni multiflori
香附	Xiangfu	Nutgrass galingale rhizome; Cyperus tuber	Rhizoma Cyperi
骨碎补	Gusuibu	Fortunes' drynaria rhizome; Davallia	Rhizoma Drynariae
党参	Dangshen	Codonopsis; Pilos asiabell root	Radix Codonopsis pilosulae
桑白皮	Sangbaipi	White mulberry root-bark	Cortex Mori
桑枝	Sangzhis	Mulberry twig	Ramulus Mori
浙贝母	Zhebeimu	Thunberry fritillary bulb	Bulbus Fritillariae thunbergii
益母草	Yimucao	Motherwort herb	Herba Leonuri
积雪草	Jixuecao	Gotu kola, pennywort	Centella asiatica
淫羊藿	Yinyanghuo	Epimedium herb	Herba Epimedii
菟丝子	Tusizi	Dodder seed	Semen Cuscutae
野菊花	Yejuhua	Wild chrysanthemum flower	Flos Chrysanthemi indici
银杏叶	Yinxingye	Ginkgo leaf	Folium Ginkgo
黄芪	Huangqi	Milkvetch root	Radix Astragali
湖北贝母	Hubeibeimu	Hupeh fritillary bulb	Bulbus Fritillariae hupehensis
番泻叶	Fanxieye	Senna leaf	Folium Sennae
蛤蚧	Gejie	Gecko; Toad-headed lizard	Tokay Gecko
越橘	Yueju	Bilberry	Vaunium uligineosium. L
槐实	Huaishi	Japanese Pagodatree pod	Fructus Sophorae
蒲黄	Puhuang	Cat-tail pollen; Bullrush pollen	Pollen Typhae
蒺藜	Jili	Puncture-vine caltrop fruit	Fructus Tribuli
蜂胶	Fengjiao	Bee glue, hive dross	Propolis
酸角	Suanjiao	Tamarind fruit	Fructus Tamarindi indicae
墨旱莲	Mohanlian	Eclipta	Herba ecliptae
熟地黄	Shudihuang	Cooked Rhubarb, Cooked Rehmannia root	Radix et Rhizoma rhei, Radix Rehanniae
鳖甲	Biejia	Turtle shell	Carapax Trionycis

Category III. Raw Materials Forbidden as Health Food Ingredients

Chinese Names	Mandarin Phonics	Common English Names	Pharmaceutical or Botanical Names
八角莲	Bajiaolian	Root of Podophyllum emodi (Wall) Ying.	Rhizoma et Radix Dysosmae
八里麻	Balima	Chinese azalea fruit	Fructus Rhododendri mollis
千金子	Qianjinzi	Caper euphorbia seed	Semen euphorbiae lathyridis
土青木香	Tuqingmuxiang	Slender dutchmanspipe root	Radix Aristolochiae
山莨菪	Shanliangdang		Rhizoma Anisodus tanguticus
川乌	Chuanwu	Mother root of monkshood	Radix Aconiti carmichaeli
广防己	Guangfangji	Fangchi root	Radixaristolochiae Fangchi

(continued)

Table 16.2. Classification of raw materials for Chinese health foods (7,26,43,44) (*Cont.*)

Chinese Names	Mandarin Phonics	Common English Names	Pharmaceutical or Botanical Names
\multicolumn{4}{l}{**Category III. Raw Materials Forbidden as Health Food Ingredients**}
马桑叶	Masangye	Coryiaria sinica maxim leaf	Folium Sinica maxim
马钱子	Maqianzi	Nux vomica	Semen Strychni
六角莲	Liujiaolian	Same as Bajiaolian	Dysosma pleiantha (Hance) Woodson
天仙子	Tianxianzi	Henbane seed	Semen Hyoscyami
巴豆	Badou	Croton seed	Semen Crotonis
水银	Suiying	Quicksilver	Mercury
长春花	Changchunhua	Madagascar periwinkle grass	Herba Cathoranthi rosei
甘遂	Gansui	Kansui root	Radix Euphorbiae kansui
生天南星	Shengtiannanxing	Jack-in-the-Pulpit tuber; Fresh arisaema tuber	Rhizoma Arisaematis consanguineum
生半夏	Shengbanxia	Fresh pinellia tuber	Rhizoma Pinelliae ternate
生白附子	Shengbaifuzi	Fresh typhonium tuber	Rhizoma Typhoni giganteum
生狼毒	Shenglangdu	Fresh unbracteolated euphorbia root	Radix Euphorbiae ebractealatae
白降丹	Baijiangdan		Hydrangyrum chloratum compositum
石蒜	Shisuan	Spider lily	Herba Lycoris radiatae
关木通	Guanmutong	Manshuriensis dutchmanspipe	Caulis Aristolochiae
农吉痢	Nongjili	Crotalaria sessiliflora	Herba Crotalariae
夹竹桃	Jiazhutao	Oleander	Folium Nerium oleander
朱砂	Zhusha	Cinnabar	Cinnabaris; Red mercuric sulfide
米壳（罂粟壳）	Miqiao (Yingsuqiao)	Poppy capsule	Pericarpium Papaveris
红升丹	Hongshengdan		Red mercuric oxide
红豆杉	Hongdoushan	Chinese yew twig and leaf	Ramulus et Folium Taxi chinesis
红茴香	Honghuixiang	Henryi fruit	Fructus Illicium Henryi diels
红粉	Hongfen	Mercuric oxide	Radix Oxydum rubrum
羊角拗	Yangjiaohao	Divaricate strophanthus root	Radix Strophanthi divaricati
羊踯躅	Yangzhengshu		Rhododendron molle G. Don
丽江山慈姑	Lijiangsanchigu	Lijiang Appendiculate cremastra pseudobulb	Pseudobulbus Cremastrae seu Pleiones, Lijinag
京大戟	Jingdaji	Peking Euphorbia root	Radix Euphorbiae pekinensis
昆明山海棠	Kunmingsanhaitang		Tripterygium hypoglaucum Hutch
河豚	Hetun	Puffer, Globefish, Blowfish	Tetraodontiforms syn. Plectognathi
闹羊花	Naoyanghua	Chinese azalea flower	Flos Rhododendri mollis
青娘虫	Qingliangchong	Mung bean blister beetle	Lytta Caraganae pallas

(*continued*)

Table 16.2. Classification of raw materials for Chinese health foods (7,26,43,44) (*Cont.*)

Category III. Raw Materials Forbidden as Health Food Ingredients			
Chinese Names	Mandarin Phonics	Common English Names	Pharmaceutical or Botanical Names
鱼藤	Yuteng	Trifoliate jewelvine root	Radix seu Herba Derridis trifoliatae
洋地黄	Yandihuang	Foxglove	Folium Digitalis
洋金花	Yangjinhuan	Hindu datura flower	Flos Daturae metel
牵牛子	Qianniuzi	Pharbitis seed	Semen Pharbitidis nil
砒石（白砒、红砒、砒霜）	Pisi (Baipi, Hongpi, Pishuang)	Arsenolitum (Arsenic trioxide, ##, Arsenic]	Arsenolitum (Arsenicum trioxidum,##,Arsenic)
草乌	Caowu	Kusnezoffii root	Radix Aconiti kusnezoffii
香加皮（杠柳皮）	Xiangjiapi (Gangliupi)	Chinese Silkvine bark	Cortex Periplocae
骆驼蓬	Luotuopeng	Common peganum grass	Herba Pegani harmalae
鬼臼	Guiqiu	Common dysosma rhizome	Rhizoma dysosme versipellis
莽草	Mangcao	Lanceleaf anisetree leaf	Folium Illicii lanceolati
铁棒槌	Tiebangchui	Pendulous monkshood root	Radix Aconiti szechenyiani
铃兰	Linglan	Lily of the valley grass	Herba Convallariae majalis
雪上一枝蒿	Xueshangyizhihao	Short stalk monkshood root	Radix Aconiti brachypodi
黄花夹竹桃	Huanghuajiazhutao	Yellow oleander leaf	Folium Thevetiae
斑蝥	Banmao	Blister beetle	Mylabris
硫磺	Liuhuang	Sulfur	Sulfer
雄黄	Xionghuang	Realgar	Arsenic Disulfide
雷公藤	Leigongteng	Common three-wing-nut root	Radix Tripterygii wilfordii
颠茄	Dianque	Common atropa	Herba Atropae belladonnae
藜芦	Lihu	Black falsehellebore rhizome and root	Rhizoma et Radix Verairi
蟾酥	Zhansu	Dried toad secretion	Venenum Bufonis gargarizans

summarizes English translations of the three categories of raw materials. The directive further specifies that a health food cannot contain more than 14 plant and animal raw ingredients. If the raw materials were not from category I, the number of the raw ingredients would be limited to 4. If the raw materials were not from either category I or II, the number of the raw materials would be restricted to 1. Furthermore, the raw materials must pass toxicological safety assessment.

If a raw material is one of national protective plant and animal items, attention needs to be paid to relevant restrictions, and some special permits may be needed. Some of these items are as follows: Desert-living cistanche (*Herba cistanchis*), Licorice root (*Radix glycyrrhizae*), Ephedra (*Herba ephedrae*), and bear gall powder.

In 2003, the MOH revised the "Standards for Toxicological Assessment of Health Foods" (17). The standards established requirements for testing samples, preparation of testing samples, and contents of animal testing. In addition to actual animal test procedures, the standard also included considerations for toxicological assessment and some general principles on selection of toxicological tests and judgment of test results (12).

According to the category of the raw ingredients used for a health food and the processing technology, animal test requirements for the finished products could range from no need for toxicological testing to four tiers of tests. The test requirements for each function claim of health foods made from different sources of ingredients are summarized in Table 16.3. The four tiers of tests consist of:

1. Tier I: Acute toxicity (LD50, integrated LD50, maximum tolerated dose)
2. Tier II: genetic toxicity (Ames test or V79/HGPRT gene mutation test, mammalian bone marrow cell chromosome aberration test, TK gene mutation test, mice sperm abnormality test, mice testicle cells chromosome aberration test, dominant lethal test, unscheduled DNA synthesis test, sex-linked recessive lethal test), 30-day feeding study; traditional teratogenicity study
3. Tier III: sub-chronic toxicity: 90 day feeding study; reproductive study; metabolism study
4. Tier IV: Chronic toxicity (including carcinogenicity study)

Table 16.3. General principles for toxicological test selection for health foods (18)

Raw Material Origin	Process/ Consumption Method	Safety Assessment Requirement	
		Raw Ingredient	Finished Products
Common foods and Category I ingredients	Processing and consumption same as tradition.	Not required (NR)	NR
	Water extracted, dosage greater than regular.	NR	Tier 1 and some Tier 2 tests (3 mutation tests & 30 days feeding), teratogenicity study (if needed).
	Non-water extracts, dosage greater than regular.	NR	Tier 1, 2, 3 tests.
Approved nutrition supplements	Raw material source, process, and quality meet relevant national requirements.	NR	NR
Category II	—	—	Tier 1 and some Tier 2 tests (3 mutation tests and 30 days feeding), teratogenicity study and Tier 3 tests (if needed).
Other ingredients (Other than common foods, and Category 1 & II ingredients)	No historical human consumption data at all.	Tier 1, 2, 3, 4 tests.	Tier 1 and tier 2, some tier 3 tests (if needed).
	Some human consumption data from a few countries.	Tier 1, 2, 3 tests; tier 4 tests if needed).	Tier 1 and tier 2, some tier 3 tests (if needed).
	Widely consumed abroad, toxicity data exist.	Tier 1, 2 tests, further tier tests (if needed).	Tier 1 and tier 2, some tier 3 tests (if needed).

Functionality Evaluation

The MOH established function evaluation procedures in 1996 (19), then revised them in 2003 (17). The new procedures added human feeding trials to some function evaluation procedures. In addition to actual evaluation procedures for each function, the procedures set general principles for the functional assessment of health foods, and the rules and requirements for evaluation trials. The evaluation is based on results of animal and/or human feeding trials.

The animal experiments use either in-bred rats or mice of the same gender. The experimental design needs to include three dosage groups and one negative control group, and it may include one positive (model) control and blank control group if needed. At least one of the three dosage groups has to be five times the recommended dose for humans in the rodent experiments or ten times the recommended dose in tests conducted in mice, and each group requires 8–12 rats or 10–15 mice. The animal feeding tests normally last 30 days, and up to 45 days if necessary. However, the animal tests for several function claims require less than 30 days. The samples submitted for functional evaluation must be the final products that have passed the safety toxicology assessment and proven to be safe for consumption. A toxicological safety assessment report and a sanitary inspection report from the same lot of the sample have to be provided. To be suitable for the animal feeding administration, the samples may be further prepared through a water extraction or concentration.

Among the 27 function claims, 20 categories are required for human feeding trials whereas 22 categories require animal tests. Seven categories, such as enhancing immune function, enhancing sleep, alleviating physical fatigue, enhancing anoxia endurance, assisting in protection against irradiation hazard, increasing bone density, and assisting in protection against liver chemical injury, require only animal feeding trials.

The human feeding trials for health food functionality evaluation, in principle, are preceded by a positive confirmation of health effects derived from the animal tests. However, five categories—reducing eye fatigue, alleviating acne, eliminating skin pigmentation, improving skin moisture, and improving skin oil content—do not include precedent animal feeding tests. A self and parallel control will be employed in the human trial design. The valid number of subjects from each group (feeding and control) at the end of the trial must exceed 50, and the drop-out rate during the trial cannot exceed 20 percent. The feeding trial duration must be at least 30 days. Prior to the feeding trial, a protocol and schedule shall be submitted to the ethical committee for approval. A signed agreement has to be obtained from the volunteer subject after he or she has been communicated with regarding every aspect of the trial. Physical conditions, syndromes, and routine blood indices of the subjects should be closely monitored before and during the trial. The age and health conditions of the subjects vary with function evaluation procedures. Table 16.4 summarizes the experimental requirements for the animal and human feeding experiments of each function evaluation.

Submission of Application for Health Food Approval Certificate

Prior to submission for application, the following reports must be obtained from an authorized laboratory: report of toxicology safety assessment, report of functionality evaluation, analytical report of active ingredient, report of product stability study, and report of sani-

Table 16.4 Experimental requirements for functionality evaluation (18)

Functions	Animal Experiment		Human Feeding Trial	
	Experiment Design	Testing Parameters	Trial Design	Testing Parameters
Enhancing immune function	Animal: mice 18-22g. Duration: 30-45 days.	Body weight, organ/body weight ratio, cell immune function, fluid immune function, mono-nuclei phagocyte function, NK cell activity.	Not required (NR).	Not applicable (NA).
Assisting blood lipid reduction	Animal: adult male rats, + 1 high lipid control. Duration: 30-45 days.	Body weight, Serum total cholesterol (TC), triglycerides (TG), high density lipoprotein cholesterol (HPL-C).	Subjects: hyperlipemia patients (>18 & <65 years) (Serum mmol/L: TC≥5.2 or TG≥1.65). Duration: 30-45 days.	Serum TC, TG, HDL-C and respective reduction rate.
Assisting blood sugar reduction	Animal: adult mice or rats, + 1 high blood sugar model control group. Duration: 30-45 days.	Body weight, fasting blood sugar, sugar tolerance.	Subjects: diabetic II patients (>18 & <65 years)(sugar content mmo/L: fast blood ≥7.8 or post meal 2-hour blood ≥11.1). Duration: 30-45 days.	Sign, fast blood sugar, urine sugar, 2-hour PP (postprandial) blood sugar, serum TC, TG, HDL-C.
Anti-oxidative function	Animals: old rats (>12 months) or adult mice (8-12 months), + 1 model control. Duration: 30-60 days.	Body weight, serum & tissue malonaldehyde (MAD), lipofasci, superoxide dismutase (SOD), glutathion peroxidase (GSH-Px).	Subjects: healthy adults (>45 & <65 years). Duration: 3-6 months.	Serum MAD, SOD, GSH-Px.
Assisting in memory improvement	Animal: weaned of adult mice or rats. Duration: 30-45 days.	Jumping test, dark avoiding test, bi-directional avoiding test, water maze test.	Subjects: with similar background. Duration: 30-45 days.	Directional memory, associating, memory, image free recollection, non-significant image re-recognition, human characteristic connection recollection, memory quality.

(continued)

Table 16.4 Experimental requirements for functionality evaluation (18) (Cont.)

Functions	Animal Experiment		Human Feeding Trial	
	Experiment Design	Testing Parameters	Trial Design	Testing Parameters
Alleviating eye fatigue	NR	NA	Subjects: children, teenagers, adults with eye fatigue. Duration: 30-45 days.	Eye syndrome improvement (eye aching, eye swelling, intolerance of light, sight cloudiness, eye dryness), clear vision durability, hyperopia ability.
Facilitating lead excretion	Animal: adult mice or rats, blank & model control groups. Duration: 30-45 days.	Body weight, blood lead, bone lead, liver tissue lead.	Subjects: children (blood lead >100ug/L) or adults (>200ug/L) with high blood lead. Duration: 30-45 days.	Blood lead, urine lead, urine calcium, urine zinc.
Thinning throat mucus	Animal: rats 150-220g. Duration: 30-45 days.	Rat cotton ball embedding test, rat foot finger swelling test.	Subject: chronic pharyngitis patients (>18 & <65 years). Duration: 15-30 days.	Throat syndrome, signs.
Assisting hypertension alleviation	Animal: rats with hypertension, 10-12 weeks. Duration: 30-45 days.	Body weight, blood pressure, heart rate.	Subjects: hypertension patients (>18 & <65 years) (SYS≥140mmHg, DIA≥90mmHg). Duration: 30-45 days.	Clinical syndrome & signs, blood pressure, heart rate.
Enhancing sleep	Animal: adult mice. Duration: 30-45 days.	Body weight, prolongation of sleeping time from barbital sodium, hypnotizing test with sub-threshold dosage of barbital sodium, latency barbital sodium hypnotizing.	NR	NA
Facilitating lactation	Animal: mother (the second time birth givers & within 3 days) and baby (<3 days old) mice or rats. Duration: 15-30 days.	Weekly body weight of mother & baby rats or mice.	Subjects: hypogalactia mothers (<35 years old). Duration: 7-15 days.	Degree of breast swelling, milk quantity, milk protein content.

Table 16.4 Experimental requirements for functionality evaluation (18) (*Cont.*)

Functions	Animal Experiment		Human Feeding Trial	
	Experiment Design	Testing Parameters	Trial Design	Testing Parameters
Alleviating physical fatigue	Animal: adult mice or rats. Duration: 30-45 days.	Swim with load test, blood lactic acid, serum carbamide, liver or muscle glycogen.	NR	NA
Enhancing anoxia endurance	Animal: adult mice. Duration: 30-45 days.	Body weight, atmospheric anoxia endurance test, survival test from sodium nitrite poisoning, anoxia test against acute brain hypoblood.	NR	NA
Assisting in protection against irradiative hazard	Animal: mice 18-22g Duration: 14-30 days pre-irradiation, continue after irradiation up to 45 days.	Body weight, white cell count of peripherial blood, DNA content of marrow cell, micronucleus test of marrow cells, blood/tissue SOD activity, serum hematolytic agent content.	NR	NA
Weight reduction	Animal: male rats 100-180g, +1 model control. Duration: 30-45 days.	Body weight, food intake, inra-fat mass (fat pads around testicle & kidney), fat/body weight ratio, foods absorption rate.	Subjects: simple obese patients (BMI≥30, male fat >25%, female fat >30%). Duration: 35-60 days.	Body weight, waist & buttocks perimeters, intra-fat content, blood uric acid, urine ketones.
Enhancing child growth and development	Animal: weaned rats. Duration: 42-56 days.	Body weight, body length, foods absorption rate.	Subjects: under developed healthy children (6-10 years old). Duration: 3-6 months.	Height, body weight, chest perimeter, upper arm perimeter, internal fat content.

(*continued*)

Table 16.4 Experimental requirements for functionality evaluation (18) (Cont.)

Functions	Animal Experiment		Human Feeding Trial	
	Experiment Design	Testing Parameters	Trial Design	Testing Parameters
Increasing bone density	Animal: protocol 1 (for calcium dominant samples): weaned rats (4 weeks old), + 1 low calcium control & 1 calcium carbonate control; protocol 2 (for other samples): female adult Wistar rats with ovary removed. Duration: 3 months for both.	Body weight, bone calcium content, bone density.	NR	NA
Alleviating nutritional anemia	Animal: newly weaned rats, + 1 low iron control. Duration: 30-45 days.	Body weight, hemoglobin, free erythrocyte protoporphyrin content.	Subjects: low hemoglobin anemia adults and children (Hb g/L: adult male ≤130, female ≤120, children under 6≤110, children of 7-18≤120) Duration: 30-120 days.	Hemoglobin, hemo-ferric globin, free erythrocyte protoporphyrin/serum iron transferring globin ratio.
Assisting in protection against liver chemical injury	Animal: adult rats or mice for protocol 1 (carbon chloride liver injury model) & protocol 2 (alcohol liver injury model), + 1 model control. Duration: 30-45 days.	Protocol 1: body weight, alanine aminotrasferase (ALT), aspartate aminotransferease (AST), liver & tissue pathological check; protocol 2: body weight, MAD, Reduced GSH, TG, liver tissue pathological check.	NR	NA
Alleviating acne	NR	NA	Subjects: male or female volunteers with acne I-III (>14 & <65 years). Duration: 30-45 days.	Number of acnes, degree of skin damaging, skin oil content.

Table 16.4 Experimental requirements for functionality evaluation (18) (Cont.)

Functions	Animal Experiment		Human Feeding Trial	
	Experiment Design	Testing Parameters	Trial Design	Testing Parameters
Eliminating skin pigmentation	NR	NA	Subjects: volunteers with facial pigmentation (>18 & <65 years). Duration: 30-45 days.	pigmentation area, pigmentation color.
Improving skin moisture	NR	NA	Subjects: 30-50 years, skin moisture ≦12. Duration: 30-45 days	Skin moisture.
Improving skin oil content	NR	NA	Subjects: 30-50 years, skin oil content ≦10 or ≧27. Duration: 30-45 days.	Skin oil content.
Regulating gastrointestinal flora	Animal: mice 18-22g. Duration: 14-30 days.	Body weight, Bifidobacteria, Lactobacilli, Enterococci, Coliforms, Clostridium perfringens.	Subjects: no gastrointestinal diseases or not using antibody within a month (<65). Duration: 14-30 days.	Bifidobacteria, Lactobacilli, Enterococci, Coliforms, Pseudobacilli, Clostridium perfringens.
Facilitating digestion	Animal: adult rats or mice. Duration: 15-45 days.	Body weight, body weight increase, food intake, food absorption rate, small intestine movement test, digestive enzymes.	Subjects: children protocol: 4-10 years old, low body weight due to inadequate; adult protocol: adults with functional dyspepsia. Duration: 30-45 days for both.	Children protocol: appetite, daily intake, partiality for certain foods, body weight, hemoglobin; Adults protocol: clinical syndromes, gastro-intestinal movement test.
Alleviating constipation	Animal: adult male mice, +1 blank & 1 model control. Duration: 7-15 days.	Body weight, small intestine movement test, time for the first bowel movement, feces weight, number of feces, feces appearance.	Subjects: < 3 bowels per week. Duration: 7-15 days.	Syndrome and sign, feces appearance, bowel movements/day, bowel movement condition.
Assisting in protection against gastric mucosa injury	Animal: Wistar or SD rats, 160-180g. Duration: 14-45 days.	Body weight, degree of gastric mucosa injury.	Subjects: chronic gastritic patients (>18 & < 65 years). Duration: 30-45 days.	Clinical syndromes, gastroscopic observation and sign.

tary inspection. Only authorized laboratories are permitted to conduct these tests and to issue reports. There are currently more than 30 such laboratories around the country authorized by the MOH. Not every laboratory can conduct all tests. Each laboratory is authorized to evaluate certain functions according to its capacity (38). The imported health (functional) foods can be tested only by the Institute of Nutrition and Food Safety, the Chinese Center for Disease Control in Beijing.

Upon completion of all laboratory tests and obtaining the reports discussed previously, a manufacturer and an importer or an entrusted agent of a foreign functional food manufacturer can submit an application for the imported health food approval certificate. In addition to the aforementioned reports, an applicant shall include the following forms, information, or materials in its application (21):

1. Hygiene Permit Application Form for Imported Health (Functional) Foods
2. Product formulation and its relevant scientific evidence
3. Name and content of active ingredient(s) and the analytical procedure for the ingredients
4. Processing techniques and flowchart
5. Product quality specification (industry standard)
6. Inspection reports (as stated previously) issued by the authorized laboratory
7. Product packaging design (including product label)
8. Product manual or description
9. An entrust contract, if an applicant is an entrustee
10. Evidence documents such as product sale permit issued by relevant authority from the manufacturer's country
11. Other relevant data in support of the approval processing

Each document should include one original copy and 13 photocopies. Along with the application, three sealed product samples of small packages should be submitted. Each page of application materials except the application form and laboratory reports issued by an authorized laboratory shall be stamped with the applicant's company seal. If an applicant is applying for more than one product, it should submit an application for each product separately.

When the application for the same health (functional) foods is submitted jointly by two or more partners, the Approval Certificate of Health (Functional) Foods must be signed jointly. However, the certificate shall be issued only to one holder who is determined by all partners to have the responsibility. In addition to all the materials required of a regular application, the joint application shall include a letter of recommendation for the sole holder of responsibility, signed by all users of the certificate.

If an application for an imported health food is submitted through an entrusted entity, an entrusted contract (the Contract) shall be submitted together. The entrusted contract must meet the following requirements:

1. Each product must include one original entrusted contract.
2. The contract must include names of the entrustee and the company that issued the contract, name of product, entrusted content, and the date the contract was signed.
3. The contract must include the seal of the entrusting company or a signature from the legal representative of the company.

4. The entrusting company in the contract must be the same as the manufacturing entity in the application.
5. The entrustee in the contract must be the same as the applicant.
6. The product name stated in the contract should be the same as the name in the application form.
7. If a valid period is specified in the contract, the application date shall be within the period.
8. If the entrustee is assigning the contract to another entity for application, a permit from the manufacturing entity needs to be submitted.
9. The Chinese translation of the contract must be notarized by a Chinese notary.

The evidence documents for imported health foods such as the permits for production and sales issued by the country or region of manufacturer must meet the following requirements:

1. One original document for each product. If the original document can't be provided, the document has to be authorized by the entity issuing the document or by a Chinese Embassy or Consulate from the production country.
2. The document shall include name of the issuing organization, the name of the manufacturing company, the product name, and the date of the document signed.
3. The document must be issued by a governmental administrative authority, an industrial association, or an inspection entity authorized by the governmental authority of the production country.
4. The document should include the issuing entity's seal or a signature from a legal representative.
5. The name of the manufacturing entity and the product should be consistent with the content in the application.
6. If the document has a specified valid period, the date for application should be within the period.
7. The Chinese translation of the document should be notarized by a Chinese notary.

Assessment Procedure of the Health Food Evaluation and Approval Committee

An application for imported health food should be submitted to the Division of Traditional Medicine Protection of the SFDA. The Division will organize a quarterly evaluation and approval committee meeting in the last two weeks of each quarter. Any applications accepted by the end of the second month of the quarter will be evaluated in the committee meeting. The committee chairman or vice-chairman will conduct the meeting, and only a meeting with more than two-thirds of the committee in attendance is valid because any approvals must be agreed upon by two-thirds of the committee. Within 30 days of the evaluation meeting, the committee will grant an approval or issue a refusal letter to applicants (3,21,31).

The Committees will evaluate each applicant based on each of the following categories:

1. Verification of the health food name: As specified in the regulation and directive for health food labeling, health food names must be accurate and scientifically sound. Names of people and places, and symbols or any exaggerating or misleading names,

are not allowed. The name of a nonprimary ingredient cannot be used as a health food product name (24-28).
2. Verification of the application forms and materials: Application Form for Health Foods or Application Form for Imported Health Foods should be completed and filed in handwritten form or by typewriter and cannot be filed by copying and pasting. The contents must be legible. For the imported health foods, all application materials in foreign languages (except for address) need to be translated into Chinese. The translated application, together with the original language, will be submitted.
3. Assessment of the health food formulation: Formulation must be accurate with supportive evidence. For a health food using fungus, probiotics, or plant or animal raw materials as ingredients, a relevant report or certificate shall be submitted. For the antifatigue or enhancing child growth and development function claim application, an analytical report for stimulant and hormones shall be included.
4. Evaluation of the manufacturing technique: The processing must be in compliance with food industry good manufacturing practice and other, related hygiene requirements. The process should not affect, damage, or change active ingredients stability or produce harmful intermediates. The processing information should include the preparation of each ingredient, the processing of the finished product, and its major technical parameters.
5. Assessment of the quality standards: An application should include a manufacturing specification covering raw materials, raw material sources, and finished product. A standard and its qualitative and/or quantitative procedure for active ingredients should be established and submitted. If national standards exist for such ingredients or finished products, they should be followed.
6. Evaluation of the toxicology safety assessment reports: Toxicological safety evaluation shall follow the Procedure for Toxicological Assessment of Health Food issued by the MOH as outlined in a previous section of this chapter, "Safety Control of Raw Materials and Finished Products."
7. Evaluation of the health food functionality assessment reports: Health food functionality assessment must follow the Procedures for Functional Assessment of Health Foods issued by the Ministry of Health. The functionality assessment can be conducted only by the agencies authorized by the MOH. The Institute of Nutrition and Food Safety is the only entity authorized to conduct functional and toxicological assessment and analysis for imported health foods. For the functionality assessment requirements, please refer to the previous section, "Functionality Evaluation."
8. Assessment of the active ingredient information: An application should include an analytical report of the active ingredient no matter whether there are single or multiple active ingredients. However, if an active ingredient cannot be identified under current technology, it can be substituted by submitting a list of ingredients and respective contents for the ingredients related to the claimed function.
9. Evaluation of the product stability data: An applicant for certificate of health food approval needs to provide the product stability data that includes information about testing procedures, data, conclusions, and so on. A stability study can be conducted in the following manner: placing a packaged product under 37–40°C and relative humidity of 75 percent for three months, and then pulling samples every month to test for quality parameters representing quality of the product. This accelerated experiment can provide an equivalent of two years of shelf life. The study should cover the products

made from three different lots. If conditions permit, one year of ambient storage stability should be conducted.

10. Evaluation of the product hygiene inspection reports: All hygiene inspection reports must come from the health administrative authorities or laboratories at a province or higher level. The Institute of Nutrition and Food Safety is the only agency that can issue a sanitary inspection report for the imported health foods. The report should cover the product samples made from the most recent three lots. All samples must comply with relevant national standards, or industry standards if no national standards are available.

11. Verification of the product label and manual: The health food product labels and manuals must comply with the relevant national standards and requirements. The Administrative Regulations for Health Foods and the Directive of Labeling Guideline for Health Foods established a series of rules for health food labeling, product manual, and advertisement (3,28–30). In essence, a food without an approval certificate cannot, according to the regulations, make any function claims on its label, in manuals, or in advertisements. The content for health food labels, manuals, and advertisements must be accurate and true and cannot imply any disease-healing propaganda. The content should include: (a) the health promotion claim and target population; (b) consumption method and recommended dosage; (c) storage method; (d) name and content of active ingredients or function-related ingredients; (e) health food approval certificate number; (f) health food symbol; and (g) other labeling contents provided by the relevant standards and requirements. The directive has set guidance for the health food label design as well (28–30).

The evaluation committee appointed by the SFDA from the expert pool holds four evaluation meetings annually, normally in the last month of each quarter (21). The pool consists of the experts from food hygiene, nutrition, toxicology, medicine, and other related professions around the country. If the evaluation committee determines that a health food from an application needs to be reexamined, reexamination should be conducted by the examination entity appointed by the SFDA. The cost of the reexamination will be borne by the health (functional) food applicant.

The SFDA shall issue an Approval Certificate of Imported Health (Functional) Foods to the imported health (functional) foods that pass examination. The products that have received an Approval Certificate of Imported Health (Functional) Foods shall have the approval letter, number, and the health (functional) foods symbol designed by the MOH marked on the package. The food hygiene control and inspection agency from an entry port shall inspect the presented Approval Certificate of Imported Health (Functional) Foods. Upon verification that the product is qualified, the agency will grant an import entry.

The holder of an Approval Certificate of Health (Functional) Foods or Approval Certificate of Imported Health (Functional) Foods can engage in technology transfer or cooperative manufacturing with other parties on the basis of the certificate. For such technology transfer, the holder and the transferee need to jointly apply to the MOH to have a duplicate of the Approval Certificate of Health (Functional) Foods issued. To receive a duplicate, the holder needs to present the existing approval certificate and a valid technology transfer contract. The duplicate certificate will be issued to the transferee who in turn has no right to make further transfer (22).

Production and Sale of Health (Functional) Foods

Before the production of a health (functional) food with an approval certificate, a food manufacturing company needs to apply for a production permit from the local provincial health administrative authority with direct jurisdiction. The provincial health authority will conduct an examination and issue a permit such as "xxx Health Food (Functional) Food" to attach onto the manufacturer's Hygiene License upon approval, before production can commence. The following materials shall be submitted when applying for the production permit:

1. The manufacturer's valid hygiene license for food production issued by the health administrative authority with direct jurisdiction
2. The original or a duplicate of the Approval Certificate of Health (Functional) Foods
3. The health (functional) foods specification and sanitary practice procedure established by the manufacturing company as included in the certificate application, and the relevant explanation for establishment of the specification and the procedure
4. In the case of technology transfer or cooperative manufacturing, the submission of a valid contract for technology transfer or cooperative manufacture signed by the holder of the Approval Certificate of Health (Functional) Foods
5. Information on the production facility conditions, production technical personnel, and quality control system
6. Quality and hygiene inspection reports for three lots of products

The health food products that are not examined and approved by the MOH cannot be manufactured under the name of health (functional) foods. The enterprises that are not examined and approved by provincial health administration authorities are not allowed to manufacture health (functional) foods.

As do any other food companies, the health food companies need to have a Good Manufacturing Practices (GMPs) and Hazard Analysis of Critical Control Point (HACCP) in place as required by the recently issued directives (34,39,40). In addition, the health food manufacturers must also follow the approved content to produce health (functional) foods and are not allowed to change the product formulation, the processing technique, or the product quality standard established by the applicant, name, label, and product manual. The production procedure and conditions should conform to the sanitary practice and other hygiene requirements for the relevant food industry. The processing technique shall be able to maintain stability of the active ingredients in the product. The processing should not cause loss, destruction, or conversion to the active ingredients, or produce harmful intermediates during the production.

The packaging used should be in good condition. The packaging materials or containers in direct contact with the health (functional) foods shall conform to the relevant hygiene standards or requirements. They should be beneficial to the active ingredients stability preservation of the health (functional) foods.

The health food dealers or retailers should ask for a copy of the Approval Certificate of Health (Functional) Foods and relevant product inspection reports. To purchase the imported health (functional) foods, they should also ask for a copy of the Approval Certificate of Imported Health (Functional) Foods and the certificate of inspection from the relevant entry port agency for food hygiene control and inspection.

General Hygiene Standards of Health Foods

In addition to the labeling provisions established for general foods and special nutritional foods by different ministries, the MOH issued the "Provisions for Labeling of Health Foods" in 1996 (20). The provisions provide some additional and specific requirements for health food labeling. They cover the requirements for raw ingredients, finished product hygiene standards, and packaging and labeling.

In terms of the function claims, health (functional) foods are classified into 27 categories as listed in Table 16.1. But based on chemical structures of active ingredients, all health foods basically fall into the following classes:

- Polysaccharides: dietary fiber, lentinan and so on
- Functional sweeteners: mono-saccharide, oligose, and poly-glycitol and so on
- Functional fats (fatty acids): poly-unsaturated fatty acids, phospholipids, choline and so on
- Free radical quenchers: superoxide dismutase (SOD), glutathione peroxidase and so on
- Vitamins: vitamin A, vitamin E, and vitamin C and so on
- Peptides and proteins: glutathione, immunoglobulin and so on
- Probiotics: lactobacillus, bifidobacteria and so on
- Trace elements: selenium, zinc and so on
- Others: octacosyl alcohol, phytosterols, saponins and so on

In general, all health (functional) foods must pass scientific evaluation through qualitative and quantitative composition analysis and functional testing in animal studies or human feeding trials. They have to contain effective functional components that have stable and noticeable body function regulation. A health (functional) food should have at least one function regulating the human body, and it should contain the corresponding functional ingredient(s) with a minimum effective dosage. If necessary, the effective ingredient(s) should be controlled within a maximum limit. The formulation and manufacturing method of health (functional) foods should be scientifically sound.

The health (functional) foods should have suitable appearance, color, odor, taste, and texture. They should be free of any unpleasant or unacceptable odor and taste. The usage of food additives or nutritional fortification substances should comply with the dosage requirements for food additives or nutritional fortification substances. The food additives used should comply with the corresponding national standards or trade standards. The raw materials and supplementary ingredients should comply with the corresponding national standards, trade standards, or relevant regulations. The health (functional) foods for infants and lactating women can contain neither stimulants nor hormones. The health (functional) foods for athletes shall not contain any prohibited drugs. Pesticides, veterinary drugs, radioactive substances, and biological toxin residue limits shall comply with the corresponding national standards.

The health (functional) foods should be in compliance with the limits established in the requirements for the national hygiene standards. Any health (functional) foods that do not have a limit in the national standards should comply with the limits of lead, arsenic, and mercury as listed in Table 16.5. Any health (functional) foods that do not have a microbiological limit in the national standard shall comply with the limits as listed in Table 16.6 as determined by physical properties of the product.

Table 16.5. Standard Limits of Heavy Metals for Health Foods (20).

Heavy Metals	Allowable Limits (≤mg/kg)	
	General Product	Specific Product
Lead	0.5	1.5 for capsules; 2.0 for solid beverages and capsules containing algae or tea.
Arsenic	0.3	1.0 for solid beverages containing algae or tea and all capsules.
Mercury	-	0.3 for solid beverages containing algae or tea and all the capsule.

Table 16.6. Microbiological Limits for Health Foods (20).

	Limits			
	Liquid		Solid or semi	Solid
Parameters	Protein ≤0%	Protein < 1.0%	Protein ≤0%	Protein < 4.0%
Total bacterial count, cfu/g or mL ≤	1,000	100	30,000	1,000
Coliform, MPN/100g or 100mL ≤	40	6	90	40
Mold, cfu/g or mL ≤	10	10	25	25
Yeast, cfu/g or mL ≤	10	10	25	25
Pathogen (enteropathogenic bacteria and pathogenic coccus)	Non detected			

Control and Supervision of Health (Functional) Foods

According to the Food Hygiene Law and the regulations and standards issued by the MOH, the health administrative authorities at various levels should strengthen controlling, monitoring, and supervision of the health (functional) foods. The MOH has a right to conduct random inspection for any approved health (functional) foods and release inspection results to the public.

Under the following circumstances, the MOH preserves the right to reexamine any approved health foods:

1. As a result of scientific advancement, the understanding of the previously approved health effect has changed.
2. The formulation, production techniques, and health function of an approved health food are suspected of alteration.
3. When tightening for supervision and monitoring of health foods is necessary.

For products that fail the re-examination or refuse the re-examination, the MOH shall revoke the Approval Certificate of Health (Functional) Foods. For the products that pass re-examination, the approval certificate will remain valid.

Several random inspections for the health food products in the China market and the health food manufacturers throughout the country have been conducted in the past through integrated efforts from multiple ministries. Quite a few health foods have been suspended from production. Some examples of violations are as follows: exaggerating effects in the product manual, using stimulants in the weight reduction health foods, use of forbidden ingredients.

Provisions of Penalty

The health administrative authorities of the local governments above a county level can impose penalties pursuant to Article 45 of the Food Hygiene Law in any of the following cases:

- Production and sale of products under the name of health (functional) foods without the approval certificate issued by the MOH as specified in the regulation
- Sale of products under the name of health (functional) foods without a permit for imported health (functional) foods
- Name, label, and manual not used in accordance with the approved contents

Article 45 of the Food Hygiene Law states that a violator who produces a food claiming to have a specific health function without an approval certificate from a health administrative authority of the State Council, or who falsifies the content in its product description and manual, will be suspended from production. The violator will lose the earnings made from the violated action and be fined as much as one to five times the amount of the earnings. If no earnings were made from the violated action a penalty of 1,000 to 50,000 Yuan dollars will be imposed. Serious violators will have their Hygiene License revoked.

Article 18 of the Provisions for Administrative Penalty over Food Hygiene further defines that an entity that produces a health food under a revoked Approval Certificate for Health Food or Approval Certificate for Imported Health Food will be treated as a violator without any certificates (2,41,42).

If healing effects are promoted in the advertisements of any health (functional) foods, or superstitious beliefs are used in health (functional) foods publications, a penalty shall be imposed pursuant to the relevant regulations in the Transitional Provisions for Food Advertising issued by the State Administration for Industry and Commerce and the MOH. A penalty shall be imposed following the relevant regulations, in case of violation of the Food Hygiene Law and any other applicable hygiene requirements.

Conclusion

More than 4,000 health food products had been approved by the end of 2002 but less than half that number had actually been put into the marketplace. Among the approved products, those regulating immune function, assisting in blood lipid reduction, and used to counter fatigue accounted for about 60 percent, whereas nutrition supplements represented about 10 percent. The massive number of health foods available in the market created problems of supervision for the health authorities and created confusion for consumers. The Chinese health officials and professionals are addressing the issues. The recently revised Technical Standards for Testing and Assessment of Health Foods is an example. The regulations are in the process of revision. The expected revision may expand or make adjustments in the following areas (10): (1) scope of the preliminary evaluation; (2) assessment and inspection of production conditions; (3) implementation of an expiration date for the approval certificate; (4) authorization of a testing laboratory and refinement of its responsibility; (5) re-examination of the approved health food products.

References

1. Zhonglian Jing, August, 2002, Development Trend of Chinese Functional Foods, presented at International Seminar on Technology and Development of Functional Food, organized by the China Institute of Food Science and Technology, Xi'an.

2. China State Council, 1995, Food Hygiene Law of the People's Republic of China, National Law, Promulgation Date: 1995-10-30; Effective Date: 1995-10-30.

3. MOH, 1996, Administrative Regulations for Health Food, Order No. 46 (Promulgation Date: 1996-03-15; Effective Date: 1996-06-01).

4. Yin Dai & Xueyun Luo, 1996, Functional Food in China, Nutritional Reviews, Vol. 54, No. 11, 21–23.

5. Weijian Weng & Junshi Chen, 1996, The Eastern Perspective on Functional Foods Based on Traditional Chinese Medicine, Nutrition Reviews. Vol. 54, No. 11, 11–16.

6. Arai, S. 2002, Global view on functional foods, Asian perspectives, British Journal of Nutrition, 88, Suppl. 2, 139–143.

7. Brian Tomlinson, Thomas Y.K. Chan, Juliana C.N. Chan, Julian A.J.H. Critchley, and Paul P.H. But, 2000, Toxicity of Complementary Therapies: An Eastern Perspective, Journal of Clinical Pharmacology 2000, 40:451–456.

8. Foreign Agricultural Service, the United States Department of Agriculture, 2001, China, People's Republic of Food and Agricultural Import Regulations and Standards—Administration Regulation for Health Foods, an unofficial English Translation, GAIN Report #CH1050, http://www.atoshanghai.org.

9. Foreign Agricultural Service, the United States Department of Agriculture, 2001, China, People's Republic of Food and Agricultural Import Regulations and Standards, Health Food Standards 2001, an unofficial English Translation, GAIN Report #CH1049, http://www.atoshanghai.org.

10. Jiangsheng Huang, November, 2002, Current Situation and Development Trend of Health Food, presented at the First Professional Symposium on Nutrition and Health Food, organized by the Division of Nutrition and Health Food, Chinese Nutrition Society, Hainan.

11. Yi Tang, November, 2002, Current Administration Status of Nutrient Supplements in China, presented at the First Professional Symposium on Nutrition and Health Food, organized by the Division of Nutrition and Health Food, Chinese Nutrition Society, Hainan.

12. Xiaoqiang Gao, November, 2002, Common Problems Encountered in Application Submission for Health Food Approval Certificate, presented at the First Professional Symposium on Nutrition and Health Food, organized by the Division of Nutrition and Health Food, Chinese Nutrition Society, Hainan.

13. Yuexin Yang, November, 2002, Development of Nutrition and Health Food Standard and Its Impact on Social Economy, presented at the First Professional Symposium on Nutrition and Health Food, organized by the Division of Nutrition and Health Food, Chinese Nutrition Society, Hainan.

14. Weixin Yan, 2003, Considerations on Function Identification of Functional Foods in China, presented at the Functional Food Forum: Science and Development in Beijing, organized by ILSI, China Focal Point.

15. MOH: 2003, Public Notice No.17, 2003 (Directive; Public Notice No. 17, 2003; Promulgation Date: 2003-06-12; Effective Date: 2003-10-10).

16. SFDA, 2003, Announcement on Issues Concerning New Application Form for Healthy Food (Directive; Document No.: GuoShiYaoJianZhu[2003]291; Promulgation Date: 2003-10-27).

17. MOH, The People's Republic of China, 2003, Technical Standards for Testing and Assessment of Health Food, WeiFaJianFa (2003) No. 42. (Promulgation Date: 2003-2-14; Effective Date: 2003-5-1).

18. Jianxian Zheng, 1999, Functional Foods, China Light Industry Publisher, Volume 2, p. 1–49.

19. MOH: Evaluation Procedures and Test Methods Used in the Functional Analysis of Healthy Food, 1996 (Directive; Document No.: WeiJianFa[1996]38; Promulgation Date: 1996-07-18; Effective Date: 1996-07-18).

20. MOH, 1996, General Hygiene Requirements for Healthy Food (Directive; Document No.: WeiJianFa[1996]38; Promulgation Date: 1996-07-18; Effective Date: 1996-07-18).

21. MOH, 1996, Operating Procedures and Technique Requirements for the Review of Healthy Food (Directive; Document No.: WeiJianFa[1996]38; Promulgation Date: 1996-07-18; Effective Date: 1996-07-18).

22. MOH, 2001, Notification on Standardizing Technique Transfer of Health Food (Directive; Document No.: WeiFaJianFa[2001]71; Promulgation Date: 2001-03-08).

23. MOH, 2001, Notification on Printing and Distributing Provisions for the Review of Healthy Food Containing Fungi and Probiotics (Directive; Document No.: WeiFaJianFa[2001]84; Promulgation Date: 2001-03-23).

24. MOH, 2001, Notification on Restricting the Manufacture of Healthy Food Using Wild Animals and Plants as Raw Materials, (Directive; Document No.: WeiFaJianFa[2001]160; Promulgation Date: 2001-06-07).

25. MOH, 2001, Notification on Restricting the Manufacture of Healthy Food Containing Licorice Roots Northwest Origin and other Herbs (Directive; Document No.: WeiFaJianFa[2001]188; Promulgation Date: 2001-06).

26. MOH, 2002, Notification on Further Standardizing the Management of Raw Materials for Healthy Food (Directive; Document No.: WeiFaJianFa[2002]51; Promulgation Date: 2002-02-28).

27. MOH, 2002, Notification on Printing and Distributing Provisions for the Review of Healthy Food Containing Enzyme Preparation (Directive; Document No.: WeiFaJianFa[2002]100; Promulgation Date: 2002-04-14).

28. MOH: Provisions for Healthy Food Labeling, 1996, (Directive; Document No.: WeiJianFa[1996]38; Promulgation Date: 1996-07-18; Effective Date: 1996-07-18).

29. MOH, 2002, Reply on Several Issues Concerning Adding Hygiene Certificate Number to Health Food Labeling (Directive; Document No.: WeiFaJianFa[2002]319; Promulgation Date: 2002-12-18).

30. MOH: Notification of Printing and Distributing Measures for Denominating Health Related Products of the Ministry of Health (Directive; Document No.: WeiFaJianFa [2001] 109; Promulgation Date: 2001-04-11).

31. MOH: Notification on Printing and Distributing Measures for Application and Review of Health Related Products (Directive; Document No.: WeiFaJianFa [1999] 150; Promulgation Date: 1999-04-13).

32. MOH: Notification on Matters Concerning Further Standardizing Supervision and Administration on Health Related Products (Directive; Document No.: WeiFaJianFa [2003] 1; Promulgation Date: 2003-01-03).

33. State Administration for Industry and Trade, Ministry of Health: Notification on Strengthening Supervision and Control over Healthy Food Advertising (Directive; Document No.: GongShangGuangZi[2000] 257; Promulgation Date: 2000-10-31).

34. MOH, 2002, Notification on Printing and Distributing HACCP Implement Guidance for Food Industry (Directive; Document No.: WeiFaJianFa[2002]174; Promulgation Date: 2002-07-19).

35. MOH, 1990, Administrative Provisions for New Resource Food Hygiene (MOH regulation; Order No. 4; Promulgation Date: 1990-07-28; Effective Date: 1990-07-28).

36. State Administration for Industry and Commerce: Transitional Provisions for Food Advertising (State Administration for Industry and Commerce regulation; Order No. 72; Promulgation Date: 1996-12-30; Revision Date: 1998-12-03).

37. MOH, 2002, Administrative Provisions for Food Additive Hygiene (MOH regulation; Order No. 26; Promulgation Date: 2002-03-28; Effective Date: 2002-03-28)

38. MOH, 2002, List of Inspection Agencies Authorized by the Ministry of Health for the Health Related Products.

39. GAQSIQ, 2003. Measures for Safety Supervision and Administration of Food Manufacturing and Processing Enterprises (GAQSIQ Regulation; Order No. 52; Promulgation Date: 2003-07-18; Effective Date: 2003-07-18).

40. State Certification and Accreditation Administration, 2002, Administrative Provisions for Hazard Analysis and Critical Control Point (HACCP) Accreditation Management System in Food Industry (Directive; Document No.: Public Notice [2002] No. 3; Promulgation Date: 2002-03-20; Effective Date: 2002-05-01).

41. MOH, 1997 Procedures for Supervision over Food Hygiene (MOH regulation; Order No. 50; Promulgation Date: 1997-03-15; Effective Date: 1997-06-01).

42. MOH, 1997, Provisions for Administrative Penalty over Food Hygiene (MOH regulation; Order No. 49; Promulgation Date: 1997-03-15; Effective Date: 1997-03-15).

43. Zhufan Xie, 2002, Classified Dictionary of Traditional Chinese Medicine, (Beijing: Foreign Language Press).

44. Hopkins Technology, LLC, 1996, Materia Medica, from "Traditional Chinese Medicine and Pharmacology," CD-ROM.

17 Report of ILSI Southeast Asia Region Coordinated Survey of Functional Foods in Asia

E-Siong Tee

Introduction

Foods are traditionally recognized as providing essential nutrients for nourishing the human body. The content of these nutrients varies greatly among the various foods, and consumers are encouraged to eat a variety of foods to meet their nutritional needs. In recent years, a great deal of attention has been given to components other than nutrients that are found in foods. A great deal of research has been undertaken on the potential health significance of these components. Foods containing such components have been termed "functional foods."

There is as yet no unanimously accepted definition of functional foods globally, although several definitions have been proposed. A generally accepted understanding is that functional foods are foods that, by virtue of physiologically active food components, provide health benefits beyond basic nutrition. No country in the world currently uses the term **functional foods** in its regulations. Even the FAO/WHO Codex Alimentarius system does not have clear guidelines or specifications for functional foods.

In the Asian region, there is increased interest and trade in functional foods. The region has tremendous potential for development of these foods. It is therefore vital that there should be intensified research and development. There should be greater interaction among countries in the region and some degree of harmonization in the development of these foods in Asia. ILSI SE Asia Region will attempt to play this role in stimulating and coordinating research and development of functional foods in the region. As an initial step, a survey of the status of functional foods in the region was conducted.

Objective of Survey

The primary objective of this survey is to establish a database of the types, regulatory requirements and standards of "functional foods" in the Asian region. This information will serve as useful input for a position paper on the Asian perspectives of functional foods that is being prepared as a collaborative activity among all ILSI Asian branches (China, Japan, India and Southeast Asia Region).

Survey Methodology

A questionnaire was sent out to regulatory agencies in all the Southeast Asian and selected Asian countries and regions as well as selected companies. In addition, publications and seminar papers by various Asian scientists and authorities on the topic were obtained to provide additional information.

Because no universally accepted definition of functional foods exists, respondents were given a generally accepted definition to provide them with a common direction for the survey. It is generally recognized that functional foods are foods that possess physiological/health benefits beyond basic nutritional functions.

To obtain a wider database, respondents were informed that "functional foods" include "health" foods, food for specific health use (FOSHU), health tonics, dietary supplements other than those mentioned in the next paragraph, and so on. These may be in the form of foods, drinks, tonics, or snacks.

Respondents were instructed to exclude drugs, traditional medicine and dietary supplements in pharmaceutical dosage forms (for example, vitamins, minerals, fatty acids, and amino acids, in pills, capsules, and tablet form) from the survey.

Respondents/Sources of Information

Representatives from the following national regulatory and research agencies and international organizations representing 11 countries and regions responded to the survey questionnaire, as listed in the table that follows.

Country	Agency/Organization
China	Institute of Nutrition and Food Safety
Indonesia	National Agency for Drug and Food Control
Japan	International Life Sciences Institute
Malaysia	Food Quality Control Division, Ministry of Health
Philippines	Bureau of Food and Drugs
Philippines	Food and Nutrition Research Institute
Singapore	Agri-Food and Veterinary Authority
Singapore	Health Promotion Board
South Korea	Food and Drug Administration
Taiwan	Department of Health
Thailand	Food and Drug Administration
Vietnam	Food Administration
Yangon	Food and Drug Administration

Definitions and Regulatory Status

The definition or usage of the term "functional foods" varies greatly among the 11 countries and regions included in the survey. Only three of them, Japan, China and Taiwan, have products that may fall wholly or partially within the scope of functional foods as defined in this survey. All three respondents, however, do not use the term "functional foods" in the regulations pertaining to these foods. The terms used are more along the lines of "health-promoting foods."

In South Korea and the Philippines, the term "functional foods" was used to include dietary supplements in pharmaceutical dosage forms. In the remaining countries (Indonesia, Malaysia, Myanmar, Singapore, Thailand, and Vietnam), there are no regulations on functional foods. All these countries recognize functional foods as an important area that needs to be given attention but have not enacted specific regulations for these foods. Some of these countries have regulations pertaining to dietary supplements in pharmaceutical dosage forms.

Functional Foods and Health-Promoting Foods

Japan is probably the first country in the world to have defined "functional foods," although the term is not used in its regulatory system. The two main functions of food have conventionally been recognized as providing nutrients (primary function) and sensory functions (tastes, flavors, and so on) (secondary function). A third or tertiary function of food was thought to be that pertaining to regulating physical conditions such as functions of bio-regulation, disease prevention, recovery from disease, regulation of biorhythm, and control of aging. Foods possessing these third functions were defined as functional foods. Intensive research was carried out and it was shown that many food ingredients have this third function. As a result, a system called "foods for specified use," or FOSHU, was introduced in 1991 under the Nutrition Improvement Law. The word "functional" was not used as it was being used in the definition of "drug" in the Pharmaceutical Affairs Law. FOSHU was one of the five categories of "Foods for Special Dietary Uses" created under this law and was defined as "foods in the case of which specified effects contributing to maintain health can be expected based on the available data concerning the relationship between the foods/food's contents and health, as well as foods with permitted labeling which indicates the consumer can expect certain health effects upon intake of these particular foods." The other four categories were single foods for the ill, milk powder for pregnant or lactating women, formulated milk powder for infants, and foods for the aged with difficulty in masticasting or swallowing.

In China, too, the term "functional food" is not used. Instead, the term "health foods" is used and is defined as "food that has special health functions, suitable for the consumption by special groups of people and has the function of regulating human body functions, but is not used for therapeutic purposes." The regulation for the Control of Health Food was issued by the Ministry of Health in November 1996.

The Health Food Control Act was enacted in 1999 in Taiwan. Under this Act, the term "health food" is used and is taken to mean foods with specific nutrient or specific health care effects as specially labeled or advertised, but not food aimed at treating or preventing human diseases.

Functional Foods and Dietary Supplements in Pharmaceutical Dosage Forms

In some countries, the term "functional foods" was used to include dietary supplements in pharmaceutical dosage forms. In South Korea, a new law called the Health Functional Food (HFF) Act was enacted in August 2002. HFF in Korea refers to "processed and manufactured goods in the form of tablets, powders, granules, liquids and pills that help enhance and preserve the health of the human body using nutritional or functional ingredients". This Act is intended to be rather similar to the Dietary Supplement Health and Education Act of 1994 (DSHEA) in the United States. Although the term "functional foods" is used in this Act, it is not within the definition used in this survey.

The FOSHU system in Japan was reorganized, in line with the introduction of a new system of Foods with Health Claims in April 2001. This system comprised the existing FOSHU and a new category of Foods with Nutrient Function Claims. It was also proposed that the restriction of food forms in FOSHU system be removed so that products in tablet and capsule forms be included in the system.

The Bureau of Food and Drugs in the Philippines has defined functional foods as "any finished, labeled, processed product that contains herbs/plant materials, marine/animal sources, amino acids, vitamins and minerals, and other substances, single or combined, which is not intended or marketed as a conventional food. It may come in different forms such as liquid, powder, granule, tablets for specific health use that is backed by scientific and relevant studies but is not intended to treat or cure any disease/condition."

According to the Indonesian National Agency for Drug and Food Control (NADFC), there is no regulation on functional foods in the country. The Agency is in the process of formulating such a regulation. It is anticipated that the proposed regulation will be along the line of foods as the definition provided was: "natural and processed foods which contain one or more ingredients which, based on empirical or scientific evidence, qualify for having physiological functions beneficial to health."

In Myanmar, there is no specific regulation on functional foods. However, a system similar to the FOSHU system in Japan was reported to be in place. Products have to be subjected to review by an approval system prior to being permitted to be on sale. No details of the system are available. However, as can be seen later in the section on types of products, it is doubtful that the system is indeed that of functional foods as defined in this survey.

Specific Regulations for Dietary Supplements

There are no functional food or health-promoting foods regulations in Malaysia, Singapore, Thailand, or Vietnam and there are no indications that such regulations will soon be introduced in these countries. These countries, however, have regulations pertaining to dietary supplements.

In Malaysia, the term "functional food" is not used in the regulatory system and there is no official definition of the term. Food products are regulated under the Malaysian Food Regulations 1985. A new regulation on nutrition labeling and claims for foods was gazetted on March 31, 2003.

There are separate regulations for dietary supplements that come under the purview of the Drug Control Authority (DCA), with the National Pharmaceutical Control Bureau serving as its secretariat. Dietary supplements have been defined as "products intended to supplement the diet, taken by mouth in forms such as pills, capsules, tablets, liquids or powders and not represented as conventional foods." Each of these products has to have an application submitted to the DCA for registration. The review process includes safety evaluation, manufacturing process, and checks on the permitted claims. Advertisements for supplements also require approval by the Medicines Advertisement Board.

In recent years, there has been an increase in a number of food products that contain extracts of various botanical or animal components. In many cases, these products are not clearly marketed as "food" or "drug." It has been difficult to determine which authority should regulate the marketing and sale of such food-drug interphase products, that is, the Food Quality Control Division or the Drug Control Authority. To enable a quick decision, a Committee for the Classification of Food-Drug Interphase Products has been formed since 2000.

In Singapore, there is currently no legal or official definition for functional foods. In general, food and supplements derived from food, such as royal jelly, bee pollen, botanical (fruit, cereal or plant-based) beverages, and protein- and carbohydrate-based powdered beverages, come under the purview of the Agri-Food and Veterinary Authority. The import and sale of these products in Singapore are governed by the Sale of Food Act and the Food Regulations. Importers of these products are required to ensure that the food products they intend to import comply with the requirements of the Food Regulations, including the labeling requirements. On the other hand, dietary supplements such as vitamins, minerals, amino acids, essential fatty acids, phospholipids, and preparations for medicinal or pharmaceutical purposes come under the purview of the Centre of Pharmaceutical Administry of the Health Sciences Authority.

The definition of dietary supplements in Thailand is as follows: "products which are directly consumed other than normal staple food and are often in the form of tablets, capsules, powder, liquid or other forms and are intended for general persons in good health," a definition that is not too different from that of Malaysia. The regulation of these products is, however, rather different from that of Malaysia and Singapore; the Food and Drug Administration Thailand regulates these under Foods for Special Dietary Uses.

Premarketing Approval System

This section describes the premarketing approval systems only in Japan and China, where the two systems are closer to the definition of "functional foods" of this survey. In the other countries where functional foods refer to dietary supplements, the premarketing approval system would be more appropriate for products in pharmaceutical dosage forms and would very likely fall under a different regulatory agency from food.

In the FOSHU system in Japan, each application had to be vetted on a case-by-case basis by a premarketing approval system set up by the Ministry of Health and Welfare (MHW). The criteria stipulated by the MHW are the following:

a. The food should be expected to contribute to the improvement of one's diet and the maintenance/enhancement of health.
b. The health benefits of the food or its constituents should have a clear medical nutritional basis.
c. Based on medical and nutritional knowledge, appropriate amounts of daily intake should be defined for the food or its constituents.
d. Judged from experience, the food or its constituents should be safe to eat.
e. The constituents of the food should be well-defined in terms of physico-chemical properties and qualitative/quantitative analytic determination.
f. There should be no significant loss of nutritive constituents of the food in comparison with the same ones normally present in similar types of foods.
g. The food should be of a form normally consumed in daily dietary patterns, rather than consumed only occasionally.
h. The product should be in the form of a usual food, but not in another form, such as pills or capsules.
i. The food and its constituents should not be those exclusively used as a medicine.

Within the MHW, the Foods for Special Dietary Use Assessment Investigation Committee investigates each product submitted for approval. The contents of the sample products are analyzed at the National Institute of Health and Nutrition. Approved products may carry a prescribed logo of the system and an approved statement indicating the specified health benefit.

Considering the reorganization of the health claim system in Japan, new applications for FOSHU should be received by the Planning Section, Food and Health Department, Medical Bureau of the Ministry of Health, Labour and Welfare and examined by the Council on Pharmaceutical Affairs and Food Hygiene. Products in conventional food forms already approved should be examined and approved by a sectional committee on newly developed foods to be established under the Sectional Committee on Food Hygiene of the Council.

In the case of China, food products that claim health functions shall be reviewed and approved by the Ministry of Health. The following requirements must be met:

a. It must be demonstrated by necessary animal and/or human functional tests that the product has definite and stable health functions.
b. All the raw materials and their products must meet the hygienic requirements of food. They must not cause any acute, sub-acute, or chronic harmful effects to human body.
c. There shall be scientific substantiation supporting the formulation and the amount of ingredients used. Functional ingredients shall be identified. If the functional ingredients cannot be identified under the present condition, it is necessary to show the names of the main raw materials related to the health functions.
d. Information shown in the label, use instruction, or advertisement shall not claim therapeutic effects.

The approved products shall be issued a Health Food Certificate with a code number. Such foods are then permitted to use the special symbol for health food stipulated by the Ministry of Health. The functional components contained therein and the health function that the product possesses should be clearly indicated on the label.

In Taiwan, no food shall be labeled or advertised as health food unless it is registered as such in accordance with the Health Food Control Act, 1999. A manufacturer or importer of a health food must submit an application supported by information such as its ingredients, specifications and methods of analysis, other relevant data and documentation, as well as label and samples of the product. Products that meet the following requirements may be given a product registration permit by the central competent authority:

a. Contain chemical entity with definite health care effect, the reasonable intake of which is supported by scientific evidence. If current technology cannot identify chemical entities with valid health care effects in health food, the ingredients with the relevant health care effects shall be enumerated or supporting literature provided to the central competent authority for re-evaluation and identification.
b. Be duly supported by scientific assessment and test of health care effects or by academic principles that they are harmless and carry definite, steady health care effects. The method by which health care effects are assessed and the method by which toxicological assessments are made shall be established by the central competent authority.

There are specific requirements for the labeling and advertising of health foods. Besides the usual particulars of the product, the approved health care effects should be clearly stated as well as the amount of intake and important message for consumption of the health food and other necessary warnings. No health food label or advertisement shall be misrepresented or exaggerated or contain content beyond the approved scope. No health food shall claim therapeutic effect.

The health care effects of health foods shall be described in any of the following ways:

a. Claiming the effect of preventing or alleviating the illness relating to nutrients when deficient in the human body if intake of the health food can make up said nutrients.
b. Claiming the impact on human physiological structure and functions by the specified nutrients or specific ingredients contained in a health food or by the food itself after the health food has been taken.
c. Furnishing the scientific evidence to support the claim that the health food can maintain or affect human physiological structure and functions.
d. Describing the general advantages of taking the health food.

Types of Products and Claims

Only foods in the Japan and China systems are further elaborated here, because the regulatory systems in these two countries are closer to the definition of "functional foods" as intended in this survey. Examples given by other countries were mixtures of fortified foods and dietary supplements.

FOSHU in Japan

In the original FOSHU system of Japan, as previously mentioned, products are in conventional food forms. These would then be in line with the scope of the definition of functional foods as used in this survey. As of April 2001, a total of 252 items were approved. Claims for seven categories of products are approved, namely, foods:

a. that promote an increase in the intestinal microflora and helps to maintain a healthy intestinal environment
b. that help people with a high cholesterol level
c. for mineral (calcium or iron) supplementation, with high absorbability
d. of low carcinogenicity
e. helpful for people with mild hypertension
f. helpful for people who are concerned with high blood glucose
g. helpful for people who are concerned about their blood triglyceride level

Functional components must be identified on the label and include:

- oligosaccharides, lactobacillus, fibre
- soy protein, chitosan
- glycoside from eucommia leaves
- calcium citrate malate (CCM), casein phospho-peptide (CCP)
- Palatinose, maltitol, green tea polyphenols
- indigestible dextrin
- diacylglycerol

However, after 2001, the system was expanded to include products in pharmaceutical dosage forms.

With this proposed change, there were also changes to the permitted claims for FOSHU products. Claims should not include diagnosis, treatment, or prevention of diseases as stated in the Pharmaceutical Affairs Law and shall be limited to the following. Reduction of disease risk claims shall not be permitted.

a. Maintenance and improvement of indices of physical conditions that can be easily evaluated (indices evaluated by healthy persons and those evaluated during medical check-ups), for example, "helps you maintain normal blood sugar levels" or "promotes decomposition of body fat"
b. Maintenance of good physical condition and/or organ function or its improvement, for example, "regulates bowel movement" or "improves absorption of calcium."
c. Improvement of subjective and temporary, but not persistent or chronic, changes in physical condition, for example, "helpful for those who feel physically fatigued."

As of September 25, 2003, a total of 396 FOSHU products have been approved. A major proportion (57 percent) of these foods are products that contain oligosaccharides, lactobacillus, bifidobacterium, and dietary fibre and are supposed to help maintain good gastro-intestinal condition. Another 16 percent of the products contain protein, peptides, dietary fibre and diacylglycerol, or plant sterol or stanol esters and are said to be "good for those who have high serum cholesterol and triglyceride."

The total health foods market in Japan for 2001 has been estimated to be U.S. $112 billion. Some 29 percent of the total are FOSHU type products, 4 percent are foods for special dietary uses, and 7 percent are foods with nutrient function claims. The bulk of this, or 60 percent, is a mixture of "other health foods." The estimated share of functional foods in the Japanese market in 2001 was therefore U.S. $32.8 billion.

Health Foods in China

A total of 3,357 products had been approved by the Ministry of Health China as of March 2002. But it was estimated that only 30 percent of these products were currently in the market. A large proportion (about 45 percent) of these products were in conventional food forms or powdered beverage whereas another 46 percent were in the form of capsules or pills. Hence only about half of the products may fit into the scope of functional foods as defined in this survey.

Slightly more than a third of these products claimed to regulate immune system in the body. Another 18 percent claimed to counteract fatigue, and a similar percentage is said to function as being regulatory for hyperlipidemia. The top 10 functions or claims of the health foods in China are as follows:

- Regulate immune system
- Reduce fatigue
- Regulate hyperlipidemia
- Delay aging
- Effect hypoxia tolerance
- Inhibit tumor development

- Improve gastrointestinal function
- Regulate blood glucose
- Improve memory
- Improve the sleep process

The health foods in China are not permitted to make the following claims:

- The prevention or treatment of disease
- Recovery of one's youthful vigour, prolongation of life, anticancer, or curing cancer
- Secret prescription from generation to generation, nourishing food, food for improvement of health and beauty, food used in Imperial Palace

Presumptive Functional Foods

Some examples of presumptive functional foods given by Myanmar are the following:

- Iodized salt
- Several brands of malted milk drinks (SMART drinks)
- Essence of chicken/fish
- Bird's nest (Bird's nest is made by the swallow, using saliva secreted by the bird. It is used in making soup and is considered a delicacy among the Chinese.)
- Formula dietary foods; high-protein formulas
- Soya milk
- Glucose powder fortified with vitamin C and calcium
- Weight-loss products
- Biscuits for children, fortified with DHA, taurine
- Low sodium salt
- Herbal products (for example, noni juice, spirulina spread and snack, coffee or tea with linzhi extract)

It can be seen that the preceding list is a mixture of fortified/enriched foods as well as some herbal products. These examples are not the same as those in the FOSHU system of Japan and are not within the scope of the definition of functional foods as used in this survey. The respondent did point out that this is a presumptive list only.

Other examples of functional foods identified by a few other countries in the survey included foods fortified or enriched with vitamins, minerals, fatty acids, fibre, bacterial cultures, and so on. Many of these would not fit into the definition of possessing physiological or health benefits beyond basic nutritional functions.

Summary and Conclusions

Representatives from national regulatory and research agencies and international organizations from 11 countries and regions in Asia responded to the survey questionnaire on functional foods in the region. In addition, publications and seminar papers by various Asian scientists and authorities on the topic were obtained to provide additional information. The countries and regions participating in the survey are China, Indonesia, Japan, Korea, Malaysia, Myanmar, Philippines, Singapore, Taiwan, Thailand, and Vietnam. The primary objective of the survey is to establish a database of the types, regulatory requirements, and standards of "functional foods" in the region.

The definition or usage of the term "functional foods" varied greatly among the respondents. Only three of them, namely, Japan, China, and Taiwan, have products that may fall wholly or partially within the scope of functional foods as defined in this survey. All of them, however, do not use the term "functional foods" in the regulations pertaining to these foods. The terms used are geared more toward "health-promoting foods." In all these, a specific regulation was enacted and a premarketing approval system was in place to review each health product submitted prior to release into the market. The FOSHU system of Japan, started in 1991, had approved 252 functional food products by April 2001, all of which were in conventional food forms. After that date, the FOSHU system was amended to include supplements in pharmaceutical dosage forms. The "health food" regulation of China has several thousand products, comprising a mixture of foods and products in pharmaceutical dosage forms. Taiwan, too, permits the sale of health food products under a specific regulation.

In South Korea and the Philippines, the term "functional foods" was used to include dietary supplements in pharmaceutical dosage forms. In the remaining countries (Indonesia, Malaysia, Myanmar, Singapore, Thailand, and Vietnam), no regulations on functional foods exist. There is a great deal of interest among the countries and regions in Asia to regulate and permit the sale of "functional foods." In some of these countries, there are existing regulations pertaining to dietary supplements in pharmaceutical dosage forms.

Of major concern to regulatory authorities in the survey is the health and safety assessment and the appropriate communications for the scientific community and the public. Another important area in which countries have voiced concern is that of the need for a premarketing approval system to ensure safety and efficacy of products. More work needs to be carried out on appropriate methodologies for scientific substantiation of claims for functional foods. The main consumer concerns include the need for quality control of products by the industry, especially controlling the standard of manufacturing practice, for example, GMP, and the importance of regular inspection by the relevant authorities.

From the survey findings, it is clear that there is great variation in the understanding on the subject. Functional foods are at very different stages and even directions of development in the region. Such developments, if allowed to persist, would be most detrimental to the development of these products in the region. A harmonized approach is needed, therefore, to the development of functional foods. It is recognized, however, that although such an approach is a desirable long-term goal, it will take some time to materialize since definition and legislative differences vary widely across the countries.

This would be beneficial to the advancement of the industry and would bring about greater consumer confidence in these products. Indeed, there is a great deal of potential for the development of functional foods in the Asian region. A large number of food products and biologically active ingredients are unique to the region and could very well be useful in promoting the well being of the population. There should, of course, be scientific substantiation of these beneficial effects. Greater interaction among countries and regions in Asia will bring about greater advancement of functional foods.

18 Germany and Sweden: Regulation of Functional Foods and Herbal Products

Joerg Gruenwald and Birgit Wobst

Abstract

Regulation and definition of functional food and herbal products is not harmonized within the European Union (EU). Health claims are possible in Sweden and are regulated by the Swedish Code of Practice. Swedish food law prohibits claims that a food can prevent, relieve, or cure diseases.

The Swedish Code of Practice allows generic claims related to well-established diet-health connections. A documentation of the effect of generic claims is not required. Eight specific relations are given, which are closely related to official nutrition recommendations. Product-specific physiological claims have been possible since 1998. The claimed effects have to be documented in at least one product-specific clinical study in humans and should be supported by scientific bibliographic documentation. The Swedish Nutrition Foundation (SNF) publishes successful applications for innovative claims as soon as the product is launched with the claim.

In Germany no special legislation exists for functional food. It is regarded as food for normal consumption and regulated by German food law. Health-related claims are not regulated by a special code but are becoming more readily accepted by German authorities if they are true, are not misleading, and claim only the benefits for good health or well-being. Generally, references to special organs or diseases are prohibited and the consumer must not be frightened.

Within Europe herbal products can be regulated under food law or as medicines. The regulation of traditional herbal medicines in Europe is under discussion. Germany has a long tradition of herbal medicine use, and most products are regulated as medicine or traditional medicine. Bibliographic applications are possible. Furthermore, botanicals are used as spices, flavorings, or teas under food law, and, more recently, herb extracts are added to yogurt or drinks as functional foods. In Sweden, botanicals can be authorized as drugs by the Medicinal Products Agency before marketing. Herbal products can also be defined as natural remedies with a simplified registration if they have been traditionally used within Europe. The situation with herbal products under food law is comparable to that in Germany.

Introduction

The modern and healthful diet should be rich in cereals, vegetables, and fruits, and low in meat and especially saturated fats. Pursuit of a healthful lifestyle, sense of well-being, and increased knowledge about nutrition-related diseases are some possible reasons for the rapid development of functional foods. Self-responsibility in health issues and prevention-oriented and healthful nutrition practices are reconciled in the concept of functional food, which can be generally regarded as a specific health-promoting effect of a food based on scientific proof. The functional food concept is being pushed rapidly by the food industry, and the legal framework is lagging behind that rapid development. Legislation is usually a long, complex process, and legislators are unlikely to embark upon such a process without good reason (Gruenwald et al. 2002).

Some approaches to the harmonization and definitions of functional food and health claims in Europe are as follows: The Council of Europe's "Guidelines concerning scientific substantiation of health-related claims for functional food" (2001); the EU discussion paper on "Nutrition Claims and Functional Claims (SANCO/1341/2001) and Draft proposal for "Regulation of the European Parliament and of the Council on Nutrition, Functional and Health Claims Made on Foods" (Working document SANCO/1832/2002); the European Commission Concerted Action on Functional Food Science in Europe (FUFOSE) and Process for the Assessment of Scientific Support for Claims on Foods (PASSCLAIM), coordinated by The International Life Sciences Institute (ILSI) Europe in order to establish a science-based approach for concepts in functional food science (Aggert et al. 1999); and an extensive discussion by the Codex Alimentarius.

The Codex Alimentarius is an international standards-setting body with the purpose of protecting the health of consumers and ensuring fair practices in the food trade. The Codex published "Codex General Guidelines on Claims" (1979, 1991), which could be used as the basis for national approaches on health claims. The essential requirements of these guidelines in relation to functional food are the following:

- It must be safe.
- It must not be misleading to the consumer.
- It must be in accordance with national food law.

These guidelines relate to claims made for a food irrespective of the food's being covered by an individual Codex standard. The guidelines are based on the principle that no food should be described or presented in a manner that is false, misleading, or deceptive or is likely to create an erroneous impression regarding its character in any respect.

The person marketing the food should be able to justify the claims made.

Definition of Claims by the Codex Alimentarius

A claim is "any representation which states, suggests or implies that a food has particular characteristics relating to its origin, nutritional properties, nature, production, processing, composition or any other quality."

Following are prohibited claims:

- Claims stating that any given food will provide an adequate source of all essential nutrients, except in the case of well-defined products for which a Codex standard regulates such claims as admissible claims or where appropriate authorities have accepted the product to be an adequate source of all essential nutrients.
- Claims implying that a balanced diet or ordinary foods cannot supply adequate amounts of all nutrients.
- Claims that cannot be substantiated.

- Claims as to the suitability of a food for use in the prevention, alleviation, treatment or cure of a disease, disorder, or particular physiological condition unless they are in accordance with the Codex standards for special dietary uses or permitted under national laws.
- Potentially misleading claims that contain incomplete comparatives and superlatives or claims as to good hygiene practice, such as "wholesome," healthful" or "sound."

Health claims must be consistent with national health policy, including nutrition policy, and support such policies. Health claims should be supported by specific consumer education.

The following information should appear on the label or labeling of the food bearing a health claim:

- Information on the target group, if appropriate.
- Information on how to use the food to obtain the claimed benefit, if appropriate.
- If appropriate, advice to vulnerable groups on how to use the food and to groups, if any, who need to avoid the food.
- Maximum safe intake of the food where necessary.

Generally, the Codex guidelines on health claims are always related to national food law:

> Any health claim must be accepted by or be acceptable to the competent authorities of the country where the product is sold. **Only health claims that support national health policy and goals should be allowed.**

A good example for different national approaches to health claims is the situation in Sweden compared to that in Germany. Sweden has a voluntary code for claims, whereas Germany has no clear definitions and relies on the interpretation of EU directives to determine whether a claim is medicinal (European Advisory Service 2002).

Sweden

In Sweden, the regulation of health claims and functional food is well organized and progressive. Before entering the European Union (EU), Sweden passed a self-regulating plan for the marketing of foods with health claims. Sweden stresses the importance of a timely and broad debate on health claims including enhanced function claims and disease risk reduction claims within the EU.

Swedish Nutrition Foundation

The Swedish Nutrition Foundation (SNF) was established in 1961. Its objective is to support scientific research within nutrition and adjacent fields. The Foundation also promotes the implementation of developments within this field of research.

The Foundation organizes an annual national or international scientific symposium and is responsible for the publication of the *Scandinavian Journal of Nutrition*. The SNF also works in an advisory capacity on, for example, functional foods and health claims used in the marketing of food products.

The SNF consists of a Board, a Research Committee, and a Nutrition Advisory Committee.

The Swedish Code of Practice

The Swedish Code on health claims, entitled "Health Claims in the Labelling and Marketing of Food Products, The Food Industry's Rule (Self-Regulating Programme)" was introduced in 1990. In 1989 the National Board of Health and Welfare and Drug Department (today the Medicinal Products Agency) decided to stop applying medicinal product legislation "to products commonly found on the dinner table" [Asp 2002]. After that, the use of health claims in the labeling and marketing of foods became possible.

The Nutritional Labeling directive of the Food Act prohibits claims that a food can prevent, relieve, or cure diseases. Otherwise, health claims that are true, not misleading, and not derogatory to other foods are allowed.

Nutritional Safety

A food product bearing a health claim must be safe. This basic requirement for all foods becomes especially relevant for functional foods, which are intended to be consumed regularly over a long period of time (Asp 2002).

According to the Council of Europe, products for which health-related claims are made should "fit into a nutritionally adequate diet" and "not be in conflict with national nutrition policies" (Council of Europe 2001).

Definition of a Health Claim in the Swedish Code

The original Swedish Code from 1990 defined a health claim as "an assessment of the positive health effects of a foodstuff, i.e. a claim that the nutritional composition of the product can be connected with prophylactic effects or the reduced risk of a diet-related disease."

Generic Claims: The Original Swedish Code

The original Swedish Code from 1990, which was revised in 1997, was limited to generic claims, related to well-established diet-health connections. These connections are closely related to the official nutrition recommendations. The eight connections around which generic claims can be used are the following:

- Obesity—energy content
- Cholesterol level in the blood—fat quality or some soluble dietary fiber
- Blood pressure—salt (sodium chloride)
- Atherosclerosis—blood cholesterol level/blood pressure; n-3 (omega -3) fatty acids in fish and fish products
- Constipation—dietary fiber
- Osteoporosis—calcium
- Caries—absence of sugars and other easily fermented carbohydrates
- Iron deficiency—iron content

These rules are applicable for all products that fulfill certain criteria regarding composition and importance for the diet. Documentation concerning the effects of the food product itself is therefore **not needed** for these kinds of claims (Asp and Trossing 2001; Asp 2002).

Product-Specific Physiological Claims

Product-specific physiological claims have been possible since June 1998, but substantiation is required through a preclearance process involving independent experts. The SNF has an advisory role defined in the Code. The responsible organizations behind the Code are the Swedish Cooperative Union, The Federation of Swedish Farmers, The Swedish Food Federation, The Swedish Federation of Trade, and The Swedish Food Retail Association.

Applicants must prepare a dossier to document that studies are scientifically sound and unobjectionable, showing the effects that are claimed. A clinical study must be performed on human subjects, and the study group must be representative of the group of people that the marketing is aimed to reach. Studies must be performed with a supply of the food product relating to its normal use during the study period; the studies must be long enough to show a lasting effect.

The company or manufacturer is responsible for the product safety according to the Swedish Food Regulations. All ingredients must be approved as food ingredients or novel foods (according to the directive on novel foods as described previously).

The SNF will check whether criteria on "functional foods" and normal foods are fulfilled.

Application

Food products referred to in the Code should be part of a normal diet and not dietary supplements or other substances in the form of capsules, tablets, dragées, powders, or the like. The products "should through their composition contribute positively to a nutritionally adequate diet."

The application requires a nutrition declaration, containing at least energy and seven nutrients (protein, carbohydrates, sugars, fat, saturated fat, dietary fiber, and sodium).

At least one human clinical intervention study on which the claim is primarily based is desired (in summary) with reference to the researchers and institutes/departments that have performed the study. Additional documentation should be submitted after consultation with, and directly to, the appointed experts.

Scientific Evaluation by the SNF

The scientific quality must be documented through a review process; publication in well-established international scientific journals with peer-review may be an alternative.

The evaluation process of the application for a product-specific physiological claim is initiated by the SNF as soon as the application is received.

Within four working weeks a panel of at least three experts will be appointed by the Research Board of the SNF (the applicant will be informed to ascertain whether any expert may be challenged on grounds of partiality).

The number of required studies will be decided on a case-by-case basis, depending on how well established the physiological effect is considered to be.

The evaluation will consider the scientific documentation in relation to the type of claim the applicant wishes to make—not the exact wording or other formulation of the claim.

The evaluation shall be completed within 90 days of the SNF's receiving the documentation. If additional time is needed, an agreement will be reached depending on the case.

If the applicant does not accept the evaluation of the expert panel, and reports this to the SNF within two weeks, a new expert panel must be appointed.

The evaluation must be carried out confidentially. The evaluation report will become public if and when the product is put on the market with a product-specific physiological claim according to the assessed application.

In labeling and marketing it is permitted to state that the product has undergone evaluation of the scientific documentation according to the Code. This must be stated in a standardized text, for example: "Health benefit with scientific documentation evaluated according to the Food Industry Rules (Self Regulating Program)."

The evaluation will also follow applicable parts of the Council of Europe's Policy Statements concerning Nutrition Food Safety and Consumer Health. Functional food should fit into a nutritionally adequate diet. It should be communicated that good health is closely linked to good nutrition and good dietary habits according to the established dietary recommendations and food guidelines. If supported by scientific evidence, functional food, as part of such a diet, could be of additional benefit to the consumer's health and well-being and contribute to the reduction of disease risks (Council of Europe 2001).

Compliance with the Code of Marketed Products

The Assessment Board for Diet-Health Information (the Board; BKH) was established in November 2001 independently from the SNF. It deals with complaints regarding whether a particular marketing and labeling action complies with the Code. After giving the manufacturer an opportunity to explain and to justify the marketing, the Board decides whether the complaint is to be upheld.

An Example of How One Product-Specific Human Study Is Considered Sufficient

The first application and scientific evaluation for a product-specific physiological claim has been completed and published by the SNF. This serves as an example of how one product-specific human study can be considered sufficient. The application concerned a product consisting of a low-fat yogurt and "müsli" enriched in beta-glucan soluble fiber from oats. The effect on blood glucose and insulin response has been studied for up to two hours following consumption of the test meals. A yogurt without "müsli" served as the control. The Glycemic Index (GI) was determined by standardized methods.

Three human studies were documented, of which two were carried out with the specific product used in the application. Only these two studies were directly relevant for the application. One study showed a statistically significant reduction of blood glucose and insulin after the meal. The SNF concluded that the documentation supported a claim that the product lowers, smoothes out, or attenuates the blood glucose level after a meal. The SNF experts pointed out that the studies were sufficient only because the positive effects of beta-glucan are well documented and approved. For an unknown food or ingredient, these studies would clearly not have been sufficient. It was concluded that the human studies submitted with the application would support a product-specific health claim concerning effects on blood glucose levels. The marketed product should be identical with the product used in the relevant study. The suggested claim is that the product "balances the blood sugar level after a meal." The experts recommended that the wording of the claim should be discussed, because "balances" very unspecifically indicates "changes." They proposed to use "reduces" or "smoothes out" instead.

Germany

Germany is one of the European countries that have not yet established specific guidelines or voluntary codes on health claims, comparable to that in Sweden. No specific legislation exists pertaining to functional foods. Functional food is regarded as food and, therefore, regulated by the General Foodstuff and Commodities Act, the law for foods for particular nutritional uses (Council directive 89/398/EEC of May 1989, as amended, implemented into German food law: Diätverordnung §4 LMBG) or legislation on novel foods. According to German food law (with some exceptions regarding dietetic foods according to "Diätverordnung"), claims in the marketing and advertising of food products are prohibited from:

- Mentioning or referring to the elimination, alleviation, or prevention of a disease.
- Making reference to medical recommendations or medical expertise.
- Showing pictures of people in medical activities or medical-profession uniforms.
- Publishing documents or information in books or magazines instructing people to treat diseases. Food products are also prohibited from looking like pharmaceutical products because the similarity could mislead consumers.

Pharmacological effects may be documented and claimed for professional groups, such as pharmacists or physicians.

Health claims on dietetic foods are not generally prohibited. There are no clear regulations. The German authorities may tolerate moderate statements on beneficial effects that are clearly not related to diseases. Health claims are often "regulated" by court decisions.

According to §18 LMBG, disease-related claims in publications are prohibited, regardless of whether the claim is true and scientifically proven. In contrast, no general restriction is given in §18 LMBG for health-related claims such as "supports the immune system or function" or "stimulates the metabolism." The wording of such claims must be chosen carefully and should not be misleading to the consumer (§17 LMBG).

German Food Authorities

In Germany, the Ministry for Consumer Protection, Nutrition and Agriculture (Bundesministerium für Verbraucherschutz, Ernährung und Landwirtschaft) in cooperation with the Federal Institute for Risk Assessment (BfR: Bundesinstitut für Risikobewertung) and BVL (Bundesamt für Verbraucherschutz und Lebensmittelsicherheit) are responsible for German food legislation. The food authorities of the federal "Länder" are responsible for monitoring the manufacturing process of food supplements and the supplements on the market. The marketing of food supplements is regulated by the German food law (law of food and consumer goods, Lebensmittel und Bedarfssgegenstände Gesetz LMBG).

What Is Food?

Foods serve the purposes of nutrition and pleasure. The German Food Law ("Lebensmittel- und Bedarfsgegenstände-Gesetz", LMBG) defines foods as:

- Substances intended to be consumed by humans in a raw, prepared, or processed state. They can be derived from plants and animals, or from modern technology. Foods can be classified as:
  – traditional foods
  – dietary foods
  – functional foods (including "probiotic and prebiotic foods")
  – novel foods

According to a new definition of food (EC Regulation No. 178/2002), the ingestion of a foodstuff and not the nutritional value is the major requirement for being regarded as food (see EU definition of food in this section).

Food supplements are foods for general consumption according to German law. They include vitamins, provitamins, enzymes, minerals, and other components of a normal diet that are isolated and then offered in concentrated form as tablets, coated tablets, or in some other form or admixed with foods (for example, beta-carotene, co-enzyme Q10).

The crucial point in food law is food safety. According to German food law, foods and supplements in foods may be placed on the market only **if they are safe.** Substances, microorganisms, and parasites may constitute a risk to health. The new BfR, therefore, assesses foods in regard to substance risks and microbial risks.

Dietary Foods

The Directives on foods for particular nutritional uses are incorporated into German food law.

Framework provisions on foodstuffs for particular nutritional uses (PARNUTS) are laid down by Council Directive 89/398/EC, as amended by European Parliament and Council Directives 96/84/EC and 1999/41/EC.

Foodstuffs for particular nutritional uses are defined by the Directive as foodstuffs that, owing to their special composition or manufacturing process, are clearly distinguishable from foodstuffs for normal consumption, are suitable for their claimed nutritional purposes, and are marketed in such a way as to indicate such suitability.

A particular nutritional use must fulfill the particular nutritional requirements of one of the following:

i. Certain categories of persons whose digestive processes or metabolism are disturbed
ii. Certain categories of persons who are in a special physiological condition and who are able to obtain special benefit from controlled consumption of certain substances in foodstuffs
iii. infants or young children in good health

Products in the preceding categories may be characterized as "dietetic" or "dietary."

The Directive prohibits the labeling, presentation, and advertising of foodstuffs for normal consumption to use the adjectives "dietetics" or "dietary" either alone or in connection with other words.

The Directive makes provision for the drafting of rules under Standing Committee procedure:

- For the labeling of foodstuffs for normal consumption
- To indicate their suitability for a particular nutritional use
- To permit reference on the labeling, presentation, and advertising to a diet or category of persons for which a PARNUTs product is intended
- To provide for derogations from the prohibition on health claims in certain clearly defined cases

However, no such harmonized rules have been adopted or proposed.

Specific provisions on foods intended for use in energy-restricted diets for weight reduction have been drawn up and published as Commission Directive 96/8/EC. This Directive applies to foods for particular nutritional uses intended for use in energy-restricted diets for weight reduction and presented as such. These are specially formulated foods that when used as instructed by the manufacturer replace the whole or part of the total daily diet.

Dietary Foods As Defined by German Food Authorities

Dietary foods are foods that correspond to the special nutritional requirements of specific groups of people. They must be suitable for the given nutritional purpose and must be clearly distinct from foods for general consumption (§1 para. 1 and 2 "Diätverordnung" [Diabetic Regulations on Dietary Foods]).

Besides its general provisions, the DiätVO also contains special provisions for specific groups of dietary foods such as foods for infants and small children. The groups of dietary foods for which special provisions have already been issued or for which individual provisions are to be issued are listed in Annex 8 § 4a para. 1 DiätVO. Examples are energy-restricted diets for weight reduction, food for diabetics, sports nutrition, nutrition for infants, and food for sodium-reduced nutrition.

There are some specific exemptions from § 18 LMBG (prohibition of health-related claims) in the dietetic food order. They are listed in § 3 LMBG. Claims are possible for dietetic food (together with a diet meal plan) for:

- newborns with neonatal diarrhea
- renal and liver cell insufficiency
- genetic metabolic disorders
- food for special nutritional purposes in the case of:
  - maldigestion and malabsorption
  - impaired food ingestion
  - diabetes mellitus
  - chronic inflammatory bowel disease
  - chronic pancreatitis
  - gout

Manufacturers are more and more frequently attempting to identify functional food or food supplements as "diets for special medicinal purposes" for people with special nutritional needs and not intended to serve as the only food source (ergänzende bilanzierte Diäten, according to §1 para 4a (2b) DiätVO).

In principle, the labeling, presentation, and advertising of medical foods may not refer to the prevention, treatment, or cure of a disease. The term "dietetic" or "dietary" may be mentioned. The purpose for the use of the product must be given, which makes it interesting for manufacturers who wish to make health claims. Claims that are currently used in Germany are "for special nutritional needs" in the following indication areas:

- coronary heart diseases
- impaired immune defenses
- rheumatoid diseases
- age-related eye diseases (for example, macular degeneration)
- menopausal disorders
- prostate and breast cancer
- expenditure of intense muscular effort
- osteoporosis
- men/women with extreme performance/occupational stress
- hair loss

Food for special nutritional purposes is still food and should be used primarily as nutrition in the case of special physiological conditions. Pharmacological effects must not predominate nutritional function. The purpose is to provide additional nutrition for people with special diseases and not for healthy people.

Typically, dietetic foodstuff has to be clearly different from a "normal" food for daily consumption. The German authorities critically regard the development of notification regarding functional food, food supplements, and drugs as food with special medical purposes.

For products that do not belong to any of the groups in Annex 8, it is required that notification of the product be made when it is first placed on the market by forwarding an example of the product label to the Federal Institute for Protection of Consumers and Food Safety (BVL). Additional scientific data is required, proving the particular nutritional use as well as compositional details of the product.

Food Supplements As Defined by German Authorities

Food supplements are foods that contain one or more nutrients in a concentrated form (for example, vitamins, minerals, and trace elements) but no energy. They are offered for sale in typical medical forms such as tablets, capsules, or coated tablets and are intended to supplement a normal diet. Food supplements are not medicinal products and, therefore, require no registration or marketing authorization.

The German BfR regards on principle food supplements as useless for healthy individuals with a normal diet. They are convinced that the body gets all the nutrients needed from a balanced diet, but point out that an unbalanced diet can be compensated for by the intake of food supplements. The specific enrichment of foods with individual nutrients may be helpful in certain situations, but those are relatively rare in Germany.

Functional Food

The term "functional food" is increasingly being used for foods with a benefit to health beyond that of strict nutrition. From the viewpoint of the German authorities, functional food should specifically influence and satisfy consumer needs for healthful foods. In contrast to their view of medical forms of nutrient concentrates as food supplements, consumers typically regard functional foods as normal food. Terms such as "designer foods" or "nutraceuticals" are sometimes used as synonyms for "functional foods."

There is no legal definition for the term "functional foods" in Germany. Hence, they can be encountered on the market both as foods for general consumption, such as probiotic yogurt, and as foods for special dietary uses, such as margarines enriched with plant sterols (see the example in the later section, "Novel Food").

"Functional foods," which correspond to novel foods within the intention of the Novel Foods Regulation or which contain novel ingredients are not, in principle, freely available on the market but must undergo a European approval procedure.

The claim (advertising) "functional foods" is regulated for foods for general consumption in the LMBG and in the regulations governing food labeling or nutrient labeling. For foods for special dietary uses, the corresponding provisions of the Diet Regulation apply. Because of the bans on misleading (§17 LMBG) and disease-related (§18 LMBG) advertising, problems may arise with the claim "functional foods."

Health Claims on Food Labels or Advertising in Germany

Following are health claims permitted for yogurt.

Yogurt with Omega-3-fatty acids:

"Healthy heart and tasty" (Herzgesund und lecker)

"Omega-3-fatty acids for healthy cholesterol levels" (Omega-3 Fettsäuren für eine cholesterinbewusste Ernährung)

"Supports natural (immunological) resistance" (unterstützt die natürlichen Abwehrkräfte)

Probiotic yogurt:

"Daily consumption of the probiotic culture balances the natural gut flora" (. . . probiotische Kultur, die bei täglichem Verzehr die Darmflora in ein natürliches Gleichgewicht bringt)

Bread:

"Healthy cholesterol levels – the desire of your heart" (Cholesterinbewusst nach Herzenslust)

". . . bread optimizes bone formation . . ." (mit ...Brot lässt sich der Aufbau von Knochenmasse . . . optimieren)

"good for bones and teeth" (gut für Knochen und Zähne)

Drinks:

> ACE-Drink: "the high dose combination of vitamin A, C and E in one beaker may actively support the body's defense system and, therefore, is a daily plus for your health" (die hochdosierte Kombination der Vitamine A, C und E, die in einem Becher enthalten ist, kann aktiv die körpereigenen Abwehrkräfte stärken und ist deshalb ein tägliches Plus für Ihre Gesundheit)

Wellness drink:

> "Natural and seductive: rose, pomegranate, hibiscus and red clover are refreshing in taste, increasing the mood and are known for their aphrodisiacal effects" (natürlich und verführerisch: Rosen, Granatapfel, Hibiskus und Rotklee sind erfrischend im Geschmack, heben die Stimmung und sind für ihre aphrodisierende Wirkung bekannt).

Health Claims: Examples of Court Decisions

Possible claim: (OLG Hamburg 2LR 2001, 147/151)

> "Stimulates the defense troops of the immune system" (regen die Abwehrtruppen des Immunsystems an)

Impossible claim: (OLG Köln ZLR 2001, 872/880)

> According to studies red wine can contribute to the prevention of . . . heart diseases" (Laut Studien kann Rotwein dazu beitragen . . . Herzkrankheiten vorzubeugen)

This claim was prohibited because of the disease-related (heart disease instead of healthy heart) advertising.

Novel Food

Novel foods are foods and food ingredients that were not used for human consumption to a significant degree within the European Community before 1997. The Novel Foods Regulation within the European Community (Regulation (EC) No 258/97 of the European Parliament concerning novel foods and novel food ingredients) classifies the following categories of foods and food ingredients:

- containing or consisting of genetically modified organisms (GMOs) within the meaning of Directive 90/220/EEC
- produced from, but not containing, GMOs (for example, paste from genetically modified tomatoes, oil derived from genetically modified rapeseed)
- with a new or intentionally modified primary molecular structure; or
- consisting of, or isolated from, micro-organisms, fungi, or algae; or

- consisting of, or isolated from, plants, or food ingredients isolated from animals, except for foods and food ingredients obtained by traditional propagating or breeding practices, and having a history of safe use; or
- to which a production process not currently used has been applied, where that process gives rise to significant changes in the composition or structure of the food or food ingredient, which affects its nutritional value, metabolic effect, or level of undesirable substances (Regulation (EC) No 258/97)

The scope of the Novel Foods Regulation does not extend to additives, flavorings, or extraction solvents for which there are other legal EU provisions to the extent that the safety level in this regulation is guaranteed.

Before novel foods and food ingredients can be placed on the market, it must be documented that they are:

- safe for the consumer
- not misleading to the consumer
- not different from products or food ingredients that they are intended to replace to such an extent that their normal consumption would be nutritionally disadvantageous for the consumer

Furthermore, according to Directive 90/220/EEC and Directive 2001/18/EC, foods that contain or consist of GMO must be subjected to an environmental risk assessment to ensure environmental safety. To protect human and animal health, food, and feed containing or produced from GMO, two new regulations were published by the European Parliament and by the Council on 22 September 2003: Regulation (EC) No 1829/2003 on genetically modified food and feed, and Regulation (EC) No 1830/2003 concerning the traceability and labeling of GMO and the traceability of food and feed products produced from genetically modified organisms and amending Directive 2001/18/EC.

The BfR (Federal Institute for Risk Assessment) is responsible for assessing the safety of novel foods in terms of health. An expert committee made up of scientists from the fields of toxicology, allergology, nutrition, microbiology, molecular genetics and genetic engineering, food chemistry and technology, and veterinary medicine advises the institute.

Novel foods can be placed on the market either by notification or by authorization procedure, depending on the type of product.

The BfR is a founding member of a European network of laboratories that will be responsible for the development and evaluation of detection methods for genetically modified ingredients in foods. Laboratories from all European member states are members of this network. It will also examine the labeling obligation for genetically modified ingredients in foods.

An Example of a Disease-Related Health Claim in Germany

Margarine enriched with plant sterols passed the Novel Foods Authorization procedure ((EC) No 258/97). The claim on the label is "for all who wish to actively lower cholesterol levels" and "for people with increased cholesterol levels." These claims are not in accordance with §18 LMBG because of the disease-related claim. On the other hand, the

European Commission decided in the Novel Food authorization procedure that a claim to reduce elevated cholesterol levels should be put on the label. Therefore, the decision of the EU was accepted in a German court decision. The manufacturer was requested to delete the word "clearly" from the label (OLG Hamburg 29.08.2002, Az.: 3U 236/01). It was decided that only a claim related to the reduction of elevated cholesterol levels was approved by the Commission. The manufacturer claimed a reduction of clearly elevated cholesterol levels, which is prohibited now by this decision.

Herbal Medicines

Some countries in Europe regulate herbal products primarily under food law (for example, UK); others regulate primarily under pharmaceutical law (for example, Germany, France) with the possibility of a simplified pharmaceutical registration.

In the absence of guidelines on nonlicensed herbal products, the EU committee on Proprietary Medicinal Products in 1992 developed a list of herbs that were banned or restricted in any EU country. This list still tends to act as a list of prohibited herbs for food use in the EU, although it has no legal value. Since then, no further EU action has been taken in the area of nonlicensed herbal products [EAS 2002].

There is a variety of different legislation procedures and classifications of botanicals all over the world. In Europe, botanicals are marketed under several different categories from full drug status to foodstuff. The different categories are as follows:

- Herbal medicinal products
- Prescription drugs
- OTC drugs (self-medication or prescription)
- Traditional medicines
- Dietary supplements, dietetic foods, foodstuffs
- Herbs as food, nutraceuticals, or food additives (flavorings)
- Cosmetics, cosmeceuticals

The most important categories are medicinal products or foods (Gruenwald 2002).

Definition of Food and Medicine within Europe

Food

EC Regulation No. 178/2002 defines "food" (or "foodstuff") as meaning: any substance or product, whether processed, partially processed, or unprocessed, intended to be, or reasonably expected to be ingested by humans. "Food" includes drink, chewing gum, and any substance, including water, intentionally incorporated into the food during its manufacture, preparation, or treatment. It includes water after the point of compliance, but does not include feed, live animals (unless they are prepared for placing on the market for human consumption), plants prior to harvesting, medicinal, cosmetics, tobacco and tobacco products, narcotic or psychotropic substances, or residues and contaminants (Gruenwald et al. 2002). Herbs could be regulated as dietary supplements, foodstuffs, or cosmetics according to national food legislation. No harmonized EU regulation exists except for flavorings and spices (Gruenwald 2002).

Medicine

EC Directive 2001/83/EC (former 65/65 EC) relating to medicinal products for human use defines a medicinal product as "any substance or combination of substances presented for treating or preventing disease in human beings." Any substance or combination of substances administered to human beings with a view to making medicinal diagnosis or to restoring, correcting, or modifying physiological functions in human beings is also considered to be a medicinal product. "Substance" in this context includes human, animal, vegetable, and chemical matter (Gruenwald et al. 2002).

Traditional Medicine within the EU

In 1996 the European Commission requested a study on the actual use of Directive 65/65 EC for herbal medicinal products. The following procedures were obtained from this study:

- Registration procedure with complete proof of safety and efficacy is possible in all member states.
- Bibliographic applications are in principle possible in all member states. The ESCOP (European Scientific Cooperative on Phytotherapy) and WHO monographs are regarded as useful bibliographic summaries by most member states.
- National or traditional medicinal products exist in many member states. Long-term experience is sometimes enough proof for efficacy, and the national authorities ensure a sufficient level of quality (Bast et al. 2002).

Traditional Medicine and Directive 2001/83/EC

A proposal for an EC Directive defines traditional herbal medicinal products as any medicinal product containing as active ingredients one or more herbal substances or one or more herbal preparations, or one or more such herbal substances in combination with one or more such herbal preparations (Gruenwald et al. 2002).

In 1999, the ad hoc Working Group on Herbal Medicinal products (HMPWG) of the European Medicines Evaluation Agency (EMEA) was established. A number of proposals and guidelines in the field of quality, safety, and efficacy of herbal medicinal products were published (EMEA/HMPWG/25/99, EMEA/HMPWG/23/99).

In 2002, the European Commission adopted a proposal amending Directive 2001/83/EC, which proposes a simplified registration procedure for herbal medicinal products with a traditional medicinal use of at least 30 years with a minimum period of 15 years within the EU (Bast et al. 2002). Since October 2003, the Council of the EU published a common position on the amendment of Directive 2001/83/EC for traditional herbal medicinal products. The long tradition of the medicinal products makes it possible to reduce the need for clinical testing, insofar as the efficacy of the medicinal product is plausible on the basis of long-standing use and experience. Preclinical tests do not seem necessary when the medicinal product on the basis of the information on its traditional use proves not to be harmful in specified conditions of use. The quality aspect of the medicinal products is independent of its traditional use. Products should comply with quality standards in relevant European Pharmacopoeia monographs or those in the pharmacopoeia of a member state. The amendment is restricted to herbals or herbals in combination with vitamins and minerals. If the common position of the European Council will be accepted without variation by the Parliament, the directive came into force in 2004.

Herbal Medicinal Products in Germany

Germany has a long tradition of herbal medicines, which can be sold outside pharmacies. Legislation is based on monographs (Commission E monographs) and acceptance of bibliographic data as proof of efficacy and safety and the use of national tradition. A positive list of traditional medicinal herbal products exists ("Traditionsliste"). Claims for those products have to be mild. A declaration that the efficacy of the product is not scientifically proven but is based on traditional use is required.

Germany food law regulates herbs as dietary supplements (LMBG as mentioned previously). The products do not need a registration or notification if the drugs have no monograph or are not generally regarded as medicines (Gruenwald 2002). To be regarded as supplements the concentration of herbs must be below the pharmacological dosage (generally, 10 percent of the pharmacological dosage is accepted by the authorities).

Sweden

Medicinal products have to be authorized by the Medicinal Products Agency (MPA) prior to marketing. They have to be registered with full documentation.

As in all Scandinavian countries, herbal products can also be defined as a "natural remedy." The MPA defines natural remedies as medicinal products, in which the active ingredient is derived from natural sources (plant, mineral, bacteria, animal). Chemical isolation or modification is not allowed. The main criterion for the classification of herbal products is the intended use and intention indicated in the marketing presentation as well as the dosage form. The registration is simplified for natural remedies for traditionally used herbs within Europe (Gruenwald 2002).

Conclusion and Perspective

There are many ongoing discussions and decisions to be made in the field of functional food and health claims in Europe. In July 2003, the European Commission adopted a draft regulation on nutrition and health claims made on food. The regulation will need the approval of the European Parliament and the Council. The regulation is anticipated to come into force by 2005. If the European proposal of health claims comes into force, "reduction of disease risk" claims will be possible that were previously prohibited. Disease-related messages, which were totally prohibited by EU legislation, will be allowed if they can be scientifically substantiated and authorized at EU levels. This means that clinical studies on functional foods will be required. Advertising slogans such as "Haribo makes children happy" will not be affected by the regulation. The situation in Germany may be clarified when the regulation comes into force, because Germany has no special program for health claims. The intended EU regulation procedure for substantiation of health claims will be comparable to that already existing in Sweden.

Regarding herbal products, the application as traditional herbal medicinal products permitting a simplified application procedure will be possible within the EU in the near future.

References

Aggert, PJ, Alexander, J, Alles, M, et al. 1999. Scientific Concepts of Functional Foods in Europe. Consensus Document. *British Journal of Nutrition* 81: S1–S27.

Asp, Nils-Georg. 2002. Health claims within the Swedish Code. *Scandinavian Journal of Nutrition* 46(6): 131–136.

Asp, Nils-Georg, and Trossing, Mari. 2001. The Swedish Code on Health-related claims in action—extended to product-specific physiological claims. *Scandinavian Journal of Nutrition* 45: 189–192.

Bast, A, Chandler F, Choy P, et al. 2002. Botanical health Products, Positioning and Requirements for Effective and Safe Use. *Environmental Toxicology and Pharmacology* 12: 195–211.

Codex Alimentarius. 1979; 1991: Codex General Guidelines on claims CAC/GL 1-1979 (Rev. 1-1991).

Council of Europe. 2001. Technical Document. Guidelines concerning scientific substantiation of health-related claims for functional food.

European Advisory Services (EAS). 2002. *Marketing food supplements, fortified & functional foods in Europe—Legislation & Practice.* 3d ed. 146 pp. Brussels. EAS.

Gruenwald, Joerg. 2001. The Regulation of Botanicals as Drugs and Dietary Supplements in Europe. *Ethnomedicine and Drug Discovery.* 241–248.

Gruenwald, Joerg, Thomas Brendler, Christof Jaenicke, Erica Smith. 2002. *Plant-based Ingredients for Functional Foods* 1st ed. Pp. 421–473. Leatherhead: Leatherhead Publishing.

General information was obtained from the home pages of: www.bfr.bund.de/, www.europa.eu.int/comm/food/fs/sc/scf/index_en.html, www.foodlaw.rdg.ac.uk/eu/news.htm.

Federal Institute for Risk Assessment (BfR: Bundesinstitut für Risikobewertung) www.bfr.bund.de (www.bgvv.de)

Federal Institute for consumer protection and food safety (BVL: Bundesamt für Verbraucherschutz und Lebensmittelsicherheit) www.bvl.bund.de

Swedish Nutrition Foundation. www.snf.ideon.se

19 Functional Foods: Australia / New Zealand

Jane L. Allen, Peter J. Abbott, Susan L. Campion, Janine L. Lewis, Marion J. Healy

Introduction

Functional foods comprise a small but increasing proportion of the food supply in Australia and New Zealand. There is currently no specific framework for regulating these foods; however, some types of functional foods are addressed by current regulatory measures. The growing emergence of a broader range of functional foods challenges the conventional food regulatory paradigm that addresses the safety, nutritional adequacy, and broader wholesomeness of the food supply.

The range of ingestible products currently in the market place represents a continuum that includes conventional foods, functional foods, dietary supplements or complementary medicines, over-the-counter medicines, and prescription medicines (Figure 19.1). At the extremes of the continuum, the products are readily recognized as either foods or therapeutic goods and are regulated accordingly. However, between these extremes there are products that are difficult to classify, often because the products contain higher levels of vitamins and minerals than are generally permitted in foods or because they contain new or modified constituents that are purported to have positive health effects.

In Australia, products that are ingested are regulated as either foods or therapeutic goods (medicines). In New Zealand, however, there are currently three regulatory categories of products: foods, medicines, and dietary supplements. Furthermore, within the broad category of dietary supplements, there are some that are generally recognized as food-type products, whereas others are more "medicinal" in nature. The current legislative framework (described in Healy et al., 2003) in which functional foods are accommodated is illustrated in Figure 19.2.

As identified in Figure 19.2, Australia and New Zealand share a common and sole set of regulatory measures for foods, whereas no such measures exist in common for dietary supplements and medicines/therapeutic goods.

A general requirement of the *Australia New Zealand Food Standards Code* (hereafter referred to as the *Food Standards Code*) is that foods entering the food supply in Australia and New Zealand must be "safe" and "suitable." The *Food Standards Code* specifies in greater detail food regulatory requirements in relation to contaminants, composition, and labeling (ANZFA, 2002a). These requirements have been developed in line with the underlying statutory objectives for developing/reviewing food standards and other regulatory measures. (Section 10, *FSANZ Act* 1991—refer to FSANZ, 1991).

Within the Australian and New Zealand framework, foods are regulated according to their intended uses either as part of the general dietary intake for nutritional purposes (that is, general-purpose foods) or for special nutritional purposes such as meal replacement and infant formulas (that is, special-purpose foods) (ANZFA, 2002a). Functional foods are not currently defined in food legislation and do not fall within either the general- or special-

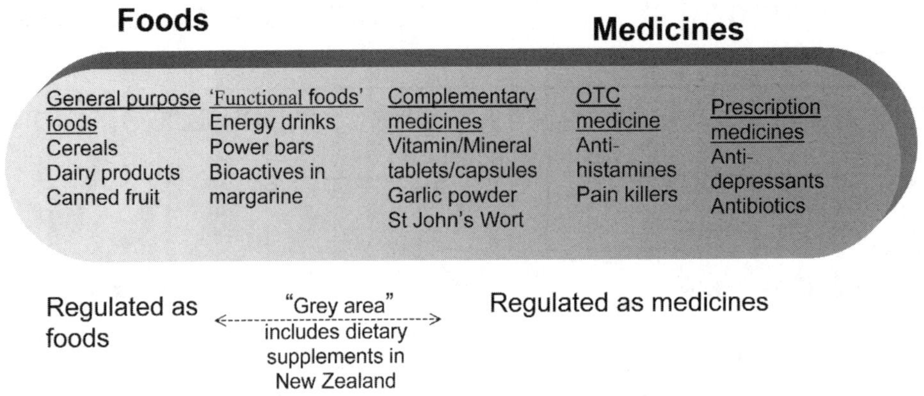

Foods   **Medicines**

General purpose foods	'Functional foods'	Complementary medicines	OTC medicine	Prescription medicines
Cereals	Energy drinks	Vitamin/Mineral tablets/capsules	Anti-histamines	Anti-depressants
Dairy products	Power bars	Garlic powder	Pain killers	Antibiotics
Canned fruit	Bioactives in margarine	St John's Wort		

Regulated as foods ←----- "Grey area" includes dietary supplements in New Zealand -----→ Regulated as medicines

Figure 19.1. Schematic placement of "functional foods" in the continuum of ingestible products, ranging from foods and medicines.

Box 19.1

Objectives of the Authority [FSANZ] in developing or reviewing food regulatory measures and variations of food regulatory measures

(1) The objectives (in descending priority order) of the Authority in developing or reviewing food regulatory measures and variations of food regulatory measures are:
 (a) the protection of public health and safety; and
 (b) the provision of adequate information relating to food to enable consumers to make informed choices; and
 (c) the prevention of misleading or deceptive conduct.
(2) In developing or reviewing food regulatory measures and variations of food regulatory measures, the Authority must also have regard to the following:
 (a) the need for standards to be based on risk analysis using the best available scientific evidence;
 (b) the promotion of consistency between domestic and international food standards;
 (c) the desirability of an efficient and internationally competitive food industry;
 (d) the promotion of fair trading in food;
 (e) any written policy guidelines formulated by the Council for the purposes of this paragraph and notified to the Authority.

NEW ZEALAND:

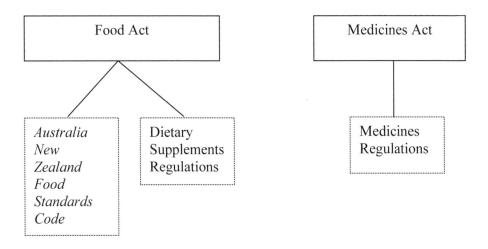

AUSTRALIA:

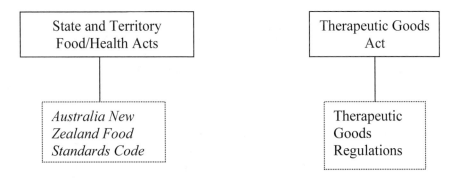

Figure 19.2. Schematic representation of legislative framework for ingestible products in Australia and New Zealand.

purpose categories because, although consumed as part of the normal diet, such foods are intended to have physiological impacts beyond meeting basic nutritional requirements.

No internationally agreed upon and recognized definition for functional foods exists as yet. As a working option FSANZ, uses the following definition:

> Functional foods are similar in appearance to conventional foods and are intended to be consumed as part of a usual diet, but have been modified to have physiological roles beyond the provision of simple nutrient requirements. (NFA, 1993)

FSANZ recognizes several characteristics that are important in identifying foods that fall in the "functional" category. First, the products should be presented as foods and avail-

Table 19.1. Current Australia/New Zealand food standards that may apply in some respects to functional foods

Safety	Nutritional Balance	Claims*, Efficacy, and Content
Novel Foods (Standard 1.5.1)	Special-Purpose Foods (such as, Standard 2.9.3)	Nutrition Information Requirements (Standard 1.2.8)
Contaminants and Natural Toxicants; Prohibited and Restricted Plants and Fungi (Standards 1.4.1; 1.4.4)	Formulated Supplementary Sports Foods (Standard 2.9.4)	Labeling of Ingredients (Standard 1.2.4)
Food additives; Processing Aids (Standards 1.3.1, 1.3.3)	Vitamins and Minerals (Standard 1.3.2)	Mandatory Warning and Advisory Statements and Declarations (Standard 1.2.3)
Formulated Caffeinated Beverages (Standard 2.6.4)		Transitional Standard Health Claims** (not currently permitted) (Standard 1.1A.2)

* The majority of nutrient content claims in Australia are currently subject to a Code of Practice, which does not apply in New Zealand. The policy aspects of the regulatory management of nutrient content claims are currently under review. (Refer to www.foodsecretariat.health.gov.au.)
** Refer to the section "Claims Relating to Functional Foods."

able as part of a normal food supply (for example, in grocery stores and specialty shops, such as health food stores). Second, functional foods are not foods that are essential for the maintenance of life—no subgroups in the population require these foods for survival. Third, functional foods should have a physiological function over and above that provided by the nutrients found in the traditional diet.

Although a specific regulatory framework for functional foods has not been fully established in Australia or New Zealand, the current regulatory measures within the *Food Standards Code* are applicable to at least some types of functional foods (see Table 19.1).

This chapter considers the regulatory challenges posed by functional foods in New Zealand and Australia. The current regulatory approach is examined and the challenges and issues in relation to functional foods are explored, using case studies when applicable. The regulatory focus is on the health and safety aspects of these foods, as well as on any associated claims.

Safety Aspects

Traditional versus Nontraditional Regulatory Approaches

A functional food is generally characterized by the increased level of a specific food ingredient. Although a significant increase in the level of specific ingredients in food products may potentially offer health benefits, it may, in some cases, also have adverse health effects. In New Zealand and Australia the traditional regulatory approach to the safety of foods has been to regard the food as safe unless a significant hazard has been identified. Thus, so-called "traditional" foods are accepted in the community because either there is no evidence of adverse health effects through long-term use or there is adequate knowledge in the community to address any identified hazard. Such foods are considered to have a history of safe consumption. Functional foods are unlikely to be regarded as "traditional" in this sense, and a more proactive approach is needed to establish safety. Thus, evidence of safety prior to the marketing of "nontraditional" foods and food ingredients is now regarded as essential.

Functional foods can raise safety issues both because of the nature of the functional ingredient itself and/or the increased level of exposure. Ingredients of traditional foods are being extracted or chemically synthesized and used in ways that alter both the form and context of their traditional use—for example, phytosterols derived from vegetable oils. Thus, in some cases consumers are being exposed to levels of food ingredients that are considerably higher than the levels to which they have been exposed in traditional foods. The form of the ingredient used in a functional food may also be different, which could influence both its bioavailability and toxicity—for example, the use of chromium picolinate compared to other forms of chromium.

Food ingredients and other bioactive substances are also increasingly being sourced from nontraditional food sources, such as microalgae in the case of docosahexaenoic acid (DHA) and tall oils in the case of phytosterols. This raises the possibility of introducing other substances with unknown toxicity into the food supply; establishing specifications for the purity and identity of the ingredient is important in these cases. Two regulatory measures that can be used to ensure the safety of functional food ingredients in New Zealand and Australia are discussed below; that is, regulation of novel foods and of nutritive substances and other bioactive substances.

Regulation of Novel Foods in Australia and New Zealand

To address concerns associated with nontraditional foods and food ingredients, a new food regulatory measure[1] was introduced in June 2001 to require so-called "novel" food and novel food ingredients to undergo a premarket safety assessment. Under this standard, "novel food" means "a nontraditional food for which there is insufficient knowledge in the broad community to enable safe use in the form or context in which it is presented." Nontraditional food in this context means "a food that does not have a history of significant human consumption in the broad community in Australia or New Zealand" (refer to Standard 1.5.1 in ANZFA, 2002a). This standard is intentionally broad in its scope and potentially encompasses a wide range of foods and food ingredients. It does not characterize novel foods in terms of physical or technological parameters but rather in terms of the inherent risk to the consumer associated with the foods. Nontraditional foods or ingredients likely to be regarded as novel are those that are new to the broad community or are presented in a way that is new, and for which the safety is unknown. It is likely that many "functional" foods or ingredients may be regarded as novel and thus require premarket safety assessments and approval.

Regulation of Nutritive and Other Bioactive Substances in Australia and New Zealand

To ensure the safety of substances added to foods for nutritional purposes, the scope of the food standards has recently been broadened to include not only vitamins and minerals but also amino acids and nucleotides. Such substances are referred to as "nutritive substances," and premarket safety assessment and approval is required before their addition to foods. A nutritive substance is defined as "a substance not normally consumed as a food in itself and not normally used as an ingredient of food, but which, after extraction and/or refinement, or synthesis, is intentionally added to food to achieve a nutritional purpose, and includes vitamins, amino acids, electrolytes and nucleotides" (refer to Standard 1.1.1 in ANZFA, 2002a). Nutritive substances are generally a single chemical entity or related entities.

Other bioactive substances, which do not necessarily have a nutritional purpose, are increasingly being used as food ingredients. These include substances that are reported to have performance-enhancing properties, for example, creatine or glucosamine sulfate (AIS, 2003). Food products containing these substances are specifically designed for individuals engaged in sporting or other high-exertion activities. Because of their specialized purpose, currently these bioactive substances are being considered only for addition to foods that fall within the definition of sports foods and are not permitted in the general food supply (refer to Table to Paragraph 2c, Standard 2.9.4, ANZFA 2002a).

Benchmarks for the Safety of Food Ingredients

For all foods, food ingredients, and substances added to foods, the purpose of a safety assessment is to confirm that there is a reasonable certainty that no harm will result from the intended use of the respective food, food ingredient, or substance added to the food (adapted from OECD, 1993). The type and level of evidence to demonstrate safety may differ depending on the particular circumstance. For foods or food ingredients for which there is some history of safe use in another country, for example, South American root tubers, the level of additional evidence required to establish safety may be less than that for a food or food ingredient that is completely new to the human diet. Specific information on the history of use of a food ingredient includes the frequency of consumption, the level of exposure for the whole population or specific subpopulation groups, and the period of use.

Similarly, for a new food ingredient that has some use in therapeutic goods, for example, creatine, there may be relevant safety information available, such as potentially adverse effects in humans, which can be used to assess its safety for use in foods. Other useful information could include the exposure levels, duration of exposure, and intended target population. It is important to note however, that prior use of an ingredient in a therapeutic good does not ensure its safety in foods; therapeutic goods generally carry a higher risk and are designed for individuals to treat or to alleviate the symptoms of disease. They are therefore usually used on the advice of a medical practitioner or health worker and are accompanied by advice on use, including safe dosage levels. There are also post-market reporting mechanisms for possible adverse effects.

In contrast, foods provide a nutritional benefit and are expected to be low risk for all sectors of the population. There is generally no information provided on levels of use and, in most cases, there are minimal restrictions on the use of food by particular subpopulation groups. The suitability and safety of adding to foods substances reported to have some therapeutic benefits needs to be considered carefully in the context of potential for misuse or over-exposure, or for misleading consumers with respect to possible benefits.

For all foods and food ingredients, FSANZ requires basic information in order to undertake a safety assessment such as, the characterization of the material, its history of use, the expected levels of dietary exposure, and known potential for adverse effects (ANZFA,

2000a). Although the emphasis of research on functional foods will primarily be on the potential benefits, there is also a need to generate adequate data to address safety concerns. In the assessment of a food or food ingredient for regulatory purposes, public health and safety are taken into account, but not potential benefits beyond essential nutrition. This is a complex issue that may need future consideration as the more traditionally held views of health benefits obtainable from foods are challenged.

Although functional foods potentially range from single chemical entities to complex foods, the initial focus of the food industry has been on particular ingredients that are reported to induce a physiological change, which may be linked to a positive health outcome. One such example is phytosterols (Application A410, ANZFA 2000b.), discussed in the box that follows.

Box 19.2

Phytosterols: A case study

Phytosterols, both esterified and nonesterified, have been a particularly prominent example of a food ingredient that is reported to have functional properties (lowering of plasma low density lipoprotein cholesterol) that may lead to positive health outcomes for the community. In this case, the marketed ingredient was well characterized in terms of composition and purity.

Potential for adverse effects was addressed by both animal and human studies, although the human studies were conducted only at the dietary intake level equivalent to the level recommended for health benefits from a single food. The longer-term human studies were limited to a three-month study at several dose levels and a 12-month study at a single dose level. Although phytosterols have been used as therapeutic agents at higher dose levels in some countries, these data were not considered appropriate to support the safety of phytosterol use in foods because of differences in chemical composition, differences in the target population, and the lack of adverse reaction monitoring for foods. There was also no evidence that phytosterols lowered plasma LDL cholesterol when used in foods other than edible oil spreads. Initial use of phytosterols and phytosterol esters was therefore restricted to one food type, namely, edible oil spreads at a level equivalent to 8% (w/w) free phytosterols.

Nutritional Issues

Regulatory Principles for Addition of Vitamins and Minerals to Foods

In Australia and New Zealand, the addition of vitamins and minerals to food has traditionally required premarket approval. The regulatory principles underpinning the approval to add vitamins and minerals to conventional foods (both general purpose and special purpose) are firmly grounded within a traditional nutrition paradigm that has dietary adequacy (including essentiality) as its basis to support physiological growth, development, and maintenance of health.

The approach to vitamin and mineral addition to general-purpose and special-purpose foods is based on the international precedent established by the Codex *General Principles for the Addition of Essential Nutrients to Foods* (Codex Alimentarius Commission). The FSANZ Regulatory Principles (Lewis, Broomhead, Jupp and Reid, 2003) are summarized as follows:

1. modified restoration (equivalent to or more than pre-processed levels)
2. nutritional equivalence of substitute foods
3. fortification (to address public health need)
4. appropriate nutrient composition of a special-purpose food

The first two principles confine the range of addition of particular vitamins and minerals to those found naturally in certain reference foods. Furthermore, principle 2 establishes maximum limits according to those found in the reference food. In all cases, and after an assessment of the risk of the proposed food contributing to excess vitamin or mineral intakes from total dietary and supplemental sources, the maximum permitted total natural and added amount of vitamins and minerals in general purpose foods is 50 percent adult Recommended Dietary Intake (RDI) per serving. The minimum amount to qualify for a vitamin or mineral content claim is 10 percent RDI per serving.

Fortification to address public health need may be either mandatory or voluntary. In the case of voluntary fortification, a further criterion, in addition to those outlined previously, is applied, namely: "only foods whose fundamental nutritional composition is consistent with authoritative dietary guidance such as national dietary guidelines are permitted voluntary fortification." This approach takes into account the potential for fortified foods to be regarded as nutritious irrespective of their base composition and the requirement that fortified foods not undermine the broad aims of government nutrition policy and advice.

Relatively little controversy exists on the assessment of risk of excess vitamin or mineral intake based on dietary intake assessments of particular fortification scenarios, particularly because safe or tolerable upper limits of vitamins and minerals are being established internationally for many of these micronutrients. However, there is much debate about the regulatory role in deciding appropriate food vehicles for vitamin and mineral addition.

Evidence of Efficacy

Vitamins and minerals are examples of nutritive substances and constituents in foods that are essential for life. They have a well-established evidence base for their essential role in maintenance of health and prevention of deficiency disease, and several scientific bodies around the world have established dietary reference values for them. The primary purpose of functional foods is not conventional nutrition but one that focuses on health enhancement or specific reduction of risk of disease, including through effects on risk factors such as serum cholesterol.

The evidence for some vitamins having a specific health enhancement or protective role is emerging but is generally not sufficiently developed for authorities to establish associated reference values for regular dietary intake. Regulators such as FSANZ have previously relied on the generally available and undisputed evidence for dietary essentiality of vitamins in order to decide on acceptable regulatory parameters for the general food supply. The challenge posed by functional foods lies in how to determine appropriate levels of evidence for other health enhancement roles as a basis for approval of the addition of vitamins to foods with a functional intent. An additional challenge is to decide an acceptable level of evidence to apply across the whole spectrum of nutritive substances, that is, including those for which evidence for dietary essentiality does not exist for example, nonessential amino acids, nucleotides, and glucuronolactone.

Nutritional Quality of Functional Foods

The nutritional quality proffered by the food "vehicle" is also an important consideration for regulators. Several precedents within the nutritional paradigm exist in the *Food Standards Code* to maintain the nutritional quality of foods, for example, application of qualifying compositional criteria to a food category before it can be approved for vitamin fortification. Functional foods potentially vary in their basic nutritional quality, including:

- those containing a highlighted biologically active substance that is consistent with the nutritional contribution of the food vehicles—for example, breakfast cereals containing beta glucan formulated for bowel health, or polyunsaturated margarines containing phytosterols
- those that are consistent with authoritative dietary guidance but have no functional link to the highlighted active substance—for example, yogurt with echinacea
- those that could be considered as inconsistent with the thrust of authoritative dietary guidance, such as vitamin fortified sugar confectionery

The issue for regulators is the extent to which the underlying nutritional quality of the foods should be regulated and whether this should depend on its conformance to authoritative dietary guidance. One of the main concerns is the potential risk of dietary distortion if food vehicles are used that are not commensurate with the traditional nutritive profile.

One approach to maintaining a distinction between functional foods and general-purpose foods is to specify the regulatory requirements separately, thus maintaining the integrity of the regulatory principles applying to general purpose and special purpose foods. It is considered that this approach will offer greater sustainability in the face of a potentially increasing array of highly fortified products.

The discussion of Formulated Caffeinated Beverages (FCBs) or "energy drinks"[2] that follows provides a case study of the approach taken in Australia and New Zealand to the regulatory management of this particular functional food (ANZFA, 2001). These products contained higher amounts of vitamins than are permitted in general-purpose foods, and thus challenged the established regulatory principles for vitamin content of general-purpose and special-purpose foods.

Claims Relating to Functional Foods

The claims that manufacturers of functional foods may want to make about their products form a continuum, as shown in Figure 19.3.

"Health claims" as such are not defined in New Zealand or Australian food standards (although certain matters are prohibited), and the definition of a health claim varies internationally. Categories 3–4 in Figure 19.3 tend to cover what are commonly called "health claims" in the Australian and New Zealand context, and this term is used later in this chapter.

Consumer knowledge is variable about the risks and benefits of new ingredients and substances added to the food supply or those that have traditionally been regulated as com-

Box 19.3

Formulated caffeinated beverages: A case study

Permission was sought for addition of certain B group vitamins to FCBs on the basis of evidence for their beneficial role in energy metabolism.

FSANZ considered the application against its regulatory objectives, the primary one being protection of public health and safety. Dietary modelling was used to estimate total vitamin intake from a diet that included FCBs; permissions were granted provided that the total intake was below each vitamin's established upper reference limit. In many cases, these permissions were above the customary limit for general-purpose foods namely, 50% of the RDI per serving—for all population groups likely to consume the product. Only maximum (as opposed to minimum) limits on vitamin content were established, and these limits were expressed per one-day quantity as determined by the manufacturer such that the higher the concentration of vitamin, the smaller the one-day quantity of FCB to be advised on the label. No controls over the underlying nutritional quality of the product were established.

Complementary risk-management strategies were applied through labeling controls to protect vulnerable consumers (for example, not suitable for children) to ensure that the product was not promoted for nutritional reasons and to minimize any potential for consumer deception.

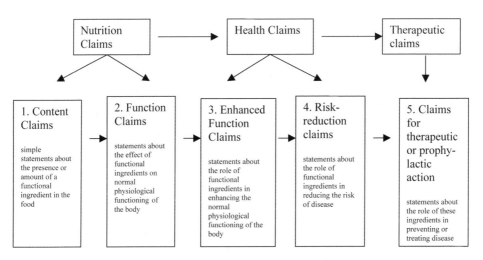

Figure 19.3. Continuum of claims. Reproduced from: *Review of Health and Related Claims*. Full Assessment Report (ANZFA, 2000c).

plementary medicines or dietary supplements. The knowledge of some ingredients, such as nonculinary herbs, is characterized by gaps in scientific knowledge about effectiveness, but there exists considerable anecdotal information and/or an established belief in the benefit of these substances and ingredients. In these cases, the mere declaration of a particular ingredient as required by current food regulatory measures, or a simple claim as to its presence in a food (without any related claim for benefit being made) may, depending on consumers' understanding, imply benefits that may not be able to be substantiated scientifically.

Is it the responsibility of the regulator to address this issue and consider ways within its legislated mandate to ensure that consumers are not deceived? One way of approaching this may be, for example, to prescribe specifications for certain ingredients to ensure that either an appropriate part of the plant is used or a minimum amount of the biologically active substance is present before a claim for presence is made. However, this approach may be problematic because of the lack of data on the range of chemical entities that may beneficially affect human health. Alternatively or additionally, should consumers be alerted via a disclaimer to the potential unreliability of these ingredients to deliver implied beneficial outcomes? What role does consumer choice, including uninformed consumer choice, play in this regard?

The current regulatory environment in Australia and New Zealand in relation to the types of claims shown in Figure 19.3 is dynamic and may change in the near future as trends in other countries and the role of food standards in controlling such claims is reviewed. Currently, some content claims are regulated through the *Food Standards Code* (for example, those related to vitamins and minerals; fatty acids, lactose, gluten, and sodium) (refer Standards 1.3.2 and 1.2.8 respectively in ANZFA, 2002a) whereas others are left to self-regulation or the general requirements of fair-trade laws (that is, that claims be truthful and not misleading). Function claims are left to the general requirements of fair-trade laws (that is, that the information should not be false or misleading)[3].

Some enhanced function and risk-reduction claims (refer to Figure 19.3) and claims for therapeutic or prophylactic action are regulated through the prohibitions in the *Food Standards Code* relating to labels on, or advertisements for, foods. However, the current standard (refer to Standard 1.1A.2 in ANZFA, 2002a) is problematic because no explicit definitions are provided for the types of claims that are permitted or prohibited and, furthermore, the current arrangements do not always capture implied claims. This has the effect of making the current regulatory measures difficult to interpret and enforce, creating confusion for the industry, consumers, and enforcement agencies.

The current arrangements can also restrict the extent to which messages about the role of foods in healthful diets, or in diets for managing particular conditions, can be conveyed to consumers. These arrangements therefore constrain the formation of partnerships between industry groups and public health or professional associations in promoting healthful diets. Such initiatives can find difficulty in promoting the specific benefits of consuming particular foods, such as reducing the risk of developing cancers or heart disease.

As part of the overall process of determining whether the current standard in relation to health claims in Australia and New Zealand should be reviewed, and potentially relaxed to permit certain claims, a pilot project was conducted in 1998. The objectives of the pilot were to try the use of a proposed management framework for health claims using a well-substantiated claim, and to address a significant public health problem. Further details are provided in the box that follows.

> Box 19.4
>
> ### Pilot health claim trial—folate and neural tube defects
>
> A claim relating to the relationship between periconceptional folate intake and reduced risk of neural tube defects (NTD) was used for the pilot, and a special exemption to the prohibition on the making of health claims was made to the *Food Standards Code* to permit the claim to be made in food labeling and advertising. The management framework used in the pilot had five components, as shown in Table 19.2. The pilot ran for a period of 12 months, and more than 100 foods were approved to use the claim, including a mixture of primary and processed foods. Although many foods were approved to use the claim, the actual number of these foods that carried the claim, or for which the claim was used in advertising, was significantly less.
>
> **Table 19.2.** Management framework used in the folate pilot. Reproduced from: *Review of Health and Related Claims*—Full Assessment Report (ANZFA, 2000c).
>
> | A Regulatory Mechanism | Substantiation and Qualification | Education and Communication | Monitoring and Evaluation | Compliance and Enforcement |
> |---|---|---|---|---|
> | A standard permitting different types of health claims and supporting interpretive guidelines/ industry code of practice. | Systems for establishing the substantiation, nutritional qualifying and disqualifying criteria, and communications effectiveness of claims. | A coordinated national nutrition education campaign in support of health claims. | A system for managing health claims that would include ongoing programs for monitoring and evaluation. | A nationally coordinated co-regulatory system for compliance and enforcement. |
>
> ### The management framework
>
> Stakeholders consulted as part of the process evaluation for the pilot considered that it adequately covered the main areas and issues relevant to managing claims and that it ensured that relevant partners were engaged in the elements of the framework that related to their needs. The expert advisory committees established for the purposes of the pilot were considered to provide a credible, independent basis for substantiating the health claim and establishing the eligibility criteria for approved foods. In addition, stakeholders considered that the co-regulatory approach achieved a balance between industry self-regulation and legally binding enforcement of the *Food Standards Code*.
>
> The process evaluation also identified several aspects of the framework that could require strengthening if it were to be used for health claims more broadly. The first of these was the need for guidelines to be developed on the requirements and scientific substantiation of claims other than those for folate and NTDs. It was also recom-

> mended that there was a need for stakeholder consultation on the rationale and merits of claims. There is also a need to clarify the interface between the Code of Practice Management Committee and the enforcement agencies for the *Food Standards Code* in relation to their respective roles and responsibilities, to avoid duplication and confusion. Furthermore, the ability to sustain public education activities in relation to claims was raised as an area that required consideration if health claims were to be permitted more broadly.
>
> The pilot health claim process demonstrated that it was possible to design a comprehensive framework for managing a health claim, utilising the resources of government, industry, and public health agencies.
>
> ### Impacts on public health
>
> The proportion of people who had heard of folate and were aware of the folate-NTD association increased during the period of the pilot. However, the ability to link this increase to the pilot *per se* was limited, due to other activities related to increasing public awareness of this issue that were occurring at the same time. Knowledge of food sources of folate increased during the pilot for both leafy green vegetables and breakfast cereals. However, there was no change in the sales of products approved to carry the folate claim during the period of the pilot (ANZFA, 2000d).

In terms of applying the findings of the folate/NTD trial more broadly, there are several issues to consider, including:

- What is an appropriate definition of a health claim?
- Related to the previous, what is an appropriate level of regulatory intervention for health claims?
- What types of evidence are needed to substantiate a health claim?
- What systems need to be put in place to ensure that claims remain valid over time?
- What is an appropriate level of public education to support claims—is the label information enough?

New standards in relation to health and nutrient claims in Australia and New Zealand are under development following ministerial guidance on the broader policy environment.[4]

Conclusion

Currently, Australia and New Zealand do not have a comprehensive regulatory framework for functional foods *per* se, and it is still not clear as to the extent to which such a framework is necessary.

Balancing the requirements for a thorough safety evaluation for functional foods against the practical limitations of testing in both animals and humans will be a challenge for the development of functional foods in the future. Similar challenges lie in the requirements for manufacturers to be able to adequately describe and promote their products, thereby allowing consumers to be adequately informed but not misled.

If a decision is taken that a more specific framework is required, then several options are available, including:

- Permissions may be spread across relevant horizontal standards, for example, novel foods, labeling requirements, vitamin and mineral permissions.
- Specific vertical (commodity) standards may be developed to accommodate different functional food types.

Account must also be taken of the binary classification of products as foods or therapeutic goods and the need to maintain the integrity of the food supply as it relates to general-purpose and special-purpose foods. Australia and New Zealand will also build on the lessons learned from regulatory systems established in other countries.

References

AIS 2003 *AIS Sports Supplement Program*. www.ais.org.au/nutrition/SuppPolicy.htm
ANZFA 2000a *Guidelines for amending the Australian Food Standards Code. Standard A19-Novel Foods*. www.foodstandards.gov.au/novelfoodsappguide
ANZFA 2000b. *Phytosterol Esters Derived from Vegetable Oils*. Full Assessment Report, Application A410. www.foodstandards.gov.au/standardsdevelopment/applications
ANZFA 2000c. *Review of Health and Related Claims*. Full Assessment Report, Proposal P153. Information Officer, FSANZ, PO Box 7186, Canberra MC, ACT 2601, Australia.
ANZFA 2000d. *Evaluating the Folate-Neural Tube Defect Health Claim Pilot*. Process Evaluation of the Management Framework. Outcome Evaluation. Information Officer, FSANZ, PO Box 7186, Canberra MC, ACT 2601, Australia.
ANZFA 2001. *Formulated Caffeinated Beverages*. Inquiry Report, Application A394. www.foodstandards.gov.au/standardsdevelopment/applications
ANZFA 2002a. *Australia New Zealand Food Standards Code*. Australia New Zealand Food Authority, Anstat, Melbourne. Also at www.foodstandards.gov.au/foodstandardscode
Codex Alimentarius Commission. *General Principles for the Addition of Vitamins and Minerals to Foods* CAC/GL 09—1987 (Rev 1989, 1991).
FSANZ 1991. The *Food Standards Australia New Zealand Act 1991*. http://scaleplus.law.gov.au (accessed March 2003)
Healy, Brooke-Taylor and Liehne. 2003. Reform of food regulation in Australia and New Zealand. Food Control. Vol. 14, No. 6, Sept. 2003, 357–366.
Lewis, Broomhead, Jupp and Reid. Nutrition considerations in the development and review of food standards, with particular emphasis on food composition. *Food Control*. Vol. 14, No. 6, Sept. 2003, 300–407.
NFA 1993. Functional Foods Policy Discussion Paper. July 1994.
OECD 1993. *Safety evaluation of foods derived from modern biotechnology: concepts and principles*. OECD Publication Service, Paris.

Glossary

AIS	Australian Institute of Sport
ANZFA	Australia New Zealand Food Authority
FSANZ	Food Standards Australia New Zealand
NFA	National Food Authority
OECD	Organization for Economic Cooperation and Development

Notes

1. Such regulatory measures are referred to as food "standards" in Australia and New Zealand.
2. Refer to Standard 2.6.4 in ANZFA 2002a.
3. The respective Commonwealth and State legislation can be referenced at http://www.austlii.edu.au (accessed March 2003).
4. Refer to www.foodsecretariat.health.gov.au.
5. ANZFA/FSANZ documents accessed from the Web site www.foodstandards.gov.au can also be obtained from the Information Officer, FSANZ, PO Box 7186, Canberra MC, ACT 2601, Australia.

20 Regulation of Functional Foods in Spain

Luis García-Diz and Jose Luis Sierra Cinos

The absence of a specific regulation on functional foods in Spain has forced the food industry marketing them to use other regulatory standards, such as those for dietary foods and/or foods for special diets, which are governed by Real Decreto (Royal Decree) 2685/1976 [1] and subsequent amendments. However, this regulation applies only to products explicitly directed at "special situations" (that is, pregnancy) or "special diets" (that is, weight loss). Some foods among those classified as "functional" have been marketed under this regulation, but this is obviously not a logical marketing channel. Only those foods intended for particular nutritional uses should be labelled and marketed as dietary foods.

Since becoming law, Real Decreto (RD) 2685/1976 has been modified on numerous occasions. The result of these modifications was to expressly state that dietary foods/foods for special diets are intended for people with certain physiological conditions, for special situations in which a controlled intake of certain food components would be beneficial, and for people who have problems with the processes of assimilation or metabolism. All these situations represent an altered physiology.

In our country, commercialization of these types of products requires a prior notification to the overseeing public health authority, except in special cases such as for low caloric value. In any case, whether the dietary products are attributed with preventive or healing properties, these should be classified under the appropriate regulation. This frames these products clearly outside the category of functional foods.

Actual situation in Spain

In general, the labelling, display, and advertising of foodstuffs is regulated by RD 1334/1999 and its subsequent modification 238/2000. These laws prohibit any claims of health benefits in the labelling and advertising of foods and foodstuffs.

In Spain, nutritional claims are governed by RD 930/1992. Only claims related to calories, macronutrients, cholesterol, fiber, and sodium are allowed under this legislation. Nutritional claims regarding vitamins and minerals are considered only when they appear in significant quantities (more than 15 percent of RDI). Foods with nutritional claims are required to bear nutrition labelling.

Frontier products regulation

The increased understanding and public awareness of the dependence of good health on eating a proper diet has led to growing numbers of "frontier products," "fantasy products," or "miracle products." Out of concern for this trend, Spanish authorities developed RD 1907/96, which attempts to regulate the advertising and commercial promotion of products, activities, and services that claim a healthful purpose.

Some articles of this RD limit the advertising and marketing of food with supposed functional properties.

Article 1. Public Health Control of Advertising

1. Public health authorities and other related governing bodies in each case, in agreement with article 27 of the General Law of Public Health, as well as the applicable special regulations in each case and those established in this Ordinance, will control the advertising and commercial promotion of the products, materials, substances, energies, or methods that declared or are presented as being useful for the diagnosis, prevention, or treatment of illness or physiological developments, weight loss, change in physical or psychological state, and the restoration, correction or modification of organic functions or other health purposes.

   (...)

Article 4. Prohibitions and Limitations on the Advertising of Claimed Healthful Purpose

(...)

Any kind of advertising, either direct or indirect and of either mass or individualized promotion, of products, materials, substances, energy or methods with claimed healthful purpose is prohibited in the following cases:

1. Advertising aimed at the prevention, treatment, or cure of communicable diseases, cancer, and other tumorous illnesses, insomnia, diabetes and other diseases of the metabolism.
2. Advertising that suggests specific slimming properties or that claim to counter obesity.
3. Advertising that claims a therapeutic usefulness for one or more illnesses, without adjusting to the requirements and exigencies anticipated in the Law of Medications.
4. Advertising that provides guarantees of relief or certain cure.

   (...)

7. Advertising that attempts to use testimonials made by public health professionals, famous or well-known people, or by real or depicted patients, as a way of eliciting consumption of the product.
8. Advertising that seeks to substitute for a normal dietary regiment or common practices of good nutrition, especially in the cases of maternity, nursing, childhood or advanced age.
9. Advertising that attributes concrete and specific preventive, therapeutic, or healing properties to certain forms, indications, or brands of nutritious products of ordinary consumption.
10. Advertising that attributes preventive properties, cures, or other properties, apart from those recognized in this special category of regulation, to nutritional products aimed at dietary or special regiments.

    (...)

12. Advertising that suggests or indicates that use or consumption of the specific product increases physical, psychic, athletic, or sexual performance.
13. Advertising that uses the term "natural" as characteristically linked with preventive or therapeutic effects.
14. Advertising of a superfluous character or that seeks to replace the value of medications or legally recognized health products.
15. Advertising of a superfluous character or that seeks to substitute for the consultation or intervention of health care professionals.
16. And, in general, advertising that makes specific preventive or therapeutic claims that are not backed by sufficient clinical tests or certified scientists and expressly recognized by the State Public Health Administration.

In the last few years the European Commission has carried out a wide study outlined in the white paper (COM (1999) 719 White Paper on Food Safety) that has given rise to a large number of regulatory initiatives on functional foods, incorporating the particular opinions of all the member countries, allowing that the positions of the European Union in this field and the opinions of each of the countries, Spain among them, are known through the statements presented in the document.

Interpretive Agreement Regarding the Advertising of Food Properties in Relation to Health, Signed by the Ministry of Health and Consumption (MSC) and the Spanish Federation of Industries of Food and Beverages (FIAB)

In current view of the reality of a market that has already accepted functional foods and the legal framework usually used to carry out their sales to the final consumer, the authorities responsible for health and consumer protection have signed an agreement with the representatives of the food and beverage industries, in which they establish acceptable interim standards awaiting the adoption of the national legislation for these products. Parts of this document precisely reflect the feeling of Spanish health administration authorities and make it possible to manage the functional characteristics of food or nutritious products on the part of the industry.

The purpose and scope of this agreement is to regulate the advertising of food properties related to health, until the current European directive with application toward functional foods is approved and adapted to the national legislation.

In Spain, at the time of signing this agreement, the advertising of food properties related to health is regulated by Real Decreto 212/92, General Standard of the labelling, display, and advertising of nutritious products, and by RD 1907/96 regarding advertising and the commercial promotion of products, activities, or services with claimed healthful purpose.

The first of these, in Article 4, literally establishes that the labelling and the means thereof should not:

1. Be of such a nature that misinforms the buyer, especially:

   (...)

1.4. Attributing to a nutritious product a preventive, therapeutic or healing properties of a human illness, excepting the applicable qualities of mineral waters and the nutritious products dedicated to special diets.

In accordance with this article, preventive, therapeutic, or healing indications of a human illness cannot be attributed to alimentary products. The alimentary products may possess physiological or nutritious properties and may declare those properties, so long as they respect the principles of RD 212/92.

To adapt the commercial reality of the last few years to the next effective legal framework forthcoming in the countries of the European Union, the signatory parts leave the following out of the scope of application of the voluntary agreement:

- Claims that attribute preventive, therapeutic, or healing properties of a human illness to a nutritious product, or that mention those properties.
- Claims relative to foods prepared for dietary or special regiments of their diverse natures that will be regulated by RD 2685/76, for which the Technical-Sanitary Regulations (TSR) are approved, as well as for the specific legislation that establishes itself via TSR in agreement with the annex of RD 1809/91 that the TSR modifies.
- The claims relative to bottled waters that will be governed by RD 1164/91 for which the TSR is approved for the elaboration, circulation, and trade of bottled drinking water.
- Also excluded from the scope of application of this document are declarations of nutritious properties (also known as "nutritional claims") of the type: "it contains vitamin A", "with calcium", ... that are regulated by RD 930/92. This law includes labelling standards for nutritious properties of food products and is carried out in accordance with that which established it.

Health claims are allowed according to the definition set forth in the following section. A concern addressed in this document is to define the boundary between these claims and preventive, therapeutic, or healing claims. Because of the different interpretations about where such a boundary is located, the object of this document is to trace the boundary in a clear and operative manner, also establishing a pursuant commission to which doubts could be presented.

Definition of Health Claims

For the purpose of this agreement, "health claims" shall be defined as all claims relative to:

- The function of one or more nutrients or constituents of a food in the human body,
- The effect of one or more food products on a person's health, or
- Healthful nutrition habits.

The agreement establishes a clarifying framework for the advertising of food properties in relation to health so that those responsible for making claims will know the limits within which claims of healthful qualities can be made. With the purpose of facilitating the interpretation of claims, an example list of prohibited claims was created, both in terms of their contents and the meaning of the individual words contained in the claims.

Because no individual reacts to similar stimuli in exactly the same physiological manner, the use of categorical words that imply a beneficial action is not acceptable. Consequently, the words "to facilitate," "to favor," "to help," and so on indicate simply that it is easier to achieve a certain objective with the specified action as a starting point, but attainment of the objective can also depend on many other factors.

In case doubts exist on whether certain claims related to health do not conform to what was established in the Interpretive Agreement, the Pursuant Commission will create a report aimed at solving the problem. The voluntary use of this Interpretive Agreement on the part of the signatory parties is done within the current legislation on the matter. The Agreement shall not serve to deny the possibility of using the jurisdictional course.

General Requirements for Reinforcing the Right of the Consumer to Truthful and Legal Information Related to Health

- When a claim is made related to health it must pass scientific tests that demonstrate the truthfulness and accuracy of the stated or implied objectives.
- The nutrient or food constituent from which the claim originates should be present (or absent in some cases) in a significant quantity for the function or property attributed to it.
- The declaration of properties in printed advertising and in labelling must be accompanied by a mention of the importance of consuming a balanced and varied diet to satisfy nutritional needs (example: a varied and balanced diet is recommended).
- It shall not be suggested that a certain trademark of a food product produces particular effects in the body when similar products produce those same effects.
- When a health claim is made, all nutritional labelling requirements of RD 930/92 must be met.

Examples of Prohibited Claims

Next, a relationship will be shown, although merely as an example (and not exhaustive), that includes certain claims that are prohibited (because the relationship is merely illustrative, in no way will it be understood that any other claim that does not appear in the relationship is allowed). This guide is a good example of the intent of the health authorities with regard to what was allowed prior to the appearance of these types of functional foods.

In the following list of suggested examples, the term or terms for which they are not considered allowed is highlighted in boldface.

Regenerative effect or action
Eliminates the need or substitutes for medication
Tonic effect or action
Relaxant effect or action
It helps to protect you against **intestinal infections**
Cures constipation
The regular consumption of X decreases **the risk of suffering . . . disease**
Eliminates the need or substitutes for a medical treatment
It improves your **natural defenses**
It protects you against **X illness**
It helps fight **osteoporosis**

It **stimulates** the immune system
It facilitates the **elimination** of cholesterol
It facilitates **correct** ossification
Fiber **prevents colon cancer**
X brand oranges are recommended by the WHO
Product X helps you **to lose weight**
The regular consumption of X is recommended **to fight against kidney stones**

(. . .)

All these examples appear literally in the agreement signed by the Spain Federation of Food and Beverage Industries; therefore, there is small possibility for the inclusion of health claims in the advertising of foods or nutritious products with functional characteristics.

Last, the composition and functions of the Tracking Commission of the agreement is detailed.

Tracking Commission of the Agreement about Interpretation of Advertised Properties of Foods in Relation with Health

The Tracking Commission will be formed by two members from the Health Administration and two from the Federation of Food and Beverage Industries.

In case doubts exist on whether certain claims related to health are in agreement with what is established in the Interpretive Agreement or not, a report from the Tracking Commission can be requested before their diffusion. If the party responsible for the claim requests it, he or she will be given audience to justify the claim before the Tracking Commission. This Commission will answer to the submitted consultation requests related to the advertising of health-related properties within a maximum period of 10 calendar days. At the end of this period, if there has not been an answer indicating necessary modifications, the advertising will be allowed to use.

In this manner, administrative silence or delay in resolution of the outlined cases will not paralyze the activity of the companies.

The agreement, accepted and ratified by the parties, was signed March 20, 1998, in the Ministry of Health and Commerce in Madrid.

Diet supplements regulation

Other legal framework in which, at this moment, these types of products seek protection is being presented to the consumer as diet supplements, using the fact that some of them increase normal levels of some of the nutrients they contain. In this case neither preventive nor therapeutic allegations can be made, because they are expressly prohibited in the standards and there are maximum allowed quantities for some nutrients.

The form of presentation (tablets, capsules, pills, and so on) and labelling of these products place them before the consumer's eyes near medications, creating a new category called "Nutraceuticals, Pharmafoods, and Biofoods."

In October 2003, RD 1275/2003 regarding diet complements was published. This assumes a translation of community standards on the same subject to Spain. It has special importance because a large number of products classified as nutraceuticals are probably protected in this legislation for their commercialization. There exists an ample range of nutrients and other elements that may be present in these nutritional supplements, including vitamins, minerals, fatty acids, amino acids, fiber, and others. In this "RD," only the standards for vitamins and minerals are established. At a later date, if the European Union has not yet established a regulation relative to other nutrients and substances used in nutritional supplements, standards that are clearly based on scientific data may be adopted. The vitamins and minerals that can be used are included in appendix I of the RD.

On the other hand, it is important that the vitamins and minerals used in the preparation have an adequate bioavailability, and for that reason only certain vitamin-like substances and mineral salts can be used. In the production of this type of preparation, other substances that have been approved by the Scientific Committee of the Human Alimentation may also be used. This RD specifies the maximum and minimum amounts of vitamins and minerals that can contain the nutritional complements. Using the reference of the RDI published in RD 930/1992, the amounts of vitamins and minerals that can contain these preparations will be such that the daily intake recommended by the manufacturer corresponds to over 15 percent of the minimum RDI and 100 percent of the maximum RDI. The labelling of this type of product must be of nutritional content, specifically prohibiting the attribution of preventive or curative properties of diseases and reference to nutritional or health claims.

While this standard is in force in the institutions of the European Union, a new standard is being studied that will provide an opening for functional foods, harmonizing the postures of the different countries. In this new agreement, only the category of "functional foods" would be regulated, definitively leaving out other alternative classifications that could be confusing for the consumer.

Community Legislation

The community directives are focused at the moment on obtaining coordination among food-related legislations that address industrial innovation, security of products, rights of the consumer, and the free movement of goods within the member states.

The European Union member states have different rules and interpretations regarding food-related legislation. Therefore, a claim that is permitted in one country may well be prohibited in another and vice versa. As a result, certain foods cannot be distributed in certain countries, which has the direct effect of creating unequal conditions of competition on the interior market in those countries. It therefore becomes a priority to adopt rules at the community scale applicable to these commercial food products.

With agreement to the community legislation there are no separate categories for functional foods or functional food ingredients, although admittedly there is the possibility that the expression "functional" could be included on the labels of some products.

Labelling regulation

Directive 2000/13/EC relative to the labelling, presentation, and advertising of foodstuffs makes expressed reference to the prohibition of attributing in the labelling and the advertising references to the prevention, treatment or cure of illnesses.

Article 2

1. The labelling and methods of labelling must not:
    (a) be such that could mislead the purchaser to a material degree, particularly:
        (i) as to the characteristics of the foodstuff and, in particular, as to its nature, identity, properties, composition, quantity, durability, origin or provenance, and the method of manufacture or production;
        (ii) by attributing to the foodstuff effects or properties which it does not possess;
        (iii) by suggesting that the foodstuff possesses special characteristics when in fact all similar foodstuffs possess such characteristics;
    (b) be subject to Community provisions applicable to natural mineral waters and foodstuffs for particular nutritional uses, attribute to any foodstuff the property of preventing, treating, or curing a human disease, or refer to such properties.

Only in the case of dietary foods directed toward a special dietary regimen, or to special alimentation, is the possibility of making reference to the physiological condition or illness for which they are dedicated obviously considered. In any case, these types of foods are considered under specific legislation regulated by Directive 89/398/EEC and modified partially by the Directives 96/84/EC, for the temporary authorization of commercialization, and 1999/41/EC, in which low sodium and celiac foods are excluded from this food group. Foods dedicated to special medical uses also are regulated separately through Directive 1999/21/EC.

The different categories of dietetic foods included in the actual legislation are:

1. Infant and child formulas
2. Processed cereal-based foods for infants and young children
3. Food intended for use in calorie-restricted diets for weight reduction
4. Dietary foods for special medical purposes
5. Foods intended to restore the physical expenditure of intense muscular effort, especially for athletes

The obvious result of this is that it is impossible to market functional foods through this standard.

Meanwhile, the community created a group of experts that has produced a white paper about alimentary safety, covering all related aspects, including new foods, and the situation and possible regulation of health claims.

White Paper on Food Safety

COMMISSION OF THE EUROPEAN COMMUNITIES
Brussels, 12 January 2000
COM (1999) 719 final

"Assuring that the EU has the highest standards of food safety is a key policy priority for the Commission. This White Paper reflects this priority. A radical new approach is proposed. This process is driven by the need to guarantee a high level of food safety."

In the document is the creation of a European Authority and its corresponding counterparts at the national level, the agencies of food safety. These new organizations will be in charge of securing a high level of protection of consumer health in the area of food safety, which will improve and maintain consumer confidence. They will also make a determination of the risks and communicate their conclusions of those risks related to the field of alimentation. For this, they will be assisted through the advice of expert panels, which will ensure independence, quality, and adequate transparency.

The creation of the European Authority and the national agencies and the beginning of the activities planned for the first half of 2002 have been completed and many of the outlined objectives have already been reached, such as work in the analysis of risks and critical points of control, the new foods (those that have been genetically modified, GMO, or algae), the information to the consumers in the processes of detection of risks, the labelling and advertising of foods and nutritious products so that the consumers make an informed choice, the inclusion of complete labelling of ingredients, the regulation of the therapeutic and dietary claims, deceitful advertising, and nutritional information. The specific plans of the commission regarding nutritional labelling are summarized in points 65, 66, and 67. The nutritional information needed to understand the new labelling of food is found in point 74.

XIV. Labelling of food
No. 65.
Act: Proposal for amending Directive 79/112/EEC on the labelling, presentation, and advertising of foodstuffs.
Objective: To specify the conditions under which "functional claims" and "nutritional claims" may be made.
Adoption by Council/ Parliament July 2002

No. 66.
Act: Proposal for amending Directive on nutrition labelling
Objective: To bring the provisions on nutrition labelling in line with consumer needs and expectations
Adoption by Council/ Parliament July 2002

No. 67.
Act: Proposal for amending Directive on misleading advertising
Objective: To clarify the scope of the Directive with regard to claims concerning, in particular, food, health, and the environment
Adoption by Council/ Parliament July 2002

XVI. Nutrition
No. 74.
Act: Proposal for Council Recommendations on European dietary guidelines
Objective: To support the member states in their development of nutrition policy at the national level. To streamline the flow of information to enable consumers to make informed choices.

Adoption by Council/ Parliament December 2001

Of these objectives, number 65 is focused on the need for defining a standard for functional foods, which is nonexistent at present in the member countries of the European Union. This is supplemented with number 14, which aims to reinforce the training of politicians and the information provided to consumers so that they can adequately understand the claims that will be made on these types of foods or food products.

Among the approved directives that apply to this type of product, there is 2000/13 EC, related to the labelling, presentation, and advertising of foodstuffs, and 2002/46/EC, related to the coordination of the laws of the Member States related to food supplements.

Directive 2000/13/EC of the European Parliament and of the Council of 20 March 2000 on the Coordination of the Laws of the Member States Relating to the Labelling, Presentation, and Advertising of Foodstuffs

(...)

(8) Detailed labelling, giving the exact nature of the product and its specific characteristics, which enables the consumer to make a choice in full knowledge of the facts, is the most appropriate, because it creates the fewest obstacles to free trade.
(9) Therefore, a list should be drawn up of all information that should, in principle, be included in the labelling of all foodstuffs.

(...)

(14) The rules on labelling should also prohibit the use of information that would likely mislead the purchaser or attribute medicinal properties to foodstuffs. To be effective, this prohibition should also apply to the presentation and advertising of foodstuffs.

(...)

Have Adopted This Directive:

Article 1

1. This Directive concerns the labelling of foodstuffs to be delivered as such to the ultimate consumer and certain aspects relating to the presentation and advertising thereof.
2. This Directive shall also apply to foodstuffs intended for supply to restaurants, hospitals, taverns and other similar mass caterers (hereinafter referred to as 'mass caterers').

(...)

Article 2

1. The labelling and methods used must not:
    (a) be such as could mislead the purchaser, particularly:
        (i) as to the characteristics of the foodstuff and, in particular, as to its nature, iden-

tity, properties, composition, quantity, durability, origin or provenance, method of manufacture or production;
(ii) by attributing to the foodstuff effects or properties which it does not possess;
(iii) by suggesting that the foodstuff possesses special characteristics when in fact all similar foodstuffs possess such characteristics;
(b) subject to Community directives applicable to natural mineral waters and foodstuffs for particular nutritional uses, attribute to any foodstuff the property of preventing, treating or curing a human disease, or refer to such properties.

(...)

The rest of the articles refer to the characteristics that must be included in the labelling in terms of regulation, obligatory contents, and form of the labelling itself. The last point states the terms under which this directive would go into effect.

Article 27

This Directive goes into effect 20 days after its publication in the *Official Journal of the European Communities*.

Article 28

This Directive is addressed to the Member States. Approved in Brussels, 20 March 2000.

The candidates to be considered functional foods do not have space in the framework of this directive, but they could opt for being covered under that for dietary products, in which foods modified for special situations are considered.

Food supplements normative

As in Spain, the other marketing alternative within the European environment would be to consider them as food supplements. The European legislation is summarized in Directive 2002/46/EC. The Directive opens the possibility to contemplate substances with nutritional or physiological effect. Reference is made to the establishment of a list of positive components, but on the other hand it defines them as products made in dosed form (capsules, tablets, chewable tablets, and so on), which definitely takes them away from the concept of food. However, this type of presentation could be adequate for "nutraceuticals," and this legislation is nearer to health and nutritional claims than is foodstuff legislation.

Directive 2002/46/EC of the European Parliament and the Council of 10 June 2002 on the Approximation of the Laws of the Member States Relating to Food Supplements

Whereas:

(1) There are an increasing number of products marketed in the Community as food containing concentrated sources of nutrients and presented with the purpose of supplementing the intake of those nutrients from the normal diet.

(...)

(4) Consumers, due to their particular lifestyles or other reasons, may choose to supplement their intake of some nutrients through food supplements.
(5) In order to ensure a high-level protection for consumers and to facilitate their choice, the products put on the market must be safe and bear adequate and appropriate labelling.

(...)

(8) Specific rules concerning nutrients, other than vitamins and minerals, or other substances with a nutritional or physiological effect used as ingredients in food supplements, should be laid out at a later stage, provided that adequate and appropriate scientific data make them become available. Until such specific Community rules are adopted and without prejudice to the provisions of the Treaty, national rules concerning nutrients or other substances with nutritional or physiological effects used as ingredients of food supplements, for which no Community specific rules have been adopted, may be applicable.

(...)

(13) Excessive intake of vitamins and minerals may result in adverse effects and therefore necessitate the setting of maximum safe levels for them in food supplements. Those levels must ensure that the normal use of the products under the instructions of use provided by the manufacturer will be safe for the consumer.
(14) When maximum levels are set, the upper safe levels of vitamins and minerals should be taken into account, as established by scientific risk assessment based on generally acceptable scientific data and intakes of those nutrients as part of the normal diet.
(15) Food supplements are purchased by consumers to supplement the intake of certain elements in the diet. In order to ensure that this aim is achieved, if vitamins and minerals are declared on the label of food supplements, they should be present in the product in significant amounts.

(...)

Article 1

1. This Directive concerns food supplements marketed and presented as foodstuffs. These products shall be delivered to the ultimate consumer only in prepackaged form.
2. This Directive shall not apply to medicinal products as defined by Directive 2001/83/EC of the European Parliament and of the Council of 6 November 2001 on the Community code relating to medicinal products for human use (4).

Article 2

For the purposes of this Directive:

(a) "food supplements" shall mean foodstuffs with the purpose of supplementing the normal diet and concentrating sources of nutrients or other substances with a nutritional or physiological effect, alone or in combination, marketed in dose form, namely forms such as capsules, pastilles, tablets, pills and other similar forms, sachets of powder, ampoules of liquids, drop-dispensing bottles, and other similar forms of liquids and powders designed to be taken in small measured unit quantities;
(b) "nutrients" shall mean the following substances:
   (i) vitamins
   (ii) minerals

This article, despite the fact that it establishes a positive list of substances, also leaves the door open for the incorporation of other substances that are not nutritious.

(. . .)

Article 6

1. For the purposes of Article 5(1) of Directive 2000/13/EC, the name for the commercial products covered under this Directive shall be "food supplement."
2. The labelling, presentation, and advertising must not attribute the properties of preventing, treating, or curing a human disease to food supplements or make reference to such properties.
3. Without prejudice to Directive 2000/13/EC, the labelling shall have the following specific information:
   (a) the names of the categories of nutrients or substances that characterize the product or an indication of the nature of those nutrients or substances;
   (b) the portion of the product recommended for daily consumption;
   (c) a warning to not exceed the stated daily recommended dosage;
   (d) a statement which says that food supplements should not be used as a substitute for a varied diet;
   (e) a statement which says that the products should be stored out of the reach of young children.

Article 7

The labelling, presentation, and advertising of food supplements shall not include any mention, stated or implied, that a balanced and varied diet cannot provide adequate quantities of nutrients in general.

(. . .)

The last published Directive that makes direct reference to the labelling of the foods and foodstuffs in the field of the European Union began as a draft document that was remitted to all the member countries for review and comment. The comments were incorporated into the document for final approval. This text shows perfectly the spirit with which the European Union approaches the topic of the functional foods and the unique points of view of Spain in the claims that it presented. The final text will be the regulatory framework in which foods and functional nutritious products will move in Spain in coming years.

SANCO/1341/2001 Discussion Paper on Nutrition and Functional Claims

Introduction:

1. As food production has become more and more complex, consumers have become increasingly interested in the information appearing on food labels. They have also become more interested in their diet, its relationship to health, and, more generally, the composition of foodstuffs they are selecting. For these reasons it is important that the information about foodstuffs and their nutritional value appearing on the label that is used for their presentation, marketing, and advertising should be clear, accurate, and meaningful.
2. The European Community has adopted detailed rules on nutritional labelling. However, this is not the case with some specific claims. There is, of course, the basic provision that claims should not materially mislead the consumer, and proper enforcement would attempt to prevent abuses in this area. However, Member States have pointed out that this general idea could be open to different interpretations and therefore is not satisfactory for dealing with some specific claims.
3. The food industry has responded to the increased consumer interest in nutrition by providing nutritional labelling on many foods and by highlighting the nutritional value of products through claims in their labelling, presentation, marketing, and advertising. Many people would argue that this evolution could be considered positive for providing relevant information to the consumer. However, for the food industry, it was also an opportunity to use nutritional claims as a marketing tool.
4. In view of the proliferation of the number and type of claims appearing on the labels of foodstuffs, and in the absence of specific provisions at the European level, some Member States have adopted legislation and other measures to regulate their use. This has resulted in different approaches and numerous discrepancies regarding both the definitions of terms used and the conditions warranting the use of claims. These discrepancies could act as barriers to guaranteeing a high level of consumer and public health protection, and could constitute obstacles to the free movement of foodstuffs and the proper functioning of the internal market.
5. For these reasons, coordination of rules regarding claims at the Community level is being advocated. For consumers, rules concerning the classification and conditions for use of nutritional and functional claims have top priority. The industry would also favor uniform rules for the entire Community for a number of reasons. In its White Paper on Food Safety (Paragraph 101, Action no. 65), the Commission proposed consideration of whether to introduce specific provisions to govern "nutrition claims" and "functional claims," to reach the dual objectives of achieving the free movement of foodstuffs between Member States and a high level of consumer protection.

General Considerations

6. Directive 2000/13/EC provides that the labelling, presentation, and advertising of foodstuffs should not mislead the consumer about characteristics of the foodstuffs, or attribute to the product effects or properties it does not possess, or suggest that the foodstuff possesses special characteristics when in fact all similar products possess such characteristics.

(...)

Definition of the Term "Claim"

12. One of the problems in talking about claims is the lack of common understanding about the terminology used. It is therefore highly important to agree on a number of relevant definitions. As a starting point, it would be useful to find a common definition for the generic term "claim."
13. The Codex Alimentarius defines "claim" in the General Guidelines on Claims CAL/GL 1-1979 (Rev. 1-1991)) as

    Any representation which states, suggests, or implies that a food has particular characteristics relating to its origin, nutritional properties, nature, production, processing, composition, or any other quality.

14. This definition shares many elements with the European Union's definition of nutrition claims (see below) but is justifiably broader. It needs to be considered carefully as a starting point in the search for a Community definition.
15. There have been many attempts to define the different types of claims (nutritional claims, nutrient function claim, enhanced function claim, and so on). However, sometimes the differences between them may in fact be insignificant. They often overlap, hence the difficulty in placing them in distinct categories. Furthermore, it is doubtful that all consumers will be able to distinguish clearly between different types of claims. Nevertheless, for reasons of clarity of the rules, it would seem necessary to provide such definitions in any future Community legal measures.

Nutritional Claims

16. **Council Directive 90/496/EEC on nutrition labelling includes a definition of "nutritional claim,"** and this definition could be used as a basis for discussion. The definition is as follows:

    "Any representation and any advertising message that states, suggests, or implies that a foodstuff has particular nutritional properties due to the energy (caloric value) it:

    - provides
    - provides at a reduced or increased rate or
    - does not provide, and/or due to the nutrients it
    - contains,
    - contains in reduced or increased proportions or
    - does not contain.
        A reference to qualities or quantities of a nutrient does not constitute a nutritional claim insofar as it is required by legislation.

17. This definition was adopted over 10 years ago for the purpose of nutritional labelling.

    Although it is true that the majority of nutrition claims concern nutrients or substances that have a nutritional function, such as protein, carbohydrates, fat, components or macronutrients, and vitamins and minerals, there is an increasing number of claims for other substances, such as fiber, antioxidants (lutein, lycopene), and lactic acid bacteria, which do not have a nutritional but rather a physiological effect. Some people argue, therefore, that the definition of nutrition claims should take this fact into account.

(...)

Different Types of Nutrition Claims

19. Among most types of nutrition claims, Codex Alimentarius has defined a "nutrient content claim" as "a nutrition claim that describes the level of a nutrient contained in a food". Examples of nutrient content claims would be "source of calcium," "high in fiber," or "low in fat." These claims could be described as "absolute" nutritional claims. On the other hand, there are "comparative" nutritional claims, which Codex has defined as "a claim that compares the nutrient levels and/or caloric value of two or more foods. Words or phrases such as "reduced," "less than," and "increased" are common in comparative nutritional claims. A compilation of Some of the most commonly found nutrition claims include:

- High /rich/excellent source of
- Without/free
- Increased/low/weak/poor
- Source of/contains/reduced/light

(...)

21. For the use of the term **"light"**, Codex Guidelines suggests following the same criteria as for the term "reduced" and to include an indication of the characteristics which make the food "light" as well. Indeed, this type of claim often refers to very different components of the food; for example, "light" could refer to less fat, less sugar, less caffeine, and so on.
22. Special consideration should be given to the use of the term "diet." It often has similar connotations to the term "light," but there is a high risk of confusion between the terms "diet" and "dietary." Of these terms, "dietary" is being reserved exclusively for foods for particular nutritional uses under Community legislation.
23. Special consideration may also be needed for all **comparative claims**. Indeed, if a comparison is made, it must be clear which products are being compared. For example, if a product bears the claim "reduced fat," the question that arises is, "reduced compared with what?" Therefore, a **reference product** is needed to compare with.

    Which product to select as a reference product then becomes the question. It could be the same brand of the same product, for example "Cream Cheese" and "Cream Cheese Light," but sometimes the same brand product does not exist.
24. All **claims relating to dietary cholesterol** merit particular attention. Indeed, some have strongly argued that, since dietary cholesterol is not a major factor in coronary heart disease and since there is a danger of confusion with blood cholesterol levels, "low cholesterol" claims, "reduced cholesterol" claims, "X% less cholesterol" claims, and "cholesterol-free" claims should not be made.

25. In many cases it has been pointed out that consumers do not understand the difference between **dietary and blood cholesterol**. Most people know that eggs are very high in dietary cholesterol, so they assume that by completely avoiding eggs, they are doing all that is needed to lower their blood cholesterol. Still, they continue to eat fatty foods that are very rich in saturated fats. This concept misleads consumers into thinking that as long as they avoid eggs or other foods containing dietary cholesterol, they can continue to eat any other foods that are high in saturated fat.

26. As far as "low sodium" and "very low sodium" claims are concerned, a difference in approach exists between Community legislation and the Codex Guidelines for the use of nutrition claims. Indeed, under Community legislation these two types of claims are to be regulated under Directive 89/398/EEC on foods intended for particular nutritional uses (dietetic products). This means that products making such claims are considered to be dietetic foods. However, under certain conditions that can be determined under procedures defined in the above Directive, foodstuffs for normal consumption that are suitable for a particular nutritional use may indicate such suitability. The Codex Guidelines do not reserve such claims for dietetic foods. It should also be noted that many people think that the terms "sodium" and "salt" are interchangeable, and therefore the conditions warranting the claims for sodium are also those for claims about salt content.

Criteria for Making Nutrition Claims

27. Considering that the regulations apply to claims, not foods, they must apply to all foods and not only to specific food types, with the exception of foods for particular nutritional uses. This exception especially applies in the case of functional claims. Some would advocate limiting such claims to a distinct category of foods they would call "functional foods", although others would reject that idea. Indeed, every food has a function and there is no adequate justification for creating a special category of "functional" foods.

28. Some maintain that since **"X% fat free" or "only X% fat" claims** can be misunderstood and potentially misleading, even if factually true, they should not be allowed. For example, a product claiming to be "80% fat free", is a product with 20% fat content, which is quite a high fat content for most products, but the claim "80% fat free" can lead consumers to assume that the product is low in fat.

29. It has been argued that claims of **"without additives" or "no additives"** should apply when the product has not been manufactured with the addition of the nutrient that is the subject of the claim, while the nutrient is usually added to similar products.

    As mentioned above, these claims are prohibited if all products of the same category are required by law to be produced without the addition of the nutrient/ingredient/other substance in question.

30. Some of the criteria used to define the term **"low"** in reference to some nutrients are explained in the Annex. However, there may still be a need to distinguish those products that are **naturally low** in a certain nutrient. In this case many prefer to express the claim in the form: "a food low in *(the nutrient)*"; the same is valid for the term "high" or "rich": "a food with high *(the nutrient)* content."

31. Under Community legislation, **claims regarding vitamins and minerals** are allowed if they are present in the product in significant amounts. Directive 90/496/EEC on

nutrition labelling contains an Annex listing the vitamins and minerals for which claims are allowed and their Recommended Daily Allowances (RDA) and states that, as a rule, 15 percent of that RDA should be considered the minimum standard for what constitutes a "significant amount." It should be noted that this list of vitamins and minerals and their respective RDAs dates back to 1990 and stems from the corresponding Codex Guidelines. The revision of this Directive is considered in the White Paper on Food Safety. In the context of this revision of the Directive, it has been suggested that the figure of 15 percent also should be revised. Indeed, some have argued that 15 percent is quite high and that, as a consequence, many foods that are generally considered as good dietary sources of some micronutrients would not qualify for a claim.

32. For **comparative nutrient claims**, a number of **general conditions** should be applied. Such general conditions should include the following:

- The foods being compared should be different types of the same food or similar foods.
- The claim should include a statement of the difference in caloric value or nutrient content.
- The following information should appear in close proximity to the comparative claim:
  –The amount of difference, expressed as a percentage (fraction, or an absolute amount);
  –The identity of the food(s) to which the food is being compared.

33. For the terms **"increased"** or **"reduced"**, Codex Guidelines propose that claims should be made only when there is a minimum 25 percent increase or reduction of the nutrient that is the subject of the claim by comparison with the equivalent standard product (for which no claim is made). As mentioned above, it is important to clarify what the standard product is.

34. The 25 percent minimum difference established by the Codex seems to pose no major problems in the cases of caloric content and macronutrients. Member States have not expressed a particular concern about this condition, even if some have adopted different criteria such as requiring a minimum increase or reduction of 33 percent, or, for reduced fat claims, a 50 percent difference. However, **specific conditions may be necessary for vitamins and minerals**.

35. As far as **micronutrients** are concerned, the **Codex Guidelines** establish the following:

"For micronutrients, a 10% difference in the Nutrient Reference Values (NRVs) between the compared foods would be acceptable, and a minimum absolute difference in the caloric value or nutrient content equivalent to the figure defined as "low" or as a "source" in the Table of Guidelines". The point made in paragraph 32 is especially relevant.

36. The terms **"more"** or **"less"** could be used when making claims for foods with changes in their caloric or nutrient content of less than 25 percent. Again, whether and how the final consumer perceives these differences must be considered.

Functional Claims

37. In its White Paper on Food Safety, the Commission described functional claims as "**claims relating to beneficial effects of a nutrient on certain normal bodily functions.**" This would cover the claims describing the physiological role of a nutrient or other substance in growth, development, and normal functions of the body. Typical examples of functional claims would state the presence of a nutrient or other substance and briefly describe its role on human physiology, for example: "High in protein. Protein helps to build and repair body tissues."

38. **Codex Alimentarius** Guidelines for the Use of Nutrition Claims define "nutritional function claim" as "a nutritional claim that describes the physiological role of the nutrient in the growth, development, and normal functions of the body." The following examples are presented:

    "Calcium aids in the development of strong bones and teeth"; "Protein helps to build and repair body tissues"; "Iron is a factor in red blood cell formation"; "Vitamin E protects the fat in body tissues from oxidation"; "Contains folic acid: folic acid contributes to the normal growth of the fetus."

39. The above definition could be used as a basis for defining functional claims in possible future Community legislation. Some people believe that rather than developing a new definition, it would be better to use the existing Codex to avoid confusion and limit potential trade disputes. The Commission services are not aware of other serious alternatives to the Codex definition.

Criteria for the Use of Functional Claims

40. Following the definition of functional claims, consideration should be given to a number of **conditions to be fulfilled for the use of such claims**.

(...)

48. Even if conditions expressed above were to be adopted, some doubt would remain about their uniform application. They would therefore maintain that a **premarketing approval** is necessary to ensure that claims are appropriate and avoid disputes at the national or intracommunity level. It has been suggested that one possible approach would be to **compile a list of approved claims for each nutrient or substance** and possibly a specific drafting of each claim. Some favor a procedure of cooperation between Member States and the Commission for compiling and updating this list. Others think that this should be done directly at the Community level and with the involvement of the future European Food Authority.

49. Another possibility could be the introduction of a **notification procedure to the competent authorities of Member States** for food labels bearing a claim. This would facilitate monitoring and allow for a quick reaction, where necessary, from the appropriate authority when new products and/or labels are placed on the market. A variant of this option could be that the **label is notified at the European level** and is valid for

the Community as a whole. This, however, would require considerable resources at the Community level.

(. . .)

Annex:

Claims and Conditions Warranting the Claims for Different Nutrients (and Other Substances)

CLAIM: *LOW CALORIES*
CONDITIONS: Codex: less than 40 kcal/100g and less than 20kcal/100ml. **Conditions in use in some Member States:** less than 50kcal /100 g and less than 20kcal/100ml

CLAIM: *CALORIE -FREE / WITHOUT CALORIES*
CONDITIONS: Codex: less than 4kcal/100ml.

CLAIM: *LOW FAT*
CONDITIONS: Codex: no more than 3g/100g and 1.5g/100ml. **Conditions in use in some Member States:** No more than 3g per 100g for solids or per 100ml for liquids.
(...)

The claims contributed by the Health Administration of Spain summarize the experience obtained after four years of the voluntary agreement signed with the Federation of Industries of Food and Beverages. In the adaptation of the next community legislation on functional foods, the inspection and sanctioning activities of the Autonomous Communities and City Councils, who have those capacities in the Spanish administrative structure, have been taken into account.

Comments of the Spanish Authorities Concerning the Discussion Paper on Nutritional and Functional Claims. (SANCO/1341/2001)

General Remarks

The Spanish authorities are grateful for the Commission's document, as it covers the current situation and suggests several approaches that can be used to draw up Community legislation to bring into accordance nutrition and functional claims made for ordinary foodstuffs.

From the Spanish authorities' point of view, these two types of claims should each be covered by a separate piece of legislation.

As for nutritional claims, these should be examined by the group of experts on labelling as part of their work on nutritional labelling, with the objective of updating and amending the existing Directive on nutrition labelling. At the same time, the different types of claims and the conditions to be met for making them are specified in the Annex.

As for functional claims, separate legislation should exist containing: a definition of "functional foodstuff" and "functional labeling," specific criteria for using and limiting claims, and a procedure for the prior authorization of claims by the appropriate authorities in the Member States.

Comments on the Contents of Specific Points

Definition of the Term "claim"

12. From our point of view, the term needs to be defined.
13. The Codex Alimentarius definition would be acceptable as a starting point for arriving at a general definition of "claim."
15. It would be good to define the different types of claims in future Community legislation.

Nutritional Claims

17. It should be possible to make nutritional claims only when they refer to the nutritional value of the product, that is, only when they relate to nutrients in the food (macronutrients: proteins, carbohydrates and fat; micronutrients: vitamins, minerals, and trace elements). We think that any claim concerning the non-nutritional effects of other substances in foodstuffs (fiber, antioxidants, lactic bacteria) should not be seen as relating to nutritional properties and should be covered by legislation separate from that governing nutrition claims.
19. Nutritional claims may be expressed in quantitative terms, so the following examples taken from the document are acceptable:
    –High/rich in
    –Without/free
    –Increased/enriched
    –Low/weak/poor
    –Contains
    –Reduced/light
21. The use of the term "light" would have to relate to a lower caloric value compared with the reference foodstuff, but a "light" product would have to keep the same characteristics, or be of the same nature, as the reference product.

For this term to be used, the reduction in the caloric value compared with the reference product would have to be quantified.

22. To avoid confusion, the terms "diet" and "dietary" should not be used in any type of claim.
23. No comparative claim should be made unless there is a reference product.
24 & 25. No claims concerning the content of cholesterol in foodstuffs should be made as nutritional claims.
26. As with claims concerning cholesterol, the claims relating to sodium content should not be made as nutrition claims (taking in account the connection between high blood pressure and cardiovascular disease).

Criteria for Making Nutrition Claims

27. Every food has a function and there is no need to create a special category of "functional" foods.

    This means that "functional claim" needs to be defined and covered by legislation separate from that governing nutritional claims.
28. The claim "X% fat free" is not acceptable because it misleads consumers; the only acceptable term is "only X% fat," that is, where the claim indicates the fat content of the product.

31. A distinction should be made between nutrients that are naturally contained in a product and nutrients that are added to foods. In the latter case, the 15 percent figure should be maintained, with a much lower value for natural sources.
33 & 34. In order to make any comparative claim, the characteristics of the reference product should be clearly indicated.

A minimum figure should be established for "reduced" / "increased" claims should when they apply to the caloric content and macronutrients, such as the 25 percent set by the Codex. However, a 30 percent minimum figure should be considered, bearing in mind that Article 1(3) of European Parliament and Council Directive 94/35/EC of 30 June 1994 defines "reduced calorie" as: "with a caloric value reduced by at least 30 percent compared with the original foodstuff or a similar product."
35. The 10 percent minimum figure laid out by Codex should be considered for micronutrients but only when it is applied to those that occur naturally.
36. The terms "more" or "less" could be applied to the same criteria as those discussed in points 33 and 34.

Functional Claims

To produce appropriate Community legislation, functional claims should be dealt with in their own directive, separate from the directive on nutritional labelling and that on nutrition claims.

37. The definition of functional claim, the term "beneficial" should be replaced by "physiological." The definition should also mention that the nutrient is not separate but forms part of the foodstuff.
38 & 39. The Codex guidelines can be used as a basis for defining functional claims, but the directive should be separate from the one on nutrition labelling.

Criteria for the Use of Functional Claims

48. Functional claims should be compiled in a list of approved general functional claims, indicating the conditions attached to their use. However, each Member State should be allowed to adapt it in accordance with their country's particular situation (public health policy).
49. Introducing a notification procedure (as well as an authorization procedure) to notify the appropriate authorities of the Member States about products for which there is an intent to make a functional claim (labelling and advertising). Notifying the Community would also be useful in generating a list of allowed claims and the conditions for using them at the Community level.

Regulation of the European Parliament and of the Council on Nutrition, Functional and Health Claims Made in Foods: Working Document SANCO/1832/2002

The final working document SANCO/1832/2002 that includes the claims presented by the Member States to the working document SANCO 1341/2001, and was still being developed by the end of the year 2002, will serve as a framework for the Spanish juridical classification in this matter.

3.1. Nutritional Claims

21. In order to present consumers and the industry with unambiguous guidelines concerning the use of nutrition claims, clear and simple rules should be set. At the international level, Codex Alimentarius has developed guidelines for the most commonly used nutritional claims (such as "low," "rich," "light," and so on). Similar criteria also exist in the laws of some Member States. The Annex to this proposal provides a list of nutrition claims and their specific conditions of use. This Annex takes into account existing provisions of some Member States, the Codex Alimentarius guidelines, and some Community provisions. In order to revise and adapt the Annex promptly, when necessary, modifications to this Annex should be adopted through the Committee procedure referred to in Article 18.

22. Under current Community legislation (Council Directive 89/398/EEC6 on the approximation of laws of the Member States relating to foodstuffs intended for particular nutritional uses) the terms "dietetic" and "dietary" are reserved for foodstuffs intended for particular nutritional uses. Following the consultations leading to the Discussion Paper on Nutrition Claims and Functional Claims, the majority opinion was that the term "diet" has the same connotation as "dietetic" and "dietary" and therefore should not be used as a term in a claim on ordinary foods. This is the opinion presented in this proposal.

23. The conditions for "low sodium" claims are to be set under Council Directive 89/398/EEC on foods for particular nutritional uses (PARNUTS). Claims on ordinary foods should be able to be made under the same conditions as those set for dietetic foods and should indicate suitability for the particular nutritional use if they fulfill the conditions set under Article 2, paragraph 3, of Council Directive 89/398/EEC. If no conditions are established for this type of claim under the PARNUTS Directive, there is a risk of giving the impression that the products bearing such claims are no longer dietetic foods, which may have legal implications.

3.2. Functional Claims

24. Functional claims shall describe well-established and widely accepted roles of nutrients in growth, development, and normal physiological functions of the body. For example: "calcium aids in the development of strong bones and teeth." Being based on long-established and noncontroversial science, these claims should not need to be assessed and approved before their use in the labelling, presentation, and advertising of foods.

    However, it is necessary to provide the possibility for national authorities to verify the scientific substantiation of a functional claim and/or its conformity with the provisions laid out in this Regulation, and, where necessary, to temporarily suspend the use of such claims and refer the matter to the Community. A summary of all decisions taken under the Community procedure will then be included in a "Register."

3.3. Health Claims

25. There are a number of claims that are generally known under the broad term of "health claims" which would describe a relationship between a category of food, a food or one of its constituents, and health.

26. According to the amended Directive on medicinal products, a medicinal product is defined as follows: "(a) Any substance or combination of substances having properties of treating or preventing disease in human beings. (b) Any substance or combination of substances which may be used in human beings for making a medical diagnosis or for restoring, correcting, or modifying physiological functions".

   Furthermore, Directive 2000/13/EC on labelling, presentation, and advertising of foods specifically prohibits attributing to foods any properties of prevention, treatment or cure of a human disease, or any reference to such properties. This proposal maintains this prohibition; however, a difference between "prevention" and "significant reduction of a risk factor for a major disease" is made. Indeed, it is known that diet and eating certain foods can make important contributions to the support and maintenance of health, and can play a role in the management of certain disease risk factors.

27. The European Parliament Resolution of March 1998 on the Green Book on the General Principles of Foodstuff Legislation from the European Union called on the Commission to propose legislation on food claims to ensure that health claims are only authorized if they are tested and confirmed by an independent body within the European Union. It also called on the Commission to continue to ban the use of claims referring to the suitability of a food for the treatment, cure, or prevention of a disease, though claims referring to the reduction of the risk of disease should be allowed if they are based on sufficient and recognized scientific findings and if they are tested and confirmed by an independent body within the European Union. Furthermore, the European Parliament Resolution of June 2001 on the White Paper on Food Safety called on the Commission to address "enhanced function claims and disease reduction claims" and consider this as a priority for legislation.

28. In this proposal, health claims include both "enhanced function claims" and "reduction of disease risk factor claims." Enhanced function claims would be those that refer to specific beneficial effects beyond the widely accepted nutritional effects of a food or food constituent on physiological functions of the body; for example, "calcium may help to improve bone density. Food A is rich in calcium." Reduction of disease risk factor claims would be those that refer to the possibility of food category, a food, or food constituent to significantly reduce the risk in the development of a human disease; for example, "sufficient calcium intake may reduce the risk of osteoporosis in later life. Food A is rich in calcium."

Article 5

General Principles of Substantiation

1. Nutrition, functional, and health claims shall be based on, and substantiated by, generally accepted scientific data.
2. The food business operator making a nutritional or functional claim is responsible for justifying and substantiating the claim being made.
3. The appropriate authority of a Member State shall be empowered to require the food business operator or the person placing the product on the market to produce the scientific work and the data establishing the compliance with this Regulation.

The use of functional claims is based on the existence of scientific data that confirm this condition and the manufacturer making the claim must justify the use of the claims through scientific substantiation.

Article 6

Prohibited Claims

1. Any health claim, other than enhanced function claims, reduction of disease risk factor claims, and other claims that make reference to general, nonspecific benefits of the nutrient or food for overall good health, well being, and normal functions of the body shall be prohibited.
2. Any nutrition, functional, or health claim making reference to psychological and behavioral functions shall be prohibited.
3. Beverages containing more than 1.2 percent by volume of alcohol shall not bear any nutrition, functional, or health claim.

Claims that make reference to psychological functions or behavior, or to health in general are expressly prohibited, as well as those that make reference to functional or health claims in alcoholic drinks.

Specific Requirements for Functional Claims

Functional claims may be used provided that, in addition to the general requirements put forth in this Regulation, the following information shall also appear on the label:

- A statement indicating the importance of a balanced diet and a healthful lifestyle;
- The quantity of the food and pattern of consumption required to obtain the claimed beneficial effect;
- In certain cases, the target population;
- In certain cases, a statement addressed to persons who should avoid using the food;
- In certain cases, a warning not to exceed quantities of the product that may present a health risk.

Food labels that contain functional claims must contain additional information for the target consumer, indicating the importance of a balanced diet and the risks associated with an exaggerated consumption of the food.

Community Procedure

1. Where the appropriate authority of a Member State has detailed grounds for considering that a functional claim may not be in conformity with the provisions set forth in this Regulation, or that the scientific substantiation of a functional claim is not satisfactory to justify the claim, that Member State may temporarily suspend the use of that claim in the labelling, presentation, and advertising of the food in question within its territory. It shall inform the other Member States and the Commission and give reasons for its decision.

If the maker of the claim does not clearly justify the functional claims, the use of claims on the label will be suspended.

Procedure for Approval of Health Claims

In addition to the general requirements set forth in this Regulation and without prejudice to Article 2 of Directive 2000/13/EC of the European Parliament and of the Council on the labelling, presentation, and advertising of foods, the only health claims that can be made are enhanced function claims and reduction of disease risk factor claims, as defined in Article 2 of this Regulation, and approved in accordance with this Chapter.

1. To obtain the approval mentioned in Article 12, an application shall be submitted to the authorities.
2. The application shall be accompanied by the following documentation:
    a) the name and address of the applicant;
    b) the food or the category of food for which the health claim is requested, along with its particular characteristics;
    c) a copy of the studies which have been conducted and any other available materials that demonstrate the health claim to be in compliance with the criteria set forth in this Regulation;
    d) a copy of other scientific studies that are relevant to the proposed health claim;
    e) a proposal for the wording, in all Community languages, of the health claim for which approval is sought including, as the case may be, specific conditions for use; and
    f) an executive summary of the application.
3. The Authorities, after having verified that the application contains the required documents mentioned in paragraph two, shall acknowledge receipt of the application to the applicant within 15 days of its receipt. The acknowledgement shall state the date of receipt of the application.
4. The Authorities shall inform the Member States and the Commission of the application without delay and shall make the application and any supplementary information supplied by the applicant available to them. It shall make a summary of the application available to the public.
5. The Authority shall publish a detailed guide concerning the information applied for in the presentation of the application.

Registry

1. The Commission shall establish and maintain a *Community Registry of Functional and Health Claims*, referred to hereafter in this Regulation as "the *Registry*."
2. The *Registry* shall be available to the public.

It will be necessary to apply for the use of new health claims through a procedure that includes the presentation of a complete report that includes technical-scientific documentation.

In July 2003 the Commission of the European Parliament published a proposal for the regulation on nutritional and health claims made on foods. In light of the technological innovation in the food sector and the demand from consumers and industry alike, it is proposed that a new legislative framework on the use of claims be established.

The proposed Regulation would allow health claims under strict conditions and following independent scientific review and Community authorization.

Specific issues of the proposal of regulation:

- In addition to the definition of "nutrients" which covers the caloric value and the "traditional" nutrients (protein, carbohydrate, fat, fiber, sodium, vitamins and minerals), it is proposed to cover also "other substances with nutritional or physiological effects" (for example, antioxidants, probiotic bacteria).
- Some consumer organizations and member states in the European Union state that products that do not have a "desirable" nutritional profile, such as candies, snacks high in salt and fat, and cookies and cakes high in fat and sugar should not be allowed to bear claims. They allege that such foods would become more attractive to consumers because of the way in which they will be labelled and advertised, and many consumers that are currently eating them in moderation would consume them in greater quantities. The concept of prohibiting the use of claims on certain foods on the basis of their "nutritional profile" is contrary to the basic principle in nutrition that there are no "good" and "bad" foods but rather "good" and "bad" diets. However, food vendors presented the products through the claims as providing a benefit, that is, as "good" or "better" products. Consumers may perceive them as such, influenced by the promotional campaigns. Therefore, some restrictions on the use of claims on foods based on their nutritional profile should be anticipated. In particular, the respective amounts of total fat, saturated fat, trans fatty acids, sugars, sodium or salt, are commonly cited as criteria for the "nutritional profile" of products. It also does not seem appropriate to attribute nutritional or health claims to alcoholic beverages, but all of these proposals are currently far from meeting with the required consensus.
- Nutritional labelling should become obligatory for all foods bearing nutritional and health claims. All health claims should also be complete in order to give a better overall picture of the food.
- The approach to communicating nutritional claims must be carefully considered. A claim that is not understood is completely useless; meanwhile a claim that is misunderstood could even be misleading. It should therefore be ensured that complicated or misleading claims that are ultimately incomprehensible to consumers shall not be used.
- Many claims already found on the market make reference to general, non-specific benefits and to general wellbeing. For example: "excellent for your body", "reinforces the body's resistance", etc. Not only are these claims vague and often meaningless, they are also unverifiable and therefore should not be allowed. Behavioral functions can be influenced by many factors besides dietary ones. Therefore, it is deemed appropriate not to allow the use of claims that make reference to behavioral functions.
- Commission Directive 96/8/EC on foods intended for use in calorie-restricted diets for weight reduction, prohibits in the labelling, presentation, and advertising of products covered by the Directive and especially those designed for weight control, any reference to the rate or amount of weight loss which may result from their use, or to a reduction in the sense of hunger or an increase in the sense of satiety. It is therefore justified that such references should also be prohibited for all foods.

Nutritional Claims

To present consumers and the industry with clear guidelines concerning the use of nutrition claims, clear and simple rules should be set. At the international level, Codex Alimentarius has developed guidelines for the most commonly used nutritional claims (such as "low," "rich," "light," and so on). For comparative claims, such as "increased" or "reduced," the question that arises is to what is the comparison being made. It is therefore necessary that the products being compared be clearly identified to the final consumer. The comparison shall be made between foods of the same category, taking into consideration a range of foods in that category and including other brands in the comparison. The difference in the quantity of a nutrient and/or caloric value should be stated and the comparison should relate to the same quantity of food.

Health Claims

This proposal for a Regulation on the use of claims maintains the prohibition on claims referring to the prevention, treatment, or cure of a human disease; however, claims referring to the reduction of the risk of disease should be allowed "if they are based on sufficient and recognized scientific findings and if they are tested and confirmed by an independent body within the European Union." In order to ensure a coordinated scientific assessment of these claims, the European Food Safety Authority (EFSA) should carry out such assessments.

In order to ensure that health claims are truthful, clear, and reliable, the Authority, in its opinion and the subsequent authorization procedure, should take into account the wording of the claims assessed. The scientific assessment should be followed by a decision by the Commission, under a regulatory procedure. In summary, the authorization procedure set forth in the proposed Regulation is as follows:

1. The applicant will submit an application to the Authority.
2. The Authority will render an opinion within three months.
3. The Authority will forward its opinion to the Commission, the Member States, and the applicant, and will make its opinion public. The public may make comments to the Commission.
4. The Commission will prepare a draft decision within three months of receipt of the opinion of the Authority.
5. The Commission will inform the applicant of the final decision taken. The final decision will be published in the Official Journal of the European Union.
6. A summary of the final decision will also be included in the "Registry."

Being based on long-established and noncontroversial science, health claims that describe the role of a nutrient or other substance in growth, development, and normal physiological functions of the body shall undergo a different type of assessment and approval prior to their use in the labeling, presentation, and advertising of foods. It is therefore proposed that to adopt a list of permitted claims describing the role of a nutrient or other substance in growth, development, and normal physiological functions of the body following the opinion of the Authority be adopted.

For the purpose of transparency and to avoid the repetition of applications of health claims that have already been assessed and for those health claims that have gone through the Community procedure, a "Registry" of such claims shall be established and regularly updated.

Conclusion

The proposed rules would contribute to a high level of protection of human health and promote the protection of consumer interests by ensuring that foods bearing nutritional and health claims are labelled and advertised in a clear and adequate manner, allowing consumers to make informed choices. The proposed rules would thus be in line with the general principles and requirements of foodstuff legislation, as stipulated in Articles 5–8 of the recently adopted Regulation (EC) 178/2002 of the European Parliament and the European Council and with Article 153 of the Treaty. The proposed rules would also take into account the importance for the food industry to have a regulatory environment that will allow it to innovate and remain competitive at the Community and international level.

The directives for the regulation of nutrition and health claims are clear. However, in case C-221/00, Austria versus the Commission, which was heard very recently, the European Court of Justice interpreted the existing labelling Directive as banning all health claims relating to human diseases. It is evident that the present legislation gives rise to important controversies.

References

COM (1999) 719 final. White Paper on Food Safety. Commission of The European Communities. Brussels, January 12, 2000.

COM (2003) 424 final 2003/0165 (COD). Proposal for a regulation of the European Parliament and of the Council on Nutrition and Health Claims made on foods. Commission of The European Communities. Brussels, July 16, 2003.

Commission Directive 94/54/CE of November 18, 1994, concerning the compulsory indication on the labelling of certain foodstuffs of particulars other than those provided for in Council directive 79/112/EEC. Official Journal L 300, 11/23/1994.

Council Directive 79/112/EEC of December 18, 1978, on the approximation of the laws of the member states relating to the labelling, presentation and advertising of foodstuffs for sale to the ultimate consumer. Official Journal L 033, 02/08/1979.

Council Directive 90/496/EEC of September 24, 1990, on nutrition labelling for foodstuffs. Official Journal L 276, 10/06/1990.

Council Directive 94/35/EC of June 30, 1994, on sweeteners for use in foodstuffs. Official Journal L 237, 09/10/1994.

Council Directive 96/21/EC of March 29, 1996, amending Commission Directive 94/54/EC concerning the compulsory indication on the labelling of certain foodstuffs of particulars other than those provided for in Directive 79/112/EEC. Official Journal L 88, 04/05/1996.

Directive 89/398/EEC of May 3, 1989, on the approximation of the laws of Member States relating to foodstuffs intended for particular nutritional uses.

Directive 96/8/EC of the European Parliament and of the Council of February 26, 1996, on foods intended for use in calorie restricted diets for weight reduction. Official Journal L 055, 03/06/1996.

Directive 96/84/EC of the European Parliament and of the Council of December 19, 1996, amending Directive 89/398/EEC on the approximation of the laws of the Member States relating to foodstuffs intended for particular nutritional uses. Official Journal L 167/23 12/06/1998.

Directive 97/4/EC of the European Parliament and of the Council of January 27, 1997, amending Directive 79/112/EEC on the approximation of the laws of the Member States relating to the labelling, presentation, and advertisement of foodstuffs. Official Journal L 043 02/14/1997.

Directive 1999/41/EC of the European Parliament and of the Council of June 7, 1999, amending Directive 89/398/EEC on the approximation of the laws of the Member States relating to foodstuffs intended for particular nutritional uses. Official Journal L 124 de 05/18/1999.

Directive 2000/13/EC of the European Parliament and of the Council of March 20, 2000, on the approximation of the laws of the Member States relating to the labelling, presentation, and advertising of foodstuffs. Official Journal L 109, 05/6/2000.

Directive 2001/83/EC of the European Parliament and of the Council of November 6, 2001, on the community code relating to medicinal products for human use. Official Journal L 311, 11/28/2001.

Directive 2002/46/EC of the European Parliament and Council of June 10, 2002, on the approximation of laws of the Member States relating to food supplements. Official Journal L 183, 07/12/2002.

FAO/WHO, Codex Alimentarius, CAC/GL 23-1997. Guidelines for the use of nutrition claims.

Real Decreto 2685/1976 of October 16. RTS on foodstuffs for dietary regimen and or special. BOE 11/26/1976.

Real Decreto 1164/1991, of July 22, for which the RTS is approved for the elaboration, circulation, and trade of waters of packed drink. BOE 07/26/1991.

Real Decreto 1712/1991 of November 29 on the Health General Registration of foods. BOE 12/04/91.

Real Decreto 1809/1991 of December 13 for the one that the RTS modifies on foodstuffs for dietary regimen and/or special. BOE 12/25/1991.

Real Decreto 212/1992 of March 6 on labeling presentation and advertisement of foodstuffs. BOE 24/03/1992.

Real Decreto 930/1992 of July 17 by which the Norma is approved of labelling about nutritious properties of the foodstuffs. BOE 08/05/1992.

Real Decreto 50/1993 of January 15 by which the official control of the foodstuffs is regulated. BOE 02/11/1993.

Real Decreto 1397/1995 of August 4 by which additional measures are approved on the official control of the foodstuffs. BOE 10/14/1995.

Real Decreto 44/1996 of January 19 by which measures are adopted to guarantee the general product safety to the consumer's disposition. BOE 22/02/1996.

Real Decreto 1907/1996 of August 2 about advertising and commercial promotion of products, activities or services with health purpose. BOE 08/06/1996.

Real Decreto 1430/1997 of September 15. Specific Sanitary Tecnical Regulation of foodstuffs focused to be used in low calorie diets for reduction of weight. BOE 9/24/1997.

Real Decreto 1334/1999 of July 31. General Norma of labelling, presentation, and advertising of foodstuffs. BOE 8/24/1999.

Real Decreto 238/2000 of February 18 modifies the general norm of labelling, presentation, and advertising of foodstuffs approved in Real Decreto 1334/1999. BOE 02/19/2000.

Real Decreto 709/2002 of July 19 by which the statute of the Spanish Agency of Alimentary Safety is approved. BOE 07/26/2002.

Real Decreto 1275/2003 of October 10 relative to diet complements BOE 10/14/2003

REGULATION (EC) No 178/2002 of The European Parliament and of the Council of 28 January2002 laying down the general principles and requirements of food law, establishing the European Food Safety Authority, and laying down procedures in matters of food safety. Official Journal. L 31 02/1/2002.

SANCO/1341/2001. Discussion paper on nutrition claims and functional claims. Prepared by Directorate General Health and Consumer Protection. http://europa.eu.int/comm/dgs/ health_consumer/index_en.htm

SANCO/1341/2001 Comments by the Spanish authorities concerning the discussion paper on nutrition claims and functional claims.

SANCO/1832/2002 Working Document. Regulation of the European Parliament and of the Council on Nutrition, Functional, and Health Claims Made on Foods

SANCO D4, Discussion paper prepared by Directorate General Health and Consumer Protection. European Commission. http://europa.eu.int/comm/dgs/health_consumer/index_en.htm

21 Functional Food Legislation in Brazil

Franco M. Lajolo

Context and Concepts

Brazil was the first country in Latin America to issue legislation for functional foods (functional food claims). The motivation for that was, on one hand, the upsurge in the market of products having nutritional and health claims on the labels and inquiries to the Ministry of Health by industries planning to produce and commercialize functional foods.

On the other hand there were on the market hundreds of the so-called *natural products* containing chemicals such as amino acids, isoflavones, fatty acids, vitamins, and parts and extracts of plants (some native, some imported) being sold in pharmaceutical forms claiming to prevent or to cure all possible human diseases, including unhappiness (ANVISA 2002)

Besides unproven claims and lack of composition and quality control, those products had serious safety problems: several had well-known pharmacological or toxicological properties but were freely sold as foods or dietary supplements. These undesirable selling practices were due to a questionable interpretation of an existing regulation issued by a Justice Court decision in favor of specific products. (The regulation allowed temporary selling of the products, provided that they were safe, until their efficiency was proven. In other words, it was a transition rule, which provided time to the producers for obtaining data on the products). The confusion was also due to the fact that most of these products were proposed as "dietary supplements" but were also sold as OTC drugs (phytoterapeutics), which, in Brazil, are submitted to specific pharmaceutical norms (ANVISA, 2002; Lajolo, 2002a). In Brazil, the basic norms for foods are contained in the legislation issued in 1969: Decreto–Lei 986 (Brasil, 1969). It defines food, establishes the need for registration of foods within the Health Authority, establishes that they should have specific standards of identity and quality, and rules on labeling and on additives and on the penalty in case the regulation is not followed.

These norms had, as their main preoccupation at that time, the concern with safety and labeling in relation to additives and to the creation of standards for food quality control. In 1997, due to the existence of new products in the market and to discussions going on within the Codex Alimentarius committees, the legislation was updated and norms for Foods for Special Dietary Uses were produced (Portaria Ministerial 1549, 17/10/1997) (Brasil, 1997). These norms defined the following categories: (1) foods for special uses; (2) foods with nutrients added; (3) salt substitutes; (4) sweeteners; and (5) vitamin and mineral supplements. (Products containing vitamins or minerals in quantities higher than the RDA were considered drugs and not food supplements.)

Several subsequent regulations established the foods in each category and specific technical norms for each type of food, including their standards of identity and quality and norms for labeling, always in accordance with Codex Alimentarius propositions. This was a very important step toward modern food legislation.

At that time (1998) in Brazil, we had norms only for foods for special dietary uses as defined by Codex Alimentarius, but not for functional food or dietary supplements.

To organize the production, importing, and commerce of those supplements and the incoming functional foods, ANVISA (National Sanitary Surveillance Agency), a kind of FDA linked to the Ministry of Health (Brasil, 1999a), set up a special committee to propose a specific legislation. The committee had representatives from Academia, Industry, International Life Sciences Institute (ILSI), and ANVISA, who studied the existing regulations in several countries all over the world, seeking a balanced position (ANVISA, 2002). The suggested legislation for Brazil was issued on previously established concepts and assumptions:

- Functional foods were not to be legally defined as a new or different category but considered as food with claims.
- They should be useful to improve the well-being of population groups and should not oppose the nutrition and health policy of the country.
- They should have their efficiency and safety proved through a science-based process that should be conducted by an expert, specially nominated technical committee.
- Claims should be true and not misleading and written in a language understandable by consumers.

The idea was that a clear distinction between **normal** or **traditional foods** as opposed to **drugs** (including phytoterapeutics) should be maintained. **Therapeutic drugs** have a distinctive purpose (cure, mitigate pain, prevent disease) and a known and accepted risk, and they are used by individuals for a short period of time, rather than by the general population. Drugs are not sold in supermarkets in Brazil.

Even if not legally included as a category and officially defined, the common understanding was that functional foods are foods, in most of the cases similar to conventional foods, consumed as part of the usual diet, able to produce physiological or metabolic effects useful to maintaining good health and reducing the risk of non transmissible chronic disorders, beyond their basic nutritional properties.

The idea was also that being **food**, a functional food can not be a **nutraceutical**—a term reserved for the bioactive compound, presented in pharmaceutical forms, usually in a concentration higher than the one in foods.

This description has become the most disseminated idea in Brazil regarding **functional foods**. It is widely used by the media, as well as in scientific meetings and scientific literature. We use mostly, in Brazil, the terms **functional foods**, **nutraceuticals**, or **bioactive compounds** so that terms such as **pharmafoods**, **medical foods**, and others generated by the media have been avoided.

The Norms and What They Say

In the last two or three years, almost all the scientific meetings held in Brazil and in other Latin American countries in the area of food science and nutrition have included in their programs discussions and presentations on functional foods, motivated by scientific interest or by support by industry or the regulatory agency, which indicates how important this topic has become.

As we have seen, functional foods have not been officially defined in Brazil, and the existing legislation basically asks for demonstration of safety of products and efficacy of claims. There are now four regulations, most of which were issued in 1999, with a more recent one having been issued in 2002 (*). These are as follows:

Resolution n. 16, of April 30, 1999 (republished on 03/12/1999):

To approve the technical regulation on procedures for registration of foods and or new ingredients (Brasil, 1999a)

Resolution n. 17, of April 30, 1999 (republished on 03/12/1999):

To approve the technical regulation establishing the basic guidelines for evaluation of risk and safety of foods (Brasil, 1999b).

(*) All regulations are available on the site http:\\anvisa.gov.br/eng/legis/index.htm

Resolution n. 18, of April 30 (republished on 03/12/1999):

To approve the technical regulation establishing the basic guidelines for analysis and proof of functional and or health claims on food labels (Brasil, 1999c).

Resolution n. 19, of April 30, 1999 (republished on 10/12/1999):

To approve the technical regulation on procedures for registration of foods with functional and or health claims on their labels (Brasil, 1999d).

Finally, the Decree n. 15 establishes the existence and composition of the Technical Scientific Advisory Commission on Functional Foods and New Foods (CTCAF), instituted by the Brazilian Sanitary Surveillance Agency, which is in charge of applying the former regulations (Brasil, 1999e).

More recently, a new regulation was issued, Resolution RDC n. 2, January, 2002, which was concerned with "Bioactive Substances and Probiotics with functional or health claims." Its aim was to make as clear as possible a distinction between **functional foods** and **bioactive compounds (nutraceuticals)**, usually sold as pills, capsules, and so on, and **OTC drugs**, mostly herbs and other phytoterapeutics (Brasil, 2002).

Establishing legislation for these new products and considering incoming tendencies was a recognition of their importance as written in the preamble of Resolutions 17 and 18, which states, "the recognition of the relationship between foods, health and disease, the upsurge of new concepts on nutrient and non-nutrient needs and their metabolic importance, their importance for establishing healthy diets, the possibility of potential heath risks, the possible confusion of consumer by misleading claims, the interest of public health . . . ."

All together, the resolutions rule on demonstration of safety and efficacy of novel foods, food ingredients and bioactive compounds, and food with claims. All these products should be registered and approved by the health authority (ANVISA, 2002).

Novel foods are those without a history of use in Brazil (genetically modified foods are not covered here), and those not having standards of quality and identity are also novel foods, as are those containing substances added in concentrations much higher than usually used or present in foods (for instance, flavonoids or carotenoids).

Foods with claims in the label are regulated by the legislation that allows both functional and health claims. Functional claims are defined as those referring to the metabolic or physiological role of a nutrient or a non-nutrient in the normal body functions. Health claims are defined as those that imply, suggest, or state the existence of a relation between the food or food component and a disease or a condition related to health.

Based on the existing regulation in Brazil, functional foods can be considered as foods bearing health claims, approved by the health authority, and a premarket approval is, therefore, mandatory.

To register a product within the Ministry of Health (or sometimes in the Ministry of Agriculture) the industry should present, in addition to the documents needed for usual foods, a "Technical Report" containing the following information: the name of the product; the recommended uses and intakes; a complete description of the origin and technological process used for production; the chemical composition and molecular characterization (when needed, with description of the analytical methodology); a copy of the label with the proposed claim and the scientific support for it; and evidence in support of the claim. The report should also contain information about the status of the product in other countries, whether it was considered safe, whether the claim was approved, and possible uses and restrictions. The report is then submitted to the assessment committee for evaluation of safety and efficacy based on the evidence presented.

How Claims May Be Used and Are Assessed

Functional claims are allowed provided that they are scientifically proven. Health properties and health claims based on these properties have more strict limits in legislation. Those on health maintenance are allowed and have been relatively common; those on health risk reduction, although possible, are very scarce, and claims on mitigation of symptoms, prevention, or cure of diseases are prohibited.

The statement on the label should be scientifically proven by research that may be done by the industry itself or by published literature available in peer-reviewed journals. The extent of the demonstration expected is proportional to the type and importance of the claim made. For instance, a nutrient or bioactive component content claim may be based only on the chemical analysis of the compound and the description of its chemical structure. But claims on "bioavailability" due to the influence of the matrix of the food may depend also or maybe only on an animal assay. A risk-reduction statement would hardly be approved without mechanistic, clinical, and/or epidemiological studies using suitable biomarkers.

The scientific efficacy and safety evidence that should be presented in the dossier of the product to get it registered within ANVISA has to include all or some of the following data: chemical composition and molecular characterization (when it applies); biochemical assays (in vitro); nutritional, physiological, metabolic, and toxicological essays with animals; and cellular, clinical, and epidemiological evidence. Demonstration of historical or traditional use of the products by indigenous populations with benefits and no reported risks may be accepted in some cases as supporting evidence, provided that they are well documented.

Besides demonstrating safety and efficacy, claims must inform the target population (if any) that the claimed effect is to be observed with normal consumption of the food and obtained in the context of the normal diet, with all possible interactions. The claim should not conflict with the nutrition and health policy of the country; and it should be true, not misleading, and reflect scientific agreement. The extent of demonstration, type and number of studies, and extent of agreement or acquisition of a real consensus, are matters of debate in Brazil, as in other countries, among regulators, academia, and industry.

Because many functional foods have supplementary informative material, the information carried by such attachments is also analyzed by the committee and cannot disagree with or go beyond what was stated on the label. This is a way to limit excessive advertising that could mislead consumers. However, the legislators' intention cannot compete with the strong and usually higher impact of the general media on consumer information and choice. This question will be the subject of a new advertising and marketing regulation that is now under discussion. In some cases, product labels should also contain clear information about restrictions of their uses to specific groups or individuals, with specific physiological conditions, as well as any adverse side effects that may occur. These rules also apply to marketing and advertising material.

A multiprofessional scientific committee of members from academia and research institutes with expertise in different areas of food science, nutrition, and toxicology makes the final assessment. The information provided by the industry throughout a "dossier" on the product is analyzed by the committee according to scientific criteria and based on the quantity and quality of the literature provided and otherwise available. A final decision on the approval or prohibition of the product is based on a consensus dialog with the industry and request of additional information before a decision may also be part of the process.

Applying the Law: The Experience of Brazil

When the Assessment Committee started its work in 1999, just before the publication of the norms, there were, in Brazil a large number of products, many of which were advertised as "natural" products and were in pharmaceutical forms, being sold as foods or food supplements and bearing unproven preventive and curative claims. Some of these products consisted of complex mixtures of vegetable extracts containing very active pharmacological compounds such as amphetamines (ephedrine) and alkaloids. At the same time, foods with claims on their labels, related to added ingredients such as n-3 fatty acids, phytosterols, fibers, oligossacharides, and so on were already being displayed on supermarket shelves. They were mostly cereal containing fibers, spreads, milk and milk products, special eggs, and energy drinks and had claims associated with intestinal function and health, osteoporosis, heart diseases, cancer, stress, menopause, immune function, and other conditions.

The legislation issued and the work of the ANVISA Committee based on the scientific evidence process allowed the organization of the market. At the end of 2001, the safety, efficacy, and adequacy of claims of almost 2,000 products were assessed but only 430 were approved and are still being produced by 115 manufacturers or importers. (Actually, only 50 products could truly be called functional foods).

The process also had a visible effect on the way applications for registry of a product by small, mid-size, or large companies were presented to the agency. Initially, to substantiate claims, scientific evidence based on encyclopedia articles, magazine articles, and other *midia* used to appear as part of the technical report on the product. After some time and dialog, only scientific documents with scientific data, papers from peer-reviewed scientific and medical journals, and good technical reports appeared. The change indicated the increasing involvement of a more technical staff and thus more quality and responsibility throughout the process, which is quite relevant. Law enforcement was also achieved through consultations between ANVISA and the industry sector and by discussions on functional food science's state of the art in scientific meetings.

In the process of re-organization of the market, hundreds of products (many imported), which would be or were actually classified in other countries as **dietary supplements**, were not approved as foods. Some of them were prohibited. Others registered as **drugs**, mostly as phytoterapeutics. Following the same line, for registration purposes, mixtures of foods with isolated compounds or with herbals and herbal extracts having pharmacological activity were considered drugs, not foods, and as such had to be analyzed, thus submitting to pharmaceutical—not food—regulations.

The claims allowed in Brazil are mostly functional claims, as can be seen in some examples below:

- Spreads with phytosterols:
  "help to maintain healthy levels of cholesterol" when associated with healthy diet and life-style
- Milk containing long chain n-3 fatty acids (EPA and DHA):
  "helps to control triglyceride levels, blood fluidity, inflammatory and immune response" . . . when associated with a healthy diet and life-style
  "enhances immune defense"
  Allegations not approved: "reduces the risk of heart diseases" and "reduces blood cholesterol"
- Milk with n-3 and n-6 fatty acids from vegetable oils:
  "low in saturated fats and enriched with n-3 and n-6 fatty acids" . . . "needed to keep cholesterol low and a healthy heart"
- Milk and other products with prebiotics:
  "contribute to a healthy intestinal flora" and "helps maintaining the intestinal flora balanced"
  Allegation not approved: "assures to your son a healthy digestive system promoting his optimal development and well being" and "has the capacity to adjust intestinal flora and help digestion"
- Probiotics (Lactobacilli and Bifidobacteria):
  "help in maintaining the intestinal flora balanced"
  "help in reducing harmful bacteria"
  "help in increasing the beneficial flora"
  Allegations not approved: "increases antibodies" and also "strengthen natural defenses against daily aggression and stress"
  Chitosan:
  "helps in reducing fat and cholesterol absorption"
  "helps in weight control and cholesterol reduction" . . . "when associated with a hypocaloric diet"
- Oat, soy flakes, wheat germ, pectin, psyllium:
  "fibers help intestinal motility"
  Allegation not approved: "reduces the risk of cardiovascular diseases and osteoporosis" (this was made only for soy)

- Chewing gums with xylitol:
    "help in lowering pH thus reducing the bacterial tooth plaque" . . . "it is not fermentable by mouth bacteria" . . . "neutralize acids that harm teeth"
    Allegation not allowed: "helps in reducing cavities"

It is important to mention that due to misuse of several claims as originally allowed, they are or will be reviewed by ANVISA.

In some situations, risk management and risk communication measures were adopted. Examples are post-marketing surveillance (for olestra and chitosan products) and suggestion of research in the country (fat spreads with phytosterols). Products in pharmaceutical forms should have, on the label, the expression "This product is not intended to treat or cure any disease"; and special precautions, when needed for the elderly, children, or pregnant women, should be clearly stated.

Probiotics (and prebiotics) have been active fields of investigation and resulted in the marketing in Brazil, as in other parts of the world, of several new foods with important potential health effects and different allegations, ranging from "gut health" to "fights cancer," that are sometimes difficult to evaluate, mostly when there is a comparison between industries. For the proper assessment of probiotics to substantiate health claims in foods, there is a need for a more systematic approach with the establishment of specific guidelines, especially for human studies, that do not exist now in Brazil. In fact this should be an international effort. Currently in Brazil, only general claims on gut health are allowed for probiotics, but these may mislead consumers because their interpretation is usually very general, such as "it is good for everything." Specific claims, based on sound scientific evidence, would be more informative to the type of consumer we have, and this will be discussed in the near future in ANVISA.

It is also important to mention an issue that raised some debate among us in Brazil. The food ingredient industry has been very active in developing and advertising functional ingredients to be used in functional foods. As ingredients, they are analyzed mostly for safety; if they are safe or recognized as GRAS in other countries, their commerce is approved, but not necessarily the claims. The norms state that the possible functional or health properties and subsequent claims have to be demonstrated not for the ingredient alone but also for the product containing the ingredient, because of the possibility that interactions, within the matrix of the foods, change or reduce them. To further clarify, an ingredient added to a cereal, for instance, may not behave in the same way as it would behave in a milk drink. What is evaluated and registered and may bear claims is then the food, not the ingredient.

Bioactive Compounds Product Regulation (Dietary Supplements)

In Brazil the dietary supplements common in other countries do not exist in food or pharmaceutical regulations. Only recently, specific norms were issued to cover what was called "bioactive compounds."

These norms, ANVISA Resolution n 2 (Brasil, 2002) were issued due to a high number of locally produced but mostly imported products, usually dietary supplements in pharmaceutical forms, with health claims of a strong international advertising appeal but with safety not well documented. Brazilian health authorities consider it important to keep a clear distinction of these products from foods.

Parallel to the recognition that many of these supplements could benefit consumer health in some situations, it was important to consider the safety aspect because many of them contained **concentrated** forms of components. Even if normally contained in foods, an increased concentration as found in pills, capsules, and so on could have adverse effects after a long-term usage or could interact with drugs, reducing their efficiency.

Bioactive compounds have been defined in the regulation as "natural or synthetic substances having a demonstrated metabolic or physiologic activity." They have been classified in seven different chemical groups: carotenoids, phytosterols, flavonoids, phospholipids, organosulfur compounds, polyphenols, prebiotics, and probiotics. Probiotics were included due to their appearance in nonfood forms and were defined as "live microorganisms having the ability to improve the microbial intestinal equilibrium and thus to produce beneficial effects to the individual's health."

Bioactive compound products should be approved and registered before going to the market, in a process identical to the one previously described for functional foods. They should be proven safe and must have an allegation (functional or health claim) explaining their actions and benefits. In the case of probiotics, they must warrant that the physiological property alleged is maintained during the shelf life asserted. It is important to stress that a bioactive compound product is not approved if it has no well-defined claims on its label.

With this regulation comes a clear separation of products containing single compounds from foods. It should be noted that plant parts, plant extracts, botanicals, or herbals are ruled under the pharmaceutical regulation and as such may not be sold in supermarkets but only in pharmacies, which is an important difference between Brazil and other countries.

Products with amino acids and peptides such as creatine, taurine, arginine, and carnitine, which are common in the United States, are not included in the category. Products with these compounds are used in physical fitness centers in Brazil to increase muscular mass and to improve performance, without much control and criteria. Reports of cases of adverse effects prompted ANVISA to contract several consultations with experts who discussed their safety and their role in exercise performance. It was then concluded that not enough scientific evidence existed to substantiate their efficacy, which was actually limited to some very specific types of athletic competition and for highly trained athletes. Doubts were also raised concerning the safety of some of these products, specifically because of the lack of data on the effects of long-term ingestion. Considering therefore their uncontrolled consumption by some population groups, and also the unproven or very limited efficacy and the doubts on their safety, these products were prohibited and can be sold only as drugs or used in enteral and parenteral nutrition products.

Soya bean proteins in different forms, obtained by different technologies, are widely used in several products in Brazil and have become more popular due to the publicity on the health properties of the isoflavones they contain. Isoflavone capsules, advertising several health benefits, have also become popular among women but have been recently considered as drugs by the health authority. This categorization was based on the fact that the efficacy for stated properties such as cancer prevention and treatment, osteoporosis reduction, use for hormone therapy replacement, and heart protection were not considered proven and could not be indicated to the general population.

Although they were considered to be able to reduce some of the symptoms of menopause, care should be taken and they should always be used under medical supervision.

In Brazil there has been a reduction of fiber consumption in the last few decades due to changes in eating habits and lowered ingestion of traditional staple foods, such as beans.

This fact, and the recognized importance of fiber for gut health, has been a motivation for the production of fibers and fiber-added foods carrying several types of claims including a single "content" claim. This has caused some discussion and led to legal problems, mostly due to the lack of a clear consensus on an adopted definition, a problem that seems to be happening all over the world.

Because not all type of fibers have the capacity of lowering cholesterol, reducing blood glucose levels, and being fermented to produce short chain fatty acids, the position adopted in Brazil has been that each product has to have a proper characterization of the fiber it contains according to the claim made. Even a statement that a product is "rich" or a "good source of fiber" depends on a definition of what is considered fiber. Several products have, for instance, undigested polysacharides that physiologically behave like fiber but legally are not defined as such.

The recent definition of the U.S. National Academy of Science that included dietary fiber and also that of "added fiber" may help to introduce more defined claims. Clearly, the need exists for an international harmonization and a single definition of dietary fiber.

In Brazil, ANVISA, which is a Federal Central Health Agency, has branches in every state and often in several cities within each state. All of them may receive local applications for registering food products. Due to this fact and to the complexity of assessing claims that involve a range of science to language problems, several training activities are constantly conducted all over the country to clearly inform the branch agencies regarding the central directions of the regulation. A decision tree was also produced to help homogenize and organize the analysis of applications and speed the decision process.

Conclusion

As a final remark to this chapter, one can say that functional foods have not been defined officially in the emerging regulatory codes in Brazil, but the functional food concept has been associated with foods having health benefits beyond those of basic nutrition. Regulation is focused on safety and efficacy, and both functional and health claims are allowed provided that they are scientifically proven. Issuing regulations for food claims allowed the production of functional food within the country and promoted or renewed research efforts in important areas of food science and nutrition. In view of market globalization, there is a need for better defining concepts and criteria and of efforts to homogenize legislation in the world.

References

ANVISA 2002 Agência Nacional de Vigilancia Sanitária (ed) Alimentos com alegaçõs de propriedades funcionais. II Seminário, Brasília, 2002 95pp.

Brasil 1969 Ministério da Saúde—Decreto-Lei n. 986, de 12 de outubro de 1969. Institui normas básicas sobre alimentos. Diário Oficial da União, Brasília, 21 de outubro de 1969.

Brasil 1997 Ministério da Saúde—Portaria n. 1549 de 17 de outubro de 1997. Diário Oficial da União, Brasília, 21 de outubro de 1997.

Brasil 1999a Ministério da Saúde—Lei n. 9.782, de 26 de janeiro de 1999. Define o Sistema Nacional de Vigilância Sanitária, cria a Agência Nacional de Vigilância Sanitária e dá outras providências. Diário Oficial da União, Brasília, 27 de janeiro de 1999.

Brasil 1999b. Ministério da Saúde—Portaria ANVS/MS n. 15, de 30 de abril de 1999. Institui a Comissão de Assessoramento Técnico-científico em Alimentos Funcionais e Novos Alimentos. Diário Oficial da União, Brasília, 05 de maio de 1999.

Brasil 1999c Ministério da Saúde—Resolução ANVS/MS n. 16, de 30 de abril de 1999. Regulamento Técnico de Procedimentos para o Registro de Alimentos e ou Novos Ingredientes. Republicada no Diário Oficial da União, Brasília, 03 de dezembro de 1999.

Brasil 1999d Ministério da Saúde—Resolução ANVS/MS n. 17, de 30 de abril de 1999. Regulamento Técnico que Estabelece as Diretrizes Básicas para Avaliação de Risco e Segurança dos Alimentos. Republicada no Diário Oficial da União, Brasília, 03 de dezembro de 1999.

Brasil 1999e Ministério da Saúde—Resolução ANVS/MS n. 18, de 30 de abril de 1999. Regulamento Técnico que Estabelece as Diretrizes Básicas para Análise e Comprovação de Propriedades Funcionais e ou de Saúde Alegadas em Rotulagem de Alimentos. Republicada no Diário Oficial da União, Brasília, 03 de dezembro de 1999.

Brasil 1999f Ministério da Saúde—Resolução ANVS/MS n. 19, de 30 de abril de 1999. Regulamento Técnico para Procedimento de Registro de Alimento com Alegações de Propriedades Funcionais e ou de Saúde em Sua Rotulagem. Republicada no Diário Oficial da União, Brasília, 10 de dezembro de 1999.

Brasil 2002 Ministério da Saúde—Resolução—RDC n. 2, de 07 de janeiro de 2002. Regulamento Técnico de Substâncias Bioativas e Probióticos Isolados com Alegação de Propriedade Funcional e ou de Saúde. Diário Oficial da União, Brasília, 09 de janeiro de 2002.

Lajolo, FM 2002. Functional foods: Latin American perspectives. *Brit. J. Nutrition* 88(2): 5145–5150.

Lajolo, FM 2002a. Alimentos funcionais, bases científicas e normativas. In Boren, AG and Costa, NMB (eds). *Alimentos geneticamente modificados.* Viçosa: Editora UFV. P53–66.

22 Codex and Its Competitors: The Future of the Global Regulatory and Trading Regime for Food and Agricultural Products

Mark Mansour

In light of the proliferation of international and regional organizations claiming an interest in food standard-setting, this chapter is designed to address the role and responsibility of the UN FAO and WHO's Codex Alimentarius Commission in this regard, as well as the developing roles of the various other organizations that have of late involved themselves in this arena. The chapter concludes with an assessment of the impact of the international jurisdictional disputes that are now developing on the future of food regulation and the global food trade.

Background

The Codex Alimentarius Commission was established in 1962 under the auspices of two United Nations organizations: the World Health Organization (WHO) and the Food and Agricultural Organization (FAO). Codex is charged with establishing international standards, guidelines, and principles for use in international food trade. The primary aim of Codex is to protect the health of the consumer and ensure fair practices in the food trade. The Codex process is based primarily on reaching international consensus. More than 165 countries, as well as numerous nongovernmental (NGOs) and intergovernmental organizations (IGOs), actively participate in the proceedings of Codex. That said, the organization can and does cast votes when the Chair believes it to be in the organization's interest, and the Executive Committee is a fairly small body that acts without the direct involvement of most of the member countries.

The full Codex Alimentarius Commission meets every two years to adopt new standards, guidelines, and recommendations and to assign work to its committees. The Codex committees consist of a number of commodity committees, for example, the Codex Committee on Cocoa Products and Chocolate, as well as general subject area, or horizontal, committees—for example, the Codex Committee on General Principles—that make recommendations to the Commission.

Since its inception, the work of Codex has been of significance to the food industry. Codex has received deserved credit for helping to improve food safety standards worldwide; indeed, throughout its first 35 years, Codex operated in relative anonymity as it developed one useful standard after another. However, the importance of Codex has increased substantially in recent years for a variety of reasons. First, international trade has become more important to the economic health of the majority of nations in the world, particularly the United States. Trade in agricultural commodities and foodstuffs has grown more than 800 percent since Codex's creation in 1962. International consensus on food quality and safety issues, which has been established through the international standard-setting work undertaken by Codex, has helped forestall a number of trade disputes.

In addition, Codex has grown in importance as a result of the establishment of the World Trade Organization (WTO) in 1995. Codex is one of three international standard-setting organizations whose health and food safety standards serve as key reference points in settling trade disputes under the WTO Agreements on Sanitary and Phytosanitary Measures and Technical Barriers to Trade (SPS and TBT Agreements). The other two organizations are the International Office of Epizootics and the International Plant Protection Convention.

The Role of Codex As an International Standard-Setting Body

When Codex was established in 1962, its primary charge was to implement the Joint United Nations FAO/WHO Food Standards Program (FSP). The FSP is geared toward protecting consumer health and ensuring fair trade practices involving food. To date, the FSP has adopted more than 4,000 standards, recommendations, and guidelines. Since the establishment of Codex, the FSP has gradually evolved into a point of reference through its articulation of standards, guidelines, and code of practice for international food trade.

Specifically, Codex standards are intended to ensure safe food for the consumer that is not adulterated and is correctly labeled and correctly presented. A Codex standard for any food or foods should be drawn up in accordance with the uniform Format for Codex Commodity Standards and contain the criteria listed in the format requirements. *See* "General Principles of the Codex Alimentarius," Section I, Codex Procedural Manual.

Codex guidelines are defined as provisions of an advisory nature that are intended to assist in achieving the purpose of Codex, which is to guide and promote the elaboration and establishment of requirements and definitions for food, assist in their harmonization, and facilitate international trade. Other advisory provisions include codes of practice and recommended measures. *See* "General Principles of the Codex Alimentarius," Section I, Codex Procedural Manual.

Practically, the difference between Codex standards and guidelines lies in the requirement of formal acceptance. (As to their stature within the WTO, there is no practical difference; both carry the same weight.) Standards are to be formally adopted by member countries and, if accepted in accordance with Codex procedure, are to be acted upon in accordance with the level of acceptance acknowledged by the member country. *See* "General Principles of the Codex Alimentarius," 4.A., Section I, Codex Procedural Manual. Guidelines are also subject to Codex requirements for elaboration and adoption, but are not subject to Codex provisions relating to formal acceptance, although they may be relied upon as advisory texts. *See* "Procedures for the Elaboration of Codex Standards and Related Texts," Section I, Codex Procedural Manual.

Role of Codex Standards and Guidelines in the World Trade Organization Dispute Settlement Procedures

By accepting the Agreement Establishing the World Trade Organization (WTO Agreement), WTO member governments agree to be bound by the rules in all of the multilateral trade agreements attached to it, which include both the SPS and the TBT Agreements. Codex texts are particularly relevant to the application of the SPS and TBT Agreements because the agreements specifically direct member governments to utilize Codex texts in taking decisions under the agreements.

Should disputes arise between member governments regarding the application of agreements such as the SPS or TBT, parties have recourse to the procedures for dispute settlement under the 1994 Dispute Settlement Understanding (DSU) (Annex 2 to the General Agreement on Tariffs and Trade 1994). Article 3 of the DSU outlines the function of the dispute settlement system, which is to preserve the rights and obligations of Members under the covered agreements and to clarify the existing provisions of those agreements in accordance with customary rules of interpretation of public international law. Thus, the source of law under consideration in dispute settlement is the texts of the agreements themselves, including any explicit references to Codex standards, guidelines, or recommendations.

In the case of a trade dispute, the WTO's dispute settlement procedures first encourage the parties involved to develop a mutually acceptable solution through formal consultation.[1] If consultation does not resolve the dispute, the parties may choose to work through other mechanisms, such as "good offices," conciliation, mediation, and arbitration.

As an alternative to formal consultation, the parties may request that an impartial panel of trade experts be selected, with the parties' approval. The role of the expert panel is to hear all sides of the dispute and then make recommendations in the form of a report to the WTO's Dispute Settlement Body (DSB) (essentially the General Council of the WTO), in which all WTO Member countries are represented. After the report is submitted and adopted by the DSB, the defending party must implement the panel's recommendations and report on its compliance. It is possible that the DSB may decide by consensus not to adopt the report, or a party may appeal the recommendations, calling for further deliberation by the panel. (It should be noted that, in practice, decisions made by the expert panels are almost always upheld.)

Under either of the previously described routes of dispute of settlement concerning the application of the SPS or TBT, the parties involved are to rely when appropriate on the applicable Codex standards, guidelines, and recommendations in reaching consensus.

The Role of Codex Guidelines in the Agreement on the Application of Sanitary and Phytosanitary (SPS) Measures

Although the WTO itself is not responsible for developing food safety standards, it does have the authority to place restrictions on the use of food safety measures as unjustified or disguised barriers to trade. The WTO accomplishes this task primarily through the SPS Agreement, although the TBT Agreement also addresses requirements and other issues not covered by the SPS Agreement. As noted earlier, the trade implications of the development of a Codex guideline on biotech labeling depend in part on the referenced role of Codex guidelines in the relevant international food safety agreements.

The role of Codex standards and guidelines in the application of SPS measures is referred to several times throughout the agreement. Perhaps the most important reference occurs in Article 3 of the SPS Agreement, which assesses the harmonization of phytosanitary standards and which reads as follows:

> (1) To harmonize sanitary and phytosanitary measures on as wide a basis as possible, Members shall base their sanitary or phytosanitary measures on international standards, guidelines, or recommendations where they exist, except as otherwise provided for in the Agreement, and in particular paragraph 3.[2]

(2) Sanitary or Phytosanitary measures which conform to international standards, guidelines, or recommendations shall be deemed to be necessary to protect human, animal, or plant life or health, and presumed to be consistent with the relevant provisions of this Agreement and of GATT 1994.

International standards, guidelines, and recommendations for food safety are further defined in Annex A to the SPS Agreement as "the standards, guidelines and recommendations established by the Codex Alimentarius Commission relating to food additives, veterinary drug and pesticide residues, contaminants, methods of analysis and sampling, and codes and guidelines of hygienic practice."

The SPS Agreement also calls for member countries to take an active role in the development of the international standards, guidelines, and recommendations that are to form the basis for phytosanitary harmonization. For example, in Article 3.4, the Agreement states that

"Members shall play a full part, within the limits of their resources, in the relevant international organizations, in particular the Codex Alimentarius Commission, the International Office of Epizootics and the international and regional organizations operating within the framework of the international Plant Protection Convention, to promote within these organizations the development and periodic review of standards, guidelines, and recommendations with respect to all aspects of sanitary and phytosanitary measures."

Next, the SPS Agreement indicates that the standards and guidelines of the relevant international organizations are to be utilized in the application of risk-assessment techniques. Article 5.1 requires that Members ensure "that their sanitary or phytosanitary measures are based on an assessment, as appropriate to the circumstances, of the risks to human, animal, or plant life or health, taking into account risk assessment techniques developed by the relevant international organizations." Further, Codex standards and guidelines may play a role in the appropriate exercise of precaution, as elaborated in Article 5.7 of the agreement, which states that:

"In cases where relevant scientific evidence is insufficient, a Member may provisionally adopt sanitary or phytosanitary measures on the basis of available pertinent information, including that from the relevant international organizations as well as from sanitary or phytosanitary measures applied by other Members."

Finally, the SPS Agreement touches upon the potential role of Codex in the WTO dispute settlement procedure. Article 11.2 encourages dispute resolution panels to seek advice from experts chosen by the panel in consultation with the parties to the dispute, either through the establishment of an advisory technical experts group or consultation with the relevant international organizations. The Agreement particularly encourages the procurement of expert advice in the resolution of disputes involving scientific or technical issues, as is often the case in food safety matters. In fact, a Codex expert consulted with the panel addressing complaints over the European Union's ban on hormone-treated beef.

See Organization for Economic Co-operation and Development (OECD), *Overviews and Compendium of International Organizations with Food Safety Activities*, para. 92, May 11, 2000.

Interaction of Various NGOs and IGOs in the Codex Process

In addition to individual governmental participation in Codex, for example, the United States Government's (USG) participation, many nongovernmental organizations (NGOs) and intergovernmental organizations (IGOs) are also active in the Codex process. IGOs are comprised of numerous countries that share common interests, such as the European Community (EC), the Association of Southeast Asian Nations (ASEAN), and the WTO, and represent the interests of those governments.

By contrast, NGOs represent the interests of consumer activist groups and private industry, for example, Greenpeace, the International Council of Grocery Manufacturers Associations, and other, similar groups. The purpose of NGO participation in Codex is to provide expert information and advice to the Commission, as well as to provide the private sector with a forum for conveying the views of their members and to play an appropriate role in ensuring the harmonization of interests among the various industries they represent. To be active in Codex, NGOs must obtain "observer status." In addition to participating as NGOs with observer status, the private sector may also participate in Codex through government delegations, which may choose to include members of the private sector, such as consumer groups and industry representatives, at their discretion.

Observer status affords an NGO the right to send a nonvoting observer to Codex sessions, receive advance copies of working documents and discussion papers, circulate to the Commission its views in writing, and participate in discussions when invited to do so by the Chairman. Additionally, NGOs with observer status may be invited to participate in meetings or seminars on subjects organized under the Joint FAO/WHO Food Standards Program that fall within its fields of interest, or submit written comments in lieu of participating at the meetings.

An overview of how various IGOs, NGOs, government delegations, and other groups participate in the Codex process is set forth in the following sections.

WHO/FAO

The World Health Organization and the Food and Agriculture Organization are the international organizations responsible for food safety–related activities. These two organizations, through their respective mandates, have the responsibility of protecting consumer health, preventing the spread of disease, and ensuring the fair trade of food products. Codex was established through the Eleventh Session of the Conference of FAO in 1961 and the Sixteenth World Health Assembly in 1963 as a vehicle for implementation of the Joint FAO/WHO Food Standard Program. Both organizations subsequently adopted the Statutes and Rules of Procedure for the Commission. The goals of this program included consumer health protection, ensuring fair food trade practices and promoting the coordination of the food standards work of various international governmental and nongovernmental organizations.

The WHO and FAO, in addition to their being the IGOs responsible for the creation of

Codex, in theory complement the work of the Commission in several ways. First, the FAO and WHO assist Codex by convening expert meetings for, and providing expert consultations to, the Commission itself. FAO and WHO provide expert scientific advice on many aspects of food quality, safety, and nutrition that is relevant to the work of Codex. The FAO/WHO Expert Consultations provide independent scientific expert advice to the Commission and its specialist Committees and Task Forces. A recent example of this is the Joint FAO/WHO Consultation on the Health Implications of Acrylamide in Foods that was held by the WHO at the end of June. A second example is the recent Joint WHO/FAO Expert Consultation on Diet, Nutrition, and the Prevention of Chronic Diseases Draft Report, which provided nutrient recommendations for the prevention of obesity, diabetes, cardiovascular disease, cancer, dental disease, and osteoporosis.

Second, in order to adopt Codex standards, countries require an adequate food law as well as technical and administrative infrastructure with the capacity to implement the standard and ensure compliance with the standard. As a result, one of the roles of the WHO and FAO is to provide assistance to developing countries to enable them to take full advantage of the Commission's work. The assistance given to the developing countries by the WHO and FAO has included helping the countries with the establishment and strengthening of food control agencies. Additionally the WHO and FAO have assisted developing countries by establishing and strengthening their national food control systems.

As is apparent from the preceding discussion, FAO and WHO are closely linked, as the parent organizations, to the Codex structure. It is important to remember, however, that they are not directly involved in the business of setting food standards. Neither organization possesses compliance or enforcement responsibility, nor does either organization have any responsibility in the risk management arena. Even the FAO/WHO expert committees and consultations responsible for risk assessment exercise their responsibilities within the framework of the Codex Alimentarius system.

More relevant to this analysis is the fact that the two organizations have been operating, over the course of the past decade, from different agendas. Increasingly, tension and disagreements between the two parent organizations, both over policy direction and standard-setting, have surfaced, and this trend will have to be watched carefully for signs of effect on the broader policy debate within and around Codex. In particular, protracted FAO and WHO in-fighting could invite further involvement in the food standards process by other interested IGOs who detect a void and seek to fill it.

European Commission

The EC has official IGO status with Codex. The EC's observer status includes all the EC institutions, namely, the European Commission, European Parliament, and the Council of Ministers. As do ASEAN and Mercosur (discussed shortly), the EC interacts with Codex to represent the interests of its member countries and the region as a whole.

ASEAN

ASEAN has official IGO status with Codex. ASEAN is comprised of the following countries: Brunei, Philippines, Cambodia, Singapore, Indonesia, Thailand, Laos, Vietnam, Malaysia, and Myanmar. The primary objectives of ASEAN are the following: (1) to provide a form to discuss Codex issues, to promote information sharing, and transparency; (2) to formulate ASEAN common positions on Codex issues of mutual interest; and (3) to promote ASEAN common positions at various Codex meetings.

OECD

The Organization for Economic Cooperation and Development has official IGO status with Codex. It was established in 1961 to promote policies designed to contribute to the development of the world economy and sound economic expansion in member countries as well as nonmember countries through the process of economic development and the expansion of world trade. The OECD currently has 29 member countries, including Australia, Japan, Canada, the United States, and all the countries belonging to the EC. In particular, the OECD has been involved in the work of Codex regarding foods derived from biotechnology; it also is a proponent of an alternative to the present global arrangement in Codex whereby trade policy dictates health objectives.

Unlike the FAO and the WHO, the OECD has no mandate in the area of food safety. However, its involvement in various activities in the areas of economics, trade, and the environment has provided it a forum through which to enter the arena of food safety, even though it has no formal mandate to do so. Through its analyses of the costs and benefits associated with food safety decisions and impact on trade of regulations, as well as its recent and well-publicized activities relating to biotechnology, OECD has expanded its reach into areas that are the established province of Codex. This could lead to redundancy and conflicting international pronouncements in the coming years, especially given the relative lack of involvement by Group of 77 members in OECD activities.

CBD

The Convention on Biological Diversity (CBD), which also has official IGO status within Codex, works to conserve biological diversity, the sustainable use of its components, and the fair and equitable sharing of the benefits arising from the utilization of genetic resources. The CBD was responsible for developing the Cartagena Protocol on Biosafety, which was adopted in January 2000. As does the CBD, the Protocol seeks to protect biological diversity from the potential risks posed by living modified organisms resulting from modern biotechnology.

WSSD

The World Summit on Sustainable Development (WSSD) is sponsored by the United Nations Commission on Sustainable Development (CSD). The CSD is an IGO whose members are elected by the Economic and Social Council from among the U.N. member states and its specialized agencies. The CSD has three primary objectives. First, it reviews international, regional, and national progress in the implementation of recommendations and commitments contained in the final documents of the United Nations Conference on Environment and Development (UNCED). Second, it elaborates policy guidance and options for future activities to follow up on UNCED and achieve sustainable development. Finally, it promotes dialogue and builds partnerships for sustainable development with governments, the international community, and the major groups outside the central government who have a major role to play in the transition toward sustainable development.

Cairns Group

The Cairns Group is an IGO comprised of 18 agricultural exporting nations. Among its members are Australia, Canada, Colombia, Malaysia, Indonesia, South Africa, Philippines, and New Zealand. The primary objective of the Cairns Group is to ensure that agriculture is at the forefront of the multilateral trade agenda. This is reflected in the group's development of several proposals, primarily for submission to the WTO, calling for the elevated importance of and improved market access for agricultural commodities and food products.

Mercosur

Mercosur has official IGO status with Codex. As is ASEAN, Mercosur is a regional organization. It is comprised of Argentina, Brazil, Paraguay, and Uruguay, who in 1991 signed the Treaty of Asuncion, in which they agreed to establish the Southern Common Market, or Mercosur. The member countries belonging to Mercosur undertake to harmonize food standards and to identify and reduce technical barriers to food trade in their region.

Andean Community

The Andean Community has official IGO status with Codex. Its member countries are Bolivia, Colombia, Ecuador, Peru, and Venezuela. Among the objectives of the Andean Community is to promote the balanced and harmonious development of the member countries under equitable conditions.

Greenpeace

Greenpeace International has official NGO status with Codex. The primary objective of Greenpeace's involvement in the Codex process is to bring about the "immediate cessation of deliberate releases of genetically engineered organisms into the environment." As part of this mandate, Greenpeace has been active in pushing for mandatory, process-based labeling of all food products containing or derived from biotechnology. This would, of course, necessitate the implementation of an identity preservation system to track foods throughout the production chain to ensure that they are free from any bioengineered components. A number of other groups work closely with Greenpeace on this and other issues.

ISO

The International Organization for Standardization (ISO) is a worldwide federation of national standards bodies. Established in 1947, it now represents more than 140 countries. ISO helps to promote the development of standardization and related activities through international agreements, which are then published in the form of international standards. The Standards relate to various intellectual, scientific, technological, and economic activities. They are documented agreements that set precise criteria that are then used consistently as rules, guidelines, or definitions of characteristics. ISO and Codex are similar in function in that they both set standards that provide the framework or template for individual country standards. For example, both NAFTA and the WTO recognize ISO and Codex standards.

The Interaction of the Various Actors and Processes

Until the past several years, a tacit understanding existed that for food safety standards, Codex was the only relevant actor on the global stage. This understanding continued after the creation of the WTO. The primary impetus for the unsettled jurisdictional situation, which today confronts industry and governments, were the simultaneous and related emergence of consumer NGOs and the contentious issue of biotechnology.

These developments coincided with a rash of U.S.-EU trade disputes over beef hormones and other issues in which Codex was embroiled. Over the past several years, as it has become clearer that some of the consumer NGOs and member countries sought to use the Codex system to expand standard-setting into the arena of social policy, a disruption in the function of the standard-setting process has resulted.

The divisions within Codex are several but may be roughly described as those parties encouraging "science based" standard setting and those organizations and entities that believe that a broader set of criteria can and should be applied to the standard-setting process.

A group of industry NGOs, some member countries (both developed and developing), and several IGOs have been active in recent years in calling for removal from the standard-setting process of any mechanism that relies on so-called "other legitimate factors" that extend into the realm of the subjective, such as animal rights, consumer right to know, and related criteria, including the highly contentious "precautionary principle." Application of these criteria, these actors fear, will serve to restrain trade without providing concomitant protection of consumer health and safety.

By contrast, a loose coalition of member states (principally the EU member states and a group of European countries aspiring to EU membership), consumer NGOs, and several country delegations (led by Norway, Switzerland, and India) insist that Codex cannot fulfill its role without taking into account these "other legitimate factors." Facilitation of trade is viewed at best as a secondary objective (the position of the EC member states and Consumers International) and at worst as an excuse by developed countries and multinational corporations to impose upon global consumers their own preferences, borne principally as a desire to make a profit (this latter accusation is generally leveled by Greenpeace, Friends of the Earth, and the so-called 49th Parallel Group). At its bottom, this is the core dispute that has stripped the gears from the Codex mechanism. If one looks at the major initiatives that have characterized the past several years' debates (biotechnology labeling, the "precautionary principle," the role of "other legitimate factors" in standard setting, and the general issues of nutrition and health claims), it is readily apparent that the failure to achieve progress stemmed from the fact that none of the initiatives involved had at bottom the requisite scientific justification to enable objective debate. The result is that all the discussions around these issues have foundered for lack of any common ground among the parties that would facilitate progress.

The issues of biotechnology, regulation of infant formula and dietary supplements, and the role of the so-called "precautionary principle" have, because of their high profile, commercial impact, and attendant controversy, accelerated the pace of these disagreements and made the workings of Codex more interesting and thus more visible. As a result, over the past several years most of the attention of outside observers has been focused on the issues at which Codex has failed to achieve consensus and has ignored the areas, primarily the technical ones, in which Codex continues to develop standards that are helpful for consumers, producers, and government alike. Indeed, when the recently empanelled Codex

Evaluation Team sought input from stakeholders, virtually every party with which it met, except for consumer activist groups, said that Codex functioned best and most smoothly when it focused on science and failed most often when it strayed into the realm of social engineering.

An irony is that part of the impetus for establishing the evaluation process was the frustration by activist groups that their efforts to secure Codex standards based on the "precautionary principle" and "other legitimate factors" were stalled as a result of what they described as a dysfunctional process. The evaluators were told plainly that the system was functioning very well until these extraneous issues were thrown into the mixture, and that any present dysfunction was owing to issues such as these that were entirely inconsistent with Codex's functions.

In a very real sense, the determination by bodies such as the regional organizations, OECD, and the WHO itself to take on issues such as supplements, health claims, nutrition, obesity policy, biotechnology, and others stems from their dissatisfaction at Codex's refusal thus far to engage in standard-setting based on the application of "other legitimate factors." That refusal, in turn, is a direct outgrowth of intense and sustained industry NGO and U.S. government activity designed to maintain Codex's focus on science. The movement of these issues, which statutorily are the core of Codex's brief, to other international bodies represents a manner of "forum shopping" by activist groups and their member countries and international organization allies, designed to ensure that some friendly organization disposes of these issues in a favorable manner. In a very real sense, then, the questions industry confronts today are a direct product of its relative success at keeping Codex focused and forcing activist groups to look elsewhere for more favorable regulatory solutions.

It is difficult to imagine any Codex committee producing a report such as that released recently by the WHO on nutrition, but it places in perspective the challenge facing industry and governments. If Codex becomes a secondary focus, and attention reverts to action at the level of the OECD, WHO, and WSSD, industry will have to respond to initiatives on multiple levels. Although none of these organizations possesses Codex's jurisdictional capacity, these organizations are "bully pulpits" for activist groups and therefore present a real risk that high-profile campaigns will result in even more public misinformation than if Codex continues to manage issues.

To complicate matters further, Codex will continue to be more relevant from a legal and regulatory perspective, because as long as it remains the ultimate arbitral mechanism, it is by definition more germane to the fulfillment of industry's regulatory requirements in key markets. The central thesis posed by those who believe that Codex has ceased to be relevant is that countries which fail to achieve their national objectives at the Codex level will act unilaterally, flouting Codex, and will do so without fear of retribution.

That theory has been challenged by the recently released report of the WTO's so-called Sardines Panel. In fact, the panel's decision is the most important test of Codex's authority in the area of international standard setting. The implications of the decision for the status and legitimacy of Codex decisions are significant. The panel ruled that the European Union had acted in a manner inconsistent with its WTO obligations in failing to adopt a Codex standard for the labeling of sardines. The EU regulation, which was more restrictive than the Codex standard, was found to be more trade restrictive than necessary to fulfill a legitimate objective. The EU announced in response that it will appeal, but no trade observer with whom we have spoken believes it has any chance of winning on the merits. Incidentally, the same observers believe that the United States would win a dispute over the moratorium on biotech authorizations, as well as on the food and feed and labeling and traceability proposals, cases that would serve as the successor litmus tests for determining the scope of Codex's authority if the sardine appeal does not resolve the issue. Regardless, today the WTO sardines report stands as evidence that Codex standards have presumptive legitimacy in the Agreement on Technical Barriers to Trade.

Biotech labeling and traceability, along with quantitative ingredient labeling and a menu of other issues, remain before Codex, and that will not change regardless of whether the OECD and the WHO become more actively involved in seeking resolution to the same questions. They are now fully involved, and there is no reason to believe that this circumstance will change. Although there is a ceiling to the OECD's potential for engagement, because of a degree of misgiving about the organization by developing countries, there is no apparent deterrent immediately discernible where the WHO is concerned, given its status as one of Codex's parent organizations. Already there are proposals for significant budget increases for the organization, and among the objectives WHO seeks to achieve is greater involvement in the Codex process. One of the ancillary goals expressed is the desire to manage the Codex delegate selection process through the WHO. It is not difficult to envision where such a process change would lead: The WHO states clearly that among the criteria for delegate participation would be a narrowly focused specialization in the food safety and environmental arenas, which could result in the marginalization of trade officials, the same ones who have proven instrumental in keeping out the "precautionary principle" and "other legitimate factors the past several years." If the involvement of these officials is curtailed before Codex activities revert to the realm of traditional standard-setting, both process and outcome relative to a range of issues can be expected to change dramatically.

Such a sequence of events would leave various affected industries redeploying resources to stem the effort to impose nonscientific standards in areas in which activists have a high public profile. The risk is that should industry decide to neglect Codex, the outcome might be worse than it now appears. At present, industry's active involvement and that of its allies among the trade and commerce community tends to keep otherwise disposed Codex committee chairpersons from disregarding science and the facts in the standard-setting process. If industry concentrates its efforts elsewhere, there will be less restriction on such behavior and thus the increased likelihood of unfavorable Codex standards. Of course, these unfavorable standards will be supported by the OECD and WHO, which will step back, having persuaded Codex to adopt their agenda. At the end, industry will have expended the same base of resources and expertise, with arguably less practical result to show for the effort.

Accordingly, the goals of those who support the concept and the reality of a firmly rooted Codex process should be clear and mutual: strengthening Codex's activities in the realm of standard-setting based upon empirical criteria. The debate regarding the role of subjective social factors is academic in nature, and although interesting, should neither form the locus of a reordering of the Codex process and philosophy nor be allowed to usher in the involvement of other organizations without appropriate authority, to usurp Codex's role. The outcome will be, ultimately, a degradation of the standard-setting process, regulatory gridlock, and a loss of confidence on the part of the public in the ability of regulators to identify risk, distinguish it from hazard, promote the public welfare, and ensure the movement of goods. If this is allowed to happen, then none of the goals envisioned by the global community at Codex's inception will have come to pass, and the past forty years will have been largely wasted.

Notes

1. Trade disputes may be brought to the WTO under one of three sets of circumstances:
    1. Violation: the failure of another contracting party to carry out its obligations under the General Agreement;
    2. Non-Violation: the application by another contracting party of any measure, whether or not it conflicts with the General Agreements, that deprives a party of some benefits it should enjoy under the General Agreement; or
    3. Any other situation.
2. Article 3.3 permits Members the opportunity to introduce measures that result in a higher level of SPS protection that would otherwise be achieved by measures based on the relevant standards, guidelines, or recommendations if there is a scientific justification or if the member finds the level of protection appropriate in accordance with (1) and (2) above. Codex standards and guidelines come into play again under the definition of "scientific justification," which requires that a member determine, based on an examination and evaluation of available scientific information, that the relevant international standard, guideline, or recommendation does not provide sufficient SPS protection.

Index

Actimel, 185
Activists, 385–387
Administrative Procedure Act, 112, 113, 127
   "significant scientific agreement" standard, 125–126
"Administrative Provisions for New Food Source Hygiene" (China), 266
"Administrative Regulations for Health Foods" (China), 263
Adverse Event Reporting, 222
Advertising
   anecdotal evidence, 158
   Brazilian regulations, 371
   corrective, 162
   deceptive/misleading
      EU regulations, 345
      FTC Act, 149–156, 161–164
   EU regulations, 350–356
   expert endorsements, 158, 163–164
   express claims, 152–153
   FOSHU products, 256–257
   FTC Act, 149–156
      violations, 161–164
   implied claims, 152–153
   limiting health claims to, 187–189
   nutrition-based claims, 49–50
   Spanish regulations, 338–343
   substantiating, 156–161
   truth in, 149, 151
   U.S. regulations, 49–50, 149–156, 161–164
AGF Vitahot, 257
Agricultural Marketing Service (AMS), 70, 72
Ajinomoto General Foods, 257
Alar, 71–72
All-Bran, 25, 43–44, 79
All claims order, 161
All products order, 161
Alternative medicines, 8

American Dietetic Association, 107
American Heart Association, 103–105
Amino acids, 325–326, 374
Amurol Confections, 185
Andean Community, 384
Androstenedione, 162
Anecdotal evidence, health claims advertising, 158
Animal studies, Chinese regulations, 274–275, 287
Animal welfare, organic standards, 69, 70, 76
Antibiotics
   organic standards, 69, 73, 76
   overprescribing, 253
Antioxidants
   and cancer, 15
   FDA interim standard, 117–118
ANVISA, 367–369
   bioactives regulation, 373–375
   health claims regulation, 371–373
"Approval Certificate of Health Foods" (China), 264–265, 276, 282–285
Aquafina, 26
Arginine, 187, 374
AriZona brand, 25
Arizona Rx Memory Mind Elixir, 186–187
"Article", FFDCA definition of, 137–138, 141–147
ASEAN, Codex interactions, 382–383
Assessment Board for Diet-Health Information (BHK), 308–309
AST, 162
Australia
   food law, 321–334
   food safety, 324–327
   functional foods regulation, 321–327, 329–334
      nutritional issues, 327–329

Australia (*continued*)
  health claims regulation, 329–334
  novel foods regulation, 325–327
  vitamins/minerals regulation, 327–329
Australia New Zealand Food Authority (ANZFA), 321–324
  folate health claim evaluation, 47–48
Australia New Zealand Food Standards Code, 321, 324, 329, 331–334
"Authoritative statement" standard
  FDAMA, 111
  health claims, 83, 101, 172
Aviva, 34

Balance Bar, 23
Batch production records, 64–65
Bayer, 8, 11
Bear gall powder, 274
Bee pollen, 297
Belgium
  health claims regulation, 233
  herbals/botanicals regulation, 243
  vitamins/minerals regulations, 242
Benecol, 30, 31, 34, 184, 189
  FOSHU-approved health claim, 254, 257
Berberine, 241
Beta-carotene supplements, 174
Beverages, fruit-based, 25
BfR, 309, 312, 315
Bifidobacterium, 254, 372
Bioactive substances/compounds, 4; *See also* Nutraceuticals
  Australia/New Zealand regulations, 325–326
  Brazilian regulations, 368, 369, 373–375
  clinical substantiation, 17–20
  form of, 325
  performance-enhancing properties, 326
  proprietary, 30–32
Biodiversity, and organic agriculture, 70
Biofoods, Spanish regulations, 342–343
Biologically active principles (BAPs), 241; *See also* Bioactive substances/compounds
Biomarkers, and health claims process, 85–86
Biotechnology
  Codex and, 383, 385–387
  traceability, 387

Black currant oil, 90, 92
Blood pressure
  and dairy peptides, 3
  potassium and, 44–45, 83, 103–105, 177
    Tropicana Pure Premium, 101–108
Books, FDA labeling restrictions, 93–94
Borage seed oil, 92
Boswellia, 19
Botanicals
  Brazilian regulations, 374
  EU regulations, 241, 242–243, 316
  Irish regulations, 243
  U.S. regulations, 12, 17–20
Bovine spongiform encephalopathy, 260
Bran Buds, 177
Bran Flakes, 79
Brazil
  advertising regulations, 371
  bioactives regulation, 373–375
  botanicals/herbals regulation, 374
  dietary supplements regulation, 373–375
  food, defined, 367
  food law, 367–370
  functional claims, 370–371
  health claims regulation, 370–373
  labeling regulations, 370–373
  performance-enhancing products, 374
Brazilian Sanitary Surveillance Agency, 369
Bristol-Meyers Squibb, 8
Bundesamt für Verbraucherschutz und Lebensmittelsicherheit (BVL), 309, 312
Bundesinstitut für Risikobewertung (BfR), 309, 312, 315
Bundesministerium für Verbraucherschutz, Ernährung, und Landwirtschaft, 309
Bureau of Competition (FTC), 149
Bureau of Consumer Protection (FTC), 149
Bureau of Economics, (FTC), 149
Bureau of Food and Drugs (Philippines), 296
BVL, 309, 312

Caffeine, 25
  approved uses, 185, 189
Cairns Group, 384
Calcium
  deficiency, Japan, 248

health claims, 189
and osteoporosis, 24
Calcium acetate, 92
Calcium citrate malate, 23–24, 257
Campbell Soup Company, 154
   fat/cholesterol claims, 177
   Tomato Soup, 188
   V8 juice structure/function claim, 183
Canada
   dietary supplements, 213
   functional foods
      market, 213–214
      regulation, 216–221
   natural health products (NHP)
      industry, 213–214
      regulation, 221–225
Canada Gazette, 218, 221, 222–224
Canadian Food and Drugs Act and Regulations, 215
Canadian Institutes of Health Research (CIHR), 224
Cancer
   and antioxidants, 15
   and fiber, 25, 43–44, 79
   and phytochemicals, 24
   and selenium, 15, 174
Canola oil, genetically modified, 237
Carbohydrates, EU nutrition labeling regulations, 230
Carcinogenicity studies, 275
Cardia Salt, 187
Cardiovascular disease. *See* Heart disease
Carnitine, 204, 374
Carotenoids, 374
CarpalCare, 176
Cartagena Protocol on Biosafety, 383
Cease and desist order, FTC, 161
Center for Food Safety and Applied Nutrition (CFSAN)
   health claims, 14
   and *Pearson v. Shalala,* 15
Centers for Disease Control and Prevention (CDC), as "scientific body", 83, 101
Central Hudson Gas & Electric Corp. v. Public Service Commission of New York, 112–113, 123–125
Centrum, 8, 9

Cereals, fiber health claims, 25, 43–44, 83, 172, 177
Certifying agents, organic foods, 72
Cheerios, 26
Chevron USA v. NRDC, 141–143
Chewing gum, 257, 373
Children, obesity/overweight in, 45
China
   food law, 263, 264, 288–289
   health foods
      defined, 295
      industry, 263
   health foods regulation, 263–265, 297–298, 300–301
      approval certificate, 264–265, 276, 282–283, 285
      assessment procedure, 283–285
      control/supervision, 288–289
      evaluation/approval process, 266
      functionality evaluation, 276–281
      hygiene standards, 287
      imported foods, 282–283
      labeling, 283, 285, 287
      packaging, 286–287
      production, 286
      safety control, 266–275
      violations of, 289
   Ministry of Health (MOH), 263–290, 298, 300–301
   premarketing approval, 298
Chinese Institute of Food Science and Technology, 263
Chinese Nutrition Society, 263
Chitosan, 372
Chlorella, 92
Cholesterol
   consumer misconceptions, 353
   and oat bran, 3, 25, 44, 79–80, 172, 177
   and phytosterols, 30, 31, 315–316, 327
Cholestin case, 185
   agency and court proceedings, 140–144
   background, 139–140
   implications of, 144–146
Chondroitin sulfate, 19, 20
Chromium, 205
Citrus juices, calcium-fortified, 23–24
Civil penalties, advertising violations, 161

"Claim", defined
  Codex, 351
  EU, 231–232
Claim Identification Number (CIN), 220
Clif Bar, 25
Clinical trials
  FOSHU products, 252
  for substantiation of health claims, 85, 158–161
  Swedish regulations, 307–309
CocaCola, functional foods, 25
Code of Conduct on Health Claims (Belgium), 233
Code of Practice (Netherlands), 233–234
Code of Practice (UK), 38, 41
Codex Alimentarius
  claim, defined, 351
  "Codex General Guidelines on Claims, 304
  consumer health, 377–378
  General Principles for the Addition of Essential Nutrients to Foods, 327–328
  Guidelines for the Use of Nutrition Claims, 353–355, 359, 364
  health claim, defined, 38
  health claims regulation, 304–305
  international authority of, 385–387
  low fat/calories claims, 356, 359
  organic foods guidelines, 70, 76
  organizational interactions, 381–387
  "precautionary principle", 385–387
  and SPS Agreement, 378–381
  WTO dispute settlement procedures, 378–379
Codex Alimentarius Commission, establishment of, 377–378
Coenzyme Q10, 92
Cognitive dysfunction, and phosphatidylserine, 16, 174
Colors, EU regulations, 228, 241
Comfrey, 155
Commercial speech, and health claims, 109, 111–112, 121–125
Commission on Sustainable Development (UN), 383
Community Registry of Functional and Health Claims, 362
Compost, organic standards for, 73

Con Agra, soy foods, 26
Conjugated linoleic acid (CLA), 28
Conseil National de l'Alimentation, 233
Constitution of the United States of America, 126–127
  First Amendment, 109, 111–112, 121–125
  Fifth Amendment, 112
Consumer Advisory, FDA, 170
Consumer advocacy, and misleading health claims, 46
Consumer communication, Tropicana Pure Premium campaign, 105–108
Consumer health
  Codex Alimentarius, 377–378
  Food Standards Program, 378
  WHO/FAO, 381–382
Consumer Health Information for Better Nutrition (FDA), 40
Consumers, dietary supplement use, U.S., 21
Consumer safety, and information disclosure, 152–156
Consumers' Association (UK), 47
Consumer satisfaction surveys, 158
Contaminants
  EU regulations, 241
  preventing, 66–67
  SPS Agreement, 380
Content claims, 101
Convention on Biological Diversity (CBD), 383
Coronary heart disease. *See* Heart disease
Cosmetics, FDA-FTC jurisdiction, 150
Coumarin, 241
Council of Better Business Bureau, National Advertising Division, 186–187
Council of Europe, "Guidelines concerning scientific substantiation of health-related claims for functional foods", 304
Court of Appeals for the District of Columbia Circuit, 109, 112
Court of Appeals for the First Circuit, FDA and food additives, 90
Court of Appeals for the Seventh Circuit, FDA and food additives, 90
Court of Appeals for the Tenth Circuit, 142–143

Cranberry juice, 22, 45
Creatine, 374
Credence claims, 157
Cultivation practices, organic foods, 73
Current Good Manufacturing Practices
 (CGMPs), 56–57
  batch production records, 64–65
  equipment, 60–61
  expiration dating, 67
  labeling, 63
  laboratory operations, 65–66
  manufacturing, 66–67
  master manufacturing records, 63–64
  packaging, 63, 66–67
  personnel, 58–59
  physical plant, 59–60
  production, 61
  quality control unit, 61–62
  raw materials, 62–63
  requirements, 57–58
  utensils, 60

Dairy peptides, and blood pressure, 3
Dannon, 185
Deception Policy Statement, FTC, 149–156, 161–164
DeFelice, Stephen, 3
Dementia, and phosphatidylserine, 16, 174
Desert-living cistanche, 274
Designer foods, German regulations, 313–316
Dextrin, FOSHU-approved health claim, 255
DHA, 32, 372
Diabetes
  adult-onset, Japan, 248
  and diet, 214
  and nutraceutical research, 204–205
Diacylglycerol, 256
Diet
  and disease, 214
  and functional foods, 303–304
Dietary Approaches to Stop Hypertension (DASH) trial, 104–105
Dietary foods, Germany, 310–312
Dietary guidelines, EU recommendations, 345
Dietary ingredients
  defined

DSHEA, 55
FFDCA, 5
exemption of food additive status, 92
Generally Recognized As Safe (GRAS), 14, 20
new, 170
DSHEA regulations, 92, 96–97
EU regulations, 237–238
FFDCA definition of, 146
quality control, 62–63
safety regulations, 30, 170–171
Dietary Supplement Act, 110
Dietary Supplement Health and Education Act (DSHEA)
  dietary ingredients, 92
    defined, 55
    new, 92, 96–97
  dietary supplements
    defined, 91, 138–139, 170
    requirements for, 94–96
  GMPs, 97
  health claims
    approved, 202
    restrictions, 13–16
  imminent hazard authority, 93, 96
  passage of, xiii, 4, 5, 8–9, 89
  purpose of, 89–91
  safety standards, 92–93
  statements of nutritional support, 94
  structure/function claims, 14, 151
  use of published literature, 93–94
Dietary supplements
  adulterated, 95–97
  background, 8–9
  consumer attitudes toward, 203
  consumer use, 21
  defined, 5
    DSHEA, 91, 138–139, 170
    FFDCA, exclusionary clause, 137, 140–147
  functional foods as, 183–187, 190
  industry, U.S., 9–11, 13–29
  market
    Canada, 213
    U.S., 8–9
  misbranded, 95–96
  and *Pearson v. Shalala,* 173

Dietary supplements (*continued*)
 products considered as, 11–13
 purchasing channels, 12
 quality control, 55–67
 statements of nutritional support, 94
 structure/function claims, 175–176
 truth in labeling, 95–96
Dietary supplements regulation, *See also*
  Food supplements regulation
 Brazil, 373–375
 Codex and, 385
 FDA, 89–98, 109
 Good Manufacturing Practices (GMPs), 97
 health claims, 13-16, 111-119; *See also*
  Pearson v. Shalala
  FDA-approved, 177–182
 Japan, 258
 legal framework, 138–139
 Malaysia, 296
 New Zealand, 321, 324, 329, 331–333
 Pearson v. Shalala, 109–130
 Pharmanex V. Shalala, 137–147
 premarket approval process, 22
 Singapore, 296–297
 Thailand, 297
 U.S., 29–34
  dietary ingredients, 92
  DSHEA enactment, 89–91
  DSHEA prohibitions, 95–96
  DSHEA requirements, 94–96
  imminent hazard authority, 93, 96
  new dietary ingredients, 92, 96–97
  use of published literature, 93–94
  violations, 96
Dietetic foods, EU regulations, 238–239, 344, 359
Diet and Health: Implications for Reducing Chronic Disease Risk, 103–105
Dioxin-like polychlorinated biphenyls (PCBs), EU regulations, 241
Dioxins, EU regulations, 241
Disclaimer, structure/function claims, 6
Disclosures
 in advertising, 152–156
 mandatory, 162
Disease
 defined, 202
 diet-related, 214

Disease claims
 conventional foods, 110
 defined, 110
Dispute Settlement Body, WTO, 379
Dispute Settlement Understanding (DSU), 379
District Court for the District of Columbia, on FDA and *Pearson v. Shalala,* 112, 113–119
District Court for Utah, 137, 141–142
Doan's Pill's, 162
Docosahexaenoic acid, 325
Dole Juices, 25
Drug, Canadian Food and Drugs Act definition, 215
Drug Identification Number (DIN), 215, 222–223
Dupont, 26

Echinacea
 efficacy of, 203
 in foods, 184, 185
 standardization of, 19
E. coli 0157, 75
Econa cooking oil, 256
Eden Foods, 26
Eggland's Best, 162–163
8th Continent™, 26
Employee training, CGMPs, 58–59
Endorsements
 by experts, 163
 by groups, 163
 deceptive, 163
 use in health claims advertising, 158
Endorsers, material connection, 163–164
Energy bars, 45
Energy drinks, 25
 Australia/New Zealand regulations, 329–330
 global market, 45
Enova, 256
Ensemble, 34, 182–183
EPA, in milk, 372
Ephedra, 22, 274
Ephedrine, 371
Epigallocatechin (EGCG), 11
Equipment, quality control, 60–61
European Authority, 345

European Commission Concerted Action on Functional Food Science in Europe (FUFOSE), 304
European Commission (EC), 42
  Codex interactions, 382
  Discussion Paper on Nutrition and Functional Claims, 350–358
  food claims legislation, 231–232
  food directives, 316
  White Paper on Food Safety, 227–228, 339, 344–349
  Working Document on Nutrition, Functional and Health Claims, 358–365
European Economic Community (EEC), 227
European Food Safety Authority (EFSA), health claims, 364–365
European Medicines Evaluation Agency (EMEA), 317
European Pharmacopoeia, 317
European Union (EU)
  advertising, misleading, 345–349
  advertising regulations, 350–356
  botanicals regulation, 316
  Concerted Action Project, 236
  contaminants regulation, 241
  dietary guidelines, 345
  dietetic foods regulation, 344
  food, defined, 316
  food additives regulation, 240–241
  food law, 227–228, 231–232
  food supplements regulation, 243–244, 347–349
  functional claims, 304, 353, 361
  GMOs regulations, 237, 238, 244, 314, 315
  GMOs threshold, 74
  health claims regulation, 41–42, 304–305
    approval procedure, 362
    general well-being claims, 363
  health claims substantiation, 236, 360–363
  herbals/botanicals regulation, 242–243, 316, 317–318
  irradiation regulations, 242
  labeling regulations, 343–356
    food, 228–229
    functional claims, 361
    GMOs, 244, 315
  nutrition, 229–231
  weight reduction foods, 363
  medicine, defined, 317
  membership, 227
  novel foods/ingredients regulations, 237–238, 314–315
  nutritional claims, 304, 351–354
  "Nutrition Claims and Functional Claims", 304
  organic foods regulation, 71, 76
  PARNUTS regulations, 238–240
  PASSCLAIM, 42
  "Regulation of the European Parliament and of the Council on Nutrition, Functional and Health Claims Made on Foods", 304
  Scientific Committee on Food (SCF), 237
  vitamins/minerals regulations, 242, 348–349, 353–354
Evening primrose oil, 90, 92
Exclusionary clause, FFDCA dietary supplement definition, 137, 140–147
Exclusivity, health claims, 30–32
Expert Consultations, WHO/FAO, 382
Experts, material connection, 163–164
Expiration dating, 67, 229

Fantasy products, Spain, 337
Fat, EU nutrition labeling regulations, 230
Fat-free claims, 353
Fat substitute, 190
FDA. *See* Food and Drug Administration
Federación de Industrias de Alimentación y Bebidas, 234
Federal Food, Drug, and Cosmetic Act (FFDCA)
  article, defined, 137–138, 141–147
  dietary supplement, defined, 137, 140–147
  exclusionary clause, 137, 140–147
  GRAS ingredients, 170
  new dietary ingredient, defined, 146
  prior market clause, 137–139, 144–147
Federal Institute for Risk Assessment, 309, 312, 315
Federal Register, FDA, 80
Federal Trade Commission (FTC)
  advertising violations, 161–164
  all claims order, 161

Federal Trade Commission (FTC) (*continued*)
  all products order, 161
  Bureau of Economics, 49–50
  cease and desist order, 161
  Deception Policy Statement, 149–156, 161–164
  Dietary Supplements: An Advertising Guide for Industry, 151
  functional foods, marketing, 149–164
  Guide Concerning Use of Endorsements and Testimonials in Advertising, 158, 163
  health claims
    advertising, 187–189
    substantiation of, 21, 156–161
  law enforcement framework, 150–152
  mission, 149
  "weight of scientific evidence" standard, 33
Federation of Swedish Farmers, 234, 307
Federation of Swedish Food Industries, 234
Feeding trials, animal/human, 276, 287
Fertilizers, organic standards for, 72–73
Fiber
  and cancer, 25, 43–44, 79
  FDA interim standard, 115
  health claims, Brazil, 372, 374–375
  and heart disease, 3, 25, 44, 79–80, 172, 177
First Amendment, and disease claims, 109, 111–112, 121–125
Fifth Amendment, 112
Fish liver oils, 241
Flavonoids, 374
Flavorings, 241
Flaxseed oil, 92
Folic acid
  and neural tube defects, 16, 47–48, 112, 174, 177
  FDA interim standard, 115–116, 118–119
  pilot health claim trial, 332–333
  and vascular disease, 16
Food additives
  defined, 5
  functional ingredients as, 190
Food additives regulations
  Australia/New Zealand, 326–327
  China, 266
  EU, 228–229, 240–241
  Japan, 248
  "nutritive value" requirement, 184, 189
  petitions for FDA approval, 89–92, 170
  safety standards, 29–30
  SPS Agreement, 380
Food Advisory Committee (FAC), UK, 235
Food and Agriculture Organization (FAO), 70, 377
  Codex interactions, 381–382
Food and Drug Administration (FDA)
  Center for Food Safety and Applied Nutrition (CFSAN), 14
  the Constitution and, 126–127
  Consumer Advisory, 170
  Consumer Health Information for Better Nutrition, 40
  Current Good Manufacturing Practices (CGMPs), 56–57
  dietary supplements regulation, 89–98, 109
  enforcement authority, 127–128
  Federal Register, 80
  FTC and, 150–152
  GMA citizen petition, 119–121
  health claims
    approved, 39–41, 43, 174, 177–182
    ranking system, 32–33, 174
  Letter to Industry, 185–186
  Pearson v. Shalala
    application of, 125–128
    conventional food labeling, 121–125
    dietary supplement labeling, 110–121
    impact of, 128–130
    implementation of, 113–121
    "significant scientific agreement" standard, 110–111
    clarification of, 173
    guidance document, 113
  Task Force on Consumer Health Information for Better Nutrition, 40–41
Food and Drug Administration Modernization Act (FDAMA), xiii
  "authoritative statement" standard, 111
  enactment of, 101
  health claims process, 83–84, 102–103, 171–172

"Food Hygiene Law of the People's Republic of China", 263, 264, 288–289
Food Industry Center (Japan), 249
Food law
   Australia, 321–334
   Brazil, 367–370
   EU, 227–228, 231–232
   Germany, 309–316
   Japan, 248-249; See also FOSHU
   New Zealand, 321–334
   Spain, 337–343
   Sweden, 305–309
   U.S., 6–8
Food Policy Statement, FTC, 150–152
Food(s)
   conventional
      health claims in labeling, 121–125
      qualified disease claims, 129–130
      "significant scientific agreement" standard, 110–111
   defined
      Brazil, 367
      Canadian Food and Drugs Act, 215
      EU, 316
   dietetic
      EU regulations, 238–239
      FDA regulations, 174–175
   hypoallergenic, 174
   ingredients. See Ingredients, food
   irradiation of, 242
   labeling. See Labeling
   marketing, FDA-FTC jurisdiction, 150
   medical, 174–175, 187, 190
      EU regulations, 239
   new/novel. See Novel foods
   organic. See Organic food
   for particular nutritional uses (PARNUTS). See PARNUTS
   reduced fat, 23
Food safety regulations
   Australia/New Zealand, 324–327
   Brazil, 370–371
   China, 266–275
   dietary ingredients, 92–93, 170–171
   food additives, 29–30
   Food Standards Australia New Zealand (FSANZ), 324–327
   FOSHU products, 252
   Germany, 310
   Japan, 248–249, 260
   organic food, 75–76
   raw materials, 16–17, 62–63
   SPS Agreement, 379–381
   Sweden, 306
   U.S., 55
   White Paper (EC), 344–349
   WHO/FAO, 381–382
Foods with Nutrient Function Claims, Japan, 295–296, 300
Foods for Special Dietary Uses, Brazil, 367
Foods for specific medical purposes (FSMP), EU regulations, 239
Foods for Specified Health Uses (FOSHU) - A Guideline, 252
Foods for Specified Health Uses. See FOSHU
Food Standards Australia New Zealand (FSANZ)
   functional foods definition, 323–324
   objectives, 322
   Regulatory Principles, 328
   safety, 324–327
Food Standards Program (United Nations FAO/WHO), 378
Food supplements regulation, See also Dietary supplements regulation
   EU, 231, 243–244, 347–349
   Germany, 309, 310, 312
Footnotes, disclosures in, 155–156
ForMor Inc., 155
Formulated Caffeinated Beverages (FCBs), 329–330
49th Parallel Group, 385
FOSHU, 38, 43, 295–296, 299–300
   advertising, 256–257
   approval process, 251–252
   categories, 250
   concept, 249–250
   definitions, 250–251
   functional ingredients, 255–256, 299
   labeling, 256, 258–259, 299–300
   marketing, 48–49, 252–253, 256–257, 259–260
   premarket approval, 256–257, 297–298
   sales by health benefit claimed, 252–255

Foundation for Innovation in Medicine, 3, 203
France
   health claims regulation, 233
   herbals/botanicals regulation, 243
Freedonia Group, 4
Free speech, and disease claims, xiii, 109, 111-112, 121-125; *See also Pearson v. Shalala*
Friends of the Earth, 385
Frontier products, Spain, 337
Fruitcal, 23
FTC v. Christopher Enterprises, Inc., 155
FTC v. Pantron I Corporation, 158
FTC v. Ruberoid Company, 161
FTC v. Warner-Lambert, 162
FTC v. Western Botanicals, 155
Functional claims
   Brazil, 370–371, 372–373
   Codex definition of, 355
   EU, 353, 359, 361
      Discussion Paper, 350–358
   Spain, 356–358
Functional foods
   background, 22
   compositional standards, 27–28
   credence claims, 157
   defined, 5, 22, 293
      FOSHU, 250–251, 295, 299–300
      FSANZ, 323–324
      General Accounting Office, 169
      Health Canada, 217
      Philippines, 294, 302
      South Korea, 294, 302
   diet and, 303–304
   as dietary supplements, 183–187, 190
   flavor problems, 28
   FOSHU, 249–258, 295
   health claims, 23–24, 171–176, 191
      FDA-approved, 177–182
   importation of, 249, 257, 282–283
   labeling, 228–229
      misleading, 46–47
   market for, 204
      Canada, 213–214
      China, 263, 289
      Japan, 300
      U.S., 22–24

   marketing, 28–29, 176–191
      general principles, 50–52
   as medical foods, 187
   nutrient function claims, 258–259
   product benefits, 23–24
   public health concerns, 45–46
   structure/function claims, 175–176, 182–183
   U.S. industry, 22–29, 33–35
"Functional Foods in Europe Consensus Document", ILSI Europe, 41–42
Functional foods regulation
   Australia, 321–334
   Brazil, 367–373
   Canada, 216–221
   China. *See* Health food regulation, China
   EU, 228–229, 304
   FOSHU, 295–300
   Germany, 303, 309, 312–316
   Japan, 247–249, 257, 258–259
   Myanmar, 301
   New Zealand, 321–334
   Spain, 337–343
   Sweden, 305–309
   U.S., 29–34, 169–176
Functional ingredients
   as food additives, 190
   FOSHU, 255–256
Function claims
   Australia/New Zealand, 331
   China, 264–265, 300–301
Fungus, Chinese regulations, 266
Furans, EU regulations, 241

Gaiatsu, 249
Gatorade, 25
General Accounting Office (GAO), functional foods definition, 169
General Agreement on Tariffs and Trade (GATT), 379
General Foodstuff and Commodities Act (Germany), 309
General Law of Public Health (Spain), 338
Generally Recognized As Safe (GRAS)
   defined, 6
   dietary ingredients, 14, 20
   food additives, 170

General Mills, 26
    "authoritative statement", 83
    functional foods, 25
    health claims proposals, 172
    oats and heart health campaign, 44, 177
General Principles for the Addition of Essential Nutrients to Foods, Codex Alimentarius, 327–328
Genetically modified organisms (GMOs)
    EU regulations, 237, 238, 244, 314, 315
    "GMO free" designation, 74
    Japanese regulations, 249
Genetic engineering, and organic foods, 73–74
Genetic toxicity tests, 275
German Food Law, 309–310
Germany
    dietary foods, 310–312
    food authorities, 309–310
    food supplements, 309, 312
    functional foods, 312–316
    health claims regulation, 303, 309–316
    herbals/botanicals regulation, 243, 318
    vitamins/minerals regulations, 242
Ginger, 19
Ginkgo biloba
    efficacy trials, 20
    in foods, 185
    standardization of, 19
Gluconolactones, 25
Glucosamine, 19, 20
Glucosamine/Chondroitin Arthritis Intervention Trial (GAIT), 20
"GMO free" designation, 74
Good Manufacturing Practices (GMPs), 56–57; *See also* Current Good Manufacturing Practices (CGMPs)
    Chinese regulations, 286
    dietary supplement regulation, 97
Grape polyphenols, 19
GRAS. *See* Generally Recognized as Safe
Green Giant, 177
Greenpeace, 384, 385
Green tea, FOSHU potential, 257–258
Grocery Manufacturers of America (GMA), citizen petition, 119–121
Grocery Manufacturers of Sweden, 234

Guide Concerning Use of Endorsements and Testimonials in Advertising, FTC, 158, 163
Guidelines for the Use of Nutrition Claims, Codex Alimentarius, 353–355, 359
Gum, chewing, 257, 373

Häagen-Dazs Company, 156
Hain Food Group, 26, 184–185
Hartman Group, 4
Hazard Analysis and Critical Control Points (HACCP), 55, 286
Health, defined, 201
Health benefits
    NREA definition of, 208
    overstating, 160
Health Canada, 217
    approved health claims, 218–219
    Interim Guidance Document on Preparing a Submission for Foods with Health Claims, 220
    product-specific health claims, 219–221
    standards of scientific evidence, 218–219
Health claims, *See also under individual claims*
    anecdotal evidence for, 158
    audience targeting, 50–51, 154
    bioactives, 4, 17–20, 30–32, 326, 374,
    context of, 51
    conventional food, 121–125
    defined, 6, 79, 176, 201
        Codex Alimentarius, 38, 304–305
        NREA, 208–209
        Spain, 340–341
        Sweden, 306
    dietary supplements, 13–16
    direct, 39, 219
    DSHEA-approved, 202
    European Food Safety Authority (EFSA), 364–365
    exclusivity, 30–32
    express, 152–153
    fat-free, 353
    FDA-approved, 39–41, 43, 174, 177–182
    and First Amendment, 109, 111–112, 121–125

Health claims (*continued*)
  foods for special dietary use, 173–174
  FOSHU products, 253–255
  functional claims, China, 264–265,
    300–301
  functional foods, 23–24, 171–176, 191
    qualification of, 154–155
  general well-being, 363
  generic, 39
  Health Canada-approved, 218–219
  history of, US, 79–80
  implied, 152–153
  indirect, 39, 219
  marketing, 13–16, 42–45, 50–52
  medical foods, 173–174, 190
  misleading, 46–47
  misrepresentation of, 149–156
  NLEA-authorized, 80–84
  nutrition-based, 49–50, 173
  preliminary studies and, 174
  product-specific, 39
    Canada, 219–221
    Sweden, 307–308
  purpose of, 101–102
  qualified, 6, 14–16, 32–33, 84–85, 174
  structure/function claims. *See* Structure/
    function claims
  substantiation of, 17–20, 39–42, 85, 109
    Canada, 218–219
    EU, 236, 360
    FTC Act, 156–161
  Tropicana Pure Premium, 101–108
  unqualified, 6, 14, 32–33
Health claims regulations, *See also* Labeling
  advantages of, 51
  advertising. *See* Advertising
  Australia/New Zealand, 329–334
  "authoritative statement", 83, 101, 172
  bandwagon approach, 52
  Belgium, 233
  Brazil, 370–373
  Canada, 215–216, 221–225
  clinical trials, 17–20, 39–42, 85, 109
  Codex Alimentarius, 304–305
  disadvantages of, 51–52
  EU, 41–42, 231–232, 304–305, 351–354,
    359–360
  approval procedure, 362
  general well-being claims, 363
  substantiation of claims, 236, 360
  FDA, 39–41, 43
    -approved, 39–41, 43, 174, 177–182
    Modernization Act (FDAMA), 83–84,
      102–103, 171–172
    ranking system, 32–33, 174
  France, 233
  FTC, 21, 150–164, 187–189
  Germany, 303, 309–316
  "ignorant, unthinking, and credulous
    consumer" standard, 109
  Japan, 38, 43, 48–49, 258–259; *See also*
    FOSHU system
  Netherlands, 233–234
  and public health, 45–47
  "reasonable consumer" standard, 109
  regulatory principle, 37
  "significant scientific agreement" standard,
    110, 111–121, 125–126, 172, 173
  Spain, 234, 338–343
  Sweden, 41, 234–235, 303, 305–309
  UK, 38, 41, 235–236
  U.S., 39–41, 43, 85–86
  "weight of scientific evidence" standard,
    32–33, 109
Health drinks, 257–258
HealthFocus International, 4
Health Food Control Act (Taiwan), 298
Health foods
  defined, China, 264, 295
  industry, China, 263
Health foods regulation
  China, 263–265, 297–298, 300–301
    approval certificate, 264–265, 276,
      282–283, 285
    assessment procedure, 283–285
    control/supervision, 288–289
    evaluation/approval process, 266
    functionality evaluation, 276–281
    hygiene standards, 287
    imported foods, 282–283
    labeling, 283, 285, 287
    packaging, 286–287
    production, 286
    safety control, 266–275

violations of, 289
Taiwan, 298
Health Functional Food (HFF), South Korea, 295
Health Products and Foods Branch, 218–219
Health Protection Branch (HPB), 215
　Food Directorate, 216
Healthy Woman, 9
HeartBar, 161, 187, 189
Heart disease
　and diet, 214
　and fiber, 3, 25, 44, 79–80, 172, 177
　and nuts, 15, 174
　and oats, 3, 25, 44, 79–80, 172, 177
　and omega-3-fatty acids, 15, 174
　and plant sterols/stanols, 172, 177, 315–316
　and potassium, 24, 44–45, 83, 103–105, 177
　and soy, 25
Heavy metals
　Chinese regulations, 287
　EU regulations, 241
Heckler v. Chaney, 127–128
Heinz, lycopene campaign, 45, 188
Herbals
　Brazilian regulations, 374
　EU regulations, 241, 242–243, 316, 317–318
　German regulations, 303, 318
　Irish regulations, 243
　medicinal, 184–185
　Swedish regulations, 303, 318
　U.S. regulations, 12, 17–20
Heterochemical Corporation v. FDA, 128
HMG-CoA reductase inhibitors, 140
Honeywell, Inc., 160
Hormone replacement therapy (HRT), 34
Hormones, and organic meat production, 69, 73
Hygiene License, China, 286, 289
Hygienic practices, 58
　SPS Agreement, 380
Hypertension
　diet and, 104–105
　potassium and, 24, 44–45, 83, 103–105, 177
Hypoallergenic foods, 174

ILSI SE Asia Region survey
　definitions/regulations, 294–299, 302
　methodology, 293–294
　objective, 293
　premarketing approval, 297–299
　products/claims, 299–301
　respondents, 294
Imagine Foods, 26
Importation
　functional foods
　　China, 282–283
　　Japan, 249, 257
　"Increased", Codex Guidelines, 354
Indonesian National Agency for Drug and Food Control (NADFC), 296
Infant Formula Quality Control Procedures, 55
Infant formulas, 174
　Codex regulations, 385
　DHA in, 32
　EU regulations, 238–239
Infomercials, 163
Information, disclosure of, 152–156
Ingredients
　dietary. *See* Dietary ingredients
　food, *See also* Dietary ingredients
　　Australia/New Zealand regulations, 325–327
　　Brazilian regulations, 373
　　form of, 325
　　functional
　　　as food additives, 190
　　　FOSHU, 255–256
　　GRAS, 5
　　proprietary, 30–32
Inspectors, organic foods, 72
Institute of Medicine (IOM)
　scientific support for health claims, 174
　Tolerable Upper Intake Levels, 33
Institute of Nutrition and Safety (China), 282
Insulin resistance, 205
Intellectual property, nutraceuticals and, 30–32
Intergovernmental organizations (IGOs), Codex process interactions, 381–387
International Association of Consumer Food Organizations, 47

International Federation of Organic Agriculture Movements (IFOAM), 70, 73, 76
International Life Sciences Institute (ILSI), 368
   China Focal Point, 263
   Europe, 41–42, 304
International Office of Epizootics, 378, 380
International Organization for Standardization (ISO), 384
International Plant Protection Convention, 378
Internet, and health claims, 186
Interstate Bakeries Corporation, 160
Ireland
   herbals/botanicals regulation, 243
   vitamins/minerals regulations, 242
Irradiation
   EU regulations, 242
   organic foods, 73
Isoflavones, 9, 374
   standardization of, 19
Italy, vitamins/minerals regulations, 242

Japan
   birth rates, 247
   food law, 248–249
   Foods with Nutrient Function Claims, 295–296, 300
   FOSHU system, 38, 43, 249–258, 295–300
     marketing, 256–257, 259–260
   functional foods
     defined, 295
     regulation, 247–249
   future regulations, 259–260
   GMOs regulations, 74, 249
   health care costs, 247
   longevity of population, 247
   Ministry of Health, Labor, and Welfare (MHLW), 247–250
     Food Sanitation Law, 248–249
     Office of Health Policy on Newly Developed Foods, 251
   nutrient function claims, 258–259
   premarketing approval process, 297–298
Japan Food Additives Association, 248
Japan Health Food and Nutrition Food Association (JHNFA), 48, 250–258

"Jelly Bean Rule", 173
Johnson & Johnson, 9
Joint Health Claims Initiative (JHCI), 235–236
   Code of Practice, 38, 41

Karada Shien, 257
Kava, 170, 185
Kellogg, 34
   Ensemble line, 182–183
   fiber campaign, xiii, 25, 43–44, 79, 176–177
   folic acid health claim, 47–48, 177, 183
   soy foods, 26
Kellogg Japan, 257
Kitchen Prescription, 184–185
Klondike Lite ice cream bars, FTC labeling challenge, 151
KPMG Consultants, 214
Kraft Foods, 23
   "authoritative statement", 83
   functional foods, 25
   soy foods, 26

Labeling, *See also* Health claims regulation
   biotech, 387
   Brazilian regulations, 370–373
   Canadian regulations, 218
   Codex guidelines, 305
   conventional food, 121–125
   dietary supplements, 95–96
   EU regulations, 343–356
     food, 228–229
     functional claims, 361
     GMOs, 244, 315
     nutrition, 229–231
     weight reduction foods, 363
   false or misleading, 46–47, 128
   FOSHU products, 256, 258–259
   and free speech, 111–112
   FTC and, 150–152
   German regulations, 313–316
   health claims restrictions, 14
   health claims, types of, 101
   health foods, China, 283, 285, 287
   "ignorant, unthinking, or credulous consumer" standard, 33

logos and, 47
organic food, 69–71, 74
quality control, 63
"reasonable consumer" standard, 33
Spanish regulations, 339–343
statements of nutritional support, 94
structure/function claims, 175–176
Swedish Code of Practice, 306, 308–309
truth in, 95–96
use of books/publications, 93–94
U.S. regulations, 4–8
Web sites and, 186
Laboratory operations, quality control, 65–66
Lactobacillus, 372
Lactobacillus casei, 253
Lactoferrin, GRAS status, 20
Lean Cuisine, 156
Lebensmittel und Bedarfsgegenstände-Gesetz (LBMG), 309–310
"Less", Codex Guidelines, 354
Licensing agreements, 31
Licorice root, 274
Linseed oil, 92
alpha-Lipoic acid, 205
Lipovitan, 258
Lipton teas, 26
Listerine, 162
Lobelia, 92
Logos
  approved health food, China, 265
  on food labels, 47
  FOSHU, 250, 257
  organic foods, 72, 74–75
Lovastatin, 140–141, 143–144, 185
"Low"
  Codex Guidelines, 353
  EC guidelines, 353
Lutein, 9, 20
Lycopene, 20, 45, 188

McIntosh, David M., 113–114
McNeil Consumer Healthcare, Benecol, 30, 31, 34, 184, 189
McNeil Nutritionals, 9
Magnesium deficiency, and diabetes, 205
Magnesium orotate, 92
Malaysia, dietary supplements regulation, 296

Malaysian Food Regulations, 296
Managed care, and nutraceuticals, 34–35
Manufacturing, quality control, 63–64, 66–67
Manure, organic standards, 73, 75, 76
Marbury v. Madison, 126–127
Margarine substitutes, 184, 372
Marketing
  and deceptive advertising, 163
  FOSHU products, 256–257, 259–260
  functional foods, 28–29, 176–191
  and health claims, 13–16
    benefits of, 43–45
    general principles, 50–52
    importance of, 42–43
    public health concerns, 45–47
  health and wellness, 34
  and information disclosure, 162–156
  Swedish Code of Practice, 308–309
  target audience, 154
  Tropicana Pure Premium, 105–108
Martek Bioscience, 32
Material connection, experts/endorsers, 163–164
3-MCPD, EU regulations, 241
Mead Johnson, 9
Meal replacement products, EU regulations, 238–239
Meat production, organic management, 73
Medical devices, FDA-FTC jurisdiction, 150
Medical-disease claim, defined, 201
Medical foods
  defined, 175
  EU regulations, 239
  functional foods sold as, 187
  German regulations, 311
  health claims for, 173–174, 190
Medicinal product, EU definition, 360
Medicinal Products Agency (MPA), Sweden, 318
Medicine, defined, EU, 317
Meiji Dairies, 254
Mercosur, 384
Methylene chloride, 154
MetRx, 162
Mevacor, 140–141
Mevinolin, 140
Mice, feeding trials, 276

Microbiological testing, Chinese regulations, 287
Micronutrients
  Codex Guidelines, 354
  EU regulations, 230, 242
Milk, organic standards for, 73
Milk products, fermented, 45
Milo milk, 257
Minerals
  Australia/New Zealand regulations, 325–329
  EU regulations, 230–231, 348–349, 353–35
  Irish regulations, 242
  Italian regulations, 242
  Spanish regulations, 343
  U.S. market, 12
Ministerio de Sonidad y Consumo, 234
Ministry for Consumer Protection, Nutrition and Agriculture (Germany), 309
Ministry of Health (Brazil), 370
Ministry of Health (China), 263–290
  "Administrative Provisions for New Food Source Hygiene", 266
  function claims, 300–301
  premarketing approval, 298
  "Standards for Toxicological Assessment of Health Foods", 274
Ministry of Health and Consumption (Spain), 339–343
Ministry of Health and Welfare (Japan), premarketing approval, 297–298
Miracle products, Spain, 337
"More", Codex Guidelines, 354
Morinaga Milk Industries, 254
Myanmar, functional foods regulation, 301
Mycotoxins, EU regulations, 241
Myocardial ischemia, 205

NAFTA, 384
National Academy of Sciences
  Diet and Health: Implications for Reducing Chronic Disease Risk, 103–105
  Institute of Medicine (IOM), 17, 174
    Tolerable Upper Intake Levels, 33
  as "scientific body", 83, 101
National Advertising Division, Council of Better Business Bureau, 186–187

National Cancer Institute, xiii, 8
  fiber health claims, 43–44, 79
  "5-A-Day" program, 24
National Center for Complementary and Alternative Medicine (NCCAM), 19–20
National Consumer Council (UK), 47
National Dietary Council (France), 233
National Institutes of Health (NIH)
  Botanical Centers, 17
  National Center for Complementary and Alternative Medicine (NCCAM), 19–20
  Office of Dietary Supplements (ODS), 19–20
  as "scientific body", 83, 101
"National List, The", 73
National Nutritional Foods Association, 9
National Nutrition Survey in Japan, The, 248
National Organic Program (NOP), 69–71
  GMO standards, 74
  organic labels, 74
National Organic Standards Board, 73
National Sanitary Surveillance Agency (ANVISA), 367–369
Natto, 254
Natural foods
  defined, 6
  U.S. market, 9, 22–23
Natural Health Products Directorate (NHPD), 217
  health claims regulation, 221–225
  nutraceuticals definition, 221
Natural Health Products (NHP), Canada
  industry, 213–214
  regulation, 221–225
Natural Health Products Research Program (NHPRP), 224
Natural product number (NPN), 222
Natural products, Brazilian market, 367, 372–373
Nestle, 23
Nestle Japan, 257
Netherlands, health claims regulation, 233–234
Netherlands Nutrition Center, 233–234

Neural tube defects
    and folic acid, 16, 47–48, 112, 174, 177
        pilot health claim trial, 332–333
New Drug Application (NDA), 204
New Zealand
    dietary supplements regulation, 321, 324, 329, 331–333
    food law, 321–334
    food safety, 324–327
    functional foods regulation, 321–327, 329–334
        nutritional issues, 327–329
    health claims regulation, 329–334
    novel foods regulation, 325–327
    vitamins/minerals regulation, 327–329
Nippon Lever Co., 257
NiteBite, 187
Nitric oxide, 187
Nongovernmental organizations (NGOs),
    Codex process interactions, 381–387
Norelco Clean Water Machine, 153–154
North American Philips Corporation, 153–154
Novartis, 34
Novartis v. FTC, 162
Novel foods regulation
    Australia/New Zealand, 325–327
    Brazil, 370
    EU, 237–238, 314–315
Nucleic acids regulation
    Australia/New Zealand, 325–326
    China, 266
Numex Corporation, 164
Nutraceutical Research and Education Act, 204, 206–212
Nutraceuticals
    defined, 3–5, 6, 201
        Brazil, 368
        DSHEA, 55
        Health Canada, 217
        NREA, 208
    EU regulations, 228–229
    German regulations, 313–316
    intellectual property, 30–32
    and managed care, 34–35
    marketing, 50–52, 203–206
    proprietary protection, 204, 205
    quality control, 55–67
    Spanish regulations, 342–343
    toxicity issues, 203, 204
    U.S. regulations, 29–34
Nutraceuticals World, 4
Nutrient/nutritional claims, 38–39
    Codex guidelines, 355, 364
    comparative, 354
    EU, 229–231, 351–354
    Japan, 258–259
    Spain, 356–358
    U.S., 49–50
Nutrients, recommended intake, 17
Nutrition
    EU Discussion Paper, 350–358
    EU recommendations, 345
Nutritional Outlook, 4
Nutritional profile, EU, 363
Nutrition Business Journal, 4
"Nutrition Claims and Functional Claims", EU, 304
Nutrition industry, growth of, 4–8
Nutrition labeling, EU regulations, 229–231
Nutrition Labeling and Education Act (NLEA), xiii, 49
    authorized health claims, 80–84
    health claims for functional foods, 171–174
    intent of, 101
    "significant scientific agreement" standard, 110
    statements of nutritional support, 94
Nutritive substances, regulation,
    Australia/New Zealand, 325–326
"Nutritive value" requirement, food additives, 184, 189
Nuts, and heart disease, 15, 174

Oats, and cholesterol, 3, 25, 44, 79–80, 172, 177
Obesity
    global prevalence of, 45
    Japan, 248
Observer status, nongovernmental organizations (NGOs), 381
Ocean Spray, 45
Office of Dietary Supplements (ODS), 19–20
Office of Natural Health Products, 221

Official Journal of the European Communities, 347
Ohtsuka Pharmaceutical, 258
Olestra, 190
Omega-3-fatty acids
　Brazilian regulations, 372
　FDA interim standard, 116–117
　German regulations, 313
　health benefits, 24
　and heart disease, 15, 174
One-A-Day, 8, 11
Oranamin C, 258
Orange juice
　calcium-fortified, 23-24; *See also* Tropicana Pure Premium
　and nutrition, 103
　US market, 106–107
Organic, definition, 70
Organic Farming Research Foundation, 75
Organic foods
　certifying agents, 72
　defined, 6, 69
　EU regulations, 71, 76
　health and safety, 75–76
　inspectors, 72
　Japanese market, 249
　labeling, 74
　U.S.
　　industry, 22–23
　　market, 69
　　regulations, 69–77
Organic Foods Production Act, 69
Organization for Economic Cooperation and Development (OECD), 383
Organosulfur compounds, 374
Orotate compounds, 92
Orphan Drug Act, 204
Osteoarthritis, 20
Osteoporosis, and calcium intake, 24
Over-the-Counter drugs
　FDA-FTC jurisdiction, 150
　market, 204
Overweight, global prevalence of, 45

Pacific Foods, 26
Package labeling. *See* Labeling

Packaging
　Chinese regulations, 286–287
　quality control, 63, 66–67
　Tropicana Pure Premium, 105–108
Pallone, Frank, 204
Panda Herbal International, Inc., 155
PARNUTS, 228, 238–240
　EU regulations, 238–240, 359
　Germany, 310–311
PASSCLAIM, 42, 236, 304
Patents, 31
Pearson v. Shalala
　background, xiii, 110–121
　CFSAN and, 15
　and dietary supplements, 173
　and FDA authority over health claims, 173
　First Amendment and food labeling, 121–125
　impact of, 128–130
　interpretation/application by FDA, 125–128
　qualified health claims, 8–85
PepsiCo., functional foods, 25
Performance-enhancing products
　Australia/New Zealand, 326
　Brazil, 374
Personnel, quality control, 58–59
Pesticides
　EU regulations, 241
　Japanese regulations, 249
　organic standards for, 71–73
　SPS Agreement, 380
Pfizer, 11
Pfizer factors, 157–158
Pharmaceuticals, market, 204
Pharmafoods, Spanish regulations, 342–343
Pharmanex I, 141–142
Pharmanex II, 138, 142–143
Pharmanex III, 137–138, 139, 143–144
Pharmanex v. Shalala, 137–138, 146–147, 185
　agency and court proceedings, 140–144
　background, 139–140
　implications of, 144–146
Philippines
　Bureau of Food and Drugs, 296
　functional foods, 294, 302

Phosphatidylserine, and cognitive dysfunction and dementia, 16, 174
Phospolipids, 374
Physical plants, quality control, 59–60
Phytochemicals
  and cancer, 24
  and cholesterol, 31, 315–316, 327
Phytonutrients, defined, 6
Phytosanitary standards, harmonization of, 379–380
Phytosterols, 325, 372, 374
  and cholesterol, 31, 315–316, 327
Phytotherapeutics, Brazil, 367, 368, 372
Pillsbury, health claims, 177
Plant Protection Convention, 380
Polycyclic Aromatic Hydrocarbons (PAH), 241
Polydextrose, 257
Polyols, EU regulations, 241
Polyphenols, 374
Pork and Beans, 177
Potassium
  and blood pressure, 24, 44–45, 83, 103–105, 177
  intake, US, 105
  and stroke, 24, 83, 177
  Tropicana Pure Premium health claim, 101–108
PowerBar, 23
Prebiotics, Brazilian regulations, 373, 374
"Precautionary principle", Codex Alimentarius, 385–387
Premarketing approval
  China, 298
  FOSHU products, 256–257, 297–298
  functional claims, 355
  ILSI SE Asia Region survey, 297–299
  U.S.
    dietary supplements, 22
    food additives, 170
Prevention, self-care survey, 203
Prior market clause, 137–139, 144–147
Proanthocyanidins, 22
Probiotics
  Brazilian regulations, 372, 373, 374
  Chinese regulations, 266

European market, 45
German regulations, 313
"Process for the Assessment of Scientific Support for Claims on Foods" (PASSCLAIM), 42, 236, 304
Process controls, quality control, 61
Procter & Gamble, Fruitcal, 23–24
Product 19, 177
Product benefits. *See* Health claims
Production, quality control, 61
Production permits, China, 286
Product labeling. *See* Labeling
Product License, natural health products, 222
Product-specific health claims, 39
  Canada, 219–221
  Sweden, 307–309
Proprietary Medicinal Products (EU), 316
Proprietary products, 30–32
Proprietary protection, nutraceuticals, 204, 205
Protein content, EU nutrition labeling regulations, 230
Protein Technologies International, 26
Psyllium, 25, 176–177, 189
Publications, FDA labeling restrictions, 93–94
Public health, and health claims regulation, 45–47

Quaker Oats Company, 25, 44, 177
Quality control
  batch production records, 64–65
  Current Good Manufacturing Practices (CGMPs), 57–67
  equipment, 60–61
  expiration dating, 67
  issues in, 55–57
  labeling, 63
  laboratory operations, 65–66
  manufacturing, 66–67
  master manufacturing records, 63–64
  packaging, 63, 66–67
  personnel, 58–59
  physical plant, 59–60
  process controls, 61
  production, 61
  raw materials, 16–17, 62–63
  testing, 65–66

Quality control (*continued*)
   unit, 61–62
   utensils, 60
Quanterra™, 11
Quigley Corp., 160
QVC, Inc., 160

Raisin Bran, 79
Raisio Group, 31
Rama ProActive margarine, 257
Rats, feeding trials, 276
Raw materials
   Chinese safety regulations, 266–275
   quality control, 16–17, 62–63
Real Decreto (RD), 337–343
Recaldent, FOSHU approval, 257
Recommended Daily Allowance
   nutrition labeling, EU regulations, 231
   vitamins/minerals, 354
Recommended Daily Intake, 328
Red Bull GmbH, 25
"Reduced", Codex Guidelines, 354
Red yeast rice, 139–140, 143–144
Reference Daily Intake, 173
Reference Daily Value, 173
"Regulation of the European Parliament and of the Council on Nutrition, Functional and Health Claims Made on Foods", EU, 304
Research, to substantiate health claims, 158–161
Research-based Dietary Ingredient Association (RDIA), 32
Restitution, to consumers, 161
Risk-assessment techniques, SPS Agreement, 380
Risk-reduction claims
   Australia/New Zealand, 331
   Canada, 216
Royal jelly, 297
Rutin, 91

Safrole, 241
St. John's Wort, 92
   efficacy, 20, 155, 203
   as a food ingredient, 184, 185
   standardization of, 19

Sanitarium, folate health claim, 47–48
Sanitary and Phytosanitary (SPS) Agreement
   risk-assessment techniques, 380
   role of Codex standards/guidelines, 378–381
Sardines Panel, WTO, 366–367
Saw Palmetto, FDA interim standard, 119
Scandinavian Journal of Nutrition, 305
Schering-Plough Healthcare Products, 160
Scientific Committee on Food (SCF), EU, 237
Secretary of Health and Human Services, 93, 96
Selenium, and cancer, 15, 174
Senate Committee on Labor and Human Resources, 89
Sewage sludge, 73
Shark cartilage, 19, 91
"Significant scientific agreement" standard, 172
   Administrative Procedure Act, 125–126
   after *Pearson v. Shalala,* 111–121
   dietary supplements, 110, 111, 126
   FDA clarification of, 173
   FDA guidance document, 113
   NLEA, 110
Silver Anvil Award for Integrated Communications, 107
Simeon Management Corp. v. FTC, 158
Singapore, dietary supplements regulation, 296–297
Site licensing, NHP companies, 222
Sitostanol ester, 30
Snow brands milk, 260
SoBe, 26
Sodium, health claims, 353, 359
Sodium nitrate, 76
Soft drinks
   functional, 45
   Japan, 257
South Korea
   functional foods, defined, 294, 302
   Health Functional Food (HFF), 295
Soyatech Inc., 34
Soy foods, 26–27, 374
   and heart disease, 25
   U.S. market, 34

Soy isoflavones, 9
   standardization of, 19
Soymilk, 26
Spain
   functional foods regulation, 337–343
   health claims regulation, 234
   vitamins/minerals regulations, 242, 343
Spanish Federation of Food and Drink Manufacturers, 234
Spanish Federation of Industries of Food and Beverages (FIAB), 339–343
Spanish Ministry of Health, 234
Specialty supplements, standardization of, 17–20
Specifications and Standards for Foods, Food Additives, etc. under The Food Sanitation Law, 248
SPINS, 4, 21
Sports nutrition products, EU regulations, 239–240
Spreads, cholesterol-lowering, 30, 31, 184, 372; *See also* Stanols; Sterols
Standard Operating Procedures (SOP), laboratory testing, 65–66
"Standards for Toxicological Assessment of Health Foods", (Chinese Ministry of Health), 274
Stanols, plant, 30, 31, 172, 177
   GRAS status, 20, 184
State Food and Drug Administration (SFDA), China, 263–265
   Division of Traditional Medicine Protection, 283–285
Stay Alert Caffeine Supplement Gum, 185
Sterols, plant, 177, 315–316
   GRAS status, 20
Stouffer Foods Corp., 156
Stroke
   potassium and, 24, 83, 177
      Tropicana Pure Premium, 101–108
Structure/function claims, 39, 101
   defined, 6, 79, 176
   DSHEA, 14, 151
   functional foods, 175–176, 182–183
   regulating, 191
   scientific evidence for, 176
Sugars, EU nutrition labeling regulations, 230

Superscript, for disclosures, 155–156
Sweden
   approved health claims, 41
   functional foods regulation, 305–309
   health claims regulation, 234–235, 303, 305–309
   herbals regulation, 318
Swedish Code of Practice, 303, 306
   labeling, 306, 308–309
   marketing, 308–309
Swedish Cooperative Union, 307
Swedish Federation of Trade, 307
Swedish Food Federation, 307
Swedish Food Retail Association, 234, 307
Swedish Nutrition Foundation (SNF), 303, 305, 307–309
Sweeteners, 172, 228, 241

D-Tagatose, and tooth decay, 172
Taisho Pharma, 258
Taiwan, health foods regulation, 298
Take Control, 30, 184, 189
Tall oils, 325
Task Force on Consumer Health Information for Better Nutrition, FDA, 40–41
Taurine, 25, 374
Tea extracts, 19
Teas, flavored, 25, 26
Technical Barriers to Trade (TBT) Agreement, role of Codex standards/guidelines, 378–379
Technical-Sanitary Regulations, Spain, 340
Technical Scientific Advisory Commission on Functional Foods and New Foods (CTCAF), 369
Testimonials, use in health claims advertising, 158, 159
Testing, quality control, 65–66
Thailand, dietary supplements regulation, 297
Theragran, 8
Therapeutic claims, Canada, 216
Therapeutic Products Directorate (TPD), 215
Thompson v. Western States Medical Center, 124
Tolerable Upper Intake Levels, NAS/IOM, 33
Tooth decay, and D-Tagatose, 172

Toxicity testing, Chinese health foods, 274–275
Traceability, biotechnology, 387
Trade agreements, 378-379; *See also* Codex Alimentarius Commission; World Trade Organization (WTO)
Trade regulations, WHO/FAO, 381–382
Traditional Chinese Medicine (TCM), 263
TraumaCal Liquid, 175, 187
Travasorb Hepatic Powder, 175, 187
Treaty of Rome, 227
Tropicana Nutrition Center, 108
Tropicana Products
  "authoritative statement", 83
  calcium-fortified citrus juices, 23–24, 34
  potassium health claim campaign, 44–45, 177
  Pure Premium campaign
    health claim process, 101–105
    marketing, 44–45, 105–108
    strategy, 103
Turmeric, standardization of, 19

Unilever, 30
United Kingdom Food Standards Agency, 242
United Kingdom (UK)
  consumer advocacy, 47
  health claims regulation, 38, 41, 235–236
  herbals/botanicals regulation, 243
United Nations (UN)
  Commission on Sustainable Development, 383
  Conference on Environment and Development (UNCED), 383
  Food and Agriculture Organization (FAO), 70, 377
United States
  Codex process interactions, 381–387
  Constitution, 126–27
    First Amendment, 109, 111–112, 121–125
    Fifth Amendment, 112
  Court of Appeals for the District of Columbia Circuit, 109, 112
  Court of Appeals for the First Circuit, 90
  Court of Appeals for the Seventh Circuit, 90

Court of Appeals for the Tenth Circuit, 142–143
Department of Agriculture (USDA)
  Agricultural Marketing Service (AMS), 70, 72
  Foreign Agricultural Service (FAS), 263
  Organic seal, 74
Department of Health and Human Services, "Healthy People 2010", 23
dietary supplements
  companies, 9–11
  consumer use of, 21
  industry, 9–11, 13–29
  market, 11–13
  regulation, FDA, 89–98
District Court for the District of Columbia, on FDA and *Pearson v. Shalala,* 112, 113–119
District Court for Utah, 137, 141–142
FDA. *See* Food and Drug Administration
food law, 6–8
food sanitation regulations, 55
food supply safety, 32
functional foods
  industry, 22–29
  products/companies, 25–27
  regulation, 169–176
health claims
  advertising, 49–50
  approved, 177–182
  process, 85–86
  qualified, 84–85
  regulation, 39–41
natural food industry, 22–23
nutraceuticals regulation, 29–34
nutrition labeling, 4–8
organic foods
  industry, 22–23
  regulation, 69–77
Secretary of Health and Human Services, 93, 96
Unither Pharma, Inc., 160–161, 189
Utensils, quality control, 60–61

V8 juice, 183
Vascular disease, and B vitamins, 16
Veterinary drugs, SPS Agreement, 380

Viactiv, 9
Victor, Stephen, 163
Vitamin B, FDA interim standard, 117–118
Vitamin B6, 16
Vitamin B12, 16
Vitamin E, efficacy of, 203
Vitamin-fortified drinks, German regulations, 314
Vitamins regulations
   Australia/New Zealand, 325–326, 327–329
   EU, 242, 348–349, 353–354
      nutrition labeling, 230–231
   Ireland, 242
   Italy, 242
   Spain, 343
   U.S., 12
Vitasoy USA, 26
Voedingscentrum, 233
Voltaire, 201

Warner Lambert, 11, 257
Warning labels, EU regulations, 229
Washington Legal Foundation (WLF), 186
Weider Nutrition International, Inc., 172
Weight reduction foods, 174
   EU labeling regulations, 363
   German regulations, 311
"Weight of scientific evidence" standard, 32–33, 109

Welch Foods, antioxidants claims, 188
Wellness drinks, German regulations, 314
Wexler, Patricia, 163
Whitaker v. Thompson, 129, 130
White Paper on Food Safety, (EC), 227–228, 339, 344–349
White Wave, 26
WonderBread, 189
World Health Organization (WHO)
   Codex Alimentarius Commission, 377–378
   Codex interactions, 381–382, 386–387
   dietary practices, 45
   organic foods guidelines, 70
World Summit on Sustainable Development (WSSD), 383
World Trade Organization (WTO)
   Dispute Settlement Body, 379
   establishment of, 378
   Sardines Panel, 366–367
Wyeth, 8, 9

Xuezhikang, 140
Xylitol, 373

"Yakult" lactic acid beverage, 253, 257
Yogurt, 254

Zbars, 187
Zhi-Tai, 140